# Microbial Biosurfactants and their Environmental and Industrial Applications

T0225268

*Editors*

**Ibrahim M. Banat**

University of Ulster
Faculty of Life & Health Sciences
Coleraine, Northern Ireland, UK

**Rengathavasi Thavasi**

Jeneil Biotech Inc.
Saukville, WI, USA

**CRC Press**
Taylor & Francis Group
Boca Raton  London  New York

CRC Press is an imprint of the
Taylor & Francis Group, an **informa** business
A SCIENCE PUBLISHERS BOOK

Cover illustrations provided by one of the editors of the book, Prof. Ibrahim M. Banat.

CRC Press
Taylor & Francis Group
6000 Broken Sound Parkway NW, Suite 300
Boca Raton, FL 33487-2742

First issued in paperback 2021

© 2019 by Taylor & Francis Group, LLC
CRC Press is an imprint of Taylor & Francis Group, an Informa business

No claim to original U.S. Government works

Version Date: 20180731

ISBN-13: 978-0-367-78055-5 (pbk)
ISBN-13: 978-1-138-19795-4 (hbk)

Library of Congress Cataloging-in-Publication Data

Names: Banat, Ibrahim M., editor. | Thavasi, Rengathavasi, editor.
Title: Microbial biosurfactants and their environmental and industrial
   applications / editors, Ibrahim M. Banat, Rengathavasi Thavasi.
Description: Boca Raton, FL : CRC Press/Taylor & Francis Group, [2018] | "A
   Science Publishers book." | Includes bibliographical references and index.
Identifiers: LCCN 2018036101 | ISBN 9781138197954 (hardback : alk. paper)
Subjects: | MESH: Surface-Active Agents | Microbiological Phenomena
Classification: LCC TP994 | NLM QV 233 | DDC 668/.1--dc23
LC record available at https://lccn.loc.gov/2018036101

**Visit the Taylor & Francis Web site at**
**http://www.taylorandfrancis.com**

**and the CRC Press Web site at**
**http://www.crcpress.com**

# Preface

Surface-active molecules, otherwise known as surfactants, are chemicals that have key impacts on several aspects of our day-to-day living, as they are part of many daily products we use. These chemical surfactants mainly originate from petrochemical or oleochemical sources and are ingredients of household cleaning agents, laundry products, cosmetics, pharmaceutical, petroleum industries, environmental cleaning products and agro food processing industry. The worldwide use of surfactants has been steadily growing over the past few decades and is projected to further increase in the future. The fact is that most chemical surfactants have detrimental effects on the environment, both in sourcing and disposal, which has led to the search for alternative compounds with lesser overall impact. This shift in attitude towards biological surfactants (biosurfactants) during the past two decades has, therefore, been mainly driven by the search for environment friendly compounds and the sustainability agenda, which are becoming mainstream strategies. Many products using surfactants as an ingredient are now under investigation to see if part or all of these compounds can be substituted with sustainable biosurfactants, that is, surfactant molecules produced principally by microorganisms using sustainable renewable feedstock. This is because of the advantages they bring including relative stability at high temperature and in adverse environments, in addition to the added advantage of being readily biodegradable in the environment if, or when, discharged.

I started working with biosurfactants some 30 years ago while I was at the Kuwait Institute for Scientific Research where I realised the significance of this research and the international interest and attention in this area of research, which was in its infancy, yet was beginning to show. This compelled me to persevere in this area of research while establishing an international network of collaborators and interested scientists working on biosurfactants. It was through this that some 15 years ago, I started collaboration with Dr. Thavasi Rengathavasi where together we published several papers related to our research on biosurfactants and established a strong collaborative relationship.

The main question one always keeps in mind is why do some microorganisms produce biosurfactants? The answer is not simple or straight forward, as these compounds appear to have several functions linked to the existence of these organisms. One of the apparent functions in microorganism that use hydrocarbon substrates for growth or exist in oily substrates is to make these substrates available for them to metabolise. Another role for biosurfactants includes the motility of bacteria in viscous environments. Some biosurfactants can also play a role in quorum sensing mechanisms of the cells, which adjust gene expressions depending on cell density or on the surrounding environment. There has also been evidence that biosurfactants play an important role in biofilm formation and maintenance mechanisms, inducing biofilm formation on the one hand and disruption of mature biofilms integrity on the other, to sustain channels for nutrient and gaseous exchanges to cells at the base of the biofilm. All the above-hypothesised functions of biosurfactants have been put forward for a number of bacteria; reasons why several yeasts, however, produce large quantities of biosurfactants remain not fully explained. One explanation for yeasts is the type of environment they grow in which may have waxes or high molecular weight hydrophobic molecules.

After many years in which interest in biosurfactants was mainly among researchers, a critical stage in biosurfactants' potential commercial exploitation was reached when international companies began to show interest and a desire to explore use in some of their products. Using biosurfactants rather than chemical surfactants have deservedly become one of the top agenda items for many companies as a result of the sustainability initiative and green programs. Potential areas for use, therefore, began expanding rapidly and their beneficial outcomes have reached a stage to be dependent on whether biosurfactants can be tailored for specific applications, and whether they can be produced at a price that makes them attractive alternatives to chemical surfactants. Some issues, however, remain to be investigated and dealt with before largescale exploitation for many biosurfactants. These include cost of production, achievable yields and safety issues of producing strains such as the potential pathogen *Pseudomonas aeruginosa*. The fact that many biosurfactants have already been used as ingredients in several commercial products is a testimony to their potential for further exploitation. We, therefore, feel that there are no major impediments to the use of biosurfactants in a wide range of products and applications within the next few years, and may expect to see an increasing range of domestic products containing such products on supermarket shelves. There are limited consumer products around the world that already contain sophorolipids biosurfactants as an ingredient. These include cleaning and detergent products in Europe and the Far East, and some over-the-counter pharmaceuticals in the Far East. There is also a drive by some manufacturers aiming towards enhancing the green credentials of their product or hoping to target their product to specific sectors such as the vegan food markets that may enhance future interest.

Perhaps a measure of the current commercial interest in biosurfactants is the number and diversity of European Union-funded research projects that I have become involved with and that have major roles in biosurfactants research. These include, for example, the FP7 KILL.SPILL project, which examines the remediation of marine oil spills; the FP7 BIOSURFING project, which focuses on the production of 'new to nature' sophorolipids with applications in the food and pharmaceutical sectors; and the Horizon 2020 MARISURF project, which is pioneering the search for new biosurfactants from marine bacteria for a range of potential applications. All of these projects have many industrial partners supporting and exploring possibilities of using these compounds in their products or services.

It is with all the above in mind that when Dr. Rengathavasi Thavasi approached me to jointly initiate and edit a book in which experts in the field are invited to contribute materials related to description, potentials and exploitation of biosurfactants, I felt compelled to accept. We, therefore, thankfully acknowledge all the authors and co-authors of each chapter of the book for their valuable contributions and patience in trying to complete this task and hope this book adds another footstep in this important research and application area.

**Ibrahim M. Banat**
**Rengathavasi Thavasi**

# Contents

# 1

# Introduction to Microbial Biosurfactants

*Rengathavasi Thavasi[1],\* and Ibrahim M. Banat[2]*

## Introduction

Microorganisms produce structurally diverse metabolites with wide range of potential applications in many industrial sectors. Among such metabolites, microbial biosurfactants (BSs) are of great importance for their structural and functional diversity and broad spectrum applications (Banat 1995a, b, Desai and Banat 1997, Rodrigues et al. 2006). BSs are basically amphiphilic surface active molecules produced by bacteria, fungi, and actinomycetes that belong to various chemical groups including glycolipids, glycolipoproteins, glycopeptides, lipopeptides, lipoproteins, fatty acids, phospholipids, neutral lipids, lipopolysaccharides (Desai and Banat 1997, Banat et al. 2010, Marchant and Banat 2012b), and glycoglycerolipids (Wicke et al. 2000). Their properties/applications include emulsification, foaming, detergency, dispersion, wetting, penetrating, thickening, microbial growth enhancement (e.g., oil-degrading bacteria), antimicrobial agents, antibiofilm agents, antiadhesive, metal sequestering, resource recovery (oil recovery) (Banat et al. 2010, Fracchia et al. 2014) and wound healing agents (Ju et al. 2016, Lydon et al. 2017, Gupta et al. 2017). These interesting properties allow BSs to have the ability to replace some of the most versatile chemical surfactants that are now in practice (Marchant and Banat 2012a). In addition, BSs are promising natural surfactants that offer several advantages over chemically synthesized surfactants, such as *in situ* production from renewable substrates, lower toxicity, biocompatibility, and complete biodegradability (Makkar et al. 2011, Banat et al. 2014).

[1] Jeneil Biotech Inc., 400 Dekora Woods Blvd, Saukville, WI 53080, USA.
[2] University of Ulster, Faculty of Life & Health Sciences, Coleraine BT52 1SA, Northern Ireland, UK.
  Email: im.banat@ulster.ac.uk
\* Corresponding author: hydrobact@gmail.com

BSs are mainly composed of a hydrophilic and a hydrophobic moiety that provide them the ability to accumulate and partition between different phases (gas, liquid and solid), thus reducing surface tension (e.g., air-water) and interfacial tensions at liquid interface (e.g., oil-water or water-oil) or liquid-solid (e.g., liquid and sands or stones, wetting property). Most BSs are either anionic or neutral; the hydrophilic moiety is made of ester, hydroxyl, phosphate or carboxyl groups or carbohydrates including mono, di- and polysaccharides or proteins/peptides. The hydrophobic part is composed of saturated or unsaturated fatty acids, hydroxyl fatty acids or fat alcohols with carbon chain length ranging from 8 to 18.

The properties of BSs depend on their chemical structure and composition which affect their critical micelle concentration (CMC) (Satpute et al. 2010). At concentration above CMC, BSs molecules form micelles, bilayers or vesicles structural arrangements. This property enables the BSs to minimize surface and interfacial tension and enhance the solubility and bioavailability of hydrophobic compounds (e.g., oil in water). CMC is usually used to measure surfactant efficiency and highly active BSs have lower CMC, i.e., less concentration is required to decrease the surface tension. Another important property that affects BSs efficiency is their hydrophilic lipophilic balance (HLB) value which indicates the type of emulsion (e.g., oil in water or water in oil) formed by a BS in reaction with hydrophobic substrates (e.g., oil). BSs with low HLB values stabilize water-in-oil emulsion, whereas BSs with high HLB value stabilize oil-in-water emulsion (Desai and Banat 1997, De et al. 2015).

Among microbial biosurfactants, the three most studied glycolipids are sophorolipids (SLs) produced by *Candida/Starmerella* yeasts, rhamnolipids (RLs) produced by *Pseudomonas* species and mannosylerythritol lipids (MELs) produced by *Pseudozyma* yeasts, all of which have been taken to commercial scale production. Other biosurfactants are still in the process of making them commercially viable and evaluation of their application in different fields. This chapter provides a general scientific view of microbial BSs as an introduction to the book and individual chapters in this book will provide more detailed information on different microbial BSs, their characteristics, properties and potential applications in different fields.

# Biosynthesis BSs

## *Physiology and genetics of BSs synthesis*

The main factor determining BSs biosynthesis is the genetic makeup of the producer organisms and exposure of microbes to certain environments (oil spill). Reports on molecular genetics and biochemistry of the biosynthesis of BSs revealed the operons responsible, and enzymes and metabolic pathways required for their extracellular production. Reports on physiology and genetics of BSs synthesis are limited, for example, lipopeptide BSs produced by genus *Bacillus,* Surfactin (Peypoux et al. 1999) and in *B. amyloliquefaciens* (Zhi et al. 2017), and lichenysin (Yakimov et al. 1995) and RLs by *P. aeruginosa* (Reis et al. 2011). Surfactin is a cyclic lipopeptide BS and a potent antimicrobial agent produced as a result of non-ribosomal biosynthesis catalyzed by a large multi-enzyme peptide synthetase system called the surfactin synthetase. Biosynthetic pathways for other lipopeptides such as iturin, lichenysin and arthrofactin are also mediated by similar enzyme complexes, but are yet to be elucidated. These non-ribosomal peptide synthetases (NRPSs) responsible for lipopeptide biosynthesis display a high degree of structural similarity among themselves even from distant microbial species (Zhi et al. 2017). RLs biosynthetic regulatory factors include quorum sensing systems proteins and environmental response, and global regulatory systems within bacterial physiology, acting either at transcriptional or post-transcriptional level (Reis et al. 2011). It has also been reported that plasmid-encoded-rhlA, B, R and I genes of rhl quorum sensing system are required for production of glycolipid biosurfactants by *Pseudomonas* species (Das et al. 2008). Ochsner et al. (1994a, b, 1995) reported that the genetics of RLs biosynthesis *in P. aeruginosa,* the rhlABR gene cluster was responsible for the synthesis of RhlR regulatory protein and a rhamnosyl transferase and that active rhamnosyl transferase complex is located in the cytoplasmic membrane, with the 32.5-kDa RhlA protein harboring a putative signal sequence, while the 47-kDa RhlB protein is located in the periplasmic region and contains at least two putative membrane-spanning domains (Ochsner et al. 1994b). Ochsner and Reiser (1995) identified another regulatory gene, *rhlI*, located downstream to the *rhlABR* gene cluster. Further, it was shown that the regulation of rhamnolipid production in *P. aeruginosa* is mediated by the *rhlR-rhlI* system involving an autoinducer. However, reposts on genetics of biosynthesis of other BSs such as alasan and emulsan, MELs, CBLs, viscosin, amphisin and putisolvin are still limited.

## *Need for producing microbe*

BSs are produced by microorganisms for various reasons, among which, the following are some examples: (i) emulsification of hydrophobic substrates to increase the bioavailability (oil to the degrading microbes) (Desai and Banat 1997, Rosenberg and Ron 1999, Thavasi et al. 2011a, b); binding to heavy metals (Herman et al. 1995), (ii) pathogenesis—RLs production in *P. aeruginosa* is controlled by cell dependent systems RhlR-RhlI and the same system controls the synthesis of virulence factors, elastase and LasA protease (Ron and Rosenberg 2001), (iii) antimicrobial activity—competence for space, food material and self-defense, e.g., lipopeptide BSs are potent antibiotics against bacteria and fungi (Yakimov et al. 1995, Peypoux et al. 1999); RLs (Stanghellini and Miller 1997) and SLs were reported as promising antimicrobial agents (Elshikh et al. 2017a), (iv) regulation of attachment and detachment of microbes from surfaces—increase of cell hydrophobicity of *P. aeruginosa* in the presence cell-bound RLs (Zhang and Miller 1994) and decrease in cell hydrophobicity in the presence of cell-bound emulsifier in *Acinetobacter* strains (Rosenberg et al. 1983), (v) biofilm formation and disruption—RLs play role maintaining channels between multicellular structures in biofilms and in dispersal of cells from biofilms (Pamp and Tolker-Nielsen 2007, Quinn et al. 2013), RLs induce the release of lipopolysaccharides and thus enhance cell surface hydrophobicity, which might favor primary adhesion of planktonic cells (Al-Tahhan et al. 2000); biofilm disruption—addition of RLs disrupts *Salmonella typhimurium* and *Bordetella bronchiseptica* biofilms (Irie et al. 2005).

# Classification of Microbial BSs

Chemical/synthetic surfactants are usually classified according to the nature of their polar groups but BSs are generally categorized mainly by their chemical composition, particularly by the molecules that form the hydrophobic and hydrophilic moieties and microbial origin. According to Rosenberg and Ron (1999), BSs can be classified into two major groups based on their molecular weight as: (i) low-molecular-mass molecules, which efficiently lower surface and interfacial tension, and (ii) high molecular-weight polymers, which are more effective as emulsion stabilizing agents. The major classes of low molecular weight BSs include glycolipids, lipopeptides and phospholipids, whereas high molecular weight BSs include polymeric and particulate surfactants like polyanionic heteropolysaccharides containing both polysaccharides and proteins.

## *Low-molecular weight BSs*

### *Glycolipids (GLs)*

GLs are the well-studied molecules among all known microbial BSs. GLs are made of mono- or disaccharides connected to fatty acids. The saccharide moieties of the glycolipids comprise glucose, mannose, galactose, trehalose, and rhamnose. The hydrophobic lipids part is saturated or unsaturated fatty acids, hydroxyl fatty acids or fat alcohols. Following section will outline GL BSs (sophorolipids, rhamnolipids, trehaloselipids, cellobiose lipids and mannosylerythritol lipids).

### *Sophorolipids (SLs)*

SLs are glycolipid biosurfactant molecules produced by yeasts, such as *Candida bombicola, Yarrowi alipolytica, Candida apicola*, and *Candida bogoriensis*. The best known producing microbe is *Candida bombicolla* (ATCC 22214). SLs are composed of a carbohydrate head, sophorose, and a fatty acid tail with generally 16 or 18 carbon atoms with saturation or un-saturation. Sophorose is a disaccharide that consists of two glucose molecules linked at 1,2 position. The sophorose molecules in SLs are acetylated in certain conditions on the 6'- and/or 6''-positions. One fatty acid hydroxylated at the terminal or subterminal (β-1) positions is β-glycosidically linked to the sophorose molecule. The fatty acid carboxylic acid group is either free known as acidic or open form (Figure 1a) or esterified generally at the 4''-position called the lactonic form (Figure 1b). The hydroxy fatty acid moiety in SLs generally has 16 or 18 carbon atoms with generally one unsaturated bond (Asmer et al. 1988). Among all other biosurfactants, SLs is the only biosurfactant that has been produced with the highest yield 705 g/L by *C. bombicola* using glucose and ethyl esters of colza oil (Marchal et al. 1997).

### *Rhamnolipids (RLs)*

RLs glycolipid biosurfactants are produced by numerous bacterial species (Abdel-Mawgoud et al. 2010). Originally, RLs were discovered (Bergström et al. 1946) as extracellular products of *Pseudomonas* sp. and characterized as a mixture of two major forms based on number of sugar molecules as mono-RLs (Figure 1c) and di-RLs (Figure 1d), and there are two major forms of RLs with four derivatives based on their fatty acid composition as: Rha-Rha-C10-C10, Rha-Rha-C10, and their mono-RLs forms as Rha-C10-C10 and Rha-C10. Recently it has been revealed that there are about 60 RLs produced at different concentrations by various *Pseudomonas* species and by bacterial species belong to other families, classes, or even phyla. For example, bacteria from genus *Burkholderia, B. thailandensis* E264 (Elshikh et al. 2017b) was found to produce RLs with longer alkyl chains than RLs produced by *P. aeruginosa*. Despite the interesting surfactant properties of RLs, pathogenic nature of the producing bacteria (*P. aeruginosa*) had some concern to using this strain in industrial scale fermentation. The highest RLs yield reported for *P. aeruginosa* to date is 112 g/L 112 h (Giani et al. 1997). As a continuous search for a safer non-pathogenic strain has paid off, and there are many recent reports available on RLs production from non-pathogenic pseudomonas strains such as *P. chlororaphis* (Gunther et al. 2005) and *P. putida* KT2440 with the highest RLs production of 90g/L (Wittgens et al. 2011). Few other bacterial genera were also reported to produce RLs, for example, *Renibacterium salmoninarum* (Christova et al. 2004), *Cellulomonas cellulans* (Arino et al. 1998), and actinobacterium *Nocardioides* sp. (Vasileva-Tonkova and Gesheva 2005). A detailed review on RLs and diversity of structures, microbial origins and their roles were discussed by Abdel-Mawgoud et al. (2010). However, a recent review of literature carried out by Irorere et al. (2017) suggested that numerous research outputs have insufficient evidence to support claims of rhamnolipid-producing strains and/or yields achieved and have recommended some rigours standards to be set for reporting new rhamnolipid-producing strains and production yields.

## Trehalose lipids (TLs)

TLs are glycolipids produced by *Mycobacterium, Corynebacterium, Rhodococcus, Arthrobacter, Nocardia*, and *Gordonia* (Christova et al. 2015). TLs are composed of disaccharide trehalose attached to a long chain of hydroxyl fatty acid known as mycolic acid at C-6 and C-6′ in dimycolates and at C6 in momomycolates (Figure 1e). Mycolic acids are long chain, α-branched-β-hydroxy fatty acids. TLs produced by different organisms differ in the size and chemical nature of mycolic acid in their number of carbon atoms and the degree of unsaturation. Recently, Kügler et al. (2014) isolated TLs from two actinomycetes, *Tsukamurella spumae* and *T. pseudospumae* and the chemical composition of the TLs isolated from these two strains were identified as 1-α-glucopyranosyl-1-α-glucopyranosid carrying two acyl chains varying from C4 to C6 and C16 to C18 at the 2' and 3' carbon atom of one sugar unit. Applications of TLs include induced differentiation of leukemia cell lines (Sudo et al. 2000) and inhibition of protein kinase activity (Isoda et al. 1997).

## Cellobiose lipids (CBLs)

CBLs are glycolipids consisting of β-cellobiose sugar moiety as the hydrophilic and fatty acid and/or acetyl as the hydrophobic moiety. Unlike MELs, CBLs do not have sugar alcohol as part of the structure but consist of various R groups as side chains.

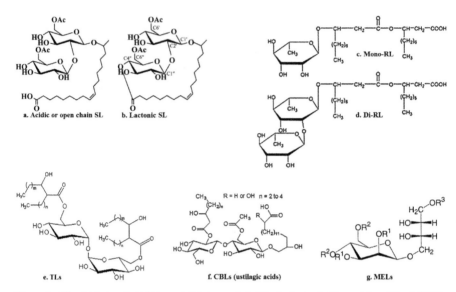

**Figure 1.** Microbial glycolipid BSs, a. Acidic/open chain SLs, b. Lactonic SLs, c. Mono-RL, d. Di-RL, e. TLs, f. CBLs and g. MELs.

CBLs producing microbes include *Ustilago maydis* (Figure 1f), *Cryptococcus humicola, Pseudozyma fusiformata, P. hubeiensis, P. aphidis, P. graminicola* and *P. flocculosa*. So far, nine different chemical forms of CBLS have been reported and discussed in detail by Kulakovskay et al. (2009) and Morita et al. (2013).

## Mannosylerythritol lipids (MELs)

MELs are another class of glycolipids produced by *Pseudozyma* sp. and *Ustilago* sp. (Figure 1g) characterized as 4-O-b-D-mannopyranosylerythritol or 1-O-b-D-mannopyranosyl-erythritol as their hydrophilic head group and fatty acyl as hydrophobic units with one or two acetyl groups at C-4' and/or C-6' of the mannose moiety. Based on their acetylation, MELs are grouped in to three forms as MEL-A di-acetylated, whereas MEL-B and MEL-C are mono-acetylated at C-4' and C-6', respectively (Morita et al. 2015). The pharma industry applications of MELs includes cell differentiation-inducing activities against human leukemia cells, rat pheochromocytoma cells, and mouse melanoma cells, and inhibit the secretion of inflammatory mediators by mast cells. In biochemical industry, MELs are used in the purification of lectins. In the environmental sector, the readily biodegradable and emulsification activity of MELs makes them suitable molecules for bioremediation applications (Yu et al. 2015).

## Fatty acids and phospholipids

Bacteria and fungi are known to produce fatty acids and phospholipids when cultured with hydrophobic substrates such as *n*-alkanes (Desai and Banat 1997). Fatty acids are produced from alkanes by microbial oxidations with surfactant properties considered as surfactants (Rehn and Reiff 1981). Besides straight-chain fatty acids, microbes produce complex fatty acids containing OH groups and alkyl branches, for example, corynomycolic acid (Kretschner et al. 1982). The hydrophilic or lipophilic balance (HLB) of surface active fatty acids depends on the hydrocarbon chain length. Fatty acids with chain length of C12 to C14 are reported to have high surface active properties (Rosenberg and Ron 1999).

Phospholipids are major components of microbial cell membranes. While growing hydrocarbon-degrading bacteria or yeast on alkane substrates, there is an increase in phospholipid concentration in the culture medium. For example, *Acinetobacter* sp. HO1-N culture was grown with hexadecane produced phospholipids (phosphatidylethanolamine as major component) rich vesicles (Kaeppeli and Finnerty 1979). Other examples of microbial phospholipids include phosphatidylethanolamine produced by *Rhodococcus erythropolis* grown on *n*-alkane and the phospholipid which lowered the interfacial tension of water and hexadecane to less than 1 mNm$^{-1}$ with a CMC of 30 mg L$^{-1}$ (Kretschner et al. 1982).

## Lipopeptides (LPs)

LPs are cyclic peptides composed of amino acids linked with fatty acids (C12–C18). LPs include the well-known lipopeptide antibiotics, surfactin from *Bacillus subtilis* (Arima et al. 1968), polymyxins isolated from *Bacillus brevis* (Marahiel et al. 1977), *B. polymyxa* (Suzuki et al. 1965), and lichenysin produced by *B. licheniformis* (Yakimov et al. 1995). LPs from other microbes include serrawettin (serratamolide) from *Serratia marcescens* NS.38 (Mutsuyama et al. 1985) and viscosin from *Pseudomonas fluorescens* (Neu et al. 1990, Bonnichsen et al. 2015), and *P. libanensis* M9-3 (Saini et al. 2008).

## Surfactin

The cyclic lipopeptide surfactin (Figure 2a), produced by *B. subtilis*, is one of the most powerful LP BS. It lowers the surface tension of water from 72 to 27.9 mN/m at concentration as low as 0.005% (Arima et al. 1968). Chemically, surfactin is composed of seven amino acids that are linked to the carboxyl and hydroxy groups on long chain fatty acids with a carbon chain length C13–C15. Surfactin is used in therapeutic and environmental applications.

## Lichenysin

A cyclic lipopeptide BS isolated form *B. licheniformis* BAS50 (Lichenysin A) (Yakimov et al. 1995). The lichenysin A (Figure 2b) isolated from *B. licheniformis* BAS50 reduced the surface tension of water from 72 to 28 mN/m with a CMC of 12 mM, which is 24 mM for surfactin. Chemical characterization of lichenysin A showed that isoleucine was the C-terminal amino acid instead of leucine and an asparagine residue was found as aspartic acid in surfactin. Application of lichenysin includes antimicrobial, antiadhesion, and antibiofilm activities (Coronel-León et al. 2016).

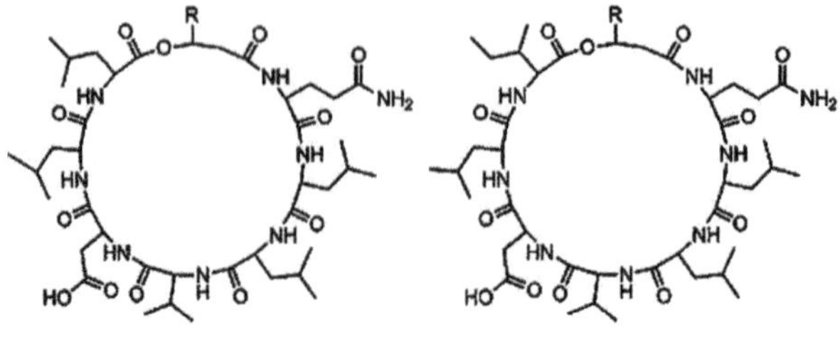

a. Surfactin        b. Lichenysin

**Figure 2.** Lipopeptide BSs, a. Surfactin and b. Lichenysin.

## Polymyxins

Polymyxins belong to the cationic cyclic lipopeptide. There are five different types isolated so far, Polymyxins A-E from *Bacillus brevis* (Marahiel et al. 1977) and *B. polymyxa* (Suzuki et al. 1965). Polymyxins are well-known, clinically used antibiotics, among the five types, polymyxin B (Figure 3a) and polymyxin E (also called colistin) have been widely used in clinical practice since 1950s. Polymyxins are highly bactericidal against gram-negative organisms and used for treating infections caused by multidrug-resistant microbes (Landman et al. 2008).

## Serrawettins

They are nonionic, aminolipid BSs, produced by *Serratia* species. Three molecular species of serrawettins were reported: *serrawettin W1*, cyclo(D-3-hydroxydecanoyl-L-seryl)2; *serrawettin W2*, D-3-hydroxydecanoyl-D-leucyl-L-seryl-L-threonyl-D-phenylalanyl-L-isoleucyllactone; and *serrawettin W3*, cyclodepsipeptide composed of five amino acids and another BSs called serratamolide (Figure 3b) was isolated from *S. marcescens* NS.38. *S. rubidaea* produces another type of serrawettin called rubiwettin R1, linked D-3-hydroxy fatty acids and RG1, b-glucopyranosyl linked D-3-hydroxy fatty acids (Mutsuyama et al. 1985, Matsuyama et al. 2011). Applications of serrawettin are antimicrobial (Wasserman et al. 1962, Matsuyama et al. 1992), biocontrol against plant oomycete pathogens (Strobel et al. 2003) and anticancer activities (Escobar-Díaz et al. 2005).

## Viscosin

It is a cyclic lipopeptide BS that contains hydrophobic amino acids bonded with fatty acid (Figure 4). Viscosin production was reported from *Pseudomonas fluorescens* (Neu et al. 1990), *P. fluorescens* SBW25 (Bonnichsen et al. 2015), and *P. libanensis* M9-3 (Saini et al. 2008). Viscosin has been explored for enhanced mineralization and emulsification of *n*-hexadecane (Bak et al. 2015), antimicrobial, and enhanced dispersion of biofilm formation by the producing microbe. Viscosin also inhibited the migration of metastatic prostate cancer cell line, PC-3M, without any known toxicity which suggests its potential biomedical applications (Bonnichsen et al. 2015). The biofilm dispersal property of viscosin was evaluated for its application in agriculture by Alsohim et al. (2014) which revealed that viscosin increased the efficiency of biofilm surface spreading over the plant root and protected germinating seedlings

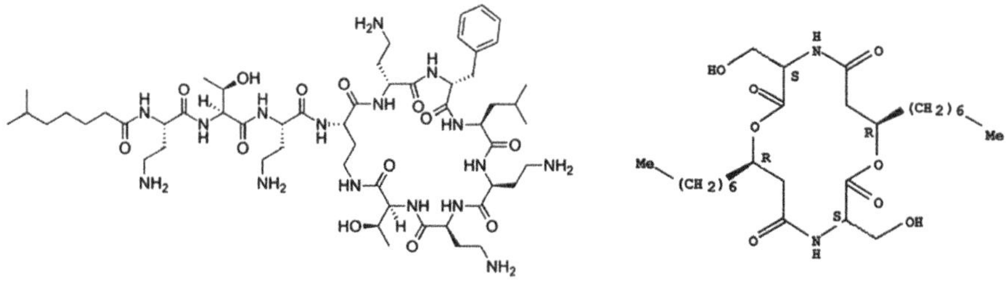

**a. Polymyxin B**                    **b. Serratamolide**

**Figure 3.** Microbial lipopeptide BSs, a. Polymyxin and b. Serratamolide.

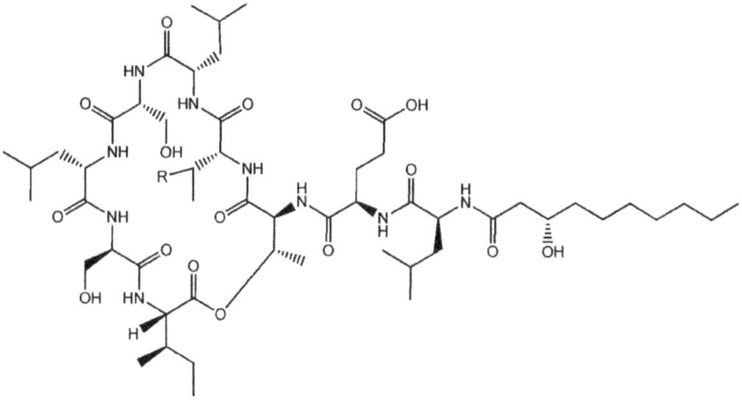

**Figure 4.** Microbial lipopeptide BS, Viscosin.

in soil from the plant pathogen Pythium. Thus, viscosin could be useful as a plant growth promoting and biocontrol agent.

## High-molecular weight BSs

A wide variety of microorganisms, including *Acinetobacter, Arthrobacter, Pseudomonas, Halomonas, Bacillus,* and *Candida* have been identified to produce polymeric BSs. Chemically, polymeric BSs are polysaccharides, proteins, lipopolysaccharides, lipoproteins and a mixture of the above. Unlike other BSs, polymeric BSs do not necessarily reduce surface tension, but they are known for their efficacy to reduce the interfacial tension between immiscible liquids and form stable emulsions. Examples of polymeric BSs are emulsan from *Acinetobacter calcoaceticus* (Rosenberg and Ron 1999), biodispersan isolated from *A. calcoaceticus* (Rosenberg and Ron 1997), liposan from *Candida lipolytica* (Cirigliano and Carman 1985) and alasan produced by *A. radioresistens* (Navon-Venezia et al. 1998). The most studied polymeric BSs are emulsan and liposan.

### Emulsan

Isolated from *A. calcoaceticus* RAG-1 by Rosenberg et al. (1979), it is a lipopolysaccharide (1000 kDa) composed of a heteropolysaccharide linked to a

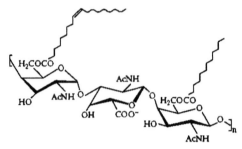

**Figure 5.** Emulsan.

fatty acid via ester and amide bonds (Figure 5). Mercaldi et al. (2008) revealed that chemically, emulsan is a complex of approximately 80% (w/w) lipopolysaccharide and 20% (w/w) high-molecular weight exopolysaccharide. Emulsan is a very effective emulsifying agent for hydrocarbons in water even at concentrations of 0.001% to 0.01%.

## *Liposan*

It is an extracellular water soluble emulsifier produced by *C. lipolytica* during stationary phase of fermentation when grown with hexadecane as the carbon substrate and is made up of 83% carbohydrates and 17% proteins. Further chemical characterization revealed that the sugar content of liposan is similar to that of the emulsifier emulsan, with a heteropolysaccharide consisting of glucose, galactose, galactosamine, and galacturonic acid (Cirigliano and Carman 1985).

## *Alasan*

Alasan is a bioemulsifier isolated from *A. radioresistens* KA53 (Navon-Venezia et al. 1995). Chemically, alasan is a complex of anionic, high molecular weight, alanine containing heteropolysaccharide and protein (Navon-Venezia et al. 1998). Emulsification activity of alasan is associated with its protein moiety which was demonstrated by the inactivation of its emulsifying activity as a result of its treatment with specific proteases (Toren et al. 2001). Alasan is 2.5 to 3 times more active even after a heat treatment at 100°C under neutral or alkaline condition (Navon-Venezia et al. 1998).

## Conclusion

Biological and physicochemical properties of microbial BSs clearly indicated their potential application in place of their chemical counter parts. Raising demand for biological/green molecules among the public and scientific community, and versatility and broad spectrum application potentials of BSs are the main driving force for recent developments in BSs science. Commercial production of BSs such as SLs, RLs and MELs changed the image of microbial BSs as commercially viable molecules with potential applications. The main factor that drives the price of BSs are downstream processing which accounts for 60% of the production cost (Satpute et al. 2010) and large scale downstream processing of SLs, RLs and MELs which are commercially available/viable as compared to the downstream processing of other BSs.

Production cost can be justified based on applications of biosurfactants, for example, medical and cosmetic application require high purity BSs which requires extensive purification process that increases the cost of the final product. On the other hand, environmental applications, such as use in bioremediation and agriculture, may be effectively achieved using crude or reasonably purified/extracted BSs where the purification cost is minimum. Development of non-GMO strains with high BSs producing property, use of renewable substrates in production medium and identification of suitable applications will encourage more industrial investments and their interest in development of such methods. Another potential approach could be the use of BSs producing strains that can grow on site/in application site as *in situ* BSs producers such as for uses in microbially enhanced oil recovery (MEOR) or in bioremediation of soil, where the potential BS producing strains can be applied with minimal nutrients and produce required amount of BS to extract the oil (in MEOR) or clean the oil (in oil spill). This approach will reduce the cost and create new avenues for BSs technology. Finally, this book is designed to provide the latest developments in BS science, focusing on applications of BSs in various fields as reported by active researchers in this area.

# References

Abdel-Mawgoud, A.M., F. Lépine and E. Déziel. 2010. Rhamnolipids: diversity of structures, microbial origins and roles. Appl. Microbiol. Biotechnol. 86(5): 1323–1336.

Alsohim, A,S., T.B. Taylor, G.A. Barrett, J. Gallie, X.X. Zhang, A.E. Altamirano-Junqueira, L.J. Johnson, P.B. Rainey and R.W. Jackson. 2014. The biosurfactant viscosin produced by *Pseudomonas fluorescens* SBW25 aids spreading motility and plant growth promotion. Environ. Microbiol. 16(7): 2267–81.

Al-Tahhan, R.A., T.R. Sandrin, A.A. Bodour and R.M. Maier. 2000. Rhamnolipid induced removal of lipopolysaccharide from *Pseudomonas aeruginosa*: effect on cell surface properties and interaction with hydrophobic substrates. Appl. Environ. Microbiol. 66: 3262–3268.

Arima, K., A. Kakinuma and G. Tamura. 1968. Surfactin, a crystalline peptide lipid surfactant produced by *Bacillus subtilis*: isolation, characterization and its inhibition of fibrin clot formation. Biochem. Biophys. Res. Commun. 31: 488–494.

Arino, S., E. Marchal and J.P. Vandecasteele. 1998. Production of new extracellular glycolipids by a strain of *Cellulomonas cellulans* (Oerskovia xanthineolytica) and their structural characterization. Can. J. Microbiol. 44: 238–243.

Asmer, H.J., S. Lang, F. Wagner and V. Wray. 1988. Microbial production, structure elucidation and bioconversion of sophorose lipids. J. Am. Oil Chem. Soc. 65: 1460–1466.

Bak, F., L. Bonnichsen, N.O. Jørgensen, M.H. Nicolaisen and O. Nybroe. 2015. The biosurfactant viscosin transiently stimulates *n*-hexadecane mineralization by a bacterial consortium. Appl. Microbiol. Biotechnol. 99: 1475–1483.

Banat, I.M. 1995a. Biosurfactants characterization and use in pollution removal: state of the art. A review. Acta Biotechnologica. 15: 251–267.

Banat, I.M. 1995b. Biosurfactants production and use in microbial enhanced oil recovery and pollution remediation: A review. Bioresour. Technol. 51: 1–12.

Banat, I.M., A. Franzetti, I. Gandolfi, G. Bestetti, M.G. Martinotti, L. Fracchia, T.J. Smyth and R. Marchant. 2010. Microbial biosurfactants production, applications and future potential. Appl. Microbiol. Biotechnol. 87: 427–444.

Banat, I.M., S.K. Satpute, S.S. Cameotra, R. Patil and N.V. Nyayanit. 2014. Cost effective technologies and renewable substrates for biosurfactants' production. Front. Microbiol. 5: 697.

Bergström, S., H. Theorell and H. Davide. 1946. On a metabolic product of *Ps. pyocyanea*, pyolipic acid, active against *Mycobacterium tuberculosis*. Ark. Chem. Miner. Geol. 23A(13): 1–12.

Bonnichsen, L., N. Bygvraa Svenningsen, M. Rybtke, T. de Bruijn, J.M. Raaijmakers, T. Tolker-Nielsen and O. Nybroe. 2015. Lipopeptide biosurfactant viscosin enhances dispersal of *Pseudomonas fluorescens* SBW25 biofilms. Microbiology. 161: 2289–2297.

Christova, N., B. Tuleva, Z. Lalchev, A. Jordanova and B. Jordanov. 2004. Rhamnolipid biosurfactants produced by *Renibacterium salmoninarum* 27BN during growth on n-hexadecane. Z. Nat. Forsch. C. J. Biosci. 59: 70–74.

Christova, N., S. Lang, V. Wray, K. Kaloyanov, S. Konstantinov and I. Stoineva. 2015. Production, structural elucidation, and *in vitro* antitumor activity of trehalose lipid biosurfactant from *Nocardia farcinica* strain. J. Microbiol. Biotechnol. 25(4): 439–447.

Cirigliano, M.C. and G.M. Carman. 1985. Purification and characterization of liposan, a bioemulsifier from *Candida lipolytica*. Appl. Microbiol. Biotechnol. 50: 846–850.

Coronel-León, J., A.M. Marqués, J. Bastida and A. Manresa. 2016. Optimizing the production of the biosurfactant lichenysin and its application in biofilm control. J. Appl. Microbiol. 120(1): 99–111.

Das, P., S. Mukherjee and R. Sen. 2008. Genetic regulations of the biosynthesis of microbial surfactants: an overview. Biotechnol. Gene. Eng. Rev. 25: 165–85.

De, S., S. Malik, A. Ghosh, R. Saha and B. Saha. 2015. A review on natural surfactants. RSC Adv. 5: 65757–65767.

Desai, J.D. and I.M. Banat. 1997. Microbial production of surfactants and their commercial potential. Microbiol. Mol. Biol. Rev. 61: 47–64.

Elshikh, M., I. Moya-Ramírez, H. Moens, S. Roelants, W. Soetaert, R. Marchant and I.M. Banat. 2017a. Rhamnolipids and lactonic sophorolipids: natural antimicrobial surfactants for oral hygiene. J. Appl. Microbiol. 123(5): 1111–1123.

Elshikh, M., S. Funston, A. Chebbi, S. Ahmed, R. Marchant and I.M. Banat. 2017b. Rhamnolipids from non-pathogenic *Burkholderia thailandensis* E264: Physicochemical characterization, antimicrobial and antibiofilm efficacy against oral hygiene related pathogens. N. Biotechnol. 36: 26–36.

Escobar-Díaz, E., E.M. López-Martín, M. Hernández del Cerro, A. Puig-Kroger, V. Soto-Cerrato, B. Montaner, E. Giralt, J.A. García-Marco, R. Pérez-Tomás and A. Garcia-Pardo. 2005. AT514, a cyclic depsipeptide from Serratia marcescens, induces apoptosis of B-chronic lymphocytic leukemia cells: interference with the Akt/NF-kappaB survival pathway. Leukemia 19(4): 572–579.

Fracchia, L., C. Ceresa, A. Franzetti, M. Cavallo, I. Gandolfi, J. Van Hamme, P. Gkorezis, R. Marchant and I.M. Banat. 2014. Industrial applications of biosurfactants. pp. 245–260. *In*: N. Kosaric and F.V. Sukan (eds.). Biosurfactants: Production and Utilization-Processes, Technologies, and Economics. Surfactant Science Series 159, CRC Press.

Giani, C., D. Wullbrandt, R. Rothert and J. Meiwes. 1997. *Pseudomonas aeruginosa* and its use in a process for the biotechnological preparation of L-rhamnose. U.S. Patent 5,658,793.

Gunther, N.W., A. Nuñez, W. Fett and D.K.Y. Solaiman. 2005. Production of rhamnolipids by *Pseudomonas chlororaphis*, a nonpathogenic bacterium. Appl. Environ. Microbiol. 71(5): 2288–2293.

Gupta, S., N. Raghuwanshi, R. Varshney, I.M. Banat, A.K. Srivastava, P.A. Pruthi and V. Pruthi. 2017. Accelerated *in vivo* wound healing evaluation of microbial glycolipid containing ointment as a transdermal substitute. Biomed. Pharmacother. 94: 1186–1196.

Herman, D.C., J.F. Artiola and R.M. Miller. 1995. Removal of cadmium, lead, and zinc from soil by a rhamnolipid biosurfactant. Environ. Sci. Technol. 29: 2280–2285.

Irie, Y., G.A. O'Toole and M.H. Yuk. 2005. *Pseudomonas aeruginosa* rhamnolipids disperse *Bordetella bronchiseptica* biofilms. FEMS Microbiol. Lett. 250: 237–243.

Irorere, V.U., L. Tripathi, R. Marchant, S. McClean and I.M. Banat. 2017. Microbial rhamnolipid production: A critical re-evaluation of published data and suggested future publication criteria. Appl. Microbiol. Biotechnol. 101: 3941–3951.

Isoda, H., D. Kitamoto, H. Shinmoto, M. Matsumura and T. Nakahara. 1997. Microbial extracellular glycolipid induction of differentiation and inhibition of the protein kinase C activity of human promyelocytic leukemia cell line HL60. Biosci. Biotechnol. Biochem. 61(4): 609–14.

Ju, L-K., S. Dashtbozorg and N. Vongpanish. 2016. Wound dressings with enhanced gas permeation and other beneficial properties. U.S. Patent N. 9,468,700 B2.

Kaeppeli, O. and W.R. Finnerty. 1979. Partition of alkane by an extracellular vesicle derived from hexadecane-grown *Acinetobacter*. J. Bacteriol. 140: 707–712.

Kretschner, A., H. Block and F. Wagner. 1982. Chemical and physical characterization of interfacial-active lipids from *Rhodococcus erythropolis* grown on n-alkanes. Applied Environ. Microbiol. 44: 864–870.

Kügler, J.H., C. Muhle-Goll, B. Kühl, A. Kraft, R. Heinzler, F. Kirschhöfer, M. Henke, V. Wray, B. Luy, G. Brenner-Weiss, S. Lang, C. Syldakt and R. Hausmann. 2014. Trehalose lipid biosurfactants produced by the actinomycetes *Tsukamurella spumae* and *T. pseudospumae*. Appl. Microbiol Biotechnol. 98: 8905–8915.

Kulakovskaya, T., A. Shashkov, E. Kulakovskaya, W. Golubev, A. Zinin, Y. Tsvetkov, A. Grachev and N. Nifantiev. 2009. Extracellular cellobiose lipid from yeast and their analogues: structures and fungicidal activities. J. Oleo. Sci. 58(3): 133–40.

Landman, D., C. Georgescu, D.A. Martin and J. Quale. 2008. Polymyxins revisited. Clin. Microbiol. Rev. 21(3): 449–465.

Lydon, H.L., N. Baccile, B. Callaghan, R. Marchant, C.A. Mitchell and I.M. Banat. 2017. Adjuvant antibiotic activity of acidic sophorolipids with potential for facilitating wound healing. Antimicrob. Agents Chemother. 61(5) pii: e02547-16.

Makkar, R.S., S.S. Cameotra and I.M. Banat. 2011. Advances in utilization of renewable substrates for biosurfactant production. Appl. Microbiol. Biotechnol. Express 1(1): 1–5.

Marahiel, M., W. Denders, M. Krause and H. Kleinkauf. 1977. Biological role of gramicidin S in spore functions. Studies on gramicidin-S negative mutants of *Bacillus brevis* 9999. Eur. J. Biochem. 99: 49–52.

Marchal, R., J. Lemal and C. Sulzer. 1997. Method of production of sophorosides by fermentation with fed batch supply of fatty acid esters or oils. US Patent 5616479.

Marchant, R. and I.M. Banat. 2012. Microbial biosurfactants: challenges and opportunities for future exploitation. Trends Biotechnol. 30(11): 558–565.

Matsuyama, T., K. Kaneda, Y. Nakagawa, K. Isa, H. Hara-Hotta and I. Yano. 1992. A novel extracellular cyclic lipopeptide which promotes flagellum-dependent and -independent spreading growth of *Serratia marcescens*. J. Bacteriol. 174: 1769–1776.

Matsuyama, T., T. Tanikawa and Y. Nakagawa. 2011. Serrawettins and other surfactants produced by *Serratia*. pp. 93–120. *In*: G. Soberón-Chávez (ed.). Biosurfactants, Microbiology Monographs 20. Springer-Verlag, Berlin Heidelberg.

Mercaldi, M.P., H. Dams-Kozlowska, B. Panilaitis, A.P. Joyce and D.L. Kaplan. 2008. Discovery of the dual polysaccharide composition of emulsan and the isolation of the emulsion stabilizing component. Biomacromolecules 9(7): 1988–1996.

Morita, T., T. Fukuoka, T. Imura and D. Kitamoto. 2013. Accumulation of cellobiose lipids under nitrogen-limiting conditions by two ustilaginomycetous yeasts, Pseudozyma aphidis and Pseudozyma hubeiensis. FEMS Yeast Res. 13: 44–49.

Morita, T., T. Fukuoka, T. Imura and D. Kitamoto. 2015. Mannosylerythritol lipids: production and applications. J. Oleo. Sci. 64(2): 133–141.

Mutsuyama, T., M. Fujita and I. Yano. 1985. Wetting agent produced by *Serratia marcescens*. FEMS Microbiol. Lett. 28: 125–129.

Navon-Venezia, S., Z. Zosim, A. Gottlieb, R. Legmann, S. Carmeli, E.Z. Ron and E. Rosenberg. 1995. Alasan, a new bioemulsifier from *Acinetobacter radioresistens*. Appl. Environ. Microbiol. 61(9): 3240–3244.

Navon-Venezia, S., E. Banin, E.Z. Ron and E. Rosenberg. 1998. The bioemulsifier alasan: role of protein in maintaining structure and activity. Appl. Microbiol. Biotechnol. 49(4): 382–384.

Neu, T.R., T. Haertner and K. Poralla. 1990. Surface active properties of viscosin: a peptidolipid antibiotic. Appl. Microbiol. Biotechnol. 32: 518–520.

Ochsner, U.A., A.K. Koch, A. Fiechter and J. Reiser. 1994a. Isolation and characterization of a regulatory gene affecting rhamnolipid biosurfactant synthesis in *Pseudomonas aeruginosa*. J. Bacteriol. 176: 2044–2054.

Ochsner, U.A., A. Fiechter and J. Reiser. 1994b. Isolation, characterization and expression in *Escherichia coli* of the *Pseudomonas aeruginosa* rhlAB genes encoding a rhamnosyltransferase involved in rhamnolipid biosurfactant synthesis. J. Biol. Chem. 269: 19787–19795.

Ochsner, U.A. and J. Reiser. 1995. Autoinducer-mediated regulation of rhamnolipid biosurfactant synthesis in *Pseudomonas aeruginosa*. Proc. Natl. Acad. Sci. USA 92: 6424–6428.

Ochsner, U.A., J. Reiser, A. Fiechter and B. Witholt. 1995. Production of *Pseudomonas aeruginosa* rhamnolipid biosurfactants in heterogeneous host. Appl. Environ. Microbiol. 61: 3503–3506.

Pamp, S.J. and T. Tolker-Nielsen. 2007. Multiple roles of biosurfactants in structural biofilm development by *Pseudomonas aeruginosa*. J. Bacteriol. 189(6): 2531–2539.

Peypoux, F., J.M. Bonmatin and J. Wallach. 1999. Recent trends in the biochemistry of surfactin. Appl. Microbiol. Biotechnol. 51(5): 553–63.

Quinn, G.A., A.P. Maloy, M.M. Banat and I.M. Banat. 2013. A comparison of effects of broad-spectrum antibiotics and biosurfactants on established bacterial biofilms. Curr. Microbiol. 67: 614–623.

Rehn, H.J. and I. Reiff. 1981. Mechanisms and occurrence of microbial oxidation of long-chain alkanes. Adv. Biochem. Eng. 19: 175–216.

Reis, R.S., A.G. Pereira, B.C. Neves and D.M. Freire. 2011. Gene regulation of rhamnolipid production in *Pseudomonas aeruginosa*—a review. Bioresour. Technol. 102(11): 6377–6384.

Rodrigues, L., I.M. Banat, G. Teixeira and O. Oliveira. 2006. Biosurfactants: potential applications in medicine. J. Antimicrob. Chemother. 57: 609–618.

Ron, E.Z. and E. Rosenberg. 2001. Natural roles of biosurfactants. Environ. Microbiol. 3(4): 229–236.

Rosenberg, E., A. Zuckerberg, C. Rubinovitz and D.L. Gutnick. 1979. Emulsifier Arthrobacter RAG-1: isolation and emulsifying properties. Appl. Environ. Microbiol. 37: 402–408.

Rosenberg, E., A. Gottlieb and M. Rosenberg. 1983. Inhibition of bacterial adherence to hydrocarbons and epithelial cells by emulsan. Infect Immun. 39(3): 1024–1028.

Rosenberg, E. and E.Z. Ron. 1999. High- and low-molecular-mass microbial surfactants. Appl. Microbiol. Biotechnol. 52: 154–162.

Saini, H.S., B.E. Barragán-Huerta, A. Lebrón-Paler, J.E. Pemberton, R.R. Vázquez, A.M. Burns, M.T. Marron, C.J. Seliga, A.A. Gunatilaka and R.M. Maier. 2008. Efficient purification of the biosurfactant viscosin from *Pseudomonas libanensis* strain M9-3 and its physicochemical and biological properties. J. Nat. Prod. 71(6): 1011–1015.

Satpute, S.K., A.G. Banpurkar, P.K. Dhakephalkar, I.M. Banat and B.A. Chopade. 2010. Methods for investigating biosurfactants and bioemulsifiers: A review. Crit. Rev. Biotechnol. 30: 127–144.

Stanghellini, M.E. and R.M. Miller. 1997. Biosurfactants their identity and potential efficacy in the biological control of zoosporic plant pathogens. Plant Dis. 81(1): 4–11.

Strobel, G.A., S.I. Morrison and M. Cassella. 2003. Methods for protection of plants from Oomyocyte pathogens by use of *Serratia marcescens* and isolates. U.S. Patent Appl. US2003/0049230 A1.

Sudo, T., X. Zhao, Y. Wakamatsu, M. Shibahara, N. Nomura, T. Nakahara, A. Suzuki, Y. Kobayashi, C. Jin, T. Murata and K.Z. Yokoyama. 2000. Induction of the differentiation of human HL-60 promyelocytic leukemia cell line by succinoyl trehalose lipids. Cytotechnology 33: 259–264.

Suzuki, T., K. Hayashi, K. Fujikawa and K. Tsukamoto. 1965. The chemical structure of polymyxin E. The identies of polymyxin E1 with colistin A and polymyxin E2 with colistin B. J. Biol. Chem. 57: 226–227.

Syldatk, C. and F. Wagner. 1987. Production of biosurfactants. pp. 89–120. *In*: N. Kosaric, W.L. Cairns and N.C.C. Gray (eds.). Biosurfactants and Biotechnology. Marcel Dekker, Inc., New York.

Thavasi, R., S. Jayalakshmi and I.M. Banat. 2011a. Effect of biosurfactant and fertilizer on biodegradation of crude oil by marine isolates of *Bacillus megaterium, Corynebacterium kutscheri* and *Pseudomonas aeruginosa*. Bioresour. Technol. 102: 772–778.

Thavasi, R., S. Jayalakshmi and I.M. Banat. 2011b. Application of biosurfactant produced from peanut oil cake by *Lactobacillus delbrueckii* in biodegradation of crude oil. Bioresour. Technol. 102: 3366–3372.

Toren, A., S. Navon-Venezia, E.Z. Ron and E. Rosenberg. 2001. Emulsifying activities of purified Alasan proteins from *Acinetobacter radioresistens* KA53. Appl. Environ. Microbiol. 67(3): 1102–1106.

Vasileva-Tonkova, E. and V. Gesheva. 2005. Glycolipids produced by Antarctic *Nocardioides* sp. during growth on n-paraffin. Process Biochem. 40: 2387–2391.

Wasserman, H.H., J.J. Keggi and J.E. Mckeon. 1962. The structure of serratamolide. J. Am. Chem. Soc. 84: 2978–2982.

Wicke, C., M. Hüners, V. Wray, M. Nimtz, U. Bilitewski and S. Lang. 2000. Production and structure elucidation of glycoglycerolipids from a marine sponge-associated *Microbacterium* species. J. Nat. Prod. 63: 621–626.

Wittgens, A., T. Tiso, T.T. Arndt, P. Wenk, J. Hemmerich, C. Muller, R. Wichmann, B. Küpper, M. Zwick, S. Wilhelm, R. Hausmann, C. Syldatk, F. Rosenau and L.M. Blank. 2011. Growth independent rhamnolipid production from glucose using the non-pathogenic *Pseudomonas putida* KT2440. Microb. Cell Fact. 10: 80.

Yakimov, M.M., K.N. Timmis, V. Wray and H.L. Fredrickson. 1995. Characterization of a new lipopeptide surfactant produced by thermotolerant and halotolerant subsurface *Bacillus licheniformis* BAS50. Appl. Environ. Microbiol. 61: 1706–1713.

Yu, M., Z. Liu, G. Zeng, H. Zhong, Y. Liu, Y. Jiang, M. Li, X. He and Y. He. 2015. Characteristics of mannosylerythritol lipids and their environmental potential. Carbohydr. Res. 407: 63–72.

Zhang, Y. and R.M. Miller. 1994. Effect of a *Pseudomonas* rhamnolipid biosurfactant on cell hydrophobicity and biodegradation of octadecane. Appl. Environ. Microbiol. 60(6): 2101–2106.

Zhi, Y., Q. Wu and Y. Xu. 2017. Genome and transcriptome analysis of surfactin biosynthesis in *Bacillus amyloliquefaciens* MT45. Sci. Rep. 7: 40976.

# 2

# Downstream Processing of Microbial Biosurfactants

*Rengathavasi Thavasi[1],* and Ibrahim M. Banat[2]*

## Introduction

The global biosurfactants (BSs) market is estimated at $4.20 billion in 2017 and projected to reach $5.52 billion by 2022, at a compound annual growth rate (CAGR) of 5.6% from 2017 to 2022 (Global biosurfactants market research report by Markets and Markets™ Inc. 2017). Increase in regulation and consumer preference towards bio-based and organic products, particularly in Europe and North America, is expected to increase BSs use in many products (Marchant and Banat 2012a). The commercial production of BSs such as sophorolipids (SLs), rhamnolipids (RLs) and mannosylerythritol lipids (MELs) changed the image of microbial BSs as commercially viable molecules with potential applications (Marchant and Banat 2012b). As their demand in the market rises, so does the demand for large scale production of cost effective sustainable renewable substrates (Banat et al. 2014) and efficient downstream processing techniques. Downstream processing of BSs accounts for 60% of the production cost (Desai and Banat 1997, Satpute et al. 2010). It is therefore an important factor that determines the final price of the produced BSs. Large scale downstream processing of SLs, RLs and MELs is commercially and economically viable. Even though simple separation processes are commercially viable, some end uses of BSs are limited by their purity, which varies significantly and many applications require further purification which increases the final product price. Medical and cosmetic application of BSs for example may require highly

[1] Jeneil Biotech Inc., 400 Dekora Woods Blvd, Saukville, WI 53080, USA.
[2] University of Ulster, Faculty of Life & Health Sciences, Coleraine BT52 1SA, Northern Ireland, UK.
  Email: im.banat@ulster.ac.uk
* Corresponding author: hydrobact@gmail.com

purified BSs while environmental and agricultural application may require low purity or crude BSs. There are many research and review articles (Mulligan and Gibbs 2004, Rau et al. 2005, Satpute et al. 2010, Winterburn et al. 2011, Weber et al. 2012, Weber and Zeiner 2015, Shah et al. 2016) providing details on downstream processing of different microbial BSs. This chapter will collectively discuss the downstream processing protocols of microbial BSs, including isolation, extraction, and purification.

## Downstream Processing of BSs

Microbial BSs are produced by fermentation process and the downstream processing of BSs starts with termination of fermentation process after reaching an acceptable product concentration in the culture broth. Specific downstream processing is required for each BS based on their critical micelle concentration (CMC), hydrophilic and lipophilic balance (HLB), chemical nature, ionic charge, solubility in extracting solvent/water, expected purity for specific application, producing microbe and location of BS (intracellular, extracellular or cell bound) (Desai and Banat 1997, Stapute et al. 2010, Najmi et al. 2018). Current BSs recovery methods have some disadvantages including:

- Higher cost of solvents usage and prolonged processing time
- Generation of large quantity of spent solvents as toxic waste and their disposal burden
- Extraction efficiency of solvents and product loss during the process, and
- Requirement of additional purification of crude BSs for specific application that require high purity BSs

Therefore, the need for new eco-friendly and cost effective downstream processes is required for efficient recovery of BSs. The following sections provide information on downstream processing of BSs that include phase separation, salt precipitation, solvent extraction, foam fractionation and chromatographic separation. A detailed list of downstream processing of BSs is provided in Table 1.

### *Phase separation*

Phase separation is a simple and inexpensive process for extraction of BSs from the culture broth. Research on phase separation was reported for SLs (Gorin et al. 1961), RLs (Mixich et al. 1997) and MELs (Rau et al. 2005). Phase separation is a result of presence of BSs higher than their solubility concentration which makes the BSs phase heavier than the water/broth. Cells remain in the aqueous phase, but the bottom BSs phase still contains cells that needs further separation such as water wash and solvent extraction.

### *Phase separation of SLs*

SLs were first reported by Gorin et al. (1961) and are produced by a non-pathogenic yeast *Candia magnoliae* (now known as *Starmerella bombicola*). SLs are the only BSs reported with volumetric yields up to 705 g/L (Marchal and Lemal 1997)

**Table 1.** Downstream processing of microbial BSs.

| Process | Biosurfactant |
|---|---|
| **Phase separation** | |
|  | Sophorolipids |
| pH mediated phase separation | Rhamnolipids |
| Temperature mediated separation | Mannosylerythritol lipids |
| **Precipitation** | |
| Ammonium sulfate precipitation | Emulsan |
|  | Biodispersan |
|  | Bioemulsifier |
|  | Rhamnolipids |
|  | Surfactin |
| Zinc sulfate precipitation | Rhamnolipids |
| Acetone precipitation | Bioemulsifer |
| Acid precipitation | Surfactin |
|  | Rhamnolipids |
|  | Trehalose lipids |
|  | Mannosylerythritol lipids |
| **Solvent extraction** | |
| Dichloromethane-methanol | Trehaloselipids |
| Ethyl acetate-hexane | Sophorolipids |
| Chloroform-methanol | Liposan |
| Ethyl acetate | Rhamnolipids |
| **Foam separation/fractionation** | |
|  | Surfactin |
|  | Rhamnolipids |
|  | Sophorolipids |
| **Crystallization** | |
|  | Cellobiose lipids |
|  | Glycolipids |
|  | Sophorolipids |
| **Recrystallization** | Sophorolipids |
|  | Rhamnolipids |
| **Ultrafiltration** | |
|  | Rhamnolipids |
|  | Surfactin |
| **Chromatography** | |
| **Adsorption–desorption and Ion exchange chromatography** | Rhamnolipids |
| **Silica gel column chromatography** | Sophorolipids |

Table adapted from Desai and Banat (1997); Mixich et al. (1997); Mulligan and Gibbs (2004); Smyth et al. (2010); Stapute et al. (2010); Weber and Zeiner (2015); Shah et al. (2016); Beuker et al. (2016); DÃaz De Rienzo et al. (2016); Dolman et al. (2017).

through fermentation. Higher concentration in the culture broth makes the SLs phase heavier and once the agitation and aeration in the fermentation process is stopped, natural SLs will self-assemble/precipitate at the bottom of the fermentor. At this point, separation of SLs is very easy and further purification requires minimal amount of solvents (ethyl acetate and hexane) to recrystallize natural SLs (Figure 1). Since natural SLs are a mixture of at least four different forms, lactonic, acidic/open chain and acetylated forms of lactonic and open chain form, for specific applications separation of at least two major forms are necessary (Thavasi, unpublished data).

**Figure 1.** Phase separation of SLs from fermentation culture broth.

## Phase separation of RLs

A detailed phase separation process for RLs was described in U.S. Patent 5,656,747 (Mixich et al. 1997). In brief, the process begins acidification of culture broth containing RLs to pH 2.5–4.0, heating the acidified broth to 90°C to 110°C, cooling the broth to 20°C–30°C and separation of the phase separated RLs by centrifugation. Final RLs product recovery reported in U.S. patent by Mixich et al. (1997) was 90%–99%, based on the quantity of RLs contained in the final fermentation broth. Though the product recovery is above 90%, acidification process should be monitored closely; a rapid acidification may hydrolyze the RLs in to sugars and fatty acids. To avoid RL hydrolysis and generation of heat, diluted acids (for example, 30% v/v, $H_2SO_4$ or HCl) could be used with close pH monitoring and high capacity cooling systems in place to maintain the temperature of the vessel at 4°C to 10°C. Controlling temperature at 4°C to 10°C facilitates the precipitation of RLs by reducing their solubility in the cold water medium. If the use of diluted acids is not desirable to avoid product dilution, concentrated acids can be used with close monitoring of pH and heat generation. This process can be optimized for specific need of the end user with few trails.

## Phase separation of MELs

MELs fermentation with *Pseudozyma aphidis* DSM 14930 reported by Rau et al. (2005) noticed formation of aggregated beads of MELs on the bottom of the sampling bottle, after 3 days of fermentation. Number of beads and their width (2–10 mm) increased with time and continued to increase as the MELs concentration reached greater than 40 g/L. Composition of the beads contained 60% MELs, small amounts of soybean oil (20%) and fatty acids (10%). At the end of fermentation (8 days), 90 g/L of MELs yield was reported (Rau et al. 2005). They also found that 93% MELs can be isolated from the culture broth by heating to 100°C for 20 minutes, followed by phase separation. Such heat induced phase separation was observed by Thavasi et al. (unpublished data) for SLs (Figure 1).

As shown by the above reports, phase separation process is a cost effective process to isolate BSs from the fermentation broth, but depend on concentration/ yield of BSs and their solubility in the water/culture medium which depend on other parameters such as pH, temperature, etc. Phase separation process can be used as the initial step in purification of BSs or it can be the final step for application that requires low purity BSs suitable for agriculture and environmental bioremediation.

## Salt precipitation

### Ammonium sulphate precipitation

Ammonium sulfate is used in precipitation of BSs such as emulsan (Rosenberg et al. 1979, Kaplan and Rosenberg 1982), alasan (Toren et al. 2001), RLs (Shah et al. 2016) and surfactin (Youssef et al. 2005) from fermentation medium. Concentration of the salt used for precipitation range from 30% to 65% saturation and the product is treated for 12 to 24 h at 4°C and BSs are separated from the salt by dialysis. This method is effective for high molecular weight BSs and not effective for BSs such as RLs. Shah et al. (2016) reported that methanol/chloroform/acetone extraction of RLs was more effective (7.50 g/L) than ammonium salt precipitation (4.90 g/L). However, chloroform is not preferred in the industrial field due to toxicity and regulatory issues. Ethyl acetate is considered as the best extracting solvent for SLs and RLs. Further, Shah et al. (2016) evaluated the zinc sulfate precipitation (40% salt concentration (w/v)) of RLs, but the recovery (5.25 g/L) is not as effective as methanol/chloroform/acetone extraction.

### Acid precipitation

Acid precipitation is a widely accepted and inexpensive method used in the recovery of crude BSs. In this method, concentrated acids like HCl and $H_2SO_4$ are used to lower the pH to 2.0 or 3.0 and the BS containing solution is allowed to stand for 8 to 12 h at 4°C. At this pH, the BSs become insoluble and separate from the culture broth at the bottom of the fermentor. Separated BS phase can be isolated from the aqueous phase by centrifugation or solvent extraction. Lowering pH must be carried out slowly with constant mixing at controlled temperature to avoid product hydrolysis or degradation. Addition of concentrated acid to water phase will increase the temperature which is detrimental to the product. BSs like surfactin (Mukherjee et al. 2006), RLs (Deziel et al. 1999, Smyth et al. 2010), SLs (Nunez et al. 2001), TLs, and MELs (Rapp et al. 1979) were isolated using this method.

## *Solvent extraction*

Solvent extraction is a common extraction method used in chemical and fermentation process. Solvents such as chloroform, methanol, acetone, ethanol, ethyl acetate, dichloromethane, butanol, pentane, *n*-hexane, acetic acid, diethyl ether, isopropanol and methyl tert-butyl ether (MTBE) are used for BSs extraction from the fermentation medium. For laboratory use, most of the above mentioned solvents are suitable while for industrial scale, several criteria should be considered before choosing a solvent, such as toxicity, extraction efficiency, partition coefficients, and solubility of the particular BS, in addition to the spent solvent's recyclability/reusability, availability and cost of the solvent.

BSs such as SLs, RLs, TLs, CBLs, MELs, and surfactin have been reported to be extracted using such solvents (Rapp et al. 1979, Desai and Banat 1997, Nunez et al. 2001, Smyth et al. 2010). Form an industrial point of view, ethyl acetate is widely used in the extraction of BSs using counter current extraction process and ethyl acetate as an accepted solvent for food grade industrial process. For example, SLs is extracted using ethyl acetate and recrystallized (Figure 2) from the concentrated ethyl acetate extract using ethyl acetate and *n*-hexane at 3:1 ratio at 4°C. If higher amount of unbound fatty acids (from the oil used in SLs production) are detected in SLs-extract, prior to recrystallization, washing 2 to 3 times with *n*-hexane can remove the fatty acids and a higher purity of natural-SLs can be achieved (Gross et al. 2017).

**Figure 2.** Recrystallized SLs from ethyl acetate extract.

## *Extractive fermentation*

Extractive fermentation is a process where product is extracted from the culture medium without terminating the process using a biocompatible extractant. This process has many advantages, such as reduction of product induced toxicity, efficient downstream processing at the end, where the product is concentrated in the extractant and needs only minimal extraction or product isolation process. Drouin and Cooper (1992) reported the production of surfactin in aqueous two-phase fermentor with *B. subtilis* ATCC 21332 in a mineral medium with 5 g/L of glucose, 0.07 M phosphate buffer (pH 6.7) and other salts. The extractant was an aqueous two-phase system of polyethylene glycol (PEG-8, 8000 mol. wt.) and dextran (D-40, 40,000 mol. wt.) used to partition the surfactant and surfactant-producing bacterial cells. It was reported that bacterial cells accumulated in the lower dextran-rich phase and surfactin in the top PEG-8 phase. These observations encourage the application of extractive fermentation methods for other BS fermentation for efficient product recovery. However, other factors such as selection of specific extractant for each BS and its biocompatibility with the producing microorganisms in addition to parameters such as the temperature, pH and salt/ionic concentration need to be optimized for better extraction results.

### Foam fractionation

In the foam fractionation process, foam containing BS is allowed to overflow from the bioreactor through a fractionation column. The collected foam is acidified to pH 2.0–3.0, the acidification precipitates BS from the foam, and at the end BS is extracted with solvents (Cooper et al. 1981, Najmi et al. 2018). In continuous process, an integrated foam recycler is employed to fractionate the foam produced and recycle the culture medium into the reactor (Winterburn et al. 2011). Foam fractionation was used as an extraction process for RLs (Beuker et al. 2016), SLs (Dolman et al. 2017) and surfactin (Makkar and Cameotra 2001). Even though foam fractionation is a solvent free innovative method for *in situ* continuous recovery of BSs, it is not a selective extraction method; surface activity properties of the target molecules play an important role in foam fractionation technique (Sarachat et al. 2010). The molecules with high surface activity can be attached to the air bubbles (foam) more efficiently than less surface active molecules. In the case of mono and di rhamnolipids produced by *P. aeruginosa*, the more hydrophobic characteristic of Rha-C10-C10 was at higher concentration ratio in the foam (more than double) than that of Rha-Rha-C10-C10 (Heyd et al. 2011). Adaptation of this method to industrial scale has many challenges: (i) changes to the current manufacturing processes, (ii) removal of anti/defoaming agent from the fermentation media (the antifoaming agent is the key to control foaming during fermentation; further, some producing microbes require large amount of aeration that results foaming), (iii) loss of fermentation medium volume and viable cells adhered to foam, (iv) difficulty to collect foam with BS, if oily substrate is used in the process (v) contamination, and (vi) separation of BS from the foam requires further extraction process/step (solvent extraction). At present, other conventional extraction processes used for BS production are user friendly and economically viable; more advanced research on foam fractionation processes is needed to address above issues and make foam fractionation an economically viable industrial process.

### Ultrafiltration

Ultrafiltration is an efficient process widely used to concentrate and purify BSs from the fermentation medium or from a concentrated extract. This process was applied to BSs such as surfactin, and rhamnolipids (Mulligan and Gibbs 1990). Fermented culture broth is passed through membranes with specific molecular weight cut-off that excludes the impurities and retain the BS which can be further processed. Critical micelle concentration (CMC) of the BS plays a main role in this process; BS form micelles at concentrations above their CMC, which allows these BS-aggregates to be retained by relatively higher molecular weight cut-off membranes. On the other hand, lower molecular weight impurities such as salts, free amino acids, peptides and small proteins are easily removed from the mixture. Mulligan and Gibbs (1990) evaluated different molecular weight cut-off membranes (MWCO) (10 to 300 kDa) for the retention of surfactin and RLs (1036 and 802 mol. Wts, respectively). They reported Amicon XM 50 (50,000 MWCO) as a suitable membrane for surfactin with a 160-fold purification (97% product purity), and 98% product recovery/ retention. In the case of RLs, YM 10 (10,000 MWCO) membrane was found suitable with 92% product recovery/retention, and 8% product loss indicating that RL micelles are smaller than 10,000 mol. wt. seeped through the membrane. RL micelles' size typically range from 80–200 nm (Gruber et al. 1993). Another study by Lin and Jiang (1997) revealed that surfactin micelles assembled in the form of supramolecules with 10,000 and 30,000 Da, and membranes with similar MWCO showed retentions of 98.8% and 97.9%, respectively. When membranes with 50,000 and 100,000 Da MWCO were used, the percentage of surfactin retention decreased significantly to 86% and 53%, respectively. It is therefore clear from above results that characterization of BS CMC, their micelle size and selection of membranes with specific MWCO are crucial for higher BS recovery. Further, removal of cells from the culture medium before ultrafiltration can prevent membrane blockage by cells and make the process more efficient and cost effective.

## *Chromatographic separation*

Chromatographic methods are mainly used for separation and purification of BS from crude mixture. Chromatographic purification of BS includes: (i) preparative-TLC, (ii) silica gel column chromatography, (iii) adsorption-desorption chromatography, (iv) ion-exchange chromatography and (v) solid phase extraction.

### *Preparative-TLC*

It is a conventional method used in separation of compounds from a mixture based on their retention value (*rf*). A silica-coated (stationary phase) glass plate with variable thickness is applied with a sample of crude BS and allowed to run in a solvent system (mobile phase). After a specific time, bands developed for each compound are visualized under UV or acid treatment (to view carbohydrates, SLs or RLs) or chromogenic reagents (Ninhydrin, to view protein, surfactin) and separated compounds are scraped and extracted further with solvents. The *rf* value obtained with a specific solvent can help to choose a specific solvent system for column chromatography and elution time of the BS.

### *Silica gel column chromatography*

Based on preparative-TLC results, eluting solvent system is selected for purification of BSs. For example, SLs can be purified in silica gel column with chloroform: methanol system in a gradient elution where acidic and lactonic SLs are separated using this method (Gross et al. 2017). In this process, the ratio of chloroform is higher at the beginning, which will elute the hydrophobic lactonic-SLs and as the methanol concentration increases, the hydrophilic acidic/open chain-SLs are eluted out at high purity. A combination of automated column chromatographic system with UV-detection systems (example, Isolera™ Spektra Systems with Accelerated Chromatographic Isolation system, Biotage USA, Charlotte, NC, USA) can make this process more efficient with less manual input.

## Adsorption and desorption chromatography

Adsorption and desorption of BS molecule to and from the stationary phase are two main processes in this method. For example, RLS are purified using Amberlite XAD 2 or XAD 16 polystyrene resins. Adsorption process begins with applying cell-free culture broth to the adsorbent column and 0.1 M phosphate buffer (pH 6.1) as the equilibrating medium. Culture broth is applied to a point where the resin becomes exhausted, which is confirmed by ultra violet (UV) absorption and when surface tension dropped below 35 mN m$^{-1}$ (Reiling et al. 1986). The resin is washed with distilled water to remove any pigments and free fatty acids. Desorption process is carried out with eluting methanol, which leads to the desorption of resin bound RLs; at the end, RLs in methanol are evaporated to obtain the crude RLs. RLs obtained using the above methods can be further purified using ion exchange chromatography. The loading capacity of the resin for RLs was 60 g/kg of XAD-2, with 60% RLs purity and 75% product recovery from the culture broth.

## Ion exchange chromatography

This method can be used as the sole process to purify BSs or as a continuation of adsorption and desorption chromatography. Reiling et al. (1986) used both adsorption and desorption (for initial purification, as described in previous section) and ion exchange chromatography (for final purification). Ion exchange chromatography was carried out in a DEAE-Sepharose CL 6B column which was equilibrated with 10 mM Tris hydrochloride buffer (pH 8) containing 10% (v/v) ethanol. RLs obtained from the adsorption column was diluted 10-fold with the same equilibrating buffer containing 20% (v/v) ethanol and applied to the column after filtration through a cotton filter. Flow rate of the column was set as 6 liters$^{h-1}$. The monitoring criterion is estimation of surface tension of the effluent; as the surface tension reaches below 35 mN m$^{-1}$, the column was washed with 2 to 3 bed volumes of 0.1 M NaCl in 10 mM Tris hydrochloride buffer containing 10% (v/v) ethanol. Then bound RLs were released by 0.8 M NaCl in the same buffer. Product recovery using this step is approximately 90%. The advantages of adsorption-desorption and ion exchange chromatography techniques include speed achieved through one/two-step recovery and high quality purified BSs output. However, this process is suitable for small scale operations to obtain high purity BSs; however, it is not economical for large scale operations, which require handling of larger volumes, recovery of product from the resin and reusability of the resin and spent solvent.

## Solid phase extraction (SPE)

SPE is a selective sample preparation method utilizing a strong anion exchange material. Behrens et al. (2016) evaluated SPE and liquid-liquid extraction (LLE) for the purification of the RLs supernatant of the *P. putida* strain KT2440 pVLT33_ rhlABC. The SPE cartridge is made of strong quaternary ammonium modified polymeric anion exchange material (Chromabond® HRXA, 3 mL volume, 200 mg adsorbent weight) and syringe filters (Chromafil® PET-20/15 MS, polyester, 0.20 μm, 15 mm). RLs are loaded onto the SPE cartridge and extracted using gravity and then negative pressure (max. 600 mbar). The cartridges are then washed with 2 mL of sodium hydroxide solution (0.1 mol/L), followed by 5 mL acetonitrile to remove the unbound materials. Cartridges are then dried using negative pressure after each washing step. The final elution of bound RLs from the stationary phase was carried out with 5 mL acetonitrile containing 2% formic acid and used for qualitative and quantitative LC-MS analysis. Sample amount and strength of the elution solution are dependent on the concentration of RLs in the supernatants and therefore conditions have to be optimized. Results from this experiment revealed substantial losses of RLs in LLE, compared to SPE, where it is only 5 to 7%. Further, acid precipitation and centrifugation of RLs isolated using SPE contained 33 to 42% which is about 6-fold higher yield of RLs than LLE extraction process. Although SPE provided higher yields and proved to be more reproducible and less time/work consuming, it still needs further research to make this process suitable for large scale operations.

## Conclusion

As discussed in this chapter, downstream processing of BSs plays a vital role in their purity, production costs and potential applications. As the demand for BSs increases, there is need for developing new downstream processes or fine tuning of existing methods. One such approach could be the utilization of BSs micelle behavior at concentrations higher that their CMC. For example, SLs phase-separate from the culture broth and precipitate at the bottom, which can be separated easily from the fermentor and further purified with minimal use of solvents. Another example is the use of ultrafiltration of BSs RLs and surfactin, where micelles played the main role in increasing the molecular size of the BSs improving their membranes' retention. This approach may be time consuming, but efficient and can be improved for better results. BSs, for example, precipitate or form micelles at certain pH which reduces the volume to be ultrafiltered and the removal of cells by centrifugation reduces cell mediated membrane blockage. The creation of a pH that induces micelles formation enhances the treatment of concentrated micelles in the ultrafiltration system. This approach will improve the efficiency of the system by reducing the volume and processing time and will make the process cost effective and solvent free or reduce the amount of solvents to be used. However, in addition to focusing on downstream processing, a serious thrust into research and development is needed as the first part of the fermentation process, i.e., (i) identification of BS over producing strains, (ii) optimization of current fermentation process to increase the yield, and (iii) identification/creation of new applications for BSs with different purity level to expand the economic impact/importance of the BSs in the market. Further, public awareness on eco-friendly technology will drive the market and encourage the industries to invest more in biobased product research.

# References

Banat, I.M., S.K. Satpute, S.S. Cameotra, R. Patil and N.V. Nyayanit. 2014. Cost effective technologies and renewable substrates for biosurfactants' production. Front. Microbiol. 5: Article 697.

Behrens, B., J. Engelen, T. Tiso, L. Blank and H. Hayen. 2016. Characterization of rhamnolipids by liquid chromatography/mass spectrometry after solid-phase extraction. Anal. Bioanal. Chem. 408(10): 2505–2014.

Beuker, J., A. Steier, A. Wittgens, F. Rosenau, M. Henkel and R. Hausmann. 2016. Integrated foam fractionation for heterologous rhamnolipid production with recombinant *Pseudomonas putida* in a bioreactor. AMB Express 6: 11.

Cooper, D.G., C.R. Macdonald, S.J. Duff and N. Kosaric. 1981. Enhanced production of surfactin from *Bacillus subtilis* by continuous product removal and metal cation additions. Appl. Environ. Microbiol. 42: 408–412.

DÃaz De Rienzo, M.A., I.D. Kamalanathan and P.J. Martin. 2016. Comparative study of the production of rhamnolipid biosurfactants by B-thailandensis E264 and *P. aeruginosa* ATCC 9027 using foam fractionation. Process. Biochem. 51(7): 820–827.

Desai, J.D. and I.M. Banat. 1997. Microbial production of surfactants and their commercial potential. Microbiol. Mol. Biol. Rev. 61: 47–64.

Déziel, E., F. Lépine, D. Dennie, D. Boismenu, O.A. Mamer and R. Villemur. 1999. Liquid chromatography/ mass spectrometry analysis of mixtures of rhamnolipids produced by *Pseudomonas aeruginosa* strain 57RP grown on mannitol or naphthalene. Biochim. Biophys. Acta Mol. Cell. Biol. Lipids. 1440: 244–252.

Dolman, B.M., C. Kaisermann, P.J. Martin and J.B. Winterburn. 2017. Integrated sophorolipid production and gravity separation. Process Biochem. 54: 162–171.

Drouin, C.M. and D.G. Cooper. 1992. Biosurfactants and aqueous two-phase fermentation. Biotechnol. Bioeng. 40: 86–90.

Global biosurfactants market research report by Markets and Markets™ Inc. https://www. marketsandmarkets.com/PressReleases/biosurfactant.asp.

Gorin, P.A.J., J.F.T. Spencer and A.P. Tulloch. 1961. Hydroxy fatty acid glycosides of sophorose from *Torulopsis magnoliae*. Can. J. Chem. 39: 846–855.

Gross, R.A., R. Thavasi, A. Koh and Y. Peng. 2017. Modified sophorolipids as oil solubilizing agents. U.S. Patent No.: US 9,650.405 B2.

Gruber, T., H. Chmiel, O. Kappeli, P. Sticher and A. Fiechter. 1993. Integrated process for continuous rhamnolipid biosynthesis. pp. 175–173. *In*: N. Kosaric (ed.). Biosurfactants: Production, Properties, Application. Marcel Dekker Inc. New York.

Heyd, M., M. Franzreb and S. Berensmeier. 2011. Continuous rhamnolipid production with integrated product removal by foam fractionation and magnetic separation of immobilized *Pseudomonas aeruginosa*. Biotechnol. Prog. 27(3): 706–716.

Kaplan, N. and E. Rosenberg. 1982. Exopolysaccharide distribution of and bioemulsifiers production by *Acinetobacter calcoaceticus* BD4 and BD413. Appl. Environ. Microbiol. 44: 1335–1341.

Lin, S.C. and H.J. Jiang. 1997. Recovery and purification of the lipopeptide biosurfactant of *Bacillus subtilis* by ultrafiltration. Biotechnol. Tech. 11: 413–416.

Makkar, R.S. and S.S. Cameotra. 2001. Synthesis of enhanced biosurfactant by *Bacillus subtilis* MTCC 2423 at 45°C by foam fractionation. J. Surfactants Deterg. 4: 355–357.

Marchal, R. and J. Lemal. 1997. Method of production of sophorosides by fermentation with fed batch supply of fatty acid esters or oils. US Patent 5616479.

Marchant, R. and I.M. Banat. 2012a. Microbial biosurfactants: challenges and opportunities for future exploitation. Trends Biotechnol. 30(11): 558–565.

Marchant, R. and I.M. Banat. 2012b. Biosurfactants: a sustainable replacement for chemical surfactants? Biotechnol. Lett. 34: 1597–1605.

Mixich, J., R. Rothert and D. Wullbrandt. 1997. Process for the quantitative purification of glycolipids. U.S. Patent No. 5656747.

Mukherjee, S., P. Das and R. Sen. 2006. Towards commercial production of microbial surfactants. Trends Biotechnol. 24: 509–515.

Mulligan, C.N. and B.F. Gibbs. 1990. Recovery of biosurfactants by ultrafiltration. J. Chem. Technol. Biotechnol. 47: 23–29.

Mulligan, C.N. and B.F. Gibbs. 2004. Types, production and applications of biosurfactants. Proc. Indian Nat. Sci. Acad. 1: 31–55.

Najmi, Z., G. Ebrahimipour, A. Franzetti and I.M. Banat. 2018. *In situ* downstream strategies for cost-effective bio/surfactant recovery. Biotechnol. Appl. Biochem. (In Press) DOI: 10.1002/bab.1641.

Nunez, A., R. Ashby, T.A. Foglia and D.K.Y. Solaiman. 2001. Analysis and characterization of sophorolipids by liquid chromatography with atmospheric pressure chemical ionization. Chromatographia. 53: 673–677.

Rapp, P., H. Bock, V. Wray and F. Wagner. 1979. Formation, isolation and characterization of trehalose dimycolates from *Rhodococcus erythropolis* grown on n-alkanes. J. Gen. Microbiol. 115: 491–503.

Rau, U., L.A. Nguyen, H. Roeper, H. Koch and S. Lang. 2005. Downstream processing of mannosylerythritol lipids produced by *Pseudozyma aphidis*. Eur. J. Lipid Sci. Technol. 107: 373–380.

Reiling, H.E., U. Thanei-Wyss, L.H. Guerra-Santos, R. Hirt, O. Käppeli and A. Fiechter. 1986. Pilot plant production of rhamnolipid biosurfactant by *Pseudomonas aeruginosa*. Appl. Environ. Microbiol. 51(5): 985–989.

Rosenberg, E., A. Zuckerberg, C. Rubinovitz and D.L. Gutnick. 1979. Emulsifier of *Arthrobacter* RAG-1: isolation and emulsifying properties. Appl. Environ. Microbiol. 37: 402–408.

Sarachat, T., O. Pornsunthorntawee, S. Chavadej and R. Rujiravanit. 2010. Purification and concentration of a rhamnolipid biosurfactant produced by *Pseudomonas aeruginosa* SP4 using foam fractionation. Bioresour. Technol. 101(1): 324–30.

Satpute, S.K., A.G. Banpurkar, P.K. Dhakephalkar, I.M. Banat and B.A. Chopade. 2010. Methods for investigating biosurfactants and bioemulsifiers: A review. Crit. Rev. Biotechnol. 30: 127–144.

Shah, M.U.H., M. Sivapragasam, M. Moniruzzaman and S.B. Yusup. 2016. A comparison of recovery methods of rhamnolipids produced by *Pseudomonas aeruginosa*. Procedia Engineering. 148: 494–500.

Smyth, T.J.P., A. Perfumo, R. Marchant and I.M. Banat. 2010. Isolation and analysis of low molecular weight microbial glycolipids: Microbiology of hydrocarbons, oils, lipids, and derived compounds. pp. 3705–3723. *In*: K.N. Timmis (ed.). Handbook of Hydrocarbon and Lipid Microbiology. Springer-Verlag, Berlin Heidelberg.

Toren, A., S. Navon-Venezia, E.Z. Ron and E. Rosenberg. 2001. Emulsifying activities of purified Alasan proteins from Acinetobacter radioresistens KA53. Appl. Environ. Microbiol. 67(3): 1102–1106.

Weber, A., A. May, T. Zeiner and A. Gorak. 2012. Downstream processing of biosurfactants. Chem. Eng. Trans. 27: 115–120.

Weber, A. and T. Zeiner. 2015. Purification of biosurfactants. pp. 129–152. *In*: N. Kosaric and F. Varda-Sukan (eds.). Biosurfactants Production and Utilization—Processes, Technologies, and Economics. CRC Press, Taylor and Francsis Group, Boca Raton, FL, USA.

Winterburn, J.B., A.B. Russell and P.J. Martin. 2011. Integrated recirculating foam fractionation for the continuous recovery of biosurfactant from fermenters. Biochem. Eng. J. 54: 132–139.

Youssef, N.H., K.E. Duncan and M.J. McInerney. 2005. Importance of 3-hydroxy fatty acid composition of lipopeptides for biosurfactant activity. Appl. Environ. Microbiol. 71: 7690–7695.

# 3

# Sophorolipids: Unique Microbial Glycolipids with Vast Application Potential[#]

*Richard D. Ashby\* and Daniel K.Y. Solaiman*

## Introduction

Glycolipids are microbially-produced amphiphilic molecules that are typically synthesized as secondary metabolites. These molecules generally accumulate at interfaces such as air and water or oil and water and have the ability to lower surface and interfacial tension. This property has driven the development of glycolipids as biobased alternatives to some synthetic surfactants that are used in cleaning, cosmetic and oral hygiene products, foods, and environmental remediation practices (Shekhar et al. 2015, Gudiña et al. 2013, Morita et al. 2013, Marchant and Banat 2012).

Sophorolipids (SLs) are yeast-derived, extracellular glycolipids that have garnered immense interest due to their large production yields and evolving commercial potential (Transparency Market Research 2014). They are composed of a sophorose sugar (2-O-β-D-glucopyranosyl-β-D-glucopyranose; GLC β-1,2 GLC) linked through a β-glycosidic linkage between the 1′ hydroxy group of the sophorose and an ω or ω-1 hydroxy group located on the fatty acid. The fatty acid chain lengths typically range from 16 to 18 carbon units and can be saturated or unsaturated. The sophorose sugar moiety may be acetylated at the C6′ and/or C6″ hydroxy group

Eastern Regional Research Center, Agricultural Research Service, U.S. Department of Agriculture, 600 East Mermaid Ln. Wyndmoor, Pennsylvania 19038 USA.
Email: Dan.Solaiman@ars.usda.gov
\* Corresponding author: Rick.Ashby@ars.usda.gov

# Mention of trade names or commercial products in this article is solely for the purpose of providing specific information and does not imply recommendation or endorsement by the U.S. Department of Agriculture. USDA is an equal opportunity provider and employer.

of each glucose residue (Kurtzman et al. 2010) resulting in non-acetylated, mono-acetylated and/or di-acetylated congeners. In addition, the carboxylic acid group of the fatty acid may be lactonized to the disaccharide ring at C4″ or remain in the open-chain conformation. Figure 1 depicts the chemical structures of di-acetylated SLs containing ω-1 hydroxy-linked oleic acid in the lactonic (Figure 1A) and free-acid forms (Figure 1B) as produced by *Starmerella* (formerly *Candida*) *bombicola* (Asmer et al. 1988, Nuñez et al. 2001). Oleic acid is known to be the favored hydrophilic component for SLs derived from *S. bombicola*; however, SLs are typically synthesized as structural mixtures of as many as 23 different variants (van Bogaert et al. 2007). Generally, variability within the chemical structure of SL occurs through the acetylation patterns at the C6′ and C6″ positions of the sophorose sugar and the configuration and length of the fatty acid moiety which is dictated by the producing strain, fermentation conditions, and feedstock.

SLs were first recognized in the early 1960s in the fermentations of *Torulopsis magnoliae* and *Candida bogoriensis* (Gorin et al. 1961, Tulloch et al. 1962). The SLs derived from *T. magnoliae* were identified as partially acetylated molecules

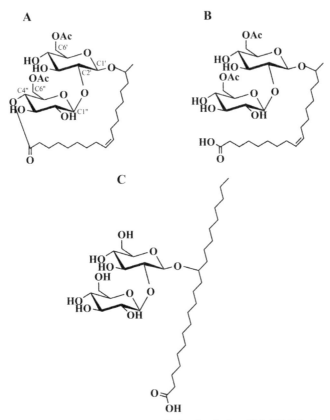

**Figure 1.** Structure of sophorolipids from *Starmerella bombicola* (i.e., 17-L-[(2′-*O*-β-glucopyranosyl-β-D-glucopyranosyl)-oxy]-9-octadecenoic acid 6′,6″-diacetate) sophorolipids in the 1′, 4″-lactone form (A) and the free acid form (B) and from *Rhodotorula bogoriensis* (i.e., 13-L-[(2′-*O*-β-glucopyranosyl-β-D-glucopyranosyl)-oxy]-docosanoic acid) (C).

containing a 2-O-β-D-glucopyranosyl-β-D-glucopyranose unit attached to 17-L-hydroxystearic acid or 17-L-hydroxyoleic acid (Tulloch et al. 1962, Tulloch and Spencer 1968) while those isolated from *C. bogoriensis* (since reclassified as *Rhodotorula bogoriensis*) contained a 13-hydroxydocosanoic acid fatty acid element (Figure 1C; Tulloch et al. 1968) which was further verified through LC/MS analysis by Nuñez et al. (2004). Since then, *T. bombicola* (since reclassified as *C. bombicola* and more recently as *S. bombicola* ATCC 22214) was identified as a SL-producer (Spencer et al. 1970); in fact, currently, *S. bombicola* is the most studied strain for SL production. This yeast strain has been recognized for its ability to produce SLs from feedstocks such as *n*-alkanes (Hu and Ju 2001a, Li et al. 2016), free fatty acids (Rau et al. 1996, Ashby et al. 2008), fatty acid esters (Asmer et al. 1988, Davila et al. 1994, Ashby et al. 2006), triacylglycerols (Asmer et al. 1988, Zhou et al. 1992, Davila et al. 1994, Zhou and Kosaric 1995, Daniel et al. 1998, Hu and Ju 2001a, Pekin et al. 2005) and alcohols (Brakemeier et al. 1998). Unfortunately, to date, the majority of triacylglycerols used for SL biosynthesis have edible applications which negatively influence their use as fermentation feedstocks. Wadekar et al. (2012) addressed this issue by reporting the synthesis of SLs from jatropha, karanja and neem oil, three low-cost inedible oils traditionally found in India. In addition, alternative raw materials, such as animal fat (Deshpande and Daniels 1995), waste restaurant oil (Fleurackers 2006), soy molasses (Solaiman et al. 2007), deproteinized whey (Daniel et al. 1998), and glycerol (Ashby et al. 2005) have also been successfully utilized in SL synthesis by *S. bombicola*. More recently, SL biosynthesis was established in *Wickerhamiella domericqiae*, and the SLs produced had 17-L-hydroxystearic acid 1′,4″-lactone 6′,6″-diacetate as the predominant chemical species (Chen et al. 2006). In 2008 and 2010, additional *Candida* strains were acknowledged to produce SLs. *C. batistae* was shown to produce mainly acidic SLs with the majority of the fatty acid side-chains representing 18-L-([2-O-β-D-glucopyranosyl-β-D-glucopyranosyl]-oxy)-oleic acid, 6′, 6″ diacetate (Konishi et al. 2008) and *C. riodocensis, C. stellata,* and *Candida* sp. Y-27208 which were also documented to produce predominantly diacetylated acidic SLs (Kurtzman et al. 2010).

## Biosynthesis of SLs

While many yeast species are now recognized as SL-producers, *S. bombicola* yet stands out as a strain of interest because of its high SL titers and ability to grow on a variety of carbon substrates. Fatty acids, particularly those with chain lengths of between 16 and 18 carbons (e.g., palmitic acid, stearic acid, oleic acid, and linoleic acid) have been identified as the most effective hydrophobic precursors for SL biosynthesis; however, related substrates such as those mentioned previously (in the introductory section) are also known to serve this purpose. Figure 2 shows the enzymatic reactions required to produce the precursor fatty acids from *n*-alkanes, triacylglycerols and their intermediates. Triacylglycerols have been well-documented as viable hydrophobic substrates for SL biosynthesis and have resulted in high titers. However, in order to be utilized for SL synthesis, the triacylglycerol must initially be hydrolyzed into its respective fatty acid components. This reaction is accomplished

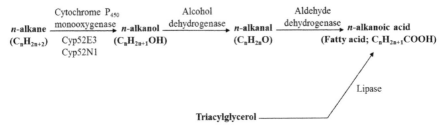

**Figure 2.** Enzymatic reactions required to produce precursor fatty acids from *n*-alkanes, *n*-alkanols, *n*-alkanals and triacylglycerols for sophorolipid biosynthesis.

through the action of lipase enzymes which may be position-specific (typically 1,3-specific; very few 2-specific lipases are known) or non-specific. The type of lipase present allows either exclusive hydrolysis at the primary positions or at random positions of the glycerol backbone, respectively. Typically, the fatty acid attached at the secondary position of the glycerol is more difficult to hydrolyze; however, 1,3-specific lipase enzymes may still function to liberate the secondary fatty acids because of spontaneous, non-enzymatic acyl migration of partial glycerides which can be promoted by acid, alkali and/or heat.

When *n*-alkanes are utilized as hydrophobic feedstocks, *S. bombicola* has the genetic capability to oxidize these alkanes to produce fatty acids that can then undergo β-oxidation to form the preferred fatty acid substrates for SL synthesis. Previous work demonstrated that cytochrome $P_{450}$ monooxygenase (CYP) enzymes are present in *S. bombicola* and function to either assimilate *n*-alkanes and/or to take part in the initial steps involved in SL biosynthesis. van Bogaert et al. (2009) identified 3 unique CYP genes in *S. bombicola* using degenerate PCR and genomic walking that coded for different gene products involved in *n*-alkane or fatty acid hydroxylation. These gene products are believed to be N-terminally anchored to the endoplasmic reticulum (Menzel et al. 1996) and were classified as members of the CYP52 family and identified as CYP52-E3, CYP52-N1, and CYP52-M1 (Nelson 1998). A subsequent study by Huang et al. (2014) further elucidated the function of each of the CYP52 gene products in SL biosynthesis. In that study, the CYP52 genes were cloned into and expressed in *Saccharomyces cerevisiae* and analyzed with a variety of alkane and fatty acid substrates using microsome proteins or whole-cell systems. Those results clarified the specific role that each CYP gene product plays in converting *n*-alkanes into SLs. CYP52-E3 and CYP52-N1 transcript levels showed clear upregulation (CYP52-N1 > CYP52-E3) when the cells were initially grown on alkanes as the sole carbon source but in each case resulting in ω-hydroxylation. These results imply the function of CYP52-N1 and CYP52-E3 are necessary to assimilate *n*-alkanes to fatty acid precursors but not to the SLs themselves.

Once primary alcohols are formed through the action of CYP52-N1 and CYP52-E3, they can be further oxidized to aldehydes by fatty alcohol oxidase (Hommel and Ratledge 1990) and finally to the fatty acid by an $NAD^+$-dependent fatty aldehyde dehydrogenase (Karunasagar 2009). In the case that no hydrophobic carbon source is present, precursor fatty acids are synthesized *de novo* from the acetyl-CoA intermediates derived from glycolysis.

In contrast to the CYP52-E3 and CYP52-N1 enzymes, the CYP52-M1 enzyme is an NADPH-dependent enzyme that has been proven to provide hydroxylated fatty acids that are essential for SL synthesis (van Bogaert et al. 2013). The catabolic pathway for SL biosynthesis from free fatty acids is shown in Figure 3. The catalytic activity of CYP52-M1 is upregulated during the stationary phase of microbial growth provided that sufficient glucose is present in the media (Roelants et al. 2013). SLs are typically synthesized as secondary metabolites and previous work has shown that the CYP52-M1 enzyme efficiently oxidizes $C_{16}$, $C_{18}$, and $C_{20}$ saturated and unsaturated fatty acids including palmitic, palmitoleic, stearic, oleic, linoleic, *cis*-9,10-epoxystearic, *trans*-9,10-epoxystearic, and arachidonic acid to both ω and ω-1 hydroxy fatty acids, further proof of its function in SL biosynthesis (Huang et al. 2014).

Experience has shown that 17-hydroxyoleic acid is the preferred fatty acid for SL synthesis in *S. bombicola* followed by 17-hydroxystearic acid, 15- and 16-hydroxypalmitic acid, and 17-hydroxylinoleic acid (Solaiman et al. 2007, Nuñez et al. 2001). Whether the hydroxy fatty acid is one of these or a less common analog, the hydroxy fatty acids are enzymatically attached to a glucose moiety at carbon 1′ through a glycosidic linkage by glucosyltransferase I (UGTA1) using UDP-glucose as the glucosyl donor resulting in a glucolipid. This reaction was initially proposed in 1982 by Breithaupt and Light in *C. bogoriensis* (Breithaupt and Light 1982) and was more recently proven in *S. bombicola* (Saerens et al. 2011, Huang et al. 2014). Another glucose moiety is enzymatically attached to the maturing SL through a glycosidic linkage between C2′ of the glucolipid and the C1″ of a second glucose molecule. This second glucosyl-transfer reaction may be catalyzed by a second glucosyltransferase enzyme (glucosyltransferase II) which has been purified but whose activities seem to be comparable to glucosyltransferase I (Esders and Light 1972). These reactions produce a non-acetylated free-acid SL which may then be acted upon by acetyltransferase enzymes (using acetyl-CoA as the acetyl donor) to produce mono-acetylated or di-acetylated products with acetyl groups attached at positions C6' and/or C6". Alternatively, the non-acetylated free-acid SL can be converted to the 1′, 4″ lactone through the action of lactone esterase and then be acetylated.

One important parameter that dictates the large-scale usefulness of any material is its cost-to-produce. A process economic model for the fermentative synthesis of SLs from two separate systems (1. glucose and oleic acid; 2. glucose and high-oleic sunflower oil) was published that demonstrated that SLs can be produced at a scale of 90.7 million kg/yr at between US$2.50 and US$3.00/kg with the raw material costs accounting for between 85% and 90% of the annual operating costs (Ashby et al. 2013). While these production numbers make SL synthesis competitive with some petroleum-based surfactants, further work was undertaken in an effort to further reduce production costs. These efforts concentrated on the use of cheaper feedstocks and improved production yields. Two feedstocks that have been tested as glucose substitutes for their viability in promoting SL synthesis were crude glycerol and soy molasses (Ashby et al. 2005, Solaiman et al. 2004). Crude glycerol is a large-volume material generated from chemical transesterification of triacylglycerols

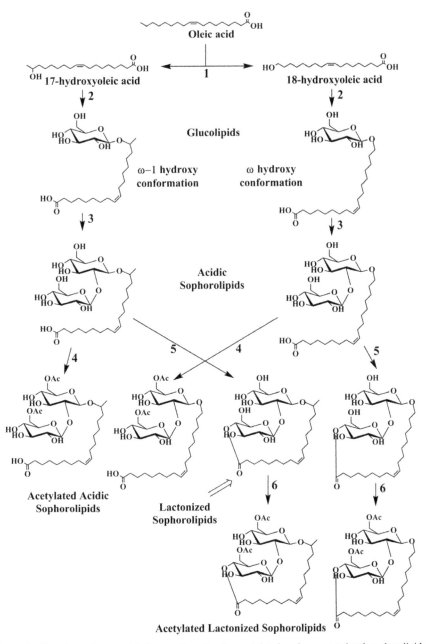

**Figure 3.** Overview of the metabolic reactions for the acetylated and non-acetylated sophorolipids derived from oleic acid in the ω and ω-1 hydroxy free-acid and lactone forms.

Reaction 1: Cofactors—$O_2$, NADPH; Enzyme—Cytochrome $P_{450}$ monooxygenase (CYP52-M1).
Reaction 2: Cofactors—UDP-glucose; Enzyme—Glucosyl Transferase 1.
Reaction 3: Cofactors—UDP-glucose; Enzyme—Glucosyl Transferase 2.
Reaction 4: Cofactors—Acetyl CoA; Enzyme—Acetyl Transferase.
Reaction 5: Cofactors—none; Enzyme—Lactone Esterase.
Reaction 6: Cofactors—Acetyl CoA; Enzyme—Acetyl Transferase.

in biodiesel production. These crude glycerol streams tend to be compositionally distinct, varying in glycerol, short-chain alcohol content, and in the amount and type of free fatty acid, mono-, di-, and triacylglycerols. This variability is dictated by the fatty acid content of the starting hydrophobic feedstock, the effectiveness of the transesterification reaction and the efficiency of the biodiesel and alcohol recovery processes. Since glycerol is easily transported into the cell and integrated into central metabolism, substituting crude glycerol for glucose did result in SL biosynthesis; however, yields were only 60% of what was obtained using glucose and oleic acid as substrates.

Soy molasses is derived from soybeans after flaking and removal of the crude soy oil through solvent extraction. After grinding and removal of the soy flour and soy protein/concentrate, the remaining material is known as soy molasses. Soy molasses is normally composed of sucrose, raffinose and stachyose which are potentially fermentable carbohydrates. By substituting soy molasses for glucose, it was found that SLs could be produced (Solaiman et al. 2004) but the sucrose was the only carbohydrate used by the yeast resulting in yields that were only 75% of those produced in the glucose/oleic acid system (Solaiman et al. 2007).

Improved yields can also enhance production economics in any biological system, especially in large-scale processes. SLs are generally regarded as having production capacities greater than other glycolipid molecules. Glucose is typically utilized as the hydrophilic substrate as it can be easily utilized for SL biosynthesis without further enzymatic conversions; however, some researchers have used combinations of glucose and other polar molecules to increase yields to over 400 g/L (Daniel et al. 1998, Pekin et al. 2005), but more commonly, the reported product's yields typically fall between 50 and 200 g/L in viable fermentation practices.

## Structural Modification of SLs (Biological or Chemical)

SLs, owing to their glycolipid structure, contain a number of sites for potential chemical modification that can and have been realized through both biological and chemical means (Pöhnlein et al. 2015). These include primary and secondary hydroxyl groups associated with the sophorose sugar and the olefinic and carboxyl groups on the fatty acid tail. By chemically or enzymatically modifying the molecules in these positions, new SL products with varying physico-chemical properties and bioactivities have been produced that are better-suited for intended applications. Generally, SLs in their native form and varying structural ratios are capable of reducing the surface tension of water from 72 mN/m to between 25 and 40 mN/m with wide-ranging critical micelle concentrations (CMC; Table 1) (van Bogaert et al. 2011a, Dengle-Pulate et al. 2013). However, recent advances have demonstrated that chemical and biological means can be utilized to tailor both the saccharide and the fatty acid tail. The hydroxy groups on the sophorose sugar and the functional groups associated with the lipid tail enable intermolecular reactions or the attachment of pendant groups to modify the properties of the SLs. Furthermore, research has shown that the fatty acid and sophorose moieties can be liberated resulting in separate novel value-added biomolecules.

Typically in SLs, the fatty acid tail is the point of greatest interest in modifying the molecules. While the acyl chains are primarily confined to lengths of between 16 and 18 carbon units for SLs produced by *S. bombicola* and 22 carbon units for SLs produced by *R. bogoriensis*, any carbon-carbon double bonds and free carboxylate groups provide points of reaction for structural alteration. Azim et al. (2006) used the terminal carboxyl group in the fatty acid tail of SLs to create a series of amino acid-conjugates. The use of carbodiimide in a coupling reaction created an amide linkage between incoming amino acids and the carboxyl group of the SLs. Thirteen conjugates were produced using this method containing such amino acids as glycine, serine, leucine, phenylalanine, aspartate and glutamate. Some of the resulting conjugates were esterified with an alkyl group to the carboxyl group of its amino acid moiety, while others had lost the carbon-carbon double bond in the alkyl chain altogether. Other studies that have exploited the carboxylate groups of SLs to produce chemically modified fatty acid side chains include work by Singh et al. (2003) who synthesized amide derivatives using lipase-catalyzed amidation reactions in the presence of tyramine, phenethylamine, and 2-(*p*-tolyl)ethylamine and by Nunez et al. (2003) who reported the lipase-catalyzed synthesis of galactose-conjugated SLs by utilizing 1,2-3,4-di-*O*-isopropylidene-D-galactopyranose.

Olefinic groups associated with the fatty acid side-chain also present reactive sites for potential derivatization. Delbeke et al. (2015) took advantage of the reactive olefinic groups in the alkyl chain of SLs to generate an aldehyde intermediate using ozonolysis. Ozonolysis resulted in the scission of the carbon-carbon double bond in the starting alkyl chain which reduced the length of the fatty acid tail from 18 carbons to 9 carbons at the location of the original double-bond. Using the aldehyde intermediate obtained through ozonolysis, reductive amination was performed in the presence of secondary amines followed by alkylation with alkyl iodide to obtain the quaternary ammonium salts which were then confirmed to have antimicrobial activity against Gram-positive bacteria.

A more recent emerging area for modifying the fatty acid side-chain of SLs is through the use of alternative hydrophobic feedstocks. It is well-known that SLs can be produced by *S. bombicola* using triacylglycerols, fatty acids, alkanes, alcohols, aldehydes and short-chain alkyl esters when the producing strain is grown under the appropriate conditions. However, these feedstocks typically result in natural SLs whose fatty acid chain length is governed by the cytochrome $P_{450}$ monooxygenase system which does not allow much variation outside chain lengths from 16 to 18 carbon atoms. van Bogaert et al. (2011b) reported two unrelated methods to circumvent the cytochrome $P_{450}$ monooxygenase enzyme. The first involved the use of pre-hydroxylated substrates (12-hydroxydodecanoic acid, 1,12-dodecanediol). Results showed that *S. bombicola* could incorporate shorter chain length fatty acids (C12) if they were already hydroxylated prior to introduction into the fermentation media. The second method utilized stearic acid mimics which could be hydroxylated by the $P_{450}$ enzyme. Dodecyl glutarate, dodecyl and tetradecyl malonate, pentenyl dodecanoate, and dodecyl pentanoate were utilized to produce unique fatty acid side-chains that could subsequently be subjected to alkaline hydrolysis to produce SLs with C5, C12 and C14 chain lengths. These new molecules reduced the surface tension of water to between 32.4 and 37.3 mN/m and had CMC values from 80 to 121 mg/L.

The sophorose sugar moiety also provides free hydroxy groups for potential modification through acylation or esterification reactions. SLs can exist in one of three natural states, either non-acetylated, or mono-acetylated, or di-acetylated at the C6' and/or C6" hydroxy locations. These acyl groups can easily be removed chemically or enzymatically to expose these free hydroxy groups for subsequent reaction. Zerkowski et al. (2006, 2010) reported the chemical derivatization of SLs to yield new amino acid-substituted sophorose moieties resulting in improved water-solubility over the parent molecules without significantly changing their surfactant properties. In that work, stearic sophoroside was used as the substrate to avoid the added complexity of the olefinic group present in the more widely studied oleic sophoroside. Existing functional groups on the amino acid units were protected and then a reactive *p*-aminobenzoic acid (PABA) linker was attached to the amino acid. These amino acids were attached to the stearic sophoroside using carbodiimide-mediated coupling reactions resulting in a series of new SLs where the sophorose moiety was chemically linked to an amino acid containing a positively charged, negatively charged, or a zwitterionic group (Figure 4).

Peng et al. (2015) utilized a cross metathesis method to attach uncharged groups onto the sophorose moiety. In order to accomplish this, the fatty acid moiety of the SL must possess an olefinic group such as in oleic acid-based SLs. A ruthenium-based catalyst $(RuCl_2[C_{21}H_{24}N_2][C_{15}H_{10}][P(C_6H_{11})_3])$ was used to catalyze the reaction of SL with alkene and *n*-alkyl acrylates. For ethylene, the 1st generation Grubbs catalyst (i.e., bis(tricyclohexylphosphine)benzylidine ruthenium(IV) dichloride) was preferred because it induced a complete conversion of the starting SL. The Grubbs reaction products were SLs where the C1' is glycosidically linked to an alkenyl group and the C4" is acylated with another alkenyl chain. Then, through an alcoholysis reaction followed by hydrogenation, SLs containing a single hydrophobic alkyl chain of 10-, 12-, or 14-carbons glycosidically bonded to C1' were produced. This approach provided a unique method to produce SLs with medium chain length (C10-C14) hydrophobic tails.

**A**                                              **B**

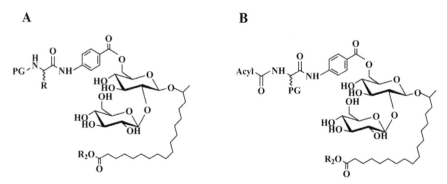

**Figure 4.** Basic structures of head group-modified sophorolipids with *para*-aminobenzoic acid (PABA)-linked amino acids (A) and acylated amino acids (B); for specifics, see Zerkowski et al. (2006). 'PG' signifies a protecting group and $R_2$ is either an ethyl or benzyl group.

Enzymatic approaches have also been explored to induce sophorose modification. Bisht et al. (1999) used *Candida antartica* lipase Novozym 435 to synthesize SL derivatives where the sophorose moiety was esterified to a hydrophobic acyl group. The starting materials used in the study were SLs esterified at the carboxylate end with methyl, ethyl, or butyl groups. Upon Novozym 435-catalysed condensation with the selected acylating agents, the primary-alcohol groups on the sophorose (at the C6' and C6" positions) of the SL esters were determined to be esterified to the corresponding acyl group (e.g., acetate, succinate, or acrylate). In addition, it was determined that minor deviations from the reaction protocol led to the formation of sophorolactones with the carboxylate group of the fatty acid esterified to the C6" hydroxy group of the sugar moiety. Similarly, Singh et al. (2003) further established the selective synthesis of an ester or amide group at carbon C6'-, C6"-, or both. The substrate for the enzymatic reactions was the ethyl ester of SLs obtained through a reaction with sodium ethoxide. Both Novozym 435 and Lipase PS-C were used to catalyze the reaction where it was found that Lipase PS-C functioned regioselectively to acylate C6", while Novozym 435 did not discriminate in performing the amidation or acylation of the C6'- and/or C6" of the SL esters. Recke et al. (2013) performed lipase-catalyzed acylations on the sugar moiety of alkyl sophorosides derived from *C. bombicola* produced in the presence of 2-dodecanol (SL-E$_{2-12}$). Novozym 435 induced the attachment of 3-hydroxydecanoic acid to both C6'- and C6" of the sophorose moiety forming a di-acylated derivative (SL-E$_{2-12}$-di-3-OH-C10). Similarly, 17-hydroxyoctadecanoic acid was esterified at C6'- or both C6'- and C6" of the alkyl sophoroside using Lipozyme IM 20 lipase (SL-E$_{2-12}$-mono-17-OH-C18 and SL-E$_{2-12}$-di-17-OH-C18). Furthermore, the investigators demonstrated that a glucose unit of the alkyl sophoroside could be removed using a snail glucuronidase to yield the corresponding alkyl glycoside, whose C6 position could then be esterified to a molecule of sebacic acid using Novozym 435 lipase.

Intermolecular reactions have also been established to produce SL-based structural derivatives. Zini et al. (2008) demonstrated the feasibility of using the lactone conformation of SLs derived from oleic acid in ring-opening metathesis polymerization reactions to produce polymeric SL molecules. These molecules were attached through the carboxylate group of the free fatty acid tail and the hydroxyl group at C4' resulting in poly(SLs), with the C18 oleic acid and sophorose alternately linked. The polymeric biomaterial had a number-average molecular weight ($M_n$) of 37,000 g/mol and a polydispersity index ($M_w/M_n$) of 2.1. The repeating units imparted an interesting asymmetric bola-amphiphilic structure which conferred unusual crystallinity properties to the novel biomaterial, in which a long-range order involving only the packing of sophorose groups was observed above the melting temperature of 123°C (close to that of low-density polyethylene) of the semicrystalline poly(SLs) ($T_g$ = 60–65°C). Subsequently, Peng et al. (2014) demonstrated the synthesis of poly(SLs) with various substituent groups at C6' and C6" of the sophorose moieties. SL-based monomers were enzymatically synthesized containing modified C6' and C6" on the sophorose moiety. These included C6'- and/or C6"-alkylated SLs synthesized enzymatically using lipase or cutinase, and C6'- and/or C6"-dehydroxylated and -iodinated/-azidated SLs obtained by chemical synthesis routes. Polymerization of these monomers in various combinations was accomplished using a second-generation Grubbs catalyst. The resultant homo- or hetero-copolymers had number-average molecular weights ($M_n$) of 36,000–84,000 g/mol and polydispersity indices of 2.2–2.9.

## Applications of SLs

### *Surfactant*

As mentioned previously, SLs consist of a polar (hydrophilic) moiety and a nonpolar (hydrophobic) moiety. As such, its amphipathic character imparts a surfactant-exhibiting surface active property. Consequently, the foremost applications of SLs being researched and developed are in areas where a surfactancy is required.

A surface activity indicator that is useful for predicting the potential application field of a surfactant is the hydrophile-lipophile balance (HLB) value first advanced by Griffin (Griffin 1949). The HLB is an index directly correlated to the relative distribution of the water-soluble (i.e., hydrophilic) and the oil-soluble (i.e., lipophilic) portions of a nonionic surfactant molecule. In a loose sense, HLB could also be viewed as the relative water solubility of a surfactant; the higher the HLB value of a surfactant, the more soluble it is in water. The range of HLB values for non-ionic surfactants is 0–20 (Griffin 1949). Figure 5 shows how HLB values of nonionic surfactants can be correlated with potential applications. A surfactant of certain HLB value would have the potential to be used as an antifoaming agent, a water-in-oil (W/O) emulsifying agent, a wetting agent, an oil-in-water (O/W) emulsifying agent, a detergent, or a solubilization agent (i.e., to facilitate the solubilization of water-insoluble substances). Griffin (1954) subsequently formulated an equation to calculate the HLB values of non-ionic surfactants. This was followed by a refined formula forwarded by Davies (1957) in which the estimation of HLBs of ionic surfactants could be compensated. The presence of ionic group(s) on a surfactant greatly expanded the HLB values to as high as 50. Experimental methods were later developed to estimate the HLB values of various surfactant types (Middleton 1968, Proverbio et al. 2003, Luan et al. 2009). The use of HLB values is widely adopted by various industries to facilitate the formulation of commercial products that require the addition of surfactants as functional ingredients.

As alluded to earlier in the chapter, SLs can assume a variety of structural configurations depending on the producing microbial strain, growth substrates,

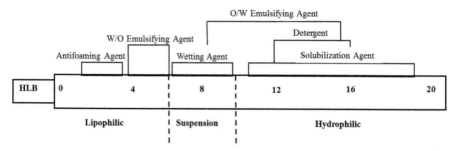

**Figure 5.** Hydrophile-lipophile balance (HLB) values and how they correlate to potential applications.

fermentation conditions, and downstream process and purification methods. Since HLB values depend heavily on the chemical structure, the assignment of a single HLB value to SL is impractical. Depending on the method of fermentative production and the actual conditions of the formulation, various HLB values have been ascribed to SLs used in particular applications. For example, Hillion et al. (1998) described the use of crude SL in its acid-form with HLB value between 4–5 as a wetting agent in the presence of divalent salt (e.g., calcium). Further, the calcium salt of SL has HLB values in the range of 4–7 and is suitable for stabilizing W/O- and also O/W-emulsions, making it useful for the preparation of cosmetic products requiring the formation of a continuous aqueous (i.e., W/O) or oil- (i.e., O/W) phase emulsion. At pH 8–12, SLs in acid or in salt forms with monovalent or divalent cations (i.e., sodium, potassium, calcium) exhibited HLB values of 13–15, making them good emulsifying agents to stabilize aqueous emulsions in cosmetic and sanitary formulations.

An effective solubilization agent requires that its HLB value falls in the range of 11–19 (Figure 5). SLs in certain structural configurations or in various modified forms can achieve these HLB values and thus can effectively function as biobased solubilization agents. Ernenwein et al. (2013) utilized the solubilization capacity of SLs to claim the preparations of hair care and skin cosmetic formulae, laundry and hard-surface cleaning products, and crop plant protection sprays. In these preparations, water insoluble ingredient(s) were solubilized by the action of SLs. In another patent application, Gross et al. (2013) described a chemical method to esterify *n*-alkyl groups of different chain lengths to the carboxyl group of SL, followed by selective lipase treatments to separately obtain di-, mono-, or non-acetylated SL-esters and SL-amides. These modified SLs were used alone or as mixtures with natural SLs to achieve the required hydrophile-lipophile balance in various O/W and W/O emulsion formulations.

The suitability of SL to use as an emulsifying agent was also studied in food science in the formulation of oxidatively stable structured lipid O/W emulsions (Xue et al. 2013). Structured lipids were synthesized via lipase-catalyzed interesterification reactions of rice bran oil and olive oil. Oil-in water emulsions of the resultant structured lipids were prepared using SL (at 0.1 wt%) as an emulsifying agent in comparison to Tween 20 (i.e., a commonly used food emulsifier). Antioxidant additives such as propyl gallate, ascorbic acid 6-palmitate or quercetin hydrate were added into the emulsion preparations to prevent peroxide formation. The results indicated that SL was as effective as Tween 20 in stabilizing the oxidatively stable O/W emulsions of the structured lipids for up to 20 days at ambient temperature. Subsequently, Koh and Gross (2016) measured the interfacial surface parameters of the alkyl esters of SL in lemon oil in water mixtures. In that study, it was concluded that among the various combinations of the components in the mixtures, the ethyl ester of SL at 0.1 wt% best stabilized a 20 wt% aqueous lemon oil mixture for up to one week without observing the separation of the oil and aqueous phases. In addition, Koh et al. (2017) tested the same series of alkyl esters of SLs in various combinations and mixtures of crude oil-related compounds with water. Although each SL derivative displayed different effectiveness in mediating the emulsification of different admixtures, the general conclusion of the authors was that various SLs performed well in comparison to the commonly used Triton X-100.

By far the most valued property of SL is its surfactant activity as a detergent for use in washing and cleaning applications. As shown in Figure 5, it requires that the molecule in its non-ionic form has an HLB value in the range of 12–16. The ionic forms of SL are expected to have HLB values > 20 as predicted based on the work of Davies (1957). The more popular parameters for predicting or indicating the suitability of a surfactant molecule for washing and cleaning applications are the critical micelle concentration (CMC) and the minimum surface tension ($\gamma_{min}$) of water in the presence of the surfactant. As with HLB values, the CMC and $\gamma_{min}$ of a SL molecule are governed by its structures and compositions, which in turn are influenced by its production and purification processes. Unlike the HLB values, however, the CMC and $\gamma_{min}$ values of any SL can be conveniently determined by experimental procedures using tensiometry. A compilation of CMC and $\gamma_{min}$ values of selected SL molecules determined under various conditions is presented in Table 1. In general, it is again obvious that the CMC and $\gamma_{min}$ values varied according to structural compositions of the biosurfactants. However, the variation found with the $\gamma_{min}$ values was not as extensive. Regardless of the range of structural changes and modification of the hydrophobic or the hydrophilic moiety of the molecule, $\gamma_{min}$ values fall into a relatively narrow range centered on mid-30's mN/m. The values at the two extremes of the $\gamma_{min}$ spectrum were for SL-E$_{2-12}$-di-3-OH-C10 ($\gamma_{min}$ = 27 mN/m) and SL-E$_{2-12}$-mono/di-17-OH-C18 ($\gamma_{min}$ = 50 mN/m; abbreviations defined earlier) (Recke et al. 2013) in which a 2-hydroxy-dodecane (instead of the hydroxy alkanoates found in naturally occurring sophorolipids) was the backbone of the hydrophobic moiety. Rosen and Dahanayake (2000a) had previously compiled a list of $\gamma$ values (note: not the $\gamma_{min}$ as reported in Table 1) of about 50 commercial nonionic surfactants at 0.001, 0.01, and 0.1 wt%. The values in their compilation ranged from the low-20s (i.e., $\gamma$ = 20.8 mN/m for [(CH$_3$)$_3$SiO]$_2$Si(CH$_3$)(CH$_2$)$_3$(OC$_2$H$_4$)$_{7.5}$OCH$_3$ at 0.1 wt%) to mid-50s (i.e., $\gamma$ = 58.2 mN/m for castor oil ethoxylate-40 (at 0.001 wt%) that contains a (OC$_2$H$_4$)$_{40}$OH hydrophilic tail). It was noted that the $\gamma_{min}$ values of the various SL compounds (Table 1) fell well within the range of the $\gamma$ values reported by Rosen and Dahanayake (2000a) for commercial nonionic surfactants, indicating that SLs are indeed functionally suitable surfactants for commercial applications. Unlike the $\gamma_{min}$ values, the CMC values of SLs varied widely (Table 1). It is very complicated to collectively summarize the reported CMC values because of the structural complexity of the SL preparations and of any salient differences in measuring methods. At the very least, however, it could be said that the reported CMC values of SLs and their derivatives fall within the range of the CMCs reported for commercial surfactants in which the lowest CMC value was 3.1 × 10$^{-6}$ M (0.00027 wt%) for C$_{16}$H$_{33}$(OC$_2$H$_4$)$_{15}$OH and the highest CMC value was 1.92 × 10$^{-3}$ M (0.027 wt%) for C$_{9-11}$H$_{19-23}$(OC$_2$H$_4$)$_8$OH (Rosen and Dahanayake 2000b). Some salient trends, however, may be gleaned from CMCs of structurally different SLs reported by individual research groups. The results of Ashby et al. (2008) for a series of lactonic SLs with different hydroxy fatty acid chain length (C$_{16}$ and C$_{18}$) and degree of unsaturation (C$_{18:0}$–C$_{18:2}$) showed that shorter chain length or higher degree of unsaturation leads to higher CMC values. Roelants et al. (2016) further demonstrated that lactonic SL forms and the acetylation of the 6'- and/or 6"-carbon of the sophorose moiety of acidic SL led to lowering of the CMC value. The data of

**Table 1.** Surfactant properties of select sophorolipid derivatives.

| SL Sample | Production Procedures | CMC (mg/L) | $\lambda_{min}$ (mN/m) | Solubility (mg/L H$_2$O) | Reference |
|---|---|---|---|---|---|
| SL-p 75% C$_{16:0}$; 92% Lc | Fed-batch: Glc+PA EtOAc extraction; Hexane precipit. | >200 | 35 | >200 | Ashby et al. 2008 |
| SL-s 76% C$_{18:0}$; 99% Lc | Fed-batch: Glc+SA EtOAc extraction; Hexane precipit. | 35 | 35 | <200 | Ashby et al. 2008 |
| SL-l 29% C$_{18:2}$; 35% C$_{18:0}$; 98% Lc | Fed-batch: Glc+LA EtOAc extraction; Hexane precipit. | 250 | 36 | >200 | Ashby et al. 2008 |
| SL-o 94% C$_{18:1}$; 100% Lc | Fed-batch: Glc+OA EtOAc extraction; Hexane precipit. | 140 | 36 | <200; ≥0.45 g/ml 70% v/v aq. EtOH | Ashby et al. 2008, Solaiman et al. 2017 |
| SL-p C$_{16:0}$; >93% Lc | Glc+$n$-hexadecane EtOAc extraction; Hexane precipit. | – | – | 0.28 g Lc SL/g EtOH; 0.017–0.025 g Lc SL/g aq. buffer; 0.010–0.025 g Ac SL/g aq. buffer or EtOH | Hu and Ju 2001b |
| SL-o + SL-s 99% Lc | Fed-batch: Glc+RO Multi-step ultrafilt. | 45.1 ± 0.1 | 33.9 ± 0.7 | – | Roelants et al. 2016 |
| SL-o + SL-s 100% Ac | Fed-batch: Glc+RO Multi-step ultrafilt. | 112 ± 7 (Mix-acetyl.) 245 ± 9 (Non-acetyl.) | 38.2 ± 0.5 40.9 ± 0.3 | – | Roelants et al. 2016 |
| SL-o $n$-alkyl C$_1$–C$_6$ esters | Esterification of SL-o | 3 µM (C$_6$) – 40 µM (C$_1$)[a] | 34 (C$_6$) – 39 (C$_1$)[a] | – | Zhang et al. 2004 |
| SL-o, GL-o 100% Ac | Glc+olive oil+OA EtOAc extract.; Chromatography; Var. enzym. rxns. | 160–170 µM (SLdiOAc, SL, GLOAc, GL) | 39.8–43.0 (SLdiOAc, SL, GLOAc, GL) | – | Imura et al. 2010 |

*Table 1 contd. ...*

*...Table 1 contd.*

| SL Sample | Production Procedures | CMC (mg/L) | $\lambda_{min}$ (mN/m) | Solubility (mg/L H$_2$O) | Reference |
|---|---|---|---|---|---|
| SL-alcohol (2 OH-C$_{12}$)[b] | Shake Flask: Glc+2-C$_{12}$H$_{25}$OH; Var. enzym. rxns. | 200 (SL-E$_{2\text{-}12}$)[b] | 31 (SL-E$_{2\text{-}12}$) | — | Recke et al. 2013 |
| | | 200 (GL-A$_{2\text{-}12}$)[b] | 31 (GL-A$_{2\text{-}12}$) | | |
| | | 200 (SL-E$_{2\text{-}12}$-di-3-OH-C10)[b] | 27 (SL-E$_{2\text{-}12}$-di-3-OH-C10) | | |
| | | 150 (SL-E$_{2\text{-}12}$-mono/di-17-OH-C18)[b] | 50 (SL-E$_{2\text{-}12}$-mono/di-17-OH-C18) | | |
| SL-s 100% ester/Ac; charged amino acid attached to sophorose head group. | Fed-batch: Glc+SA EtOAc extraction Hexane precipit. Various chemical rxns. | 90 µM SL-s (ethyl ester) | 48 | — | Zerkowski et al. 2006 |
| | | 6 µM (pH 5.8) – 24 µM (pH 9.1) SL-hydroxy-proline; (+) charge | 37-42 | | |
| | | 110 µM (pH 2.5) – 51 µM (pH 5.8) SL-N-propionyl-glutamic; (–) charge | 43-49 | | |
| | | 14–17 µM (pH 2.5–9.1) SL-glutamic; (zwitterion) | 39-41 | | |

**Abbreviations:** Critical Micelle Concentration (CMC), Minimum surface tension ($\lambda_{min}$), Sophorolipid (SL), Glucolipid (GL), Glucose (Glc), Palmitic acid (PA), Stearic acid (SA), Oleic acid (OA), Linoleic acid (LA), Rapeseed oil (RO), Lactone form (Lc), Acid form (Ac), Ethyl acetate (EtOAc), Acetyl group (OAc). [a]Estimated values based on data in Figs. 3 and 4 of Zhang et al. (2004). [b]Lipid moiety is 2-dodecanol. SL-E$_{2\text{-}12}$, 6''-mono-acetylated sophorose; GL-A$_{2\text{-}12}$, glucose as the hydrophilic moiety; SL-E$_{2\text{-}12}$-di-3-OH-C$_{10}$, 6',6''-di-(3-OH-C$_{10}$)-esters of SL-E$_{2\text{-}12}$; SL-E$_{2\text{-}12}$-mono/di-17-OH-C$_{18}$, 6'-mono(40%) and 6',6''-di(60%)-(17-OH-C$_{18}$)-esters of SL-E$_{2\text{-}12}$.

Zhang et al. (2004) also demonstrated that to an extent, the length of the alkyl chain in the acyl group of their series of SL-esters significantly influenced the CMC values of the molecules; the longer the alkyl chain, the lower the CMC value became. As mentioned earlier, Zerkowski et al. (2006) attached various charged amino acids onto the sophorose head group of ethyl ester of SL-stearic acid (Figure 4) and showed that considerable lowering of CMC values could be achieved in many of the charged SL derivatives. Because the structure and composition of SLs can be made to vary considerably depending on the details of the fermentation process and also by further chemical and/or enzymatic derivatization, the surfactant properties of SLs could span a broad spectrum of activity matching those of commercial surfactants.

## *Antimicrobial activities of SLs*

Apart from their most promising application as a biosurfactant, the next most exciting value-added property of SLs is their antimicrobial activity. Collectively, many biosurfactants have been found to be active in antibacterial, antiviral, antifungal, anti-tumor (Callaghan et al. 2016), anti-inflammatory, and immune-modulatory activities (Fracchia et al. 2012). Antimicrobial activity of different SLs and their derivatives discussed in this section are listed in Table 2. Since the first brief note by Ito et al. (1980), the antimicrobial activity of SL has been the focus of many research studies. With the more recent trend in the industry to move into biobased ingredients (APUA 2011), the promise of antimicrobial cleaning formulae based on SLs thus becomes a highly attractive proposition. Ito et al. (1980) tested the growth-inhibitory effect of SLs against yeasts such as *Candida* (7 different strains), *Pichia* (9 strains), *Debaryomyces* (3), *Saccharomycopsis lipolytica*, and *Lodderomyces inconspicua*, and found that SL only inhibited the growth of *C. lipolytica* and *P. farinose* at 0.4 mM concentration, whereas *C. parapsilosis*, *P. ohmeri* and *P. scolyti* were resistant to SL even at a concentration of 0.8 mM; the rest of the yeast species tested experienced only partial growth inhibition. Lang et al. (1989) expanded the testing of SL antimicrobial activity to include various Gram-positive and Gram-negative bacteria in addition to yeasts (i.e., *Glomerella cingulata* and *C. albican*). Furthermore, unlike Ito et al. (1980), Lang et al. (1989) specifically defined the molecular species of SL being tested as either di-acetylated (diOAc), or mono-acetylated (6″-OAc), or non-acetylated lactone of SL-oleic acid, separately or in a mixture. Results showed that the Gram-positive bacteria (i.e., *B. subtilis* and *S. epidermidis*) were susceptible to SL inhibition, whereas the Gram-negative *P. aeruginosa* and the 2 yeast species tested were resistant to SL. Furthermore, their data demonstrated that the degree of growth inhibition against the susceptible Gram-positive bacteria (i.e., *B. subtilis* and *S. epidermidis*) was in the order of diOAc > 6″-OAc > non-acetylated sophorolipid. Krivobo et al. (1994) carried out a concentration-dependent growth inhibitory study of SL against five classes of microorganisms (i.e., yeast (7 strains), Gram-positive (2 strains) and Gram-negative (2 strains) bacteria, phytopathogenic fungi (5 strains), and dermatophytic fungi (7 strains)). Their results were in general agreement with the findings of Ito et al. (1980) and Lang et al. (1989), that the Gram-positive bacteria (in this case, *Staphylococcus aureus* (at $IC_{50} < 0.1$ g/l of SL)) were more susceptible to SL growth inhibition than the Gram-negative bacteria and the yeasts/fungi (up to

**Table 2.** Antimicrobial activity of sophorolipids.

| SL Sample | Test Microorganism(s) | Gram Reaction | Inhibition Concentration | Assay Method Description | Reference |
|---|---|---|---|---|---|
| SL-o (C$_{18:1}$; Lc; diOAc) | Various species of *n*-alkane utilizing *Candida* (7 strains); *Pichia* (9 strains); *Debaryomyces* (3 strains); *Saccharomycopsis lipolytica*; *Lodderomyces inconspicua* | NA (yeast) | 0.4 mM → Complete inhibition of *C. lipolytica* and *P. farinose*; partial inhibition of all others except the following: 0.8 mM → n.i. to *C. parapsilosis*; *P. ohmeri*; and *P. scolyti* | A$_{650\,nm}$ of cultures after 1–3 d growth | Ito et al. 1980 |
| SL-o (C$_{18:1}$; Lc; diOAc) | *Bacillus subtilis*; *Staphylococcus epidermidis* | Gram (+) | 6 µg/ml (MIC) | | |
| | *Streptococcus faecium*[a] | Gram (+) | 15 µg/ml (MIC) | | |
| | *Glomerella cingulate* | NA (yeast) | n.i. | OD$_{546\,nm}$; MIC = conc. inhibiting 50% of final maximum cell growth | Lang et al. 1989 |
| SL-o (C$_{18:1}$; Lc; 6"-OAc) | *Bacillus subtilis*; *Staphylococcus epidermidis* | Gram (+) | 25 µg/ml (MIC) | | |
| | *Streptococcus faecium*[a] | Gram (+) | 29 µg/ml (MIC) | | |
| | *Glomerella cingulate* | NA (yeast) | 50 µg/ml (MIC) | | |
| SL-o (C$_{18:1}$; Lc; Non-OAc) | *Bacillus subtilis* | Gram (+) | 500 µg/ml (MIC) | | |
| | *Staphylococcus epidermidis*; *Streptococcus faecium*[a] | Gram (+) | n.i. | | |
| | *Glomerella cingulate* | NA (yeast) | n.i. | OD$_{546\,nm}$; MIC = conc. inhibiting 50% of final maximum cell growth | Lang et al. 1989 |
| SL-o (C$_{18:1}$; Lc; diOAc, 6"-OAc, or Non-OAc) | *Pseudomonas aeruginosa* | Gram (–) | n.i. | | |
| | *Propionibacterium acnes* | Gram (+) | +/– inhibition | Cell growth at 2 mg/ml of SL | |
| | *Candida albicans* | NA (yeast) | n.i. | | |

| | | | | |
|---|---|---|---|---|
| SL (olive oil as lipid substrate; undefined compositions) | Candida albicans; C. glabrata; C. parapsilosis; C. tropicalis (2 strains); Cryptococcus neoformans (2 strains) | NA (yeast) | 0.18 (C. parapsilosis) – 2 (Cryptococcus neoformans) g/l (IC$_{50}$) | 96-well microtitration plate method; IC$_{50}$ = concentration inhibiting 50% of cell growth (with 95% confidence boundaries) | Krivobo et al. 1994 |
| | Escherichia coli; P. aeruginosa | Gram (–) | 0.52 g/l (P. aeruginosa) – n.i. (E. coli) (IC$_{50}$) | | |
| | Staphylococcus aureus (2 strains) | Gram (+) | < 0.1 g/l (IC$_{50}$) | | |
| | Drechslera sp.; Fusarium oxysporum; Geotrichum candidum; Pleomorphomonas oryzae; Verticillium lecanii | (phytopath.) | 0.78 (Drechslera sp.) – n.i. (V. lecanii) g/l (IC$_{50}$) | | |
| | Epidermophyton floccosum; Microsporum canis; Trichophyton mentagrophytes (3 strains); T. rubrum; T. tonsurans | (dermatophyte) | 1.26 (M. canis) – 3.52 (T. tonsurans) g/l (IC$_{50}$) | 96-well microtitration plate method; IC$_{50}$ = concentration inhibiting 50% of cell growth (with 95% confidence boundaries) | Krivobo et al. 1994 |
| SL (canola oil lipid substrate; Ac) | B. subtilis; S. aureus; Streptococcus mutans; Propionibacterium acnes | Gram (+) | IC = 4, 1, 1, 0.5 ppm, respectively | IC = the lowest test concentration at which cell growth (OD$_{600}$) or number of colony was first observed | Kim et al. 2002 |
| | Botrytis cineria | (phytopath.) | 50% growth inhibition at 100 ppm | | |
| SL (canola oil lipid substrate; Lc) | B. subtilis; P. acnes | Gram (+) | IC = 16 and 4 ppm, respectively | | |
| SL-o (C$_{18:1}$; Lc/Ac (i.e., non-lactonic)/diOAc/6-OAc) | Streptococcus agalactiae, B. subtilis, Micrococcus luteus, E. coli, Moraxella sp., Salmonella choleraesuis, Alcaligenes latus; Ralstonia eutropha | Gram (+)/ Gram (–) | 4.891 mg/ml (Moraxella sp.); others varying degrees of inhibition to n.i. | % growth inhibition (measured by OD$_{400}$) of culture at 24 h in presence of 4.891 mg/ml SL | Gross and Shah 2002 |

*Table 2 contd. ...*

*...Table 2 contd.*

| SL Sample | Test Microorganism(s) | Gram Reaction | Inhibition Concentration | Assay Method Description | Reference |
|---|---|---|---|---|---|
| SL-o (C$_{18:1}$; *n*-Alkyl (C$_1$-C$_6$) esters; Ac/diOAc/6'-OAc/Non-OAc) | *C. albicans*; *C. antartica*; *C. tropicalis* | NA (yeast) | 100% inhibition of *C. albicans*, *C. antartica*, and *C. tropicalis* by various acetylated and non-acetylated SL esters | % inhibition (measured by OD$_{400}$) of culture at 48 h in presence of 5 mg/ml SL | Gross and Shah 2014 |
| SL (C$_{18:1}$ & C$_{18:2}$; Lc; non-OAC; variously obtained using different sugar substrate)[b] | *Rhodococcus erythropolis*, *B. subtilis*; *S. agalactiae*; *S. epidermis* | Gram (+) | 0.006 mg/ml (xylose-SL, *R. erythropolis*); 0.024 mg/ml (lactose- and arabinose-SL, *B. subtilis*) | MLD$_{50}$ = 50% lethal dose based on broth microdilution method using 96-well microtiter plates | Shah et al. 2007 |
|  | *Moraxella sp.*; *P. putida*; *Enterobacter aerogenes*; *E. coli* | Gram (−) | 0.024 mg/ml (lactose- and arabinose-SL, *Moraxella sp.*) |  |  |
| Commercial SL (Unspecified structural species)[c] | *Filobasidiella neoformans*; *C. tropicalis* | NA (yeast) | 1–15 mg/ml (MIC) | MIC = 50% inhibition of cell growth (OD) | Kulakovskaya et al. 2014 |
| SL-p (C$_{16:0}$), SL-s (C$_{18:0}$), or SL-o (C$_{18:1}$; all in Lc form | *Listeria monocytogenes*; *Salmonella spp.* | Gram (+)/Gram (−) | 0.1% (w/v) (*L. monocytogenes*) | 1 min incubation; agar dilution method | Zhang et al. 2016a |
| SL-o (C$_{18:1}$; Lc) | *Escherichia coli* O157:H7 | Gram (−) | 1% (w/v) | 24 h incubation; agar dilution method | Zhang et al. 2016b |
| SL-o (C$_{18:1}$; Lc) | *Lactobacillus acidophilus*, *L. fermentum*, *Streptococcus mutans*, *S. salivarius*, *S. sobrinus* | Gram (+) | ≥ 1 mg/ml (all *Lactobacilli*) ≥ 50 µg/ml (all *Streptococci*) | 100% growth inhibition (OD$_{600}$); microtiter plate | Solaiman et al. 2017 |

| | | | | | |
|---|---|---|---|---|---|
| SL (C18:1, 92% & C18:2, 6%; diOAc Lc) | S. mutans, S. oralis, S. sanguinis, A. naeslundii | Gram (+) | 195 mg/ml (S. mutans, S. sanguinis, N. mucosa); 97.5 mg/ml (S. oralis, A. naeslundii) | Serial Dilution Method; MIC = lowest concentration at which no bacterial growth occurred (Resazurin) | Elshikh et al. 2017 |
| | N. mucosa | Gram (−) | | | |
| + Tetracycline + Ciprofloxacin + Chlorhexidine | Same as above | | 0.03 mg/ml (All 5 strains) | Same as above | |
| SL (C18:1, > 90% non-OAc, acid form) | E. faecalis | Gram (+) | E. faecalis: 2–4 mg/ml w/kan.; < 0.016–1 mg/ml w/Cef. | Serial Dilution Method; MIC = lowest concentration where OD600 was not significantly different between culture and blank control | Lydon et al. 2017 |
| + Kanamycin + Cefotaxime | P. aeruginosa | Gram (−) | P. aeruginosa: < 0.25–16 mg/ml w/ kan.; < 0.062–0.5 mg/ml w/Cef | | |
| SL-p (C16:0), SL-s (C18:0), or SL-o (C18:1); all in Lc form | B. licheniformis, B. pumilus, B. mycoides, E. faecium, S. xylosus, A. viridans, S. cohnii, S. equorum | Gram (+) | 4.88 µg/ml SL-s (B. licheniformis, B. pumilus, B. mycoides, S. xylosus, P. luteola); 19.5 µg/ml SL-p, -s, or -o for the rest of organisms | Agar Dilution Method; MIC = lowest concentration at which no colony appeared | Solaiman et al. 2016 |
| | P. luteola, E. cloacae, V. fluvialis, E. sakazakii | Gram (−) | | | |

**Abbreviations:** Lactone form (Lc), Acid form (Ac), Acetylation (at 6'- and/or 6''-carbon of sophorose moiety; OAc), No inhibition (n.i.), Not applicable (NA). [a] ATCC strain 10541 is now reclassified as *Enterococcus hirae*. Previously, also classified as *S. faecalis* Andrewes and Horder. [b] Sugars tested included: glucose, fructose, xylose, ribose, lactose, mannose, arabinose, and galactose. [c] "Sopholiance S" from *C. bombicola* using glucose and methyl rapeseed/ate as substrates.

$IC_{50}$ of 3.52 g/L for the dermatophyte *Trichophyton tonsurans*). Among the yeasts and fungi tested, data indicated that *C. parapsilosis* yeast was the only susceptible species ($IC_{50}$ 0.18 g/l).

Kim et al. (2002) compared the antimicrobial activity of chromatographically separated lactonic and acidic SL forms against several bacterial species and a fungal strain (*Botrytis cineria*). Aside from the previously reported high sensitivity of Gram-positive bacteria (in this case, *B. subtilis* and *Propionibacterium acnes*) to SLs, data obtained by Kim et al. (2002) showed that the acidic form of SL was more inhibitory (up to 4 × for *B. subtilis* and 8 × for *P. acnes*) than the lactonic SL. This finding was contrary to the latter consensus summary of a review by Delbeke et al. (2016) which stated that the lactonic SLs generally performed better in antimicrobial tests than the acidic forms. In a patent application, Gross and Shah (2002) summarized the growth inhibitory effects of SL in the following forms: lactone, acidic, di-OAc, and 6'-OAc, against several Gram-positive and Gram-negative bacteria. Their results summarily showed that, unexpectedly, the Gram-negative *Moraxella* sp. was the most susceptible organism even when compared to the Gram-positive *B. subtilis* and *Streptococcus agalactiae*. Furthermore, and again contrary to general consensus (Delbeke et al. 2016), their results also showed that the non-lactonic (i.e., acidic) form of their SLs exhibited better growth inhibitory activity when compared to other structural forms. Shah et al. (2007) reported the isolation of SLs having different sugar head groups using monosaccharides other than glucose as the sugar substrate in the fermentation process. Based on LC/MS and the absence of OAc, it was concluded that the isolated SLs contained the monosaccharide used as the feedstock. When the isolated SLs were tested against a panel of bacteria very similar to those used in the study by Gross and Shah (2002), the xylose-SL (i.e., SL isolated from the fermentation where xylose was the sole saccharide substrate) was most inhibitory to the Gram-positive *R. erythropolis* at an $MLD_{50}$ of 0.006 mg/ml, followed by the lactose- and arabinose-SLs ($MLD_{50}$ of 0.024 mg/ml) against the Gram-positive *B. subtilis* and the Gram-negative *Moraxella* sp. Gross and Shah (2014) then synthesized by chemical and enzymatic methods a series of *n*-alkyl esters (C1-C6) of SLs having various degrees of acetylation at the 6'- and/or 6''-carbon of the sophorose head group. Using these ester derivatives, they showed that the growth of several *Candida* species as monitored by $OD_{400}$ was completely inhibited by some of these SL-alkyl esters at a high concentration of 5 mg/ml. Different *Candida* species appeared to respond differently to the panel of SL-alkyl esters. For example, *C. albicans* was susceptible to the $C_1$-ester, $C_2$-ester-diOAc and $C_2$-ester-6'-monoOAc, whereas *C. tropicalis* was best inhibited by the $C_1$- and $C_6$-esters, and the $C_2$-ester-diOAc derivatives. Kulakovskaya et al. (2014) tested a commercial preparation of SLs (i.e., "Sophoroliance" having unspecified details of structural composition) against two strains of yeast, i.e., *Filobasidiella* (*Cryptococcus*) *neoformans* and *C. albicans*, and determined the MIC (defined as 50% inhibition of cell growth based on optical density measurement) values as 1–15 mg/ml.

The antimicrobial property of SL has also been examined in a target specific application against clinically and environmentally relevent microorganisms. Sleiman et al. (2009) studied the antibacterial activity of SLs against clinically relevant bacterial stains (i.e., *E. coli*, *S. aureus*, *K. pneumoniae*, *P. mirabilis*, *P. aeruginosa* and *S. pneumoniae*), and concluded that in comparison to the standard antibiotics (i.e., ceftazidime and cefotaxime), SLs did not effectively inhibit the growth of these bacteria. Additionally, the antimicrobial activity of SLs towards environmentally harmful algae was verified (Sun et al. 2004, Lee et al. 2008). In the study by Sun et al. (2004), SLs at a concentration between 10–20 mg/l effectively arrested the motility of 90% of the cells of *Alexandrium tamarense*, *Heterosigma akashiwo* and *Cochlodinium polykrikoides* species by variously causing ecdysis, lysis or swelling of the cell. Lee et al. (2008) conducted a field mitigation spraying study against a harmful *Cochlodinium* algal bloom using a mixture of SLs (5 mg/l) and yellow clay (1 g/l), and observed a 95% removal of the bloom in 30 min after spraying. The accompanied ecological studies showed that the bacterioplankton, heterotrophic protists, and zooplankton were adversely affected to a lesser extent than the standard application of the yellow clay alone, indicating that SLs could minimize the negative effect of algal bloom mitigation to the pelagic or ocean ecosystem. In an antibacterial survey of microorganisms selected from a previously isolated collection of bacteria from salted animal hides and skins (Birbir and Ilgaz 1996, Aslan and Birbir 2011, 2012), it was shown that SLs produced from glucose and stearic acid (SL-s) exhibited a slightly better antimicrobial activity than SLs obtained using glucose and oleic acid (SL-o) or palmitic acid (SL-p) (Table 2) (Solaiman et al. 2016). In that study, the target microorganisms were selected to represent three (3) bacterial categories, i.e., Gram-positive endospore-forming (*B. licheniformis*, *B. pumilus* and *B. mycoides*), Gram-positive (*E. faecium*, *S. xylosus*, *A. viridans*, *S. cohnii* and *S. equorum*), and Gram-negative (*P. luteola*, *E. cloacae*, *V. fluvialis* and *E. sakazakii*). The results not only agreed with documented consensus conclusions that Gram-positive bacteria were generally more susceptible to the antimicrobial action of SLs than the Gram-negative organisms, but also showed for the first time that among the Gram-positive bacteria, the endospore-forming species were the more sensitive target for SL action (Solaiman et al. 2016). Two studies by Zhang et al. (2106a, b) showed that SLs exhibited antimicrobial activity toward food pathogens, i.e., *E. coli* O157:H7, a *Salmonella* spp. and *Listeria monocytogenes*. A significant reduction of bacterial counts was observed with all SLs tested (i.e., SL-o, SL-p, SL-s) at a concentration of 0.1% wt/v against the Gram-positive *L. monocytogenes* (Zhang et al. 2016b). A higher concentration of SL-o (1% wt/v) was required, however, to inhibit the growth of the Gram-negative *E. coli* O157:H7 (Zhang et al. 2016a). The antibacterial action of SLs against representative bacteria involved in oral hygiene/ tooth decay (i.e., caries development) was reported in studies by Solaiman et al. (2017) and Elshikh et al. (2017). Solaiman et al. (2017) demonstrated the growth inhibition and the cell morphology of three oral *Streptococci* (i.e., *S. mutans*, *S. salivarius* and *S. sobrinus*) and two oral *Lactobacilli* (i.e., *L. acidophilus* and *L. fermentum*) in the presence of SL. It was theorized by Solaiman et al. (2017) that in the process of caries development, the biofilm-forming oral *Streptococci* (notably *S. mutans*) laid down a protective polysaccharide matrix for the acid-producing oral *Lactobacilli* to thrive, leading to a high local concentration of acid that undermined the integrity of the hard structures (such as the enamel and bone) of the tooth (Tanzer et al. 2001, Kleinberg 2002). The results of that study showed that SL was more reactive towards the oral *Streptococci* than the *Lactobacilli*, causing growth inhibition at a minimal concentration of 50 µg/ml for the former versus 1 mg/ml for the latter (Table 2). This finding was of particular significance because the growth of (and the consequent biofilm formation by) the primary culprit, i.e., the *Streptococci*, could be inhibited at a low concentration of SL in which the *Lactobacilli* (some species of which are beneficial probiotics) were not affected. Elshikh et al. (2017) confirmed the conclusions made by Solaiman et al. and further showed that lactonic SLs were effective against *S. oralis, Actinomyces naeslundii,* and *Neisseria mucosa,* the additional common oral bacteria. In that same study, it was revealed that lactonic SLs can interfere with biofilm formation and when used in combination with popular antibiotics (i.e., tetracycline HCl, ciprofloxacin, chlorhexidine) can improve the minimum inhibitory concentrations (MIC) of the antibiotics to as low as 0.03 mg/ml (Table 2). Additional work by Lydon et al. (2017) showed the adjuvant-type activity of acidic SLs when used with kanamycin and cefotaxime against both Gram-positive (*Enterococcus faecalis*) and Gram-negative (*Pseudomonas aeruginosa*) bacterial strains in promoting wound healing.

## Conclusion

In conclusion, SLs are unique biomolecules that can be synthesized in large yields, are environment friendly, and whose amphiphilic nature imparts many distinct uses including surfactancy and antimicrobial activities among others. Due to the many functional groups present on the SL molecules, research efforts have focused on the prospects of using chemical and/or biological means to modify the SL molecules to induce property variation and using the resulting derivative molecule(s) for specific applications. By doing so, it is expected that new possibilities will soon be developed that will compete with and perhaps replace petroleum-based analogues.

## References

Alliance for the Prudent Use of Antibiotics (APUA). 2011. White Paper on Triclosan. http://emerald.tufts. edu/med/apua/consumers/personal_home_21_4240495089.pdf.

Ashby, R.D., A. Nuñez, D.K.Y. Solaiman and T.A. Foglia. 2005. Sophorolipid biosynthesis from a biodiesel co-product stream. J. Amer. Oil Chem. Soc. 82: 625–630.

Ashby, R.D., D.K.Y. Solaiman and T.A. Foglia. 2006. The use of fatty acid esters to enhance free acid sophorolipid synthesis. Biotechnol. Lett. 28: 253–260.

Ashby, R.D., D.K.Y. Solaiman and T.A. Foglia. 2008. Property control of sophorolipids: Influence of fatty acid substrate and blending. Biotechnol. Lett. 30: 1093–1100.

Ashby, R.D., A.J. McAloon, D.K.Y. Solaiman, W.C. Yee and M. Reed. 2013. A process model for approximating the production costs of the fermentative synthesis of sophorolipids. J. Surfact. Deterg. 16: 683–691.

Aslan, E. and M. Birbir. 2011. Examination of Gram-positive bacteria on salt-pack cured hides. J. Am. Leath. Chem. Assoc. 106: 372–380.

Aslan, E. and M. Birbir. 2012. Examination of Gram-negative bacteria on salt-pack cured hides. J. Am. Leath. Chem. Assoc. 107: 106–115.

Asmer, H.-J., S. Lang, F. Wagner and V. Wray. 1988. Microbial production, structure elucidation and bioconversion of sophorose lipids. J. Am. Oil Chem. Soc. 65: 1460–1466.

Azim, A., V. Shah, G.F. Doncel, N. Peterson, W. Gao and R. Gross. 2006. Amino acid conjugated sophorolipids: A new family of biologically active functionalized glycolipids. Bioconjugate Chem. 17: 1523–1529.

Birbir, M. and A. Ilgaz. 1996. Isolation and identification of bacteria adversely affecting hide and leather quality. J. Soc. Leath. Technol. Chem. 80: 147–153.

Bisht, K.S., R.A. Gross and D.L. Kaplan. 1999. Enzyme-mediated regioselective acylations of sophorolipids. J. Org. Chem. 64: 780–789.

Brakemeier, A., D. Wullbrandt and S. Lang. 1998. *Candida bombicola*: Production, of novel alkyl glucosides based on glucose/2-dodecanol. Appl. Microbiol. Biotechnol. 50: 161–166.

Breithaupt, T.B and R.J. Light. 1982. Affinity-chromatography and further characterization of the glucosyltransferases involved in hydroxydocosanoic acid sophoroside production in *Candida bogoriensis*. J. Biol. Chem. 257: 9622–9628.

Callaghan, B., H. Lydon, S.L.K.W. Roelants, I.N.A. van Bogaert, R. Marchant, I.M. Banat and C.A. Mitchell. 2016. Lactonic sophorolipids increase tumor burden in Apc[min+/–] mice. PLoS ONE 11(6): e0156845.

Chen, J., X. Song, H. Zhang, Y.B. Qu and J.Y. Miao. 2006. Production, structure elucidation and anticancer properties of sophorolipid from *Wickerhamiella domercqiae*. Enz. Microb. Technol. 39: 501–506.

Daniel, H.-J., M. Reuss and C. Syldatk. 1998. Production of sophorolipids in high concentration from deproteinized whey and rapeseed oil in a two stage fed batch process using *Candida bombicola* ATCC 22214 and *Cryptococcus curvatus* ATCC 20509. Biotechnol. Lett. 20: 1153–1156.

Davila, A.-M., R. Marchal and J.-P. Vandercasteele. 1994. Sophorose lipid production from lipidic precursors: Predictive evaluation of industrial substrates. J. Ind. Microbiol. 13: 249–257.

Davies, J.T. 1957. A quantitative kinetic theory of emulsion type. I. Physical chemistry of the emulsifying agent. pp. 426–438. *In*: J.H. Schulman (ed.). Gas/Liquid and Liquid/Liquid Interfaces. Proceedings of 2nd International Congress Surface Activity. Butterworth Scientific Publications, London, UK.

Delbeke, E.I.P., B.I. Roman, G.B. Marin, K.M. van Geem and C.V. Stevens. 2015. A new class of antimicrobial biosurfactants: Quaternary ammonium sophorolipids. Green Chem. 17: 3373–3377.

Delbeke, E.I.P., M. Movsisyan, K.M. van Geem and C.V. Stevens. 2016. Chemical and enzymatic modification of sophorolipids. Green Chem. 18: 76–104.

Dengle-Pulate, V., S. Bhagwat and A. Prabhune. 2013. Microbial oxidation of medium chain fatty alcohol in the synthesis of sophorolipids by *Candida bombicola* and its physicochemical characterization. J. Surf. Deterg. 16: 173–181.

Deshpande, M. and L. Daniels. 1995. Evaluation of sophorolipid biosurfactant production by *Candida bombicola* using animal fat. Bioresour. Technol. 54: 143–150.

Elshikh, M., I. Moya-Ramirez, H. Moens, S. Roelants, W. Soetaert, R. Marchant and I.M. Banat. 2017. Rhamnolipids and lactonic sophorolipids: Natural antimicrobial surfactants for oral hygiene. J. Appl. Microbiol. (in press). DOI:10.1111/jam.13550.

Ernenwein, C., R. Reynaud, A. Guilleret, L. Podevin, A. Rannou and F. Lafosse. 2013. Biosolubilizer. World Patent Application # WO2013182759 A1.

Esders, T.W. and R.J. Light. 1972. Glucosyl- and acetyltransferases involved in the biosynthesis of glycolipids from *Candida bogoriensis*. J. Biol. Chem. 247: 1375–1386.

Fracchia, L., M. Cavallo, M.G. Martinotti and I.M. Banat. 2012. Biosurfactants and bioemulsifiers biomedical and related applications—present status and future potentials. pp. 325–370. *In*: D.N. Ghista (ed.). Biomedical Science, Engineering and Technology. Intechopen Publishers, Rijeka, Croatia.

Fleurackers, S. 2006. On the use of waste frying oil in the synthesis of sophorolipids. Eur. J. Lipid Sci. Technol. 108: 5–12.

Gorin, P.A.J., J.F.T. Spencer and A.P. Tulloch. 1961. Hydroxy fatty acid glycosides of sophorose from *Torulopsis magnolia*. Can. J. Chem. 39: 846–855.

Griffin, W.C. 1949. Classification of surface-active agents by HLB. J. Soc. Cosmetic Chem. 1: 311–326.

Griffin, W.C. 1954. Calculation of HLB values of non-ionic surfactants. J. Soc. Cosmetic Chem. 5: 249–256.

Gross, R.A. and V. Shah. 2002. Antimicrobial properties of various forms of sophorolipids. U.S. Patent Application # 60/424,271.

Gross, R.A., R. Thavasi, A. Koh and Y. Peng. 2013. Modified sophorolipids as oil solubilizing agents. U.S. Patent Application # 20130331466 A1.

Gross, R.A. and V. Shah. 2014. Method for neutralizing fungi using sophorolipids and antifungal sophorolipids for use therein. U.S. Patent # 8,796,228.

Gudiña, E.J., V. Rangarajan, R. Sen and L.R. Rodrigues. 2013. Potential therapeutic applications of biosurfactants. Trends Pharmacol. Sci. 34: 667–675.

Hillion, G., R. Marchal, C. Stoltz and F. Borzeix. 1998. Use of a sophorolipid to provide free radical formation inhibiting activity or elastase inhibiting activity. U.S. Patent # 5,756,471.

Hommel, R. and C. Ratledge. 1990. Evidence for two fatty alcohol oxidases in the biosurfactant-producing yeast *Candida (Torulopsis) bombicola*. FEMA Microbiol. Lett. 70: 183–186.

Hu, Y. and L.-K. Ju. 2001a. Sophorolipid production from different lipid precursors observed with LC-MS. Enz. Microb. Technol. 29: 593–601.

Hu, Y. and L.-K. Ju. 2001b. Purification of lactonic sophorolipids by crystallization. J. Biotechnol. 87: 263–272.

Huang, F.-C., A. Peter and W. Schwab. 2014. Expression and characterization of CYP52 genes involved in the biosynthesis of sophorolipid and alkane metabolism from *Starmerella bombicola*. Appl. Environ. Microbiol. 80: 766–776.

Imura, T., Y. Masuda, H. Minamikawa, T. Fukuoka, M. Konishi, T. Morita, H. Sakai, M. Abe and D. Kitamoto. 2010. Enzymatic conversion of diacetylated sophoroselipid into acetylated glucoselipid: Surface-active properties of novel bolaform biosurfactants. J. Oleo. Sci. 59: 495–501.

Ito, S., M. Kinta and S. Inoue. 1980. Growth of yeasts on n-alkanes: Inhibition by a lactonic sophorolipid produced by *Torulopsis bombicola*. Agric. Biol. Chem. 44: 2221–2223.

Karunasagar, I. 2009. Bioremediation in the marine environment. pp. 173–174. *In*: H.W. Doelle, J.S. Rokem and M. Berovic (eds.). Biotechnology—Volume IX: Fundamentals in Biotechnology. Encyclopedia of Life Support Systems (EOLSS), Oxford UK.

Kim, K., D. Yoo, Y. Kim, B. Lee, D. Shin and E.-K. Kim. 2002. Characteristics of sophorolipid as an antimicrobial agent. J. Microbiol. Biotechnol. 12: 235–241.

Kitamoto, D., H. Isoda and T. Nakahara. 2002. Functions and potential applications of glycolipid biosurfactants—from energy-saving materials to gene delivery carriers. J. Biosci. Bioengin. 94: 187–201.

Kleinberg, I. 2002. A mixed-bacteria ecological approach to understanding the role of the oral bacteria in dental caries causation: An alternative to *Streptococcus mutans* and the specific plaque hypothesis. Crit. Rev. Oral Biol. Med. 13: 108–125.

Koh, A. and R. Gross. 2016. A versatile family of sophorolipid esters: Engineering surfactant structure for stabilization of lemon oil-water interfaces. Coll. Surf. A: Physicochemical and Engineering Aspects 507: 152–163.

Koh, A., A. Wong, A. Quinteros, C. Desplat and R. Gross. 2017. Influence of sophorolipid structure on interfacial properties of aqueous-arabian light crude and related constituent emulsions. J. Am. Oil Chem. Soc. 94: 107–119.

Konishi, M., T. Fukuoka, T. Morita, T. Imura and D. Kitamoto. 2008. Production of new types of sophorolipids by *Candida batistae*. J. Oleo. Sci. 57: 359–369.

Krivobok, S., P. Guiraud, F. Seigle-Murandi and R. Steiman. 1994. Production and toxicity assessment of sophorosides from *Torulopsis bombicola*. J. Agri. Food Chem. 42: 1247–1250.

Kulakovskaya, E., B. Baskunov and A. Zvonarev. 2014. The antibiotic and membrane-damaging activities of cellobiose lipids and sophorose lipids. J. Oleo Sci. 63: 701–707.

Kurtzman, C.P., N.P.J. Price, K.J. Ray and T.M. Kuo. 2010. Production of sophorolipid biosurfactants by multiple species of *Starmerella (Candida) bombicola* yeast clade. FEMS Microbiol. Lett. 311: 140–146.

Lang, S., E. Katsiwela and F. Wagner. 1989. Antimicrobial effects of biosurfactants. Fat Sci. Technol. 91: 363–366.

Lee, Y.-J., J.-K. Choi, E.-K. Kim, S.-H. Youn and E.-J. Yang. 2008. Field experiments on mitigation of harmful algal blooms using a sophorolipid-yellow clay mixture and effects on marine plankton. Harm. Algae 7: 154–162.

Li, W., J. Li and X. Song. 2016. Alkane utilization, the expression and function of cytochrome P450 in sophorolipid synthesis in *Starmerella bombicola* CGMCC 1576. Res. Rev. J. Microbiol. Biotechnol. 5: 58–63.

Luan, F., H. Liu, Y. Gao, Q. Li, X. Zhang and Y. Guo. 2009. Prediction of hydrophile–lipophile balance values of anionic surfactants using a quantitative structure–property relationship. J. Coll. Interface Sci. 336: 773–779.

Lydon, H.L., N. Baccile, B. Callaghan, R. Marchant, C.A. Mitchell and I.M. Banat. 2017. Adjuvant antibiotic activity of acidic sophorolipids with potential for facilitating wound healing. Antimicrob. Agents Chemother. 61: e02547–16.

Marchant, R. and I.M. Banat. 2012. Microbial biosurfactants: Challenges and opportunities for future exploitation. Trends Biotechnol. 30: 558–565.

Menzel, R., E. Kargel, F. Vogel, C. Bottcher and W.H. Schunck. 1996. Topogenesis of a microsomal cytochrome P450 and induction of endoplasmic reticulum membrane proliferation in *Saccharomyces cerevisiae*. Arch. Biochem. Biophys. 330: 97–109.

Middleton, J.J. 1968. A titration method for the determination of the HLB's of emulsifiers. J. Soc. Cosmetic Chem. 19: 129–136.

Morita, T., T. Fukuoka, T. Imura and D. Kitamoto. 2013. Production of mannosylerythritol lipids and their application in cosmetics. Appl. Microbiol. Biotechnol. 97: 4691–4700.

Nelson, D.R. 1998. Cytochrome P450 nomenclature. Meth. Mol. Biol. 107: 15–24.

Nuñez, A., R. Ashby, T.A. Foglia and D.K.Y. Solaiman. 2001. Analysis and characterization of sophorolipids by liquid chromatography with atmospheric pressure chemical ionization. Chromatographia 53: 673–677.

Nunez, A., T.A. Foglia and R. Ashby. 2003. Enzymatic synthesis of a galatopyranose sophorolipid fatty acid-ester. Biotechnol. Lett. 25: 1291–1297.

Nuñez, A., R. Ashby, T.A. Foglia and D.K.Y. Solaiman. 2004. LC/MS analysis and lipase modification of the sophorolipids produced by *Rhodotorula bogoriensis*. Biotechnol. Lett. 26: 1087–109.

Pekin, G., F. Vardar-Sukan and N. Kosaric. 2005. Production of sophorolipids from *Candida bombicola* ATCC 22214 using turkish corn oil and honey. Eng. Life Sci. 5: 357–362.

Peng, Y., D.J. Munoz-Pinto, M. Chen, J. Decatur, M. Hahn and R.A. Gross. 2014. Poly(sophorolipid) structural variation: Effects on biomaterial physical and biological properties. Biomacromolecules 15: 4214–4227.

Peng, Y., F. Totsingan, M.A.R. Meier, M. Steinmann, F. Wurm, A. Koh and R.A. Gross. 2015. Sophorolipids: Expanding structural diversity by ring-opening cross-metathesis. Eur. J. Lipid Sci. Technol. 117: 217–228.

Pöhnlein, M., R. Hausmann, S. Lang and C. Syldatk. 2015. Enzymatic synthesis and modification of surface-active glycolipids. Eur. J. Lipid Sci. Technol. 117: 145–155.

Proverbio, Z.E., S.M. Bardavid, E.L. Arancibia and P.C. Schulz. 2003. Hydrophile-lipophile balance and solubility parameter of cationic surfactants. Coll. Surf. A: Physicochem. Eng. Aspects 214: 167–171.

Rau, U., C. Manzke and F. Wagner. 1996. Influence of substrate supply on the production of sophorose lipids by *Candida bombicola* ATCC 22214. Biotechnol. Lett. 18: 149–154.

Recke, V.K., M. Gerlitzki, R. Hausmann, C. Syldatk, V. Wray, H. Tokuda, N. Suzuki and S. Lang. 2013. Enzymatic production of modified 2-dodecyl-sophorosides (biosurfactants) and their characterization. Eur. J. Lipid Sci. Technol. 115: 452–463.

Roelants, S.L.K.W., K.M.J. Saerens, T. Derycke, B. Li, Y.-C. Lin, Y. van de Peer, S.L. de Maeseneire, I.N.A. van Bogaert and W. Soetaert. 2013. *Candida bombicola* as a platform organism for the production of tailor-made biomolecules. Biotechnol. Bioengin. 110: 2494–2503.

Roelants, S.L.K.W., K. Ciesielska, S.L. de Maeseneire, H. Moens, B. Everaert, S. Verweire, Q. Denon, B. Vanlerberghe, I.N.A. van Bogaert, P. van der Meeren, B. Devreese and W. Soetaert. 2016. Towards the industrialization of new biosurfactants: Biotechnological opportunities for the lactone esterase gene from *Starmerella bombicola*. Biotechnol. Bioengin. 113: 550–559.

Rosen, M.J. and M. Dahanayake. 2000a. How the adsorption of surfactants changes the properties of interfaces and related performance properties. Chapter 2. *In*: M.J. Rosen and M. Dahanayake (eds.). Industrial Utilization of Surfactants. Principles and Practice. AOCS Press, Champaign, IL, USA.

Rosen, M.J. and M. Dahanayake. 2000b. How surfactants change the internal properties of the solution phase and related performance properties. Chapter 3. *In*: M.J. Rosen and M. Dahanayake (eds.). Industrial Utilization of Surfactants. Principles and Practice. AOCS Press, Champaign, IL, USA.

Saerens, K.M.J., S.L.K.W. Roelants, I.N.A. van Bogaert and W. Soetaert. 2011. Identification of the UDP-glucosyltransferase gene UGTA1, responsible for the first glucosylation step in the sophorolipid biosynthetic pathway of *Candida bombicola* ATCC 22214. FEMS Yeast Res. 11: 123–132.

Shah, V., D. Badia and P. Ratsep. 2007. Sophorolipids having enhanced antibacterial activity. Antimicrob. Agents Chemother. 51: 397–400.

Shekhar, S., A. Sundaramanickam and T. Balasubramanian. 2015. Biosurfactant producing microbes and their potential applications: A review. Crit. Rev. Environ. Sci. Technol. 45: 1522–1554.

Singh, S.K., A.P. Felse, A. Nunez, T.A. Foglia and R.A. Gross. 2003. Regioselective enzyme-catalyzed synthesis of sophorolipid esters, amides, and multifunctional monomers. J. Org. Chem. 68: 5466–5477.

Sleiman, J.N., S.A. Kohlhoff, P.M. Roblin, S. Wallner, R. Gross, M.R. Hammerschlag, M.E. Zenilman and M.H. Bluth. 2009. Sophorolipids as antibacterial agents. Ann. Clin. Lab. Sci. 39: 60–63.

Solaiman, D.K.Y., R.D. Ashby, A. Nuñez and T.A. Foglia. 2004. Production of sophorolipids by *Candida bombicola* grown on soy molasses as substrate. Biotechnol. Lett. 26: 1241–1245.

Solaiman, D.K.Y., R.D. Ashby, J.A. Zerkowski and T.A. Foglia. 2007. Simplified soy molasses-based medium for reduced-cost production of sophorolipids by *Candida bombicola*. Biotechnol. Lett. 29: 1341–1347.

Solaiman, D.K.Y., R.D. Ashby, M. Birbir and P. Caglayan. 2016. Antibacterial activity of sophorolipids produced by *Candida bombicola* on Gram-positive and Gram-negative bacteria isolated from salted hides. J. Am. Leath. Chem. Assoc. 111: 358–363.

Solaiman, D.K.Y., R.D. Ashby and J. Uknalis. 2017. Characterization of growth inhibition of oral bacteria by sophorolipid using a microplate-format assay. J. Microbiol. Meth. 136: 21–29.

Spencer, J.F.T., P.A.J. Gorin and A.P. Tulloch. 1970. *Torulopsis bombicola* sp. n. Antonie van Leeuwenhoek 36: 129–133.

Sun, X.-X., J.-K. Choi and E.-K. Kim. 2004. A preliminary study on the mechanism of harmful algal bloom mitigation by use of sophorolipid treatment. J. Exper. Mar. Biol. Ecol. 304: 35–49.

Tanzer, J.M., J. Livingston and A.M. Thompson. 2001. The microbiology of primary dental caries in humans. J. Dent. Edu. 65: 1028–1037.

Transparency Market Research. 2014. Microbial biosurfactants market (rhamnolipids, sophorolipids, mannosylerythritol lipids (MEL) and other) for household detergents, industrial & institutional cleaners, personal care, oilfield chemicals, agricultural chemicals, food processing, textile and other applications—global industry analysis, size, share, growth, trends and forecast, 2014–2020.

Tulloch, A.P., J.F.T. Spencer and P.A.J. Gorin. 1962. The fermentation of long-chain compounds by *Torulopsis magnolia*. Can. J. Chem. 40: 1326–1338.

Tulloch, A.P., J.F.T. Spencer and M.H. Deinema. 1968. A new hydroxy fatty acid sophoroside from *Candida bogoriensis*. Can. J. Chem. 46: 345–348.

Tulloch, A.P. and J.F.T. Spencer. 1968. Fermentation of long-chain compounds by *Torulopsis apicola*. IV. Products from esters and hydrocarbons with 14 and 15 carbon atoms and from methyl palmitoleate. Can. J. Chem. 46: 1523–1528.

van Bogaert, I.N.A., K. Saerens, C. de Muynck, D. Develter, W. Soetaert and E.J. Vandamme. 2007. Microbial production and application of sophorolipids. Appl. Microbiol. Biotechnol. 76: 23–34.

van Bogaert, I.N.A., M. Demey, D. Develter, W. Soetaert and E.J. Vandamme. 2009. Importance of the cytochrome P450 monooxygenase CYP52 family for the sophorolipid-producing yeast *Candida bombicola*. FEMS Yeast Res. 9: 87–94.

van Bogaert, I.N.A., J.X. Zhang and W. Soetaert. 2011a. Microbial synthesis of sophorolipids. Proc. Biochem. 46: 821–833.

van Bogaert, I., S. Fleurackers, S. van Kerrebroeck, D. Develter and W. Soetaert. 2011b. Production of new-to-nature sophorolipids by cultivating the yeast *Candida bombicola* on unconventional hydrophobic substrates. Biotechnol. Bioengin. 108: 734–741.

van Bogaert, I.N.A., K. Holvoet, S.L.K.W. Roelants, B. Li, Y.-C. Lin, Y. van der Peer and W. Soetaert. 2013. The biosynthetic gene cluster for sophorolipids: a biotechnological interesting biosurfactant produced by *Starmerella bombicola*. Mol. Microbiol. 88: 501–509.

Wadekar, S.D., S.B. Kale, A.M. Lali, D.N. Bhowmick and A.P. Pratap. 2012. Jatropha oil and karanja oil as carbon sources for production of sophorolipids. Eur. J. Lipid Sci. Technol. 114: 823–832.

Xue, C.-L., D.K.Y. Solaiman, R.D. Ashby, J. Zerkowski, J.H. Lee, S.-T. Hong, D. Yang, J.-A. Shin, C.-M. Ji and K.-T. Lee. 2013. Study of structured lipid-based oil-in-water emulsion prepared with sophorolipid and its oxidative stability. J. Am. Oil Chem. Soc. 90: 123–132.

Zerkowski, J.A., D.K.Y. Solaiman, R.D. Ashby and T.A. Foglia. 2006. Head group-modified sophorolipids: Synthesis of new cationic, zwitterionic, and anionic surfactants. J. Surfact. Deterg. 9: 57–62.

Zerkowski, J.A., D.K.Y. Solaiman, R.D. Ashby and T.A. Foglia. 2010. Charged sophorolipids and sophorolipid containing compounds. U.S. Patent # 7,718,782.

Zhang, L., P. Somasundaran, S.K. Singh, A.P. Felse and R. Gross. 2004. Synthesis and interfacial properties of sophorolipid derivatives. Coll. Surf. A: Physicochemical and Engineering Aspects 240: 75–82.

Zhang, X., X. Fan, D.K.Y. Solaiman, R.D. Ashby, Z. Liu, S. Mukhopadhyay and R. Yan. 2016a. Inactivation of *Escherichia coli* O157:H7 *in vitro* and on the surface of spinach leaves by biobased antimicrobial surfactants. Food Cont. 60: 158–165.

Zhang, X., R. Ashby, D.K.Y. Solaiman, J. Uknalis and X. Fan. 2016b. Inactivation of *Salmonella* spp. and *Listeria* spp. by palmitic, stearic, and oleic acid sophorolipids and thiamine dilauryl sulfate. Front. Microbiol. 7: 2076.

Zhou, Q., V. Kleckner and N. Kosaric. 1992. Production of sophorose lipids by *Torulopsis bombicola* from safflower oil and glucose. J. Am. Oil Chem. Soc. 69: 89–91.

Zhou, Q. and N. Kosaric. 1995. Utilization of canola oil and lactose to produce biosurfactant with *Candida bombicola*. J. Am. Oil Chem. Soc. 72: 67–71.

Zini, E., M. Gazzano, M. Scandola, S.R. Wallner and R.A. Gross. 2008. Glycolipid biomaterials: solid-state properties of a poly(sophorolipid). Macromolecules 41: 7463–7468.

# 4

# Antimicrobial Applications of Rhamnolipids in Agriculture and Wound Healing

*Soroosh Soltani Dashtbozorg,[1] Krutika Invally,[2]
Ashwin Sancheti[2] and Lu-Kwang Ju[2],\**

## Introduction

The focus of this chapter is on the applications of rhamnolipids (RLs) that are based on their antimicrobial activities and/or other biological effects. RLs are a group of nontoxic, biodegradable and environment friendly biosurfactants (BSs). The RL congeners produced by microbial fermentation have the general molecular structures that include one (mono-RL) or two rhamnose (di-RL) residues linked to one or two β-hydroxyl fatty acids (Déziel et al. 1999, Benincasa et al. 2002). Accordingly, RLs are commonly classified into four groups depending on their respective numbers of rhamnose and fatty acid residues. The most representative structures of these four RL groups are L-rhamnosyl-β-hydroxydecanoate (R-C10), L-rhamnosyl-β-hydroxydecanoyl-β-hydroxydecanoate (R-C10-C10), L-rhamnosyl-L-rhamnosyl-β-hydroxydecanoate (R-R-C10), and L-rhamnosyl-L-rhamnosyl-β-hydroxydecanoyl-β-hydroxydecanoate (R-R-C10-C10). Around 60 RL congeners of different chain lengths and/or extents of saturation have been reported to be produced by *Pseudomonas aeruginosa*, *Pseudoxanthomonas* sp., *Myxococcus* sp.,

---

[1] ChromaTan Corporation, 200 Innovation Blvd., Suite 260B, State College, Pennsylvania 16803, USA.
[2] Department of Chemical and Biomolecular Engineering, The University of Akron, Akron, OH, USA 44325-3906.
\* Corresponding author: lukeju@uakron.edu

*Enterobacter* sp., *Burkholderia* sp., *Nocardioides* sp., *Acinetobacter calcoaceticus, Renibacterium salmoninarum, Cellulomonas cellulans, Tetragenococcus koreensis* and other species of bacteria (Abdel-Mawgoud et al. 2010).

As surfactants, RLs can reduce the surface tension of water from 72 to lower than 30 mN/m. RLs can also reduce the interfacial tension between kerosene and water from 43 to 1 mN/m (Parra et al. 1989, Nitschke et al. 2005). Critical micelle concentration (CMC) and hydrophilic/lipophilic balance (HLB) are two other properties commonly used for the characterization of surfactants. Different CMC values have been reported for RLs of different molecular structures. For example, Parra et al. reported a CMC value of 5 mg/L for the di-RL R-R-C10-C10 and 40 mg/L for the mono-RL R-C10-C10 (Parra et al. 1989). Typically, the RL with shorter fatty acid moiety has a larger CMC value (Pinzon-Gomez 2009). HLB reflects the hydrophilicity/hydrophobicity of surfactants. In general, surfactants with higher HLB values can be more suitable for uses as detergents and oil-in-water emulsifiers or solubilizers (Noordman et al. 2002). HLB values reported for RLs are in the range of 10–24 (Noordman et al. 2002, Abalos et al. 2004). It has been reported that RLs can give lower minimum surface tension of water (26–29 mN/m) as compared to the other well-known glycolipid BS, sophorolipids (SLs) (33–37 mN/m). The two BSs, RLs and SLs, can give lower surface tensions than some common synthetic surfactants such as sodium dodecyl sulfate or dodecylbenzene. RLs and SLs also have much lower CMC values as compared to these synthetic surfactants. RLs are therefore considered to be more surface active than SLs and some common synthetic surfactants (Pinzon-Gomez 2009).

Another important property of RLs is their pH sensitivity. The fatty acid moiety renders RLs anionic at pH higher than the pKa values. Accordingly, the surface activity and solubility of RLs depend strongly on pH (Zhang and Miller 1992). Lebron-Paler et al. measured the pKa value for a mixture of mono-RLs. They found the pKa value to differ at RL concentrations below and above the CMC. The pKa value was $4.28 \pm 0.16$ when RL concentrations were lower than the CMC but it increased to $5.50 \pm 0.06$ at higher concentrations (Lebrón-Paler et al. 2006).

RLs have good physicochemical properties and exhibit several promising industrial applications. For example, RLs can be hydrolyzed to produce the rare sugar L-rhamnose which is used for industrial production of high-quality flavor compounds (Lang and Wullbrandt 1999, Chayabutra et al. 2001, Trummler et al. 2003). Rhamnose production via this route avoids the toxic waste generation involved with the current methods of rhamnose production. RLs may find applications in the enhanced oil recovery. The conventional recovery method extracts almost 30% of the oil in the reservoir. Surfactants can be added to improve oil recovery during the water flooding or steam injection processes. RLs can replace the synthetic surfactants or be used as a co-surfactant in the surfactant enhanced oil recovery. The use of RLs in microbial enhanced oil recovery has also been proposed (Wang et al. 2007). In addition, RLs have biological effects on a wide range of organisms. Some of their applications based on the antimicrobial activity and other biological effects are reviewed in detail in this chapter.

## Antimicrobial Activity of RLs

The antimicrobial activity of RLs on various bacteria, yeast and fungi has been well documented (Lang et al. 1989, Kitamoto et al. 1993, Kim et al. 2000, Abalos et al. 2001, Haba et al. 2003, Benincasa et al. 2004, Rodrigues et al. 2006). The effectiveness of RLs against different microorganisms is compared according to the minimum inhibitory concentration (MIC) values, as given in Table 1. MIC represents the lowest concentration of an antimicrobial agent needed to inhibit the growth of the test microorganism after incubation for a specified time. For the values given in Table 1, the activity against bacteria was determined by using a liquid culture medium which was incubated for 24 h at 37°C (Woods and Washington 1995) while the activity against yeast and fungi was determined on a solid medium (Sabouraud agar plates) which was incubated for 72 h at 25°C (Haba et al. 2003).

Figure 1 is shown for easy comparison of the antimicrobial activity of RLs against different groups of microorganisms, i.e., Gram-positive bacteria, Gram-negative bacteria, yeast and fungi. For each group, the (cumulative) percentage of samples inhibited for growth (y-axis) at the increasing RL concentration tested (x-axis) was calculated and plotted (Figure 1). With the limited number of tests done on yeast, RLs are clearly the least effective in inhibiting yeast growth. For bacteria, the Gram-positive species are generally more susceptible to RLs than the Gram-negative species (Sotirova et al. 2008). It is well known that while the cell wall of Gram-negative bacteria has a thinner layer of peptidoglycan than that of the Gram-positive bacteria, it contains lipopolysaccharides (LPS) and porin channels in the outer membrane which are absent in Gram-positive bacteria. This outer membrane comprising of LPS functions as an effective permeability barrier. RLs can cause release of LPS leading to increased permeability of the bacterial cell (Al-Tahhan et al. 2000). It is proposed that RL-induced release of LPS causes increase in cell surface hydrophobicity which may have a detrimental effect on the bacteria. On the other hand, this RL-related increase in hydrophobicity of bacterial cells may also be managed for beneficial effects when applying RLs in removal of hydrophobic contaminants from the soil as explained in the later section (Soil Remediation). As for the antifungal activity of RLs (Figure 1), the %-inhibition appears to be lower than those of antibacterial activity in concentrations lower than 20–30 mg/L. However, at higher concentrations, RLs are significantly more effective in controlling growth of fungi than Gram-negative bacteria. For example, at 32 mg/L, RLs inhibited the growth of 57% of the tested fungi but only 30% of the Gram-negative bacteria; further, at 260–280 mg/L, RLs inhibited growth of 90% of the fungi but only about 60% of the Gram-negative bacteria. On the other hand, only in the concentration range of 30–90 mg/L, RLs appear to be more effective against fungi than Gram-positive bacteria, i.e., inhibiting the growth of higher percentages of tested fungi than Gram-positive bacteria. At lower (< 30 mg/L) and higher (> 90 mg/L) concentrations, RLs are less effective against fungi than Gram-positive bacteria. For example, at 250 mg/L, RLs inhibited growth of all of the tested Gram-positive bacteria but about 90% of the fungi. The difference is nonetheless much less significant, compared to the clearly lower percentages of inhibited Gram-negative bacteria (~ 60%) and yeast (~ 30%).

**Table 1.** Minimum inhibitory concentrations (MIC) of RLs reported for (I) Gram-positive bacteria, (II) Gram-negative bacteria, (III) yeast and (IV) fungi.

| Gram-Positive Bacteria | MIC (mg/L)* |
|---|---|
| *Arthrobacter oxydans* ATCC 8010 | 16[A], 128[B] |
| *Bacillus cereus* ATCC 11778 | 4[C], 64[A, B] |
| *Bacillus subtilis* ATCC 6051 | 128[D] |
| *B. subtilis* ATCC 6633 | 8[C], 16[B], 64[A] |
| *Clostridium perfringens* ATCC 486 | 128[B], 256[A] |
| *Micrococcus luteus* ATCC 9341 | 128[C] |
| *M. luteus* ATCC 9631 | 32[A], 64[B] |
| *Mycobacterium phlei* ATCC 41423 | 16[A], 128[B] |
| *Staphylococcus aureus* ATCC 6538 | 8[C], 32[B], 128[A] |
| *S. aureus* ATCC 29213 | 128[D] |
| *Staphylococcus epidermidis* ATCC 12228 | 8[A], 32[B], 256[C] |
| *S. epidermidis* (clinical sample) | 128[D] |
| *Streptococcus faecalis* ATCC 10541 | 4[C], 64[A] |
| *Streptococcus pneumoniae* (clinical sample) | 128[D] |
| **Gram-Negative Bacteria** | **MIC (mg/L)\*** |
| *Alcaligenes faecalis* ATCC 8750 | 32[A], 64[B] |
| *Bordetella bronchiseptica* ATCC 4617 | 128[A, B, C] |
| *Citrobacter freundii* ATCC 22636 | 64[B], > 256[A] |
| *Enterobacter aerogenes* CECT 689 | 4[B, C], > 256[A] |
| *Enterococcus faecalis* ATCC 29212 & a clinical sample | > 512[D] |
| *Escherichia coli* ATCC 8739 | 32[A], 64[B] |
| *E. coli* ATCC 10536 | 250[C] |
| *E. coli* ATCC 25922, a clinical sample, & a K12 clinical sample | > 512[D] |
| *Klebsiella pneumoniae* CECT 17832 | 0.5[B] |
| *K. pneumoniae* & a clinical sample | > 512[D] |
| *Proteus mirabilis* CECT 170 | 8[C], > 256[A, B] |
| *Pseudomonas aeruginosa* ATCC 9727 | 32[C], 256[B], > 256[A] |
| *P. aeruginosa* ATCC 27853 | > 512[D] |
| *Salmonella typhimurium* ATCC 16028 | 16[C], 128[A, B] |
| *Serratia marcescens* CECT 274 | 8[B], 16[A] |
| **Yeast** | **MIC (mg/L)\*** |
| *Candida albicans* ATCC 10231 | 32[C], > 256[A, B] |
| *Rhodotorula rubra* CECT 1158 | > 256[A, B, C] |
| *Saccharomyces cerevisiae* ATCC 9763 | > 256[A, B, C] |
| **Fungi** | **MIC (mg/L)\*** |
| *Alternaria alternate* | 4[C] |
| *Aspergillus niger* ATCC 14604 | **16[A], 64[C], > 256[B] |

*Table 1 contd. ...*

*...Table 1 contd.*

| Fungi | MIC (mg/L)* |
|---|---|
| *Aureobasidium pullulans* ATCC 9348 | **32[A], 32[C], > 256[B] |
| *Botrytis cinerea* | **18[A], 170[B] |
| *Cercospora kikuchii* | 50[E] |
| *Chaetomium globosum* ATCC 6205 | **32[A], 32[C], 64[B] |
| *Colletotrichum gloeosporioides* | **65[A], 276[B] |
| *Colletotrichum orbiculare* | 50[E] |
| *Fusarium solani* | 75[B] |
| *Gliocladium virens* ATCC 9645 | **16[A], 32[B, C] |
| *Penicillium chrysogenum* CECT 2802 | **32[A], 32[C], > 256[B] |
| *Penicillium funiculosum* CECT 2914 | 16[B], 64[C], **128[A] |
| *Phytophthora capsici* | 10[F] |
| *Phytophthora cryptogea* | 25[F] |
| *Phytophthora sojae* | 8-20 [G, H] |
| *Rhizoctonia solani* | **18[A], 109[B] |

**Note:** * The sources of MIC values are indicated by their attached superscript letters: A—(Abalos et al. 2001); B—(Haba et al. 2003); C—(Benincasa et al. 2004); D—(Lotfabad et al. 2010); E—(Kim et al. 2000); F—(De Jonghe et al. 2005); G—(Dashtbozorg et al. 2015); and H—(Miao et al. 2015).

** The values immediately following "**" are for a particular RL mixture comprising of R-R-C10-C10, R-C10-C10, R-R-C10-C12, R-C10-C12, R-C12:1-C10, R-C12:2 and R-C8:2; all others are for fermentation-produced RL mixtures.

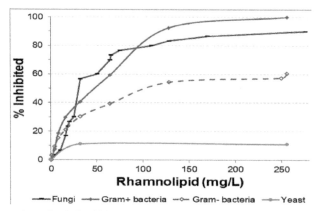

**Figure 1.** Comparison of antimicrobial activities of RLs against Gram-positive bacteria, Gram-negative bacteria, yeast and fungi. Cummulative % of tested samples inhibited the growth at increased RL concentrations (estimated according to MIC study results given in Table 1).

The antimicrobial property of RLs enables them to be used for protection of crops or target organisms, animals and humans against microbial pathogens. In this chapter, the applications of RLs are focused on two aspects: (i) effect of RLs on plant pathogens and their application in agriculture, and (ii) pharmaceutical applications focused on the use of RLs in promoting wound healing and their effects on skin cells and antimicrobial activity against wound infection bacteria, especially the Gram-positive bacteria (Benincasa et al. 2004, Vatsa et al. 2010). For the wound healing-related applications of RLs, some parameters such as wound closure rate have been measured (Piljac and Piljac 1994, Stipcevic et al. 2006, Bjarnsholt et al. 2008). There are also studies showing the effects of RLs on skin cells such as keratinocytes and fibroblasts cells (Stipcevic et al. 2006).

## Application of RLs in Agriculture

Environmental concerns over the use of synthetic agrochemicals have influenced the pursuit for bio-based environment friendly chemicals. RLs, which show excellent biodegradability as well as low aquatic toxicity, are particularly effective in rupturing the zoospores of certain plant pathogenic fungi at a very low concentration level (Miao et al. 2015). Zoospore is one of the life stages in some fungi (oomycetes), which is characterized by formation of unicellular, motile cells that are most commonly implicated in spread of the pathogen. RLs have been reported to be effective against several species of fungi and can be used as a wide-spectrum antifungal agent for the protection of different commercially important crops as shown in Table 2.

Although the antimicrobial activity of RLs offers the primary motivation, RLs can have three different acting mechanisms when used as agricultural biopesticide. First, RLs may stimulate the plants' immune activities against pathogens (Nitschke et al. 2005). Second, rhamnolipids have antifouling property that may be used to reduce undesirable attachment of microbes, pests and their spores and eggs on plants. Finally, RLs have antimicrobial properties that can be used to either kill or inhibit the

**Table 2.** Minimum inhibitory concentrations (MIC) of RLs required to inhibit the growth of different plant pathogenic fungi.

| Fungi | MIC (mg/L) | Disease/Plants Affected | Reference |
| --- | --- | --- | --- |
| *Botrytis cinerea* | 170 | Gray rot in grapes<br>Dry eye rot on apples | (Haba et al. 2003) |
| *Cercospora kikuchii* | 50[***] | Leaf spot and blight seed stain in soybeans | (Kim et al. 2000) |
| *Colletotrichum gloeosporioides* | 276 | Anthracnose in strawberries | (Haba et al. 2003) |
| *Colletotrichum orbiculare* | 50[***] | Infection in cucumber plants | (Kim et al. 2000) |
| *Fusarium solani* | 75 | Foot, root, stem and fruit rot of cucurbits and green bean root rot | (Haba et al. 2003) |
| *Phytophthora capsici* | 10[***] | Blight root rot in peppers | (Kim et al. 2000) |
| *Phytophthora cryptogea* | 25[***] | Brown root rot in chicory | (De Jonghe et al. 2005) |
| *Phytophthora sojae* | 20[***] | Soybean root rot | (Dashtbozorg et al. 2015) |
| *Rhizoctonia solani* | 109 | Eye spot in wheat and other cereals | (Haba et al. 2003) |

**Note:** *** refers to MIC values for zoospores of respective fungi.

growth of microbial pathogens. These mechanisms are described in more detail in the Mechanisms of RLs as Biopesticide section. There have also been reports for use of RLs in soil remediation and bioremediation of organic and inorganic pollutants so as to improve the crop productivity from the treated soil. These applications are described in the Soil Remediation section. Finally, the application of RLs on soy plant fungal pathogen controlling is described as a focused example in the Application of RLs in Controlling *Phytophthora sojae* section.

In addition, RLs have other attractive properties for agricultural uses. Under Section 408 of the Federal Food, Drug and Cosmetic Act (FFDCA), U.S. Environmental Protection Agency (EPA) is required to regulate the amount of pesticide residues remaining in the food or feed commodities as the result of the pesticide treatment/application. As per Code of Federal Regulations (CFR), tolerance for a given pesticide is defined as the maximum residual amount of the pesticide that is allowed to remain in the food or feed. U.S. EPA, however, has established an exemption from the requirement of a tolerance for RLs used as pesticide and fungicide for agricultural purposes (CFR). Further, RLs are biodegradable compounds (Maier and Soberón-Chávez 2000). For example, Wen et al. (2009) compared the degradation of RLs, citric acid and ethylene diamine tetraacetic acid (EDTA) in soils pre-incubated for 28 days to stimulate microbial activity. Each compound was compared at an initial concentration of 2 mmol/kg soil; RLs were also evaluated at a higher concentration of 10 mmol/kg. After 20 days, citric acid was degraded by 70%, EDTA by 14% and RLs by 36% and 29%, respectively, at 2 and 10 mmol/kg initial concentrations. The time required for RL degradation is generally shorter than those for the synthetic surfactants or other agrochemicals used in the agriculture industries (Maier and Soberón-Chávez 2000).

For agricultural use of RLs, the soil adsorption properties of RLs are also important factors to consider. In a recent study, the adsorption properties were studied by Dashtbozorg et al. (2016). The effects of soil pH and RL congeners' structures and concentrations were evaluated. The RLs used in that study had 8 groups of congener structures, including both mono- and di-RLs: RR-C10-C8 & RR-C8-C10 (molecular weight: 622.7, 5.0 wt%), R-C10-C8 & R-C8-C10 (MW: 476.6, 6.8%), RR-C10-C10 (MW: 650.8, 46.4%), R-C10-C10 (MW: 504.7, 31.0%), RR-C12-C10 & RR-C10-C12 (MW: 678.0, 4.0%), R-C10-C12 (MW: 532.7, 1.0%), RR-C10-C12:1 (MW: 676.0, 3.9%), and R-C10-C12:1 (MW: 530.7, 1.9%). All of these main congeners identified contained two β-hydroxyl fatty acids (HFAs) and at least one of the HFAs was β-hydroxydecanoic acid (C10) while the other HFA residue differed in chain length (C8, C10 or C12) and the C12 residue could be saturated or unsaturated with one double bond. RLs were found to adsorb much more preferentially on soil at a lower pH 4.5 than at pH 6.5. At pH 4.5, the adsorbed RLs reached about 2800 μg/g soil at about 200 μg/ml equilibrium aqueous-phase concentration and the adsorbed amount could still increase; at pH 6.5, the RL adsorption plateaued at about 1700 μg/g soil with an equilibrium aqueous-phase concentration of about 800 μg/ml. At the low pH, the adsorption isotherm could be approximated by a Freundlich (power-law) equation. At pH 6.5, the adsorption isotherm showed a multistage profile. The less hydrophilic congeners adsorbed more preferentially, in the following order: R-C10-C12 > R-C10-C12:1 > RR-C10-C12:1 > RR-C10-C12 > R-C10-C10 > RR-

C10-C10 > R-C8-C10 > RR-C8-C10. The order shows that the HFA chain length affected the adsorption more strongly than the number (1 or 2) of rhamnose residues. The above adsorption preference was seen more clearly with dilute solutions but diminished with increasing aqueous-phase concentration. The multistage adsorption isotherm and the diminishing adsorption selectivity to different congeners were explained by the formation of micelles and aggregates in solution and on solid surface (Dashtbozorg et al. 2016).

# Mechanisms of RLs as Biopesticide

## *Effect of RLs on plants' immune system*

Plants can recognize molecular patterns associated with the presence of microorganisms (Boller and Felix 2009). This recognition may activate an immune response which helps defend the plant from infection by phytopathogenic microorganisms. Cells of the plants' innate immune system are primarily responsible for recognizing and responding to the pathogens. Specific proteins, i.e., pattern recognition receptors, are synthesized for this purpose. The function of these proteins is the identification of microbe-associated molecular patterns (MAMPs). The identification of these molecular patterns can trigger complex signaling pathways which lead to transcriptional activation of defense-related genes and formation of antimicrobial metabolites in plant cells (Mackey and McFall 2006). RLs are considered as a new class of MAMPs involved in non-specific immunity in plants (Vatsa et al. 2010). Varnier et al. (2009) showed that three different congeners of RLs (R-C10-C10, R-R-C10-C10 and R-R-C14-C14) triggered a strong immune response in grapevine and helped to protect the plant against the fungal pathogen *Botrytis cinerea*. The immune response elicited by RLs in the grapevine included events of cell signaling like $Ca^{2+}$ influx, generation of reactive oxygen species and mitogen-activated protein kinase (MAPK) activation. The exact mechanism for induction of immune response by RLs is not completely clear. It has not been established whether specific receptors in plasma membrane are required for triggering the immune response or if the membrane disturbances themselves induce the immune response (Jourdan et al. 2009, D'apos et al. 2010).

## *Antifouling activity of RLs*

Biofouling refers to the formation of biofilm and attachment of different microorganisms on the surface in an aqueous environment. Biofilm formation consists of three major steps. The first is a biochemical step where biomolecules such as proteins, lipids and polysaccharides are attached to the surface. Second, microfouling occurs and small microorganisms such as bacteria are attached to the surface. Finally, larger organisms such as algae are attached to the surface causing macrofouling (Raya et al. 2010). It is confirmed that RLs have antifouling and anti-adhesive properties against several microorganisms (Mukherjee et al. 2006, Rodrigues et al. 2006, Das et al. 2008). For example, RLs at concentrations of 10 and 200 mg/L were demonstrated to significantly reduce the initial attachment of

*P. aeruginosa, P. putida, E. coli* and *B. subtilis* on glass and octadecyltrichlorosilicane-modified glass surfaces (Sodagari et al. 2013). Studies have shown that RLs are not only effective in preventing and slowing down the formation of biofilm but can also accelerate the dispersion phase of developed biofilm for certain bacterial species (Velraeds et al. 1996, Velraeds et al. 1998, Robert et al. 2001). For example, RLs were shown to inhibit the biofilm formation by *Bacillus pumilus* at concentrations over 1.6 mM (Dusane et al. 2010). RLs also caused significant disruption of the *in vitro* biofilm structure of *Bordetella bronchiseptica*, which is a small Gram-negative bacterium that can cause infectious bronchitis in dogs and other animals (Irie et al. 2005). Quinn et al. (2013) further studied the dispersion of pre-established single-species biofilms and a biofilm of mixed marine bacteria by RLs (and, separately, another plant-derived surfactant mixture). Ten (10) µl of a 2 g/L RL solution were added to biofilms pregrown for 2 days in wells of 96-well plates (Nunc™, Thermo Fisher Scientific). After incubation for 2 more days, the change of biofilm biomass was estimated by a crystal violet adhesion assay. The RLs reduced the *S. aureus* biofilm biomass by $85.6 \pm 3.9\%$, *B. subtilis* biomass by $88.4 \pm 5.8\%$, and *Micrococcus luteus* biomass by $74.5 \pm 6.6\%$, respectively. A similar experiment with the mixed marine bacterial biofilm showed a biomass reduction of $69.0 \pm 3.9\%$. More recently, Elshikh et al. (2017) studied the action of RLs on oral hygiene related pathogens in three different modes. RLs reduced the biofilm formation of *Streptococcus sanguinis* and *Streptococcus oralis* when these cultures were co-incubated or pre-coated in wells of 96-well plates. RLs also disrupted pre-existing biofilms of these cultures, with 80% biomass reduction for *S. sanguinis* and 65% for *S. oralis*. Also investigated were the effects of RLs on biofilms in wells with or without stainless steel coupons (2 cm²) (Chebbi et al. 2017). *Bacillus licheniformis* CAN55, *Staphylococcus capitis* SH6 and *P. aeruginosa* W10 were studied in both single and mixed bacterial cultures. The biofilm development medium used was generally the Nutrient Broth medium added with 20 g/L glucose. In wells without metal coupons, the pre-incubation (24 h) of cultures in media with RLs caused significant inhibition of biofilm formation (after phosphate buffer rinse and fresh medium addition) in the following 24 h. For *B. licheniformis* and *S. capitis*, the inhibition was about 60% at the lowest RL concentration tested, i.e., 0.04 g/L, and increased up to 90% at higher concentrations. For *P. aeruginosa* W10, the inhibition was insignificant at RL concentrations lower than 0.78 g/L but increased to 90% at concentrations higher than about 3 g/L. Pre-coating of RLs in wells (by incubation at 60°C overnight) was also shown to reduce subsequent biofilm formation (considered as the antiadhesive effect of RLs). For *B. licheniformis*, about 85% reduction was found with RLs at 0.09 g/L and above; for *S. capitis*, more than 60% reduction was found at 0.78 g/L and above. RLs also disrupted up to 99% of the biofilms pregrown for 2 days in wells with stainless steel coupons. The required RL concentration was 0.1 g/L for *B. licheniformis*, 0.5 g/L for *S. capitis*, and 1 g/L for a mixed culture of the three bacteria studied, respectively.

It is proposed that RLs can disrupt biofilms and disperse microbial cells by creating pores and channels in the biofilms (Schooling et al. 2004). Owing to their surfactant property, RLs can also react with the exopolymeric substances secreted by the biofilm-forming bacteria. This makes RLs effective as anti-adhesive and biofilm disrupting agents (Dusane et al. 2010). While many studies have documented the effects of RLs in inhibiting biofilm formation and disrupting pre-existing biofilms, the efficacy is dependent on the organisms and substrate surfaces involved. For example, Rienzo et al. (2015) did similar studies as Chebbi et al. (2017) on the antiadhesive and biofilm disruption effects of RLs (1 g/L and 5 g/L) in the presence and absence of caprylic acid (0.8% v/v). But the two studies differed in the bacterial species or strains tested, the surface substrate for biofilm formation, and the liquid medium for bacterial cultivation. Specifically, Rienzo et al. (2015) studied the biofilms of *B. subtilis* NCTC 10400, *S. aureus* ATCC 9144, *P. aeruginosa* PAO1 and *E. coli* NCTC 10418 on glass coverslips in the Luria–Bertani medium. Under these conditions, Rienzo et al. (2015) observed no significant effects of RLs in either antiadhesion or disruption of existing biofilms, contrary to the significant effects observed by Chebbi et al. (2017) and many others. It is important to perform at least laboratory studies in simulated environments to enhance successful field implementation of RLs for antifouling activities.

## *Antimicrobial effect of RLs*

Antimicrobial properties of RLs have been studied on various microorganisms. The minimum concentration of RLs required to inhibit visible growth of different microorganisms has been discussed in the Antimicrobial Activity of Rhamnolipids section (Tables 1 and 2) of this chapter. In this section, more information on the mode of action of RLs against phytopathogenic fungi and their importance to the agricultural applications is discussed.

RLs are amphiphilic and structurally similar to cell membranes of living cells. They can interact with the cell membrane lipid bilayer and cause cell lysis (Stanghellini and Miller 1997, De Jonghe et al. 2005, Varnier et al. 2009). It is proposed that surfactants can disrupt the cell membrane by modifying the lipid organization, changing the integral protein arrangement and disturbing the overall equilibrium of the cell (Manaargadoo-Catin et al. 2015). RLs exhibit antifungal activities against phytopathogenic fungi and can be used for plant protection (De Jonghe et al. 2005). Many of the phytopathogenic fungi commonly spread infection by forming zoospores which do not have cell wall and are enclosed only by cell membrane (Stanghellini and Miller 1997). RLs act against zoospore-producing plant pathogens through direct lysis of the zoospores via intercalation of RLs within the plasma membranes (Varnier et al. 2009, Canaday and Schmitthenner 2010). The zoosporic growth of different phytopathogenic fungi like *Phytophthora* sp., *Pythium* sp., *Plasmopara* sp. and *Colletotrichum* sp. is strongly affected by RLs (Varnier et al. 2009). Ju and coworkers (2016) investigated the effect of RLs against *Phytophthora sojae*. The effects of RLs on the mycelia and zoospores of this fungus were studied and compared. They found that RLs at 100 and 1000 mg/L inhibited the mycelial growth by up to ~ 30%. Moreover, RLs were much more effective in killing the zoospores; just about 20 mg/L RLs were sufficient to stop the motility and largely rupture the zoospores almost instantaneously (Dashtbozorg et al. 2015, Miao et al. 2015). Some plant pathogens have a strong cell wall and RLs may not be as effective on them as on the zoospores. The differences have been reflected in the MIC values shown earlier in Tables 1 and 2 and Figure 1. Nevertheless, as described earlier,

RLs have strong antifungal effects even against the mycelial growth of many fungi such as *Pythium myriotylum* and *Botrytis cinerea* (Perneel et al. 2008). There are commercially available RLs-based biofungicides (e.g., Zonix from Jeneil Biotech Inc., Saukville, WI, USA; http://www.proptera.com/) (Thavasi et al. 2014). The use of RLs in controlling fungal diseases, including those spread by zoospores, is an example that supports the biopesticidal applications of microbial BSs.

## *Soil remediation*

Presence of both organic pollutants like hydrocarbons and inorganic pollutants like heavy metals can decrease the lands' productivity. RLs can be used to enhance the removal of these pollutants in soil remediation and bioremediation applications. RLs are very effective in solubilizing and emulsifying hydrocarbons and are considered as effective BSs for bioremediation (Benincasa 2007, Wang et al. 2007, Nguyen et al. 2008, Whang et al. 2008). Figure 2 describes the different classes of contaminants which can be remediated using RLs (Mulligan 2005).

A significant amount of research has been conducted on the application of RLs for bioremediation, biodegradation and hydrocarbon removal from the soil. It has been reported that RLs can remove two to three times more amounts of organic materials from the soil as compared to water wash (Neto et al. 2009). Zhang and Miller demonstrated that RLs could enhance, by solubilization, the dispersion of octadecane in water by four orders of magnitude (Zhang and Miller 1992). Rahman and coworkers observed that RLs could promote total degradation of C8-C11 alkanes and 83% to 98% degradation of C12-C21 alkanes in petroleum sludge and contaminated soil (Rahman et al. 2002, Rahman et al. 2003). Two mechanisms are proposed by which RLs can improve degradation of organic contaminants. The first mechanism hypothesizes that RLs can increase the dispersion of organic compounds in aqueous medium, thus increasing the bioavailability to microorganisms present in the soil (Zhang and Miller 1992). The second mechanism suggests that RLs can cause the cell surface of the degrading soil microorganisms to become more hydrophobic, thus improving the attachment of these microorganisms with the organic compounds (Zhang and Miller 1994).

RLs are also capable of removing heavy metals from the soil. RLs have the ability to form metal complexes with multivalent cations such as cadmium, lanthanum, lead and zinc (Tan et al. 1994, Herman et al. 1995). There are several theories which are proposed to explain the mode of action of RLs in bioremediation of heavy metals. It is postulated that RLs can decrease the interfacial tension between metal and soil surfaces causing detachment of the metal from soil into the aqueous solution, thereby increasing the metal mobility (Mulligan et al. 2001, Wang and Mulligan 2004). It is also proposed that heavy metals can form complexes with RLs or get incorporated into the micelles, causing an increase in mobility of metal ions in the soil (Mulligan 2005). Another proposed mechanism for removal of heavy metals by RLs is by anion exchange. The negatively charged RLs can compete with the metal oxides for adsorption sites, thereby suppressing attachment and increasing metal mobility (Nivas et al. 1996). Table 3 provides some examples on how RLs can be used for removal of contaminants from soil.

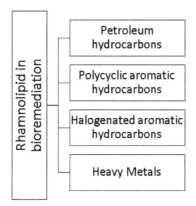

**Figure 2.** RLs in bioremediation of different classes of contaminants.

**Table 3.** Use of RLs for contaminant removal and soil remediation.

| Rhamnolipid-producing Microorganism | Applications of RLs | Reference |
|---|---|---|
| *Pseudomonas* sp. DS10-129 | RLs enhanced *ex-situ* remediation of gasoline-contaminated soil | (Rahman et al. 2002) |
| *Pseudomonas aeruginosa* UG2 | RLs enabled 23–59% removal of hydrocarbon mixture from sandy loam soil, degree of removal depending on type of hydrocarbon and concentration of rhamnolipids used | (Scheibenbogen et al. 1994) |
| *Pseudomonas marginalis* | RLs solubilized polycyclic aromatic hydrocarbons like phenanthrene | (Maier and Soberón-Chávez 2000) |
| *P. aeruginosa* | RLs enhanced mineralization of polychlorinated biphenyls like 4,4-chlorobiphenyl by 213 times as compared to the control | (Robinson et al. 1996) |
| *Pseudomonas cepacia* | RLs enhanced degradation of polychlorinated biphenyl Aroclor 1242. Almost complete degradation achieved for congeners having up to three chlorine atoms | (Fiebig et al. 1997) |
| *P. aeruginosa* BS2 | RLs mediated removal of 92% cadmium and 88% lead within 36 h | (Juwarkar et al. 2007) |

## Application of RLs in Controlling *Phytophthora sojae* as a Major Soybean Plant Pathogen

RLs have received attention to be used for controlling plant pathogens. One of the most common plant pathogens is *Phytophthora sojae*, which can cause severe damage to the stem and root of plants, especially soybean plants (Schmitthenner 1985). *Phytophthora* pathogens are classified as oomycetes (water molds). *P. infestans* and *P. sojae* are the two most common *Phytophthora* species. These species cause enormous damage to the crops, particularly tomato, potato and soybean (Nowicki et al. 2012a). *P. infestans* was responsible for the potato blight disease that caused the

Irish famine of 1845 and led to the death of millions of people (Nowicki et al. 2012b). The *Phytophthora* pathogens are longstanding and hard to control chemically.

*Phytophthora* species grow as hyphae and can reproduce sexually and asexually. Sexual reproduction happens through the differentiated female and male structures, i.e., oogonia and antheridia. Haploid nuclei (gametes) form by meiosis in these structures. Fertilization occurs to produce diploid oospores, which can survive in the soil for a long time and germinate into hyphae when the conditions are suitable (Tyler 2007). Asexual reproduction is favored in wet conditions (Ho and Hickman 1967). Three kinds of asexual spores can form during their life cycle, i.e., sporangia, zoospores and chlamydospores. Sporangia can either directly germinate into hyphae or produce 10–30 zoospores. Zoospores can produce either secondary zoospores or hyphae. Zoospores have two flagella for swimming; they are produced in the wet flooding conditions. Zoospores are the most important agent for the spreading of *Phytophthora* infection. Zoospores are chemotactically attracted towards the roots of the host plant (Morris and Ward 1992, MacGregor et al. 2002). On reaching the root, the zoospore immobilizes, encysts and inserts a germ tube into the root and starts to spread (Walker and van West 2007). After germination, they may spread on the stem and kill the plant eventually. The life cycle of *P. sojae* in the soybean field is described in Figure 3.

The cost of damages from different *Phytophthora* species in the United States was estimated at more than 10 billion dollars. *P. sojae* infects rather limited types of plants such as soybean (Spratling and Mortai 2009). Nonetheless, it has been reported that the damage cost from *P. sojae* alone was $1–2 billion per year (Tyler 2007). Oomycetes including *Phytophthora* are difficult to control. Most of the fungicides cannot kill them completely. Oospores can survive in the debris of the infected plants and germinate later to spread the infection. Moreover, because the infected plants and oospores can be buried in soil, full treatment of the infected sites

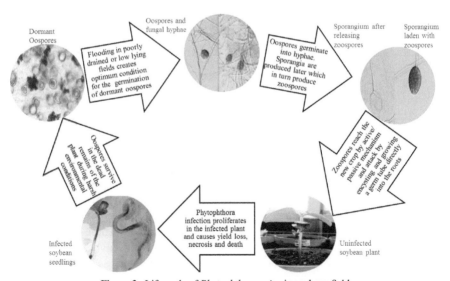

**Figure 3.** Life cycle of *Phytophthora sojae* in soybean field.

is not possible economically. It has also been reported that *Phytophthora* and most of the other oomycetes are capable of adapting to new environments. They can resist against chemical fungicides as well (Schmitthenner 1985, Schettini 1991, Fry and Goodwin 1997, Chamnanpunt et al. 2001).

Economic estimation was made for application of RLs as an antizoospore agent to control the *P. sojae* infection in soybean field (Dashtbozorg et al. 2016). Dorrance and McClure (2001) observed that *P. sojae* caused the most damage to soybean seeds during the first five days after seeding. Protecting the seeds during this initial period is therefore the most efficient way. The bulk soil volume around a germinating seed and roots during this 5-day period was estimated at 70 cm$^3$, which needs to be protected against zoospores (Dashtbozorg et al. 2016). For an average bulk soil porosity of 40% (v/v), the soil volume includes 28 cm$^3$ void space which would be filled with water in wet field conditions and zoospores can approach and attack the seed and roots through these water-filled channels. As mentioned earlier, 8–20 mg/L RLs in water was found to be sufficient to disrupt and kill the zoospores immediately (Dashtbozorg et al. 2015, Miao et al. 2015). The RL amount required in the 28 cm$^3$ water was therefore up to 0.56 mg. RL adsorption was studied in soil samples obtained from a soybean farm in Ohio. A soil pH of 6.5 was chosen for the economic estimation, based on the reported desirable soil pH of 6.3–7.0 for soybean production (Adams and Foy 1984, Schmitt 1989, Peters et al. 2005). According to the adsorption isotherm, approximately 170 µg RLs would adsorb on 1 g soil (dry weight) in equilibrium with a 20 mg/L RL solution. The 100 g dry soil in that 70 cm$^3$ volume to protect would therefore adsorb 17 mg RLs. Approximately 18 mg (~ 17 + 0.56) RLs per seed would be required for protection during this vulnerable early stage (Dashtbozorg et al. 2016). Ohio Field Crop Enterprise Budgets estimated that 175,000 seeds are planted per acre (Ward 2015). The amount of RLs required would be ~ 3.15 kg per acre to protect 175,000 seeds. Müller and Hausmann (2011) reported that the RL production cost per kg could range from $0.56 to $1.33 depending on the fermentation substrate used. Considering the highest production cost ($1.33/kg), the cost of RLs would be $4.2/acre. Ohio Field Crop Enterprise Budgets estimated that the seed cost and chemicals (pesticide, insecticide and fungicide) per acre were $75 and $34, respectively, and the variable cost was $225 (Ward 2015). The estimated RL cost ($4.2/acre) corresponded to 5.6% of the seed cost, 12% of the chemicals cost, or 1.9% of the total variable cost. The estimate supported feasibility of coating seeds with RLs for protection against zoospores during the vulnerable initial period after seeding.

## Other Pesticidal Applications of RLs

RLs have been tested for a wide range of organisms beyond the bacteria, yeast and fungi described in the previous sections. Studies conducted on RLs and their pesticidal activities indicated their broad-spectrum pesticidal potential. Some of the pesticidal applications of RLs are summarized in Table 4.

The pesticidal activity of RLs on organisms is often hypothesized to be based on its surface-active nature. In the study with green peach aphids, the mortality of aphids increased from 50% to 100% when the RL concentration was increased from

**Table 4.** Pesticidal applications of RLs.

| Pest[Ref.] | RL Concentration used (g/L) | Conclusion | Remarks |
|---|---|---|---|
| *Heterosigma akashiwo* (microalga)[A] | 0.004 | Cells turned round in 10 min at 4 mg/L and lysed at 8 mg/L | Tested with actively growing cells |
| Algae[B] | 0.05 & 0.1 | At 0.1 g/L, no algal growth; at 0.05 g/L, algal regrowth after 6 weeks | |
| Amoeba[B] | 0.25 | Lysed in 5 min | |
| Root knot nematode eggs[B] | 0.25 | Lysed after 7 days | Changed egg color to brown first |
| Nematodes[B] | 7.5 | 93% dead after 7 days | Potato, tomato and sugar beet nematodes |
| Spider mites (lemon tree)[B] | 12.5 | Death in < 15 min | |
| Green peach aphids (*Myzus persicae*)[C] | 0.1 | 100% mortality after 24 h | Conducted in green house |
| *Aedes aegypti* (yellow fever mosquito)[D] | 1 | 100% larvae mortality in 18 h | Killed both larvae and mosquitoes |
| Mosquito larvae[B] | 0.2 | Total death of larvae in 160 min | |
| Mosquito eggs[B] | 0.1 | Prevented eggs from hatching | |
| White flies (tomato plant)[B] | 1 | Full control in < 6 min | Flies stuck to leaves and died |
| Household flies[B] | 30 | 30% survival after 10 min | Conducted on flies confined in Petri dish |
| Spiders[B] | 25 | 0% survival in < 15 min | Conducted in naturally infested area |
| Red ants [B] | 5 | Effective: ant free even 2 weeks after treatment | Drenched infested mound with rhamnolipid solution |
| Grasshoppers[B] | 5 | All killed in 10 min | |

**Notes:** The superscript following pest name in the first column refers to the reference: A—(Wang et al. 2005); B—(Awada et al. 2014); C—(Kim et al. 2011); and D—(Silva et al. 2015).

40 mg/L to 100 mg/L (Kim et al. 2011). The cuticle membrane of the aphid, which is composed of complex mixture of alkanes, wax esters, fatty acids and phospholipids, was found to be dehydrated and damaged when observed through a scanning electron microscope. Mosquito eggs, larvae and the insect can also be potentially controlled with RLs. With 400–600 mg/L RLs, it took 48 h to kill the mosquito larvae population; when the concentration was increased to 800–1000 mg/L, the exposure time required to achieve 100% mortality decreased to 24 h (Silva et al. 2015). Exposure to RLs at 1000 mg/L damaged the cuticle membrane of the mosquito, thereby killing it.

RLs have also been proposed to be effective against harmful algal bloom species such as *Heterosigma akashiwo*. At concentrations above 1.3 mg/L, RLs altered the morphology of the algae. Lower concentrations of RLs were needed to affect the algae when the cells were in the lag phase, compared to when in the logarithmic growth phase. When algal cells in the early logarithmic growth phase were subjected to 4 mg/L RLs, the cells expanded and lysed within 1.5 h; the same cells lysed immediately when exposed to 6 mg/L RLs (Wang et al. 2005). During the middle logarithmic growth phase, the cells required longer time to lyse in the presence of similar concentrations of RLs. Observations by transmission electron microscopy revealed that the cells subjected to RLs showed signs of plasma membrane damage which allowed penetration of RLs to damage other organelles. Use of RLs as pesticide is still in a very nascent stage. It is a field with enormous future potential.

## *RLs application in wound healing*

A wound is defined as an injury that breaks the skin. Physical or thermal damages can result in tearing the skin (Nidhi et al. 2011). Wound healing is the biological process by which damaged tissues are regenerated. The wound healing process has different stages. Various types of cells work together to replace the damaged tissues (Rothe and Falanga 1989, Shakespeare 2001). One function of the normal skin is to protect tissues from the invasion of potential pathogens (Bowler et al. 2001). The wound area and damaged skin with exposed live tissues provide a warm, moist and nutritious environment for the growth and colonization of other pathogens (Bowler et al. 2001). *S. aureus*, *P. aeruginosa*, *Streptococcus pyogenes*, *Proteus*, *Clostridium* and other coliform species are the common pathogenic bacteria causing wound infections. These bacteria can be detrimental to the healing process. Poor management of wound infections can result in cellulites (cell inflammation). Eventually, wound infections may lead to bacteremia and septicemia when certain pathogens enter into the blood stream and the situation can be fatal (Gilliland et al. 1988). The antimicrobial activity of RLs against these or similar wound-infecting bacteria has been described in the Antimicrobial Activity of Rhamnolipids section with MIC values (Table 1).

## *Effects of RLs on skin cells*

Pantazaki and Choli-Papadopoulou (2012) investigated the effect of two RL mixtures on the viability of human fibroblast cells *in vitro*. RLs used in this study were produced by *Thermus thermophilus* HB8 with sunflower seed oil and sodium gluconate, respectively, as substrate. Both contained mixtures of mono- and di-RLs with both saturated and unsaturated fatty acid residues (Pantazaki et al. 2010). Each RL mixture was tested at 0.1–0.5 g/L concentrations. Exposure time of fibroblast cells to RL solutions was varied from 24 to 96 h. RLs had clear effects on cell morphology even during the first 24 h. Generally, fibroblast cells changed from their original polygonal shape to a fusiform shape. At these tested concentrations, RLs also decreased cell viability; the effect was stronger with the mixture produced with sun flower oil as substrate. This suggests that effects of different compositions and/ or certain congener(s) in the RL mixtures may be more cytotoxic to fibroblast cells. Effects of rhamnose were also tested at 0.25–0.5 g/L concentrations. Contrary to RLs, rhamnose stimulated the cell viability and did not affect the morphology.

Unlike the use of RL mixtures in the above study by Pantazaki and Choli-Papadopoulou (2012), Stipcevic et al. (2005) examined the effects of a pure di-RL R-R-C10-C10 on keratinocyte growth in both serum free and serum containing media and the effects on growth of fibroblast skin cells in serum containing media. The serum containing media had 10% fetal calf serum (FCS). At high concentrations, i.e., 500 mg/L and 1 g/L, the di-RL showed a cytotoxic effect on both fibroblast and keratinocyte cells and disrupted the membranes of these cells after brief exposure. This cell membrane damaging effect of RL is consistent with what has been described in earlier sections of this chapter. For example, the hemolytic ability of RLs was thought to be associated with the insertion of two fatty acid residues into the phospholipid bilayer of cell membrane (Johnson and Boese-Marrazzo 1980).

At 10–200 mg/L concentrations, the di-RL inhibited the proliferation of fibroblast cells (in the serum containing medium) and keratinocyte cells in the serum free medium. The fibroblast cell viability decreased by 61% at 200 mg/L RL; the percentages of viability decrease were smaller at lower RL concentrations. At 1 mg/L di-RL still had negative effects on the fibroblast cells, but no apparent effects on the keratinocyte cells in serum free medium. In addition to the above membrane damaging effect, the negative effects may also be related with inhibited DNA synthesis. Earlier studies showed that RLs at a concentration of 0.1 g/L can inhibit the DNA synthesis in a model human epidermoid carcinoma cell line (A 431) *in vitro*, in the absence of fetal calf serum (Piljac and Piljac 1994).

The above observations of Stipcevic et al. (2005) are more or less consistent with those of Pantazaki and Choli-Papadopoulou (2012). However, for keratinocytes grown in serum containing medium, which had a higher calcium concentration (1.2 mM, versus 0.2 mM in the serum-free medium), Stipcevic et al. (2005) found that the di-RL at 10–100 mg/L concentrations stimulated the proliferation of keratinocyte cells. The highest increase in cell proliferation and viability was observed at 50 mg/L RL concentration. Stipcevic et al. (2005) noted that the serum-containing medium with 1.2 mM calcium is the favorable medium for keratinocyte differentiation and is closer to the situation in full thickness wounds, where blood vessels are disrupted and epidermal cells are exposed to serum components. This would support positive effects of the di-RL in wound healing applications.

In a very recent report, Shen et al. (2016) showed that at 10–30 mg/L concentrations, the di-RL R-R-C10-C10 can selectively kill myofibroblast cells without causing any significant toxicity to the fibroblast cells. Fibroblast cells are the principal cells of connective tissue. They are responsible for the production of collagen and extracellular matrix. Fibroblast cells play a critical role in cell repair process by differentiating into myofibroblasts which produce an inflammatory response to injury (Baum and Duffy 2011), such as expression of contractile apparatus like α-smooth muscle actin and secretion of collagen propolypeptides to repair the open wound (Shen et al. 2016). The myofibroblast cells are expected to disappear via apoptosis when a wound heals and closes normally. If this does not happen, myofibroblasts can continue to proliferate and remodel collagen fibers, causing formation of hypertrophic scars or keloids. The selective killing of myofibroblast cells demonstrated by di-RL in cell culture system and its potent effect against scar formation on rabbit ear hypertrophic scars indicate the potential application of RLs in wound healing process. Further, Shen et al. (2016) suggested that so far there has been no other compound that inhibits myofibroblasts without causing severe toxicity to surrounding physiological cells.

## *Use of RLs in wound care*

Various methods have been developed for wound management and wound dressings. Dressings are classified in different ways, e.g., according to their functions, materials and physical forms. Dressings are fabricated for different functions. For instance, dressings can be designed and fabricated to remove necrotic tissues and improve the wound debridement process. Dressings can also be designed to allow moisture evaporation and provide a dry environment around the wound (occlusive). Furthermore, they can be used to inhibit the growth of pathogens and serve as platforms for delivery of antibacterial agents. Finally, they can be designed and applied as absorbent agents to absorb the fluid or wound exudate. According to their functions, wound dressings can be classified as debridement, antibacterial, occlusive, absorbent and adherent (Boateng et al. 2008).

Hydrocolloid, alginate and collagen are the most common materials used in fabrication of wound dressings. Wound dressings are commercially available in ointment, film, foam and gel forms (Pachence 1996, Queen et al. 2004, Falabella 2006, Boateng et al. 2008). Hydrogels are insoluble hydrophilic materials made from polymers. Hydrogels can be applied as amorphous gels or as elastic solid sheets and films. To form hydrogels, the polymeric compounds need to be crosslinked and to entrap water physically. Alginate is a natural anionic polysaccharide comprised of mannuronic and guluronic acid units. The positive effects of alginate during wound healing have been investigated in several studies (Lansdown 2002, Blaine 1947, Oliver and Blaine 1950, Gilchrist and Martin 1983, Blair et al. 1988, Blair et al. 1990, Doyle et al. 1996, Thomas et al. 2000). Sodium alginate is very commonly used in food and pharmaceutical industries. Sodium alginate is water soluble. When a viscous sodium alginate solution is exposed to multivalent cations, commonly calcium, the alginate becomes crosslinked and forms an insoluble, highly porous calcium alginate matrix. With increasing alginate concentration and crosslinking extent, the pore size and swelling property of alginate matrix is reduced. The reduction restricts permeation of solutes that are originally present in the sodium alginate solution and now entrapped in the crosslinked matrix. This property is very useful for controlling the slow or delayed release of entrapped drugs (Badwan et al. 1985, Aslani and Kennedy 1996, Shu and Zhu 2002, Wang and He 2002, Dong et al. 2006). Gelatin is another commonly used natural polymer, produced from hydrolysis of insoluble fibrous collagen. Collagen is a protein which is the major constituent of skin and bones. Gelatin has excellent biocompatibility and biodegradability. Hence, gelatin has been used widely in the medical industry, for example, as a plasma expander, in wound dressings, adhesive and absorbent pads, and in biopharmaceutical formulations for controlled drug release (Neumann et al. 1981, Ikada and Tabata 1998, Fukunaka et al. 2002, Changez et al. 2004). However, gelatin does not have satisfactory mechanical strength and is brittle (when dry), so it is not used as a sole ingredient.

Ju et al. (2016) recently developed gelatin and alginate based hydrogels for wound dressing. The hydrogels are loaded with RLs and nanometer-sized, hollow gas vesicles. The gas vesicles, produced by a cyanobacterial culture, are naturally walled with thin, gas-permeable protein layers. Including gas vesicles in the wound dressings is to enhance gas exchanges, i.e., oxygen supply and carbon dioxide removal, without accelerated moisture evaporation during the wound healing process. Moreover, loading the hydrogels with RLs is to provide slow release of RLs for the protection of the wound area against pathogens and for the potential beneficial effects on wound healing as observed in the following study.

Stipcevic et al. (2000) reported the application of an ointment containing 1% w/w di-RL R-R-C10-C10 on a patient suffering from a venous ulcer. They observed that both the collagen layer and fibrotic lesions disappeared after 41 days of treatment. In a later study, Stipcevic et al. (2006) examined the effects of the di-RL R-R-C10-C10 on the process of cutaneous wound healing using Sprague–Dawley rats. Prior to the wound healing study, a subcutaneous multi-dose study was performed on Swiss–Webster mice to assess the toxicity of RL. RL solutions were injected subcutaneously once daily for 1 week at different sites of the back side of the animal. The mice tolerated the injected RL well up to the highest level tested, i.e., 120 mg/(kg day). Wound healing study was then done on rats with full-thickness burn wounds covering 5% of the total body surface. The RL was added to Eucerin ointment, containing 71.5% white petrolatum, 23.8% lanolin and 4.7% cholesterol, at three RL concentrations: 0% (control), 0.1% and 1% (w/w). The ointments were also prepared with and without 1% antiseptic chlorhexidine hydrochloride. The distance between wound edges were measured at days 14, 21, 28, 35 and 45. For the control group, full closure was not achieved without standard antiseptic but, with the antiseptic agent, it was achieved after 45 days. Fastest rate of closure was observed for those treated with 0.1% RL, with or without the antiseptic. For example, on day 21, wounds treated with 0.1% di-RL were closed 32% more than the control. Treatment with 1% di-RL was less effective than 0.1%. The collagen content in burn wounds treated with the di-RL was also found to be lower by 47.5%, compared to the control. Fast wound closure observed for those treated with 0.1% RL in this study suggests the potential application of RLs in wound care application which reduces the time required for wound closure, thus promoting the fast recovery of patients from burn wounds.

# Summary

Increasing global demand on green natural products brought the attention of industries towards microbial BSs, for their ecofriendly, non-toxic and high performance in specific applications. As a microbial BS, RLs have well documented, antimicrobial, wound healing and skin care activities that make them ideal molecules to develop RLs-based products for above applications. The antimicrobial activities promote the use of RLs in many potential applications such as in agriculture and wound care. In agriculture, RLs can be developed as effective biopesticide. Cost estimation suggested the economic feasibility of usage of RLs, potentially as a seed coating, to control *P. sojae* infection on soybean plants. Pesticidal activities of RLs on other organisms such as algae, mosquitoes, flies, ants, spiders and grasshoppers support the idea of active development of RLs-based algaecide, insecticides and broad-spectrum biopesticide. In addition to the anticipated antimicrobial function, RLs are found to have biological effects on wound healing and skin cells. The anti-fibrotic property to reduce excessive scarring in wound healing, demonstrated with the di-RL R-R-C10-C10, by selective inhibition of myofibroblasts, indicates their potential application in would healing and skin care as RLs-containing ointment and wound dressing products. Further research focusing on RLs with above mentioned properties may lead to the development of new RLs-based products with less/non-toxic products.

## Acknowledgments

We acknowledge the financial support of Ohio Soybean Council (Projects 15-2-09 and 16-R-16) to our related study.

## References

Abalos, A., A. Pinazo, M.R. Infante, M. Casals, F. García and A. Manresa. 2001. Physicochemical and antimicrobial properties of new rhamnolipids produced by *Pseudomonas aeruginosa* AT10 from soybean oil refinery wastes. Langmuir 17: 1367–1371.

Abalos, A., M. Viñas, J. Sabaté, M.A. Manresa and A.M. Solanas. 2004. Enhanced biodegradation of casablanca crude oil by a microbial consortium in presence of a rhamnolipid produced by *Pseudomonas aeruginosa* AT10. Biodegradation 15: 249–260.

Abdel-Mawgoud, A.M., F. Lépine and E. Déziel. 2010. Rhamnolipids: Diversity of structures, microbial origins and roles. Appl. Microbiol. Biotechnol. 86: 1323–1336.

Adams, F. and C.D. Foy. 1984. Physiological effects of hydrogen, aluminum, and manganese toxicities in acid soil. pp. 57–97. *In*: Soil Acidity Liming. American Society of Agronomy, Crop Science Society of America, Soil Science Society of America.

Al-Tahhan, R.A., T.R. Sandrin, A.A. Bodour and R.M. Maier. 2000. Rhamnolipid-induced removal of lipopolysaccharide from *Pseudomonas aeruginosa*: Effect on cell surface properties and interaction with hydrophobic substrates. Appl. Environ. Microbiol. 66: 3262–3268.

Aslani, P. and R.A. Kennedy. 1996. Studies on diffusion in alginate gels. I. Effect of cross-linking with calcium or zinc ions on diffusion of acetaminophen. J. Control. Release. 42: 75–82.

Awada, S.M., M.M. Awada and R.S. Spendlove. 2014. Compositions and methods for controlling pests with glycolipids. U.S. Patent # 8,680,060.

Badwan, A.A., A. Abumalooh, E. Sallam, A. Abukalaf and O. Jawan. 1985. A sustained release drug delivery system using calcium alginate beads. Drug Dev. Ind. Pharm. 11: 239–256.

Baum, J. and H.S. Duffy. 2011. Fibroblasts and myofibroblasts: what are we talking about? J. Cardiovasc. Pharmacol. 57: 376–379.

Benincasa, M., J. Contiero, A. Manresa and I.O. Moraes. 2002. Rhamnolipid production by *Pseudomonas aeruginosa* LBI growing on soapstock as the sole carbon source. J. Food Eng. 54: 283–288.

Benincasa, M., A. Abalos, I. Oliveira and A. Manresa. 2004. Chemical structure, surface properties and biological activities of the biosurfactant produced by *Pseudomonas aeruginosa* LBI from soapstock. Antonie Van Leeuwenhoek 85: 1–8.

Benincasa, M. 2007. Rhamnolipid produced from agroindustrial wastes enhances hydrocarbon biodegradation in contaminated soil. Curr. Microbiol. 54: 445–449.

Bjarnsholt, T., K. Kirketerp-Moller, P.O. Jensen, K.G. Madsen, R. Phipps, K. Krogfelt, N. Hoiby and M. Givskov. 2008. Why chronic wounds will not heal: A novel hypothesis. Wound Repair Regen. 16: 2–10.

Blaine, G. 1947. Experimental observations on absorbable alginate products in surgery: Gel, film, gauze and foam. Ann. Surg. 125: 102–14.

Blair, S.D., C.M. Backhouse, R. Harper, J. Matthews and C.N. McCollum. 1988. Comparison of absorbable materials for surgical haemostasis. Br. J. Surg. 75: 969–971.

Blair, S.D., P. Jarvis, M. Salmon and C. McCollum. 1990. Clinical trial of calcium alginate haemostatic swabs. Br. J. Surg. 77: 568–570.

Boateng, J.S., K.H. Matthews, H.N.E. Stevens and G.M. Eccleston. 2008. Wound healing dressings and drug delivery systems: A review. J. Pharm. Sci. 97: 2892–2923.

Boller, T. and G. Felix. 2009. A renaissance of elicitors: perception of microbe-associated molecular patterns and danger signals by pattern-recognition receptors. Annu. Rev. Plant Biol. 60: 379–406.

Bowler, P.G., B.I. Duerden and D.G. Armstrong. 2001. Wound microbiology and associated approaches to wound management. Clin. Microbiol. Rev. 14: 244–269.

Canaday, C.H. and A.F. Schmitthenner. 2010. Effects of chloride and ammonium salts on the incidence of *Phytophthora* root and stem rot of soybean. Plant Dis. 94: 758–765.

Chamnanpunt, J., W.X. Shan and B.M. Tyler. 2001. High frequency mitotic gene conversion in genetic hybrids of the oomycete *Phytophthora sojae*. Proc. Natl. Acad. Sci. U S A 98: 14530–5.

Changez, M., V. Koul, B. Krishna, A.K. Dinda and V. Choudhary. 2004. Studies on biodegradation and release of gentamicin sulphate from interpenetrating network hydrogels based on poly(acrylic acid) and gelatin: *In vitro* and *in vivo*. Biomaterials 25: 139–146.

Chayabutra, C., J. Wu and L.K. Ju. 2001. Rhamnolipid production by *Pseudomonas aeruginosa* under denitrification: Effects of limiting nutrients and carbon substrates. Biotechnol. Bioeng. 72: 25–33.

Chebbi, A., M. Elsikh, F. Haque, S. Ahmed, S. Dobbin, R. Marchant, S. Sayadi, M. Chamkha and I.M. Banat. 2017. Rhamnolipid from *Pseudomonas aeruginosa* strain W10; as antibiofilm/antibiofouling products for metal protection. J. Basic Microbiol. 57: 364–375.

Code of Federal Regulations, Title 40, Protection of Environment, Part 63. Office of the Federal Register. U. S. Government Printing Office. Washington D.C.

D'apos, J., K. De Maeyer, E. Pauwelyn and M. Hofte. 2010. Biosurfactants in plant-*Pseudomonas* interactions and their importance to biocontrol. Environ. Microbiol. Rep. 2: 359–372.

Das, P., S. Mukherjee and R. Sen. 2008. Antimicrobial potential of a lipopeptide biosurfactant derived from a marine *Bacillus circulans*. J. Appl. Microbiol. 104: 1675–1684.

Dashtbozorg, S.S., S. Miao and L.K. Ju. 2015. Rhamnolipids as environmentally friendly biopesticide against plant pathogen. Environ. Prog. 28: 404–409.

Dashtbozorg, S.S., J. Kohl and L.K. Ju. 2016. Rhamnolipid adsorption in soil: Factors, unique features, and considerations for use as green antizoosporic agents. J. Agric. Food Chem. 64: 3330–3337.

De Jonghe, K., I. De Dobbelaere, R. Sarrazyn and M. Hofte. 2005. Control of *Phytophthora cryptogea* in the hydroponic forcing of witloof chicory with the rhamnolipid-based biosurfactant formulation PRO1. Plant Pathol. 54: 219–226.

Déziel, E., F. Lépine, D. Dennie, D. Boismenu, O.A. Mamer and R. Villemur. 1999. Liquid chromatography/ mass spectrometry analysis of mixtures of rhamnolipids produced by *Pseudomonas aeruginosa* strain 57RP grown on mannitol or naphthalene. Biochim. Biophys. Acta—Mol. Cell Biol. Lipids 1440: 244–252.

Dong, Z., Q. Wang and Y. Du. 2006. Alginate/gelatin blend films and their properties for drug controlled release. J. Memb. Sci. 280: 37–44.

Dorrance, A.E. and S.A. McClure. 2001. Beneficial effects of fungicide seed treatments for soybean cultivars with partial resistance to *Phytophthora sojae*. Plant Dis. 85: 1063–1068.

Doyle, J.W., T.P. Roth, R.M. Smith, Y.Q. Li and R.M. Dunn. 1996. Effect of calcium alginate on cellular wound healing processes modeled *in vitro*. J. Biomed. Mater. Res. 32: 561–568.

Dusane, D.H., Y.V. Nancharaiah, S.S. Zinjarde and V.P Venugopalan. 2010. Rhamnolipid mediated disruption of marine *Bacillus pumilus* biofilms. Colloids Surfaces B Biointerfaces 81: 242–248.

Elshikh, M., S. Funston, A. Chebbi, S. Ahmed, R. Marchant and I.M. Banat. 2017. Rhamnolipids from non-pathogenic *Burkholderia thailandensis* E264: Physicochemical characterization, antimicrobial and antibiofilm efficacy against oral hygiene related pathogens. N. Biotechnol. 36: 26–36.

Erwin, D.C. and O.K. Ribeiro. 1996. Phytophthora Diseases Worldwide. APS Press, St. Paul.

Falabella, A.F. 2006. Debridement and wound bed preparation. Dermatol. Ther. 19: 317–325.

Fiebig, R., D. Schulze, J.C. Chung and S.T. Lee. 1997. Biodegradation of polychlorinated biphenyls (PCBs) in the presence of a bioemulsifier produced on sunflower oil. Biodegradation 8: 67–75.

Fry, W.E. and S.B. Goodwin. 1997. Re-emergence of potato and tomato late blight in the United States. Plant Dis. 81: 1349–1357.

Fukunaka, Y., K. Iwanaga, K. Morimoto, M. Kakemi and Y. Tabata. 2002. Controlled release of plasmid DNA from cationized gelatin hydrogels based on hydrogel degradation. J. Control. Release. 80: 333–343.

Gilchrist, T. and A.M. Martin. 1983. Wound treatment with sorbsan—an alginate fibre dressing. Biomaterials 4: 317–20.

Gilliland, E.L., N. Nathwani, C.J. Dore and J.D. Lewis. 1988. Bacterial colonisation of leg ulcers and its effect on the success rate of skin grafting. Ann. R Coll. Surg. Engl. 70: 105–8.

Haba, E., A. Pinazo, O. Jauregui, M.J. Espuny, M.R. Infante and A. Manresa. 2003. Physicochemical characterization and antimicrobial properties of rhamnolipids produced by *Pseudomonas aeruginosa* 47T2 NCBIM 40044. Biotechnol. Bioeng. 81: 316–322.

Herman, D.C., J.F. Artiola and R.M. Miller. 1995. Removal of cadmium, lead, and zinc from soil by a rhamnolipid biosurfactant. Environ. Sci. Technol. 29: 2280–2285.

Ho, H.H. and C.J. Hickman. 1967. Asexual reproduction and behavior of zoospores of *Phytophthora Megasperma* var. sojae. Can. J. Bot. 45: 1963–1981.

Ikada, Y. and Y. Tabata. 1998. Protein release from gelatin matrices. Adv. Drug Deliv. Rev. 31: 287–301.

Irie, Y., G.A. O'Toole and M.H. Yuk. 2005. *Pseudomonas aeruginosa* rhamnolipids disperse *Bordetella bronchiseptica* biofilms. FEMS Microbiol. Lett. 250: 237–243.

Johnson, M.K. and D. Boese-Marrazzo. 1980. Production and properties of heat-stable extracellular hemolysin from *Pseudomonas aeruginosa*. Infect. Immun. 29: 1028–33.

Jourdan, E., G. Henry, F. Duby, J. Dommes, J.P. Barthélemy, P. Thonart and M. Ongena. 2009. Insights into the defense-related events occurring in plant cells following perception of surfactin-type lipopeptide from *Bacillus subtilis*. Mol. Plant Microbe. Interact. 22: 456–468.

Ju, L.K., S.S. Dashtbozorg and N. Vongpanish. 2016. Wound dressings with enhanced gas permeation and other beneficial properties. U.S. Patent # 9,468,700 B2.

Juwarkar, A.A., A. Nair, K.V. Dubey, S.K. Singh and S. Devotta. 2007. Biosurfactant technology for remediation of cadmium and lead contaminated soils. Chemosphere 68: 1996–2002.

Kim, B.S., J.Y. Lee and B.K. Hwang. 2000. *In vivo* control and *in vitro* antifungal activity of rhamnolipid B, a glycolipid antibiotic, against *Phytophthora capsici* and *Colletotrichum orbiculare*. Pest Manag. Sci. 56: 1029–1035.

Kim, S.K., Y.C. Kim, S. Lee, J.C. Kim, M.Y. Yun and I.S. Kim. 2011. Insecticidal activity of rhamnolipid isolated from *Pseudomonas* sp. EP-3 against green peach aphid (*Myzus persicae*). J. Agric. Food Chem. 59: 934–938.

Kitamoto, D., H. Yanagishita, T. Shinbo, T. Nakane, C. Kamisawa and T. Nakahara. 1993. Surface active properties and antimicrobial activities of mannosylerythritol lipids as biosurfactants produced by *Candida antarctica*. J. Biotechnol. 29: 91–96.

Lang, S., E. Katsiwela and F. Wagner. 1989. Antimicrobial effects of biosurfactants. Fett/Lipid. 91: 363–366.

Lang, S. and D. Wullbrandt. 1999. Rhamnose lipids—biosynthesis, microbial production and application potential. Appl. Microbiol. Biotechnol. 51: 22–32.

Lansdown, A.B.G. 2002. Calcium: A potential central regulator in wound healing in the skin. Wound Repair Regen. 10: 271–285.

Lebrón-Paler, A., J.E. Pemberton, B.A. Becker, W.H. Otto, C.K. Larive and R.M. Maier. 2006. Determination of the acid dissociation constant of the biosurfactant monorhamnolipid in aqueous solution by potentiometric and spectroscopic methods. Anal. Chem. 78: 7649–7658.

Lotfabad, T.B., H. Abassi, R. Ahmadkhaniha, R. Roostaazad, F. Masoomi, H.S. Zahiri, G. Ahmadian, H. Vali and K.A. Noghabi. 2010. Structural characterization of a rhamnolipid-type biosurfactant produced by *Pseudomonas aeruginosa* MR01: Enhancement of di-rhamnolipid proportion using gamma irradiation. Colloids Surfaces B Biointerfaces 81: 397–405.

MacGregor, T., M. Bhattacharya, B. Tyler, R. Bhat, A.F. Schmitthenner and M. Gijzen. 2002. Genetic and physical mapping of avrla in *Phytophthora sojae*. Genetics 160: 949–959.

Mackey, D. and A.J. McFall. 2006. MAMPs and MIMPs: Proposed classifications for inducers of innate immunity. Mol. Microbiol. 61: 1365–1371.

Maier, R.M. and G. Soberón-Chávez. 2000. *Pseudomonas aeruginosa* rhamnolipids: Biosynthesis and potential applications. Appl. Microbiol. Biotechnol. 54: 625–633.

Manaargadoo-Catin, M., A. Ali-Cherif, J.L. Pougnas and C. Perrin. 2015. Hemolysis by surfactants—A review. Adv. Colloid. Interface Sci. 228: 1–16.

Miao, S., S.S. Dashtbozorg, N.V. Callow and L.K. Ju. 2015. Rhamnolipids as platform molecules for production of potential anti-zoospore agrochemicals. J. Agric. Food Chem. 63: 3367–3376.

Morris, P. and E. Ward. 1992. Chemoattraction of zoospores of the soybean pathogen, *Phytophthora sojae*, by isoflavones. Physiol. Mol. Plant Pathol. 40: 17–22.

Mukherjee, S., P. Das and R. Sen. 2006. Towards commercial production of microbial surfactants. Trends Biotechnol. 24: 509–515.

Muller, M.M. and R. Hausmann. 2011. Regulatory and metabolic network of rhamnolipid biosynthesis: Traditional and advanced engineering towards biotechnological production. Appl. Microbiol. Biotechnol. 91: 251–264.

Mulligan, C.N., R.N. Yong and B.F. Gibbs. 2001. Surfactant-enhanced remediation of contaminated soil: A review. Eng. Geol. 60: 371–380.

Mulligan, C.N. 2005. Environmental applications for biosurfactants. Environ. Pollut. 133: 183–198.

Neto, D.C., J.A. Meira, E. Tiburtius, P.P. Zamora, C. Bugay, D.A. Mitchell and N. Krieger. 2009. Production of rhamnolipids in solid-state cultivation: Characterization, downstream processing and application in the cleaning of contaminated soils. Biotechnol. J. 4: 748–755.

Neumann, P.M., B. Zur and Y. Ehrenreich. 1981. Gelatin-based sprayable foam as a skin substitute. J. Biomed. Mater. Res. 15: 9–18.

Nguyen, T.T., N.H. Youssef, M.J. McInerney and D.A. Sabatini. 2008. Rhamnolipid biosurfactant mixtures for environmental remediation. Water Res. 42: 1735–1743.

Nidhi, K., S. Indrajeet, M. Khushboo, K. Gauri and D.J. Sen. 2011. Hydrotropy: A promising tool for solubility enhancement: A review. Int. J. Drug Dev. Res. 3: 26–33.

Nitschke, M., S.G. Costa and J. Contiero. 2005. Rhamnolipid surfactants: An update on the general aspects of these remarkable biomolecules. Biotechnol. Prog. 21: 1593–1600.

Nivas, B.T., D.A. Sabatini, B.J. Shiau and J.H. Harwell. 1996. Surfactant enhanced remediation of subsurface chromium contamination. Water Res. 30: 511–520.

Noordman, W.H., J.H.J. Wachter, G.J. de Boer and D.B. Janssen. 2002. The enhancement by surfactants of hexadecane degradation by *Pseudomonas aeruginosa* varies with substrate availability. J. Biotechnol. 94: 195–212.

Nowicki, M., M.R. Foolad, M. Nowakowska and E.U. Kozik. 2012. Potato and tomato late blight caused by *Phytophthora infestans*: An overview of pathology and resistance breeding. Plant Dis. 96: 4–17.

Oliver, L.C. and G. Blaine. 1950. Hæmostasis with absorbable alginates in neurosurgical practice. Br. J. Surg. 37: 307–310.

Pachence, J.M. 1996. Collagen-based devices for soft tissue repair. J. Biomed. Mater. Res. 33: 35–40.

Pantazaki, A.A., M.I. Dimopoulou, O.M. Simou and A.A Pritsa. 2010. Sunflower seed oil and oleic acid utilization for the production of rhamnolipids by *Thermus thermophilus* HB8. Appl. Microbiol. Biotechnol. 88: 939–951.

Pantazaki, A.A. and T. Choli-Papadopoulou. 2012. On the *Thermus thermophilus* HB8 potential pathogenicity triggered from rhamnolipids secretion: Morphological alterations and cytotoxicity induced on fibroblastic cell line. Amino Acids 42: 1913–1926.

Parra, J.L., J. Guinea, M.A. Manresa, M. Robert, M.E. Mercadé, F. Comelles and M.P. Bosch. 1989. Chemical characterization and physicochemical behavior of biosurfactants. J. Am. Oil Chem. Soc. 66: 141–145.

Perneel, M., L. D'Hondt, K. De Maeyer, A. Adiobo, K. Rabaey and M. Hofte. 2008. Phenazines and biosurfactants interact in the biological control of soil-borne diseases caused by *Pythium* spp. Environ. Microbiol. 10: 778–788.

Peters, J., P. Speth, K. Kelling and R. Borges. 2005. Effect of soil pH on soybean yield. In: Proc. 2005 Wisconsin Fertil. Chem. Assoc. Disting Serv. Award. Madison, WI.

Piljac, G. and V. Piljac. 1994. Pharmaceutical preparation based on rhamnolipid. U.S. Patent # 5,455,232.

Pinzon-Gomez, N.M. 2009. Rhamnolipid biosurfactant production from glycerol: New methods of analysis and improved denitrifying fermentation. Ph.D Dissertation. The University of Akron, Akron, USA.

Queen, D., H. Orsted, H. Sanada and G. Sussman. 2004. A dressing history. Int. Wound J. 1: 59–77.

Quinn, G.A., A.P. Maloy, M.M. Banat and I.M. Banat. 2013. A comparison of effects of broad-spectrum antibiotics and biosurfactants on established bacterial biofilms. Curr. Microbiol. 67: 614–623.

Rahman, K.S.M., I.M. Banat, J. Thahira, T. Thayumanavan and P. Lakshmanaperumalsamy. 2002. Bioremediation of gasoline contaminated soil by a bacterial consortium amended with poultry litter, coir pith and rhamnolipid biosurfactant. Bioresour. Technol. 81: 25–32.

Rahman, K.S.M., T.J. Rahman, Y. Kourkoutas, I. Petsas, R. Marchant and I.M. Banat. 2003. Enhanced bioremediation of n-alkane in petroleum sludge using bacterial consortium amended with rhamnolipid and micronutrients. Bioresour. Technol. 90: 159–168.

Raya, A., M. Sodagari, N.M. Pinzon, X. He, B.M.Z. Newby and L.K. Ju. 2010. Effects of rhamnolipids and shear on initial attachment of *Pseudomonas aeruginosa* PAO1 in glass flow chambers. Environ. Sci. Pollut. Res. 17: 1529–1538.

Rienzo, D., P. Stevenson, R. Marchant and I.M. Banat. 2015. Antibacterial properties of biosurfactants against selected Gram-positive and negative bacteria. FEMS Microbiol. Lett. 363.

Robert, J.M.I., A. Toguchi and R.M. Harshey. 2001. *Salmonella enterica* Serovar Typhimurium swarming mutants with altered biofilm-forming abilities: Surfactin inhibits biofilm formation. J. Bacteriol. 83: 5848–5854.

Robinson, K.G., M.M. Ghosh and Z. Shi. 1996. Mineralization enhancement of non-aqueous phase and soil-bound PCB using biosurfactant. Water Sci. Technol. 34: 303–309.

Rodrigues, L., I.M. Banat, J. Teixeira and R. Oliveira. 2006. Biosurfactants: potential applications in medicine. J. Antimicrob. Chemother. 57: 609–18.

Rothe, M. and V. Falanga. 1989. Growth factors. Their biology and promise in dermatologic diseases and tissue repair. Arch. Dermatol. 125: 1390–8.

Scheibenbogen, K., R.G. Zytner, H. Lee and J.T. Trevors. 1994. Enhanced removal of selected hydrocarbons from soil by *Pseudomonas aeruginosa* UG2 biosurfactants and some chemical surfactants. J. Chem. Technol. Biotechnol. 59: 53–59.

Schettini, T.M. 1991. Insensitivity to metalaxyl in California populations of *Bremia lactucae* and resistance of California lettuce cultivars to downy mildew. Phytopathology 81: 64.

Schmitthenner, A.F. 1985. Problems and progress in control of *Phytophthora* root rot of soybean. Plant Dis. 69: 362–368.

Schooling, S.R., U.K. Charaf, D.G. Allison and P. Gilbert. 2004. A role for rhamnolipid in biofilm dispersion. Biofilms 1: 91–99.

Shakespeare, P. 2001. Burn wound healing and skin substitutes. Burns 27: 517–22.

Shen, C., L. Jiang, H. Shao, C. You, G. Zhang, S. Ding, T. Bian, C. Han and Q. Meng. 2016. Targeted killing of myofibroblasts by biosurfactant di-rhamnolipid suggests a therapy against scar formation. Sci. Rep. 6: 37553.

Shu, X.Z. and K.J. Zhu. 2002. The release behavior of brilliant blue from calcium-alginate gel beads coated by chitosan: the preparation method effect. Eur. J. Pharm. Biopharm. 53: 193–201.

Silva, V.L., R.B. Lovaglio, C.J. Von Zuben and J. Contiero. 2015. Rhamnolipids: Solution against *Aedes aegypti*? Front. Microbiol. 6: 88.

Sodagari, M., H. Wang, B.M.Z. Newby and L.K. Ju. 2013. Effect of rhamnolipids on initial attachment of bacteria on glass and octadecyltrichlorosilane-modified glass. Colloids Surfaces B Biointerfaces 103: 121–128.

Sotirova, A.V., D.I. Spasova, D.N. Galabova, E. Karpenko and A. Shulga. 2008. Rhamnolipid-biosurfactant permeabilizing effects on gram-positive and gram-negative bacterial strains. Curr. Microbiol. 56: 639–644.

Stanghellini, M.E. and R.M. Miller. 1997. Biosurfactants: Their identity and potential efficacy in the biological control of zoosporic plant pathogens. Plant Dis. 81: 4–12.

Stipcevic, T., T. Piljac, J. Piljac, T. Dujmic and G. Piljac. 2000. Use of rhamnolipids in wound healing, treatment and prevention of gum disease and periodontal regeneration. U.S. Patent # 7,129,218 B2.

Stipcevic, T., T. Piljac and R.R. Isseroff. 2005. Di-rhamnolipid from *Pseudomonas aeruginosa* displays differential effects on human keratinocyte and fibroblast cultures. J. Dermatol. Sci. 40: 141–143.

Stipcevic, T., A. Piljac and G. Piljac. 2006. Enhanced healing of full-thickness burn wounds using di-rhamnolipid. Burns 32: 24–34.

Tan, H., J.T. Champion, J.F. Artiola, M.L. Brusseau and R.M. Miller. 1994. Complexation of cadmium by a rhamnolipid biosurfactant. Environ. Sci. Technol. 28: 2402–2406.

Thavasi, R., R. Marchant and I.M. Banat. 2014. Biosurfactant applications in agriculture. pp. 313–325. *In*: N. Kosaric and F.V. Sukan (eds.). Biosurfactants Production and Utilization-Processes, Technologies, and Economics. CRC Press, Florida, USA.

Thomas, A., K.G. Harding and K. Moore. 2000. Alginates from wound dressings activate human macrophages to secrete tumour necrosis factor-α. Biomaterials 21: 1797–1802.

Trummler, K., F. Effenberger and C. Syldatk. 2003. An integrated microbial/enzymatic process for production of rhamnolipids and L-rhamnose from rapeseed oil with *Pseudomonas* sp. DSM 2874. Eur. J. Lipid Sci. Technol. 105: 563–571.

Tyler, B.M. 2007. *Phytophthora sojae*: Root rot pathogen of soybean and model oomycete. Mol. Plant Pathol. 8: 1–8.

Varnier, A.L., L. Sanchez, P. Vatsa, L. Boudesocque, A. Garcia-Brugger, F. Rabenoelina, A. Sorokin, J.-H. Renault, S. Kauffmann, A. Pugin, C. Clement, F. Baillieul and S. Dorey. 2009. Bacterial rhamnolipids are novel MAMPs conferring resistance to *Botrytis cinerea* in grapevine. Plant Cell Environ. 32: 178–193.

Vatsa, P., L. Sanchez, C. Clement, F. Baillieul and S. Dorey. 2010. Rhamnolipid biosurfactants as new players in animal and plant defense against microbes. Int. J. Mol. Sci. 11: 5095–5108.

Velraeds, M.M.C., H.C. Van Der Mei, G. Reid and H.J. Busscher. 1996. Inhibition of initial adhesion of uropathogenic *Enterococcus faecalis* by biosurfactants from *Lactobacillus isolates*. Appl. Environ. Microbiol. 62: 1958–1963.

Velraeds, M.M.C., B. Van De Belt-Gritter, H.C. Van Der Mei, G. Reid and H.J. Busscher. 1998. Interference in initial adhesion of uropathogenic bacteria and yeasts to silicone rubber by a *Lactobacillus acidophilus* biosurfactant. J. Med. Microbiol. 47: 1081–1085.

Walker, C.A. and P. Van West. 2007. Zoospore development in the oomycetes. Fungal Biol. Rev. 21: 10–18.

Wang, K. and Z. He. 2002. Alginate-konjac glucomannan-chitosan beads as controlled release matrix. Int. J. Pharm. 244: 117–26.

Wang, Q., X. Fang, B. Bai, X. Liang, P.J. Shuler, W.A. Goddard and Y. Tang. 2007. Engineering bacteria for production of rhamnolipid as an agent for enhanced oil recovery. Biotechnol. Bioeng. 98: 842–853.

Wang, S. and C.N. Mulligan. 2004. An evaluation of surfactant foam technology in remediation of contaminated soil. Chemosphere 57: 1079–1089.

Wang, X., L. Gong, S. Liang, X. Han, C. Zhu and Y. Li. 2005. Algicidal activity of rhamnolipid biosurfactants produced by *Pseudomonas aeruginosa*. Harmful Algae 4: 433–443.

Ward, B. 2015. Soybean production budget. In: 2015 Ohio Enterprise Budgets, Farm Management Enterprise Budgets, Department of Agricultural, Environmental, and Development Economics; http://aede.osu.edu/research/osu-farm-management/enterprise-budgets.

Wen, J., S.P. Stacey, M.J. McLaughlin and J.K. Kirby. 2009. Biodegradation of rhamnolipid, EDTA and citric acid in cadmium and zinc contaminated soils. Soil Biol. Biochem. 41: 2214–2221.

Whang, L.M., P.W.G. Liu, C.C. Ma and S.S. Cheng. 2008. Application of biosurfactants: Rhamnolipid, and surfactin, for enhanced biodegradation of diesel-contaminated water and soil. J. Hazard Mater. 151: 155–163.

Woods, G.L. and J.A. Washington. 1995. Antibacterial susceptibility tests: Dilution and disk diffusion methods. pp. 1327–1341. *In*: P.R. Murray, E.J. Baron, M.A. Pfaller, F.C. Tenover and R.H. Yolken (eds.). Manual of Clinical Microbiology. ASM Press, Washington.

Zhang, Y. and R.M. Miller. 1992. Enhanced octadecane dispersion and biodegradation by a *Pseudomonas* rhamnolipid surfactant (biosurfactant). Appl. Environ. Microbiol. 58: 3276–3282.

Zhang, Y. and R.M. Miller. 1994. Effect of a *Pseudomonas* rhamnolipid biosurfactant on cell hydrophobicity and biodegradation of octadecane. Appl. Environ. Microbiol. 60: 2101–2106.

# 5

# Microbial Production and Applications of Mannosylerythritol, Cellobiose and Trehalose Lipids

*Chandraprasad Madihalli*[1,2] *and Mukesh Doble*[2,*]

## Introduction

Biosurfactants (BSs) are surface-active molecules which demonstrate not only reduction in surface tension of water but also versatile biochemical functions. BSs are actively produced and secreted by different classes of bacteria, fungi, or yeast and actinomycetes from various carbon sources, namely, sugars, oils, hydrocarbons and agro industrial wastes. They are amphiphilic in nature consisting of both hydrophilic and hydrophobic moieties. Unlike their chemical counterparts, BSs exhibit lower toxicity, environmental compatibility, higher biodegradability, higher selectivity and are stable under varying physical conditions (Desai and Banat 1997). Based on chemical nature, BSs are classified into lipopeptides, fatty acids, polymeric and glycolipids. BSs find useful applications in environmental remediation which include enhanced oil recovery, oil spillage control, pollutant degradation and solubilisation of water insoluble compounds (Banat et al. 2000). Further, they find potential applications in food, cosmetic, detergent industries and recently in gene and drug delivery applications (Marchant and Banat 2012a).

---

[1] Department of Biotechnology, BMS College of Engineering, Bull Temple Road, Bangalore, Karnataka, India 560019.
[2] Bioengineering and Drug Design Lab, Department of Biotechnology, Bhupat and Jyoti Mehta School of Biosciences, Indian Institute of Technology Madras, Chennai, India 600036.
  Email: chandraprasadms.bt@bmsce.ac.in
* Corresponding author: mukeshd@iitm.ac.in

Glycolipids have dominated the BS industry and few of them have even been commercialized (Marchant and Banat 2012b). Usage of renewable resources for higher productivity and multifaceted biochemical functions are the key salient features (Kitamoto et al. 2002). Glycolipid BSs include rhamnolipids (RLs), mannosylerythritol lipids (MELs), trehalose lipids (TLs), cellobiose lipids (CBLs) and sophorolipids (SLs) (Marchant et al. 2012b). These are low molecular weight compounds consisting of sugar or sugar alcohols and lipids esterified to sugars.

## Microbial Source and Structural Diversity of Glycolipid BSs

Glycolipid BSs consist of various types of sugars, their derivatives and esterified fatty acids. Mannosylerythritol lipids (MELs) consist of hydrophilic moiety in the form of 4-O-β-D-Mannopyranosyl-D-erythritol and hydrophobic moiety in the form of fatty acids and/or acetyl group which are further classified into four types, namely, MEL-A, B, C and D based on the presence and absence of acetyl groups (Figure 1). MEL-A has two acetyl groups linked to C4 and C6 positions of mannose, while MEL-B and MEL-C have one acetyl group linked to C6 and C4 positions, respectively. MEL-D consists of deacetylated mannose esterified to fatty acids at C2 and C3 positions. The MELs also vary in the length of fatty acid chain and type of saturation (saturated or unsaturated or both) as well (Smyth et al. 2010). It depends on the carbon source from which they are produced. In general, MELs are composed of fatty acids in the range of C6-C18 including unsaturated fatty acids. In addition to predominant types, there are structural variants of MELs, namely, mono/

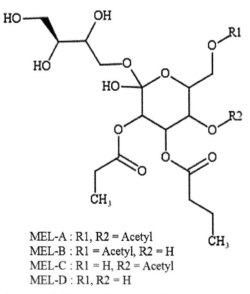

MEL-A : R1, R2 = Acetyl
MEL-B : R1 = Acetyl, R2 = H
MEL-C : R1 = H, R2 = Acetyl
MEL-D : R1, R2 = H

**Figure 1.** Structure of different MELs produced by *Pseudozyma* sp.

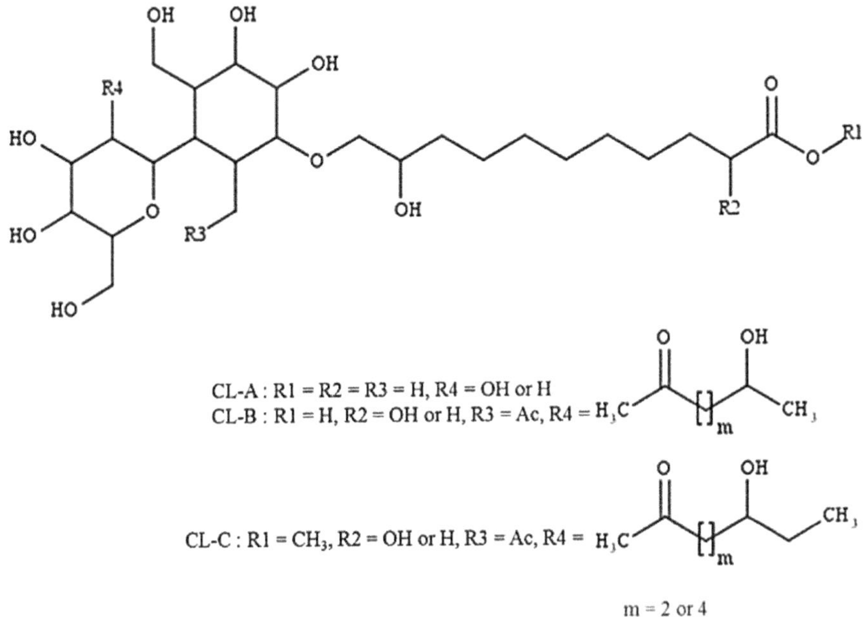

CL-A : R1 = R2 = R3 = H, R4 = OH or H
CL-B : R1 = H, R2 = OH or H, R3 = Ac, R4 = H₃C

CL-C : R1 = CH₃, R2 = OH or H, R3 = Ac, R4 = H₃C

m = 2 or 4

**Figure 2.** Structure of different cellobiose lipids.

tri-acylated MEL, Mannosylribitollipid (MRL), Mannosylarabitol lipid (MAL) and Mannosylsorbitol lipid (MSL). Biological syntheses of these compounds depend on producer strain, media used and growth conditions.

Cellobiose lipids (CBLs) are another class of glycolipids which consist of β-cellobiose sugar moiety esterified to fatty acids (Figure 2). Based on the presence of acetyl or other functional groups, CBLs are classified into three main types, namely, CBL-A, B and C (Morita et al. 2013b). All three types of CBLs have C16 fatty acids esterified to C1 position of the β-glucose unit (Figure 2). Unlike MELs, CBLs do not have sugar alcohol as part of the structure but consist of various R groups as side chains.

Another class of glycolipids called trehaloselipids (TLs) chemically consists of two glucose units linked through α, α,-1,1-glycosidic linkage (trehalose) to which the lipids are esterified. Broadly, two classes of TLs exist, namely, (1) non-ionic and (2) ionic, the production of which depends on strain physiology and growth conditions. Non-ionic TLs are represented by trehalose esterified with mono, di and trimycolates (Figure 3a) while ionic ones include succinyl trehalose lipid (STL) and trehalose tetraesters (Figure 3b). The TLs produced by *Rhodococcus erythropolis* are succinoyltrehalose lipid 1 and 2 differing in the number of succinoyl units linked to trehalose. The well-known TL is trehalose 6, 6'-dimycolate (cord factor) which is a cell wall component of Mycobacteria. The two long chain α-mycolic acid units are esterified to C6 position of each glucose (Franzetti et al. 2010).

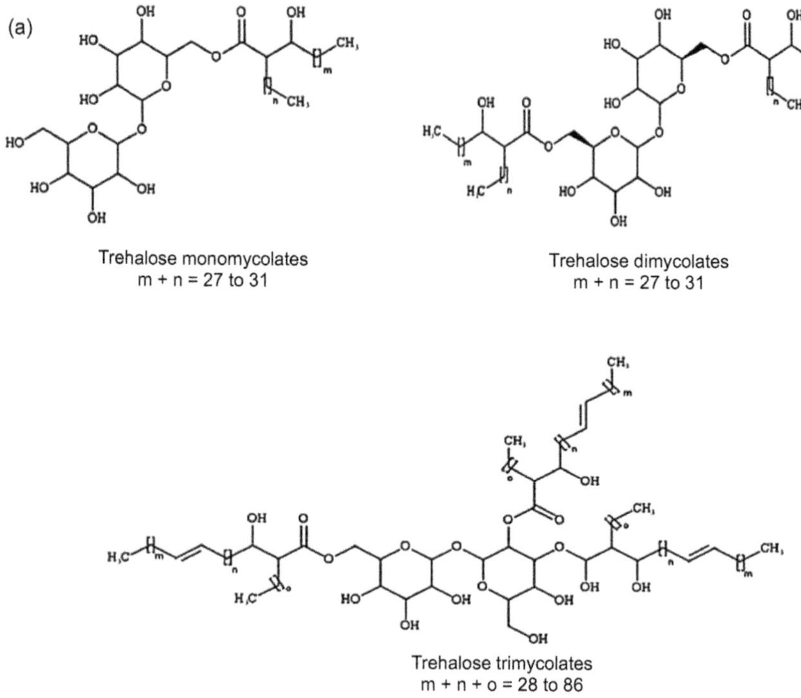

(a)

Trehalose monomycolates
m + n = 27 to 31

Trehalose dimycolates
m + n = 27 to 31

Trehalose trimycolates
m + n + o = 28 to 86

**Figure 3a.**  Structure of non-ionic trehalolipids (Kuyukina and Ivshina 2010a).

(b)

m = 5 to 9

R = either OC $(CH_2)_m$ CH$_3$ + OC $(CH_2)_2$ CH$_3$
or OC $(CH_2)_m$ CH$_3$
m = 6–12

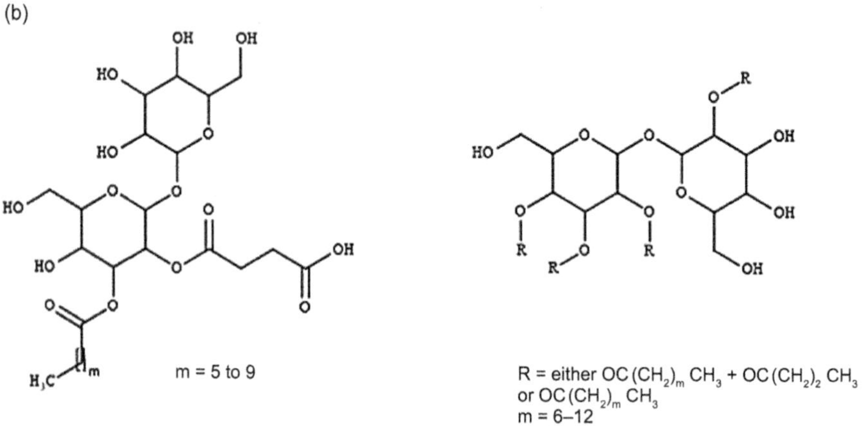

**Figure 3b.**  Structure of anionic trehalolipids (Kuyukina and Ivshina 2010a).

## Physico-chemical properties of glycolipid BSs

BSs reduce surface tension as well as interfacial tension between two immiscible liquids, solids or gases, thereby stabilizing the emulsions. Most glycolipid BSs exhibit surface and interfacial tension-reducing properties and possess other physicochemical activities such as dispersing, frothing, solubilising and wetting ability (Table 1).

Among MELs, the lower critical micelle concentration (CMC) and higher surface-active properties were exhibited by MEL-A and MEL-B while MEL-C has a higher CMC and lesser surface activity. The differences in their behaviour are due to the type of fatty acids present in each of them. MEL-A is produced predominantly by *P. antarctica*. The CMC of MEL-A is around $4.0 \times 10^{-6}$ µM with CMC (surface tension of water at CMC) of 28 mN/m while MEL-C has a CMC of $6.0 \times 10^{-6}$ µM with CMC of 25.1 mN/m. The MELs also exhibit varying emulsifying properties which depend on hydrophilic lipophilic balance (HLB) values, and the aqueous and oil phases (Boyd et al. 1972). Among MELs, MEL-A showed higher emulsifying activity towards soybean oil and *n*-tetradecane when compared to chemical surfactant, Tween 80. It also stabilizes the emulsions during storage. In a ternary system of water/MEL-A/n-decane, W/O microemulsion are formed without any of the

**Table 1.** Physico-chemical properties of glycolipid BSs.

| Biosurfactant | Microbial Source | Surface Tension | | Interfacial Tension mN/m | Reference |
|---|---|---|---|---|---|
| | | Critical Micellar Concentration (µM) | Surface Tension at CMC (mN/m) | | |
| **Mannosylerythritol lipid** | | | | | |
| MEL-A | *P. antarctica* | 2.7 | 28.4 | 2.1 | (Kitamoto et al. 2002) |
| MEL-B | *P. antarctica* | 4.5 | 28.2 | 2.4 | (Kitamoto et al. 2002) |
| MEL-A | *P. crassa* | 5.2 | 26.5 | – | (Fukuoka et al. 2008) |
| MEL-C | *P. graminicola* | 4.0 | 24.2 | – | (Morita et al. 2008) |
| MEL-C | *P. hubeiensis* | 6.0 | 30.7 | – | (Konishi et al. 2008) |
| **Trehalose lipid** | | | | | |
| TL-1 | *Rhodococcus erythropolis* | 4 (mg/L) | 36 | 17 | (Lang and Philp 1998) |
| TL-2 | *Rhodococcus erythropolis* | 4 (mg/L) | 32 | 14 | (Lang et al. 1998) |
| Trehalose-2,2,3,4-tetraester | *Rhodococcus erythropolis* | 15 (mg/L) | 26 | < 1 | (Lang et al. 1998) |
| Na salts of STL-1 and STL-2 | *Rhodococcus erythropolis* | $9.6 \times 10^{-4}$ | 30.8 | (0.2%) | (Lang et al. 1998) |
| STL-1 | *Rhodococcus* sp. SD74 | $5.6 \times 10^{-6}$ | 19.0 | – | (Tokumoto et al. 2009) |
| STL-1 (Na salt) | *Rhodococcus* sp. SD74 | $7.7 \times 10^{-6}$ | 23.7 | – | (Tokumoto et al. 2009) |

co-surfactants in the presence of MEL-A (Kitamoto et al. 2009). In the case of TLs, the surface active properties are independent of the type of surfactant over varying physical conditions such as pH, salt concentration and temperature. At a CMC of 4 mg/L, the surface tension reduced from 36 to 32 mN/m and at the same concentration, the interfacial tension reduced from 17 to 14 mN/m in water/n-hexadecane system.

## Production of glycolipid BSs

BSs are widely produced by various classes of bacteria, yeasts and fungi. The type of BS produced depends on the organism, its genetic constitution and the substrate provided (Table 2). Many of the BS producers can make use of industrial wastes as substrates. In this regard, many strains that are capable of producing BSs are isolated from contaminated soils, effluents and waste water resources. The two basic strategies adopted worldwide in increasing the BS production include (i) employing agro-industrial wastes as fermentation media to decrease the use of expensive substrates and (ii) optimization of the process in large scale for maximizing the yield and design cost-effective recovery strategies (Saharan et al. 2011).

Glycolipid BSs are gaining much interest due to their notable physico-chemical activities having potential applications in various sectors. So their industrial scale production with increased yield is necessary. Following sections will provide information on fermentation production of glycolipid BSs, MELs, TLs and CBLs.

## Production of MELs

Production of MEL from basidiomycetous yeast, *Pseudozyma* sp. has been reported in both shake flask culture as well as large scale fermenter while minimum study is performed with the latter (Arutchelvi and Doble 2011). Both the methods involved seed culture production using a growth media consisting of 4% glucose, 0.3% NaNO$_3$, 0.03% KH$_2$PO$_4$, MgSO$_4$.7H$_2$O and 0.1% yeast extract (pH 6.2), incubating at 25°C, on a reciprocal shaker (150 rpm) for 2 days (Kitamoto et al. 1992, Konishi et al. 2007). The seed culture at 1% concentration is inoculated into the above basal medium with 4% olive oil as carbon source and incubated at 28°C, 220 rpm for 7 days. The glycolipid BS was extracted from fermented supernatant using two volumes of ethyl acetate. The organic solvent fraction was concentrated by evaporating the solvent using a rotavapor and crude MELs obtained was suspended in chloroform (Konishi et al. 2007, Morita et al. 2009a, Smyth et al. 2010). Purification of different MELs was performed by gradient elution using chloroform:acetone solvent mixture (Fukuoka et al. 2007, Rau et al. 2005). Irrespective of the process type, the MELs produced vary from organism to organism wherein one of them may be predominate or only one type of MEL produced. For example, strains of *P. antarctica* produces all three types of MELs with MEL-A being the predominant. As reported elsewhere, more than 70% of the total MEL yield consists of MEL-A. Table 2 summarizes other organisms, conditions and yield of MEL obtained.

**Table 2.** MELs producers, substrates used and yield, modified from Morita et al. (2015).

| Microbial Producers | Carbon Sources | Glycolipids | Yield (g/L) |
|---|---|---|---|
| *Pseudozyma aphidis* | Soybean oil and glucose | MEL-A (main), MEL-B, MEL-C | 165[1,2,3] (Morita et al. 2013b) |
| | Soap stock and whey permeate/molasses | Mixture of MEL | 90[e] (Dziegielewska and Adamczak 2013) |
| *Pseudozyma antarctica* | Soybean oil | MEL-A (main), MEL-B, MEL-C | 40[3] |
| | Soybean oil | MEL-B (purified) | 10 |
| | n-Alkane | MEL-A (main), MEL-B, MEL-C | 140[3,4] |
| | Glucose | Mono-acylated MEL (purified) | 1.3 |
| | Soybean oil | Tri-acylated MEL (purified) | ND |
| | Waste frying oil | Mixture of MEL | 107.2[5] (Dziegielewska et al. 2013) |
| | Cellulose from Avicel and Wheat straw | SHF | 4 and 1.4 (Faria et al. 2014) |
| | Cellulose from Avicel and Wheat straw | SSF | 2.9 and 1.1 (Faria et al. 2014) |
| | Cellulose from Avicel and Wheat straw | Fed batch | 4.5 and 2.5 (Faria et al. 2014) |
| | Xylan | Fed batch | 2.0 (Faria et al. 2015) |
| *Pseudozyma churashimaensis* | Soybean oil | MEL variant (purified) | 3.8[3] |
| *Pseudozyma crassa* | Oleic acid and glucose | Diastereomer MEL-A (main), MEL-B, MEL-C | 4.6[3] |
| *Pseudozyma graminicola* | Soybean oil | MEL-A, MEL-B, MEL-C (main) | 9.6[3] |
| *Pseudozyma hubeiensis* | Soybean oil | MEL-A, MEL-B, MEL-C (main) | 9.6[3] |
| *Pseudozyma hubeiensis* | Soybean oil | MEL-A, MEL-B, MEL-C (main) | 76.3[3,4,5] |
| *Pseudozyma parantarctica* | Soybean oil | MEL-A (main), MEL-B, MEL-C | 106.7[3,4] |
| | Glucose | Mono-acylated MEL (purified) | 1.2 |
| | Soybean oil | Tri-acylated MEL (purified) | 22.7 |
| | Olive oil and mannitol | MAL (purified) | 18.2 |
| | Olive oil and arabitol | MML (purified) | ND |
| | Olive oil and ribitol | MRL (purified) | ND |

*Table 2 contd. ...*

*...Table 2 contd.*

| Microbial Producers | Carbon Sources | Glycolipids | Yield (g/L) |
|---|---|---|---|
| *P. antarctica* | Soap stock and whey permeate/molasses | Mixture of MEL | 40[5] |
| *Pseudozyma siamensis* | Safflower oil | MEL-B, MEL-C (main) | 18.5[1] |
| *Pseudozyma tsukubaensis* | Soybean oil | Diastereomer MEL-B | 73.1[4,5] |
| *Ustilago cynodontis* | Soybean oil | MEL-C | 1.4 |
| *Ustilago maydis* | Sunflower oil | MELs and cellobiose lipids | 30[6] |
| *Ustilago scitaminea* | Sugarcane juice | MEL-B | 25.1 |
| *Pseudozyma graminicola* CBS 10092 | Soybean oil | MEL-A, MEL-B, MEL-C (main) | 10 |

[1] As a mixture of MELs
[2] Feeding using resting cells
[3] Large scale productions with jar-fermenter
[4] As a mixture of MELs and cellobiose lipids
[5] Crude forms
[6] MEL as major and Cellobiose lipid as minor
ND: no data

## Effect of source of carbon and nitrogen on MEL production

The carbon and nitrogen sources are the vital nutrients for the growth and development of MEL producers. Carbon sources comprise of water soluble sugars, hydrocarbons, vegetable oils and agro-industrial wastes. Sugars such as glucose and sucrose which are water soluble, when used for MEL production, simplify the process of purification but leads to poor product yield when compared to other sources. The type of the carbon sources employed in turn affect the fatty acid content of MEL with very least effect on the hydrophilic group. The vegetable oils tested for the production of MELs include soybean, safflower, palm, corn, olive, rapeseed and coconut, out of which the soybean oil is widely used for increased production. For example, *P. parantarctica* JCM 11752 and *P. rugulosa* NBRC 10877 have produced the highest yield of MEL using soybean oil as carbon source while use of other oils resulted in less yield (Marques et al. 2009).

When vegetable oils were employed, the type of MEL produced were esterified with varying amounts of fatty acids when compared to the ones esterified upon using soluble sugars such as glucose and sucrose. In such instances, the MEL-A predominantly produced by *P. antarctica* strains had medium to long chain fatty acids linked to mannose moiety and contributing to increased hydrophobicity of the surfactant (Kim et al. 2006, Kitamoto et al. 1992). In addition, different MELs with varying proportions of saturated and unsaturated fatty acids are produced when using vegetable oils. Apart from vegetable oils, water insoluble substrates, namely, hydrocarbons (*n*-alkanes) in the range of C4 to C20 have been utilized as carbon source for MELs production. With intermittent feeding of *n*-octadecane (6% v/v), the resting cells of *P. antarctica* produced up to 140 g/L of MELs (Kitamoto et al. 2001). Recently, production of MELs has been attempted by using agro-industrial waste such as waste frying oil, cellulose from Avicl (commercial cellulose powder) and wheat straw and honey, with the highest yield being reported from waste frying oil (107 g/L) (Dziegielewska et al. 2013). Such renewable carbon sources can be a suitable carbon source for sustainable production of BSs at low production cost.

In addition to carbon source, nitrogen equally affects the MELs yield. While the use of ammonium nitrate and ammonium sulphate resulted in low yield, sodium nitrate added a 0.3% increase in yield and is considered as the best source of nitrogen. In addition, other nitrogen sources used in MELs production include urea and yeast extract. With olive oil and sodium nitrate as C and N sources respectively, *U. scitaminae* NBRC 32730 produced 12.8 g/L of MEL (Morita et al. 2009a).

## Production of TLs

### Biosynthetic pathway

According to Lang and Philp (1998), the TLs biosynthesis involves independent synthesis of carbohydrate and fatty acid moieties followed by esterification. Glucose-6-phosphate and trehalose UDP-glucose function as precursors for synthesis of carbohydrate (Lang et al. 1998). Trehalose-6-phosphate synthetase (TPS) is the key enzyme which synthesises final sugar residue by linking the two glycopyranosyl units at C1 and C1' positions. In the case of *M. tuberculosis*, the formation of mycolic residues (carbohydrate) takes place by Claisen-condensation catalyzed by the TPS enzyme. The final product α,α'-trehalose 6,6'–dicorynomycolates is formed by the stepwise esterification of completely formed lipid moiety (corynomycolic acid) to α,α'-trehalose (Kretschmer and Wagner 1983). The *n*-alkane acts as the initial substrate for the production of mono and dimycolate in the pathway. Hitherto, sufficient work is not done to unravel the complete biosynthetic pathway leading to TL formation, especially production from different carbon substrates.

Most of the studies report the use of *n*-alkanes as carbon source with other mineral salts which include $KH_2PO_4$, $K_2HPO_4.3H_2O$, $(NH_4)_2SO_4$ and $MgSO_4.7H_2O$ and further supplemented with salts of Ca, Mn, Ni, Zn and Fe. The cultures were incubated in the medium (pH 7.0) at 30°C at 180 rpm for 7 days. To purify the product, liquid-liquid extraction technique was adapted wherein the fermented broth after sonication was extracted with methyl tert-butyl ether. After separating and evaporating the organic solvent, the crude extract was suspended in methanol:water (3:1 v/v) mixture. The residual *n*-alkane fraction was removed by *n*-hexane and the methanol was evaporated to obtain crude TLs (Christova et al. 2015).

### Effect of C and N sources and growth conditions on TLs production

*Rhodococcus* sp. predominantly produces TLs by utilising *n*-alkanes. The glycolipid formed is cell bound and not secreted in to the surrounding medium. Several sp. of *Rhodococcus* utilizes a range of hydrocarbons (*n*-alkanes) with the least growth being observed when supplied with short-chain *n*-alkanes ($C_5$–$C_7$) (Lang et al. 1998). The range of *n*-alkanes tested for TLs production is *n*-octane ($C_8$) to *n*-heptadecane ($C_{17}$). In a study, it was observed that the TLs production is directly proportional to the chain length of *n*-alkanes wherein the surfactant concentration reached a maximum of 3.1 g/L with carbon source in the form of *n*-hexadecane (Tuleva et al. 2008). Naphthalene and diesel oil were shown to be the best hydrocarbons for optimal production of TLs by *R. ruber* and *R. erythropolis* (Haddadin et al. 2009). For the first time, an extracellular TL was reported when *n*-hexadecane was used alone as carbon source by the *Rhodococcus* sp. SD-74 (Tokumoto et al. 2009). Apart from hydrophobic *n*-alkanes, the production of TLs is also reported with water soluble substrates. Upon utilisation of glycerol, *R. erythropolis* ATCC 4277 was able to release TLs into the surrounding medium (Pacheco et al. 2010).

By utilising *n*-hexadecane as carbon substrate, growth kinetics studies were performed on *R. wratislaviensis* which indicated that TLs formation is growth associated and the substrate depletion is associated with increase in the biomass (Tuleva et al. 2008). Similar scenario of substrate consumption, product formation and biomass production was observed with *R. ruber* and *R. erythropolis* strains (Rapp et al. 1979, Philp et al. 2002, Haddadin et al. 2009). In addition to carbon, nitrogen also affected the TLs production. When *R. erythropolis* DSM432215 was cultured under nitrogen limited conditions, it favored the growth as well as accumulation of TLs (Ristau and Wagner 1983). Unlike production of MELs, till 2008 no attempts were made to utilize cost-effective and renewable resources such as vegetable oils for TLs production to make the process economically feasible. The extracellular glycolipid BS produced by *R. eryhtropolis* 16 LM.USTHB upon cultivation using residual sunflower frying oil was able to reduce water surface tension to 31.9 mN/m (Sadouk et al. 2008). Ruggeri et al. (2009) were able to culture *Rhodococcus* sp. BS32 and produce TLs utilizing rapeseed oil.

In addition, several optimization studies have been reported in literature. When *R. erythropolis* SD-74 was cultured under neutral pH in a very high phosphate buffer, it produced 40 g/L of TLs (Uchida et al. 1989). In addition, response surface methodology, a statistical tool was employed to enhance the yield of TLs by three folds (Mutalik et al. 2008). In another experiment, use of central composite design increased the yield of TLs by five folds in *Gordonia* sp. BS29 culture (Franzetti et al. 2009).

## *Production of CBLs*

CBLs, also known as ustilagic acid, are predominantly produced by *Ustilago* sp. The initial attempts were made to produce CBLs in *Ustilago* sp. using glucose as carbon source with both growing as well as resting cells and later by employing various renewable resources. Coconut oil (2% v/v) was the first renewable source used in CBLs production to obtain a specific yield of 0.79 g/g (Frautz et al. 1986).

### *Extraction and purification of CBLs*

The widely used method of extraction here is very similar to that of MELs, extraction with ethyl acetate followed by purification in silica gel column chromatograph using chloroform/acetone solvent mixture (Morita et al. 2011). Another method used in the purification of CBL was solvent extraction wherein the culture broth was treated with methanol after lyophilisation for 4–5 days at 5°C. After solvent evaporation, the product was suspended in deionized water and kept for 24 h at 5°C. The resulting precipitate was separated by filtration, washed three times with deionized water and suspended in methanol (Kulakovskaya et al. 2010). Qualitative analysis was carried out using thin layer chromatography for comparison of Rf values. Acid degradation of the glycolipid was performed to analyse the sugar components. In another example, purified CBLs were treated with HCl-Methanol reagent overnight at room temperature. After quenching the reaction with water, the methyl ester derivatives were extracted with *n*-hexane. The fraction containing sugar molecules were neutralized with NaOH and then analysed in HPLC (Morita et al. 2011).

# Applications of Glycolipid BSs

Low molecular weight glycolipids exhibit many interesting functional behaviour such as reducing the surface and interfacial tension of water, emulsification and de-emulsification properties, foaming potency, solubilization, self-assembling and pore-forming abilities. In addition, glycolipids exhibit versatile biological properties such as antimicrobial, haemolytic, antiviral, anti-carcinogenic and immune-modulating activities (Banat et al. 2014, De Rienzo et al. 2014, 2015, 2016). Physico-chemical properties of glycolipids are responsible for their stability at extreme conditions which include pH, salinity and temperature. Owing to their biodegradability, low or non-toxic nature, mild production conditions and being bio-based product, glycolipid BSs are believed to effectively replace the non-biodegradable, toxic and synthetic ones in the near future (Fracchia et al. 2014). Depending on their physico-chemical properties, glycolipids find useful applications in pharma, food, petrochemical industries and for drug or gene delivery. Apart from above applications, glycolipid BSs can be well exploited for solubilising, mobilising and biodegrading water insoluble hydrocarbons (Ines and Dhouha 2015).

## *Applications of MELs*

MELs being bio-based surfactants, exhibit versatile biochemical interactions and excellent interfacial properties which make them right candidates for application in many fields. MELs not only exhibit surface tension-reducing and interfacial activity but also versatile biochemical functions (Yu et al. 2015). The varying structural features with respect to number of acetyl groups, type of fatty acids (saturated and unsaturated) and their proportions render MELs with diverse functional properties mainly in the fields of pharma and environmental remediation (Morita et al. 2013c). MEL finds applications as oilfield chemicals, personal care and food processing products, household detergents, industrial and chemicals, and as textile products (Figure 4).

## *Biological activities of MELs*

The important biological activities of MELs include antimicrobial, anti-inflammatory, anti-carcinogenic, induction of cell differentiation and binding to proteins (Morita et al. 2015). Gram positive bacteria are readily inhibited by MEL-A and B whereas they exhibit very weak and no activity against Gram-negative bacteria and fungi, respectively. It has been reported that the mannopyranosyl moiety of MEL is the

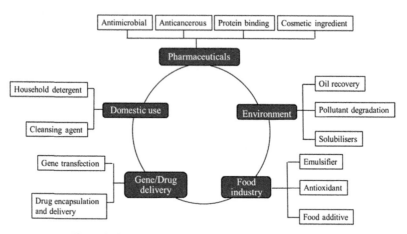

**Figure 4.** Overview of applications of MELs in various sectors.

most effective in acting against microbes and any further chemical modification of this sugar moiety might increase its effectiveness (Kitamoto et al. 1993). In addition, both MELs exhibit ability to inhibit cell proliferation and induce cell-differentiation in human leukemia cells such as K562 (Isoda and Nakahara 1997), HL60 (Isoda et al. 1997) and the KU812 (Kitamoto et al. 2002). At a concentration of 5 and 10 μM, the MEL-A and MEL-B show inhibitory activity against HL60 cells by inducing functional changes to their morphology. Studies reveal that the differentiation of granulocytes is induced by both MELs at a concentration of 10 mg/mL and inhibit the activity of phospholipid and $Ca^{2+}$-dependent protein kinase C in HL60 cells (Isoda and Nakahara 1997). Other mechanisms of differentiation by MELs include inhibition of serine/threonine phosphorylation of 30 kDa protein in HL60 cells as well as tyrosine phosphorylation of 55-, 65-, 95-, 135-kDa proteins in the K562 cells (Kitamoto et al. 2002). Through phosphate cascade systems, the MELs interfere directly with the signal transduction in cytosolic regions. The MELs exhibit ability to induce neuronal growth and their differentiation in rat pheochromocytoma PC-12 cells. Also, MEL-A exhibits anti-cancer property wherein upon treatment of the B16 (mouse melanoma) cells with MEL-A, typical symptoms of apoptotic behaviour were observed. In addition, the hall marks of apoptotic cell, namely, condensation of chromatin, fragmentation of DNA and the arrest of cells at the sub-G1 phase occurs at 10 μM concentration of MEL-A. MEL-A cause enhanced production of melanin in addition to inducing tyrosinase activity. The triggering of this apoptotic cell death was later revealed to be through a signalling pathway involving protein kinase ($C_{\alpha} PKC_{\alpha}$) (Zhao et al. 2001).

   In addition to these biological interactions, MELs also exhibit their potential in increasing the rate of gene transfection. CLSM (confocal laser scanning microscopy) and FITC (Fluorescein isothiocyanate) studies indicate that anti-sense DNA encapsulated in MEL-A conjugated to cationic liposome (cholesteryl-3h-carboxyamido ethylene-N-hydroxyethylamine) was able to localise to the nucleus of transfected cells. Complete transfer of encapsulated DNA was achieved with the fusion of membrane between cationic liposomes and the target cells induced by

MEL-A. In the same study, MEL-B and MEL-C failed to increase the transfection efficiency (Inoh et al. 2004). In another study, MEL-A was used as a ligand to bind the serum immunoglobulin (Ig). MEL-A can bind to the human immunoglobulin G (HIgG) with similar affinity as that of bovine ganglioside GMI. The amount of MEL-A attached to poly-HEMA (2-hydroxy ethyl methacrylate) for further binding to HIgG followed a dose dependent behaviour. Absorbance studies further suggested that the amount of MEL-A bound to HIgG increased with increase in its amount and finally reaching a plateau. The binding mechanism studies indicated that MEL-A preferably binds to the Fab region of the HIgG antibody and not to its Fc region, thereby exhibiting high affinity (Im et al. 2001).

## Environmental applications of MELs

Properties such as low toxicity, biodegradability and ability to enhance the solubility of poorly soluble compounds make MELs to be very promising BS for industrial applications. Rising concern about the use of synthetic surfactants in various sectors is anticipated to boost the demand for MELs, particularly due to their low toxicity (Yu et al. 2015). The growth of the global MELs market is mainly driven by the rising demand for environment-friendly products, promise for use in wide range of applications and as a natural ceramide substitute.

Solubilisation or degradation of water insoluble hydrocarbons is a big problem which drives the pollution levels in the environment significantly. BSs being amphiphiles are able to alter the physico-chemical conditions at the interfaces affecting the distribution of the chemicals amongst the phases. For example, soil contaminated with hydrocarbons is mainly composed of six phases, namely, soil particles, water, bacteria, immiscible liquid, air and solid hydrocarbons. The hydrocarbons further exist in different phases adsorbed or absorbed to soil particles, solubilized in the water phase, as a free or insoluble phase and absorbed to cell surfaces. BSs when added to this system can interact with both the abiotic particles and the bacterial cells and alter the equilibrium composition (Banat et al. 2010, Perfumo et al. 2016). Beal and Bets (2000) have reported the initial role of RLs in the biodegradation of organic contaminants. The study showed an increase in the cell-surface hydrophobicity of a BS-producing strain when compared to the non-BS producing strain upon culturing with hexadecane. Further, the solubility of hexadecane increased from 1.8 to 22.8 mg/L (Yu et al. 2015). Enhanced biodegradation by the BS producing strain is attributed to two mechanisms, namely, (1) ability to increase the solubility of the substrate which in turn helps substrate attachment to cell surface and (2) increasing the cell surface hydrophobicity, which allows hydrophobic substrates to associate easily with the microorganisms (Zhang and Miller 1992, Zhang et al. 1997, De Almeida et al. 2016). Similar to RLs, MELs are also glycolipid BSs with similar amphiphilic structure but their lower critical micellar concentration (CMC) and higher production can be advantageous over the former. MELs are already known to effectively solubilise and utilise *n*-alkanes ranging from $C_{12}$–$C_{18}$ with a yield of 0.87 g/L with 6% (v/v) *n*-octadecane after 7 days of incubation. The highest yield (of 140 g/L) was obtained with intermittent feeding of the same substrate (Table 2). Even MELs could degrade kerosene and petroleum refinery by product, up to an extent of

87% within 15 h (Hua et al. 2004). Sajna et al. (2015) observed supplementation of *Pseudomonas putida* with strain of MELs producing *Pseudozyma* sp. NII 08165 which enhanced the degradation of crude oil wherein the maximum degradation was achieved when supplemented with 2.5 mg/L of BS. Degradation of $C_{10}$–$C_{24}$ up to ~ 46% was observed in a culture broth consisting of *P. putida* supplemented with *Pseudozyma* sp. NII 08165.

Apart from hydrophobic *n*-alkanes, amphiphilic toxicants including phenols could pose potential threat to humans as well as environment. Several bioremediation technologies developed for phenolics include (i) biopiles (ii) enhanced natural attenuation (iii) sequential A/O treatment and (iv) composting. Among these, remediation through composting is the efficient one, but phenolic toxicants can exert detrimental effect on microbes by disrupting their membranes and enzyme system (Liu et al. 2012a). Previous studies reported that rhamnolipid treatment given prior to degradation has increased the adsorption of phenol by *P. simplicissimum* (Liu et al. 2012b). In another study, mono-rhamnolipid favoured the growth of the organisms by inhibiting the phenol toxicity to cells indicating its ability to repair membrane. Since MELs have similar amphiphilic structure to that of ceramide-3 (essential component of intracellular lipids of stratum corneum), MELs are able to assemble into various crystalline structures in aqueous solution (Arutchelvi et al. 2008) and so can be used for repairing the damaged cell membrane. The mild production conditions, high yield and desirable interfacial properties of MELs are the added advantages over RLs and thus can efficiently enhance the adsorption of phenol by the microorganisms (Morita et al. 2013a). In addition to the degradation of hydrophobic compounds, MELs also enhance recovery of oil from spill sites. Andrade et al. (2016) compared the efficiency of MEL-B with surfactin, a lipopeptide BS, through oil displacement test wherein the former showed a higher clear zone with light, medium and heavy oils when compared to the latter BS. Highest clear zone (6.78 cm²) was obtained with MEL-B treated with heavy oil when compared to that treated with surfactin (4.49 cm²). Overall, the study proved that MEL-B is better than surfactin for microbially enhanced oil recovery. Similar to other glycolipid BSs, MELs are also thought to have a potential role in oil pollution control, oil storage tank clean-up and enhanced oil-recovery (Safdel et al. 2017).

Madihalli et al. (2016) recently reported a new application of MEL-A, namely as a pour point depressant (PPD) of hydrocarbon fuels and biodiesel. The study demonstrated the ability of MEL-A in preventing the crystallisation of engine fuels including biodiesel at reduced temperatures, thus improving their low temperature fluidity. When MEL-A was blended with biodiesel and commercial and refinery grade diesel samples at a concentration of 0.3% (above the CMC), it was able to reduce the cloud point (the lower temperature at which fuel turns turbid and forms gel) of them by 3.2, 9.0, and 1.5°C, respectively (Figures 5 and 6). XRD and IR studies unravelled the possible mechanism of action of MEL-A. MEL-A physically associates with the fatty acid methyl esters of biodiesel and delays the nucleation, an onset for crystallisation. MEL-A being a bio-based product can be a sustainable and eco-friendly alternative to existing polymeric synthetic pour point depressants that are currently being commercially used (Madihalli et al. 2016).

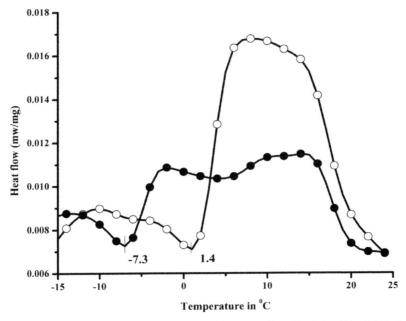

**Figure 5.** Differential Scanning Calorimetry (DSC) melting curves for unblended and blended biodiesel. Containing 0.3% MEL-A: (○) biodiesel and (●) biodiesel/MEL-A. Reprinted with permission from Madihalli et al. (2016). Copyright 2016 American Chemical Society.

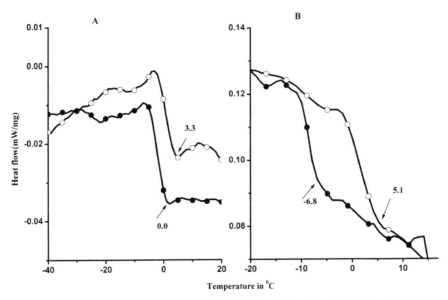

**Figure 6.** DSC melting curve for unblended and blended diesel. Containing 0.3% MEL-A of (A) CGD and (B) RGD: (○) diesel and (●) diesel/MEL-A. Reprinted with permission from Madihalli et al. (2016). Copyright 2016 American Chemical Society.

## Cosmetic applications

*Skin moisturisation*: The similarity of MELs to ceramide-3 and their ability to form various liquid crystalline phases including the lamellar phase impart them their skin care properties. Morita et al. (2009b) with 3-D cultured human skin model, TESTSKIN™, evaluated the moisturising activity of MELs on the basis of cell viability (Imokawa et al. 1989). The human skin cells were initially damaged by treating them with sodium dodecyl sulphate (SDS) and then the effect of MELs to repair them was compared with ceramide-3 (Figure 7). MELs, preferably the diastereomeric form of MEL-B, was more effective than ceramide-3 and exhibited excellent water retention properties (Sugibayashi et al. 2004).

Ceramide-like properties possessed by MELs play a role in protecting and repairing the hair fibres. Electron microscopic studies of hair repair by both MEL-A and MEL-B showed similar effect in repairing the damaged hair which is naturally performed by ceramides present in hair cuticle (Morita et al. 2010a). Increase in overall strength and sustained average friction coefficient of damaged hair were observed when treated with MELs suggesting their use as cosmetic ingredient in hair care products. In addition to the cell repairing abilities, MELs also have the ability to activate cells and induce their differentiation. The fibroblasts and papilla cells

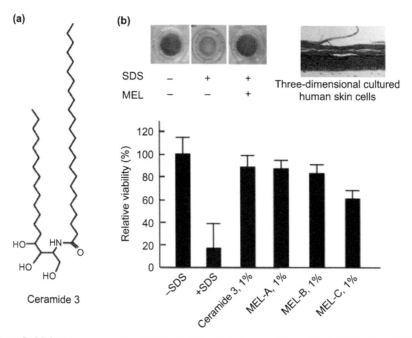

**Figure 7. Moisturizing property of MELs.** (a) Chemical structure of ceramide-3. (b) Relative viability of the cultured skin cells treated with SDS. The cultured skin cells is treated with 1% SDS, washed to remove traces of SDS, and then retreated with different MELs. MTT assay (colorimetric method) was done to predict cell viability. Ceramide-3 was used as the positive control.–SDS: non-treated with SDS, +SDS: treated with SDS. Reprinted with permission from Morita et al. (2009b). Copyright 2009.

were significantly activated upon treatment with MEL-A leading to the formation of follicles and hair growth through trans-differentiation of the adult epidermis. It was observed that relative to control, MEL-A treated cells exhibited more than 150% viability (Morita et al. 2010a). Anti-oxidative property of MELs was studied using 1,1-diphenyl-2-picryl hydrazine (DPPH) free radical- and superoxide anion-scavenging assays. All the three MELs tested showed anti-oxidant activity *in vitro*, but at lower levels when compared to arbutin, one of the well-known anti-oxidant. Of the three MELs, MEL-C showed the highest rate of DPPH and super oxide anion-scavenging activities (50.3% at 10 mg/mL and > 50% at 1 mg/mL, respectively).

Further investigations involved determining the ability of MEL-C to scavenge the peroxide and superoxide radicals from $H_2O_2$ induced oxidative stress in the skin fibroblasts. It is reported that MEL-C showed better protective activity (30.3%) when compared to arbutin (13%) at a concentration of 10 µg/mL (Takahashi et al. 2012). In another study, MEL-A was used to prevent the crystallization of pseudo-ceramide (1,3-bis-(N-(2-hydroxyethyl)-palmitoylamino)-2-hydroxypropane, PC-104) molecules as well as stabilize the emulsions. Pseudo-ceramide resembles, structurally, the ceramide component of the stratum corneum and thus it can be used to develop skin care cosmetics to overcome the scarcity of ceramide purified from animal sources. But the disadvantage of its usage is that it tends to crystallise even at a concentration of 1% leading to the destabilization of pharmaceutical formulation. Kim et al. (2014) showed that the MELs structurally being asymmetric, physically associates with PC-104 and prevents the crystallization. In addition, MELs could stabilize the emulsion for up to 4 weeks when mixed in equal proportion (1/1 w/w).

## *Applications of TLs*

Like other glycolipid BSs, TLs are finding numerous applications because of their unique multifunctional features. Structurally, they are very diverse which make them an attractive group of compounds for a variety of industrial, biotechnological and biomedical applications. Further, their use as green alternative and eco-friendly product adds up to their advantages. Figure 8 summarises the various applications of TLs.

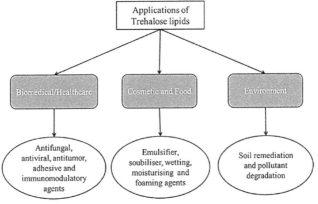

**Figure 8.** Overview of TLs applications.

## Environment applications

Generally, environmental contamination arises due to pollution caused by the insoluble organic hydrocarbons. Such compounds tend to bind to soil particles and remain insoluble, thus leading to toxicity. Bioavailability to microorganisms which can degrade them is the only alternative and ecofriendly way to remove these toxicants from soil. Glycolipid BSs, in particular, TLs, are proved to be one of the promising bio-based chemical entities that could solubilize and promote the degradation of these soil toxicants. Broadly, two mechanisms have been reported for BS-enhanced soil cleaning wherein, the first one is to increase the contact angle between the soil and the hydrophobic containment which happens below the CMC of the surfactant and the other is solubilisation of the hydrocarbon that occurs above CMC of the surfactant (Kuyukina and Ivshina 2010b). Initial studies by Page et al. (1999) on *Rhodococcus* strain H13-A revealed that compared to synthetic counterpart, Tween 80, the BS from strain H13-A is 35-fold more effective in solubilizing the polycyclic aromatic hydrocarbons (PAHs) and transferring them into aqueous phase. Another study also showed that TLs from some *Rhodococcus* strains are very effective in oil recovery from the oil contaminated sites such as sands and oil shales (Ivshina et al. 1998). Kuyukina et al. (2005) reported that TLs from *R. ruber* IEGM 231 exhibited 1.9–2.3 times more oil-removal activity than its synthetic counterpart Tween 60. TL produced by *R. ruber* using dodecane is more effective in oil removal in cold conditions, compared to the TLs produced using hexadecane. The latter is more effective at high temperatures (22–28°C). In addition, the TL BS from *R. ruber* is less toxic than the purified BSs from other bacterial strains and synthetic surfactants (Kuyukina et al. 2010a).

## Biomedical applications

BSs are considered to be safer chemicals for medical applications since they are bio-based products. The important biological properties of BSs include antimicrobial, antiviral, anti-adhesive, anticancer and immunomodulation properties (Fracchia et al. 2012, Fracchia et al. 2015). Preliminary studies on toxicity of TLs revealed that they are less irritating than SDS and thus could be used in cosmetic applications (Marques et al. 2009). The two succinoyl-TLs, namely, STL-1 and STL-2, have been investigated and found to induce cell differentiation into monocytes instead of proliferation in the human promyelocytic leukemia cells (Isoda et al. 1997). Four structural analogues of STL-3 (succinoyltrehalose lipid-3) when tested for their ability to prevent cell proliferation and induce differentiation in the same cell type, indicated the importance of structure of the hydrophobic moiety in the BS (Isoda et al. 1997). The TLs-BS *Rhodococcus erythropolis* possess high physicochemical stability and their hydrophobic nature make them ideal candidates for use in food and pharmaceutical industries, and in cosmetics applications (Lang et al. 1998). In addition, TLs have shown de-emulsification properties which can be used for breaking both w/o and o/w emulsions usually formed during oil recovery and processing (Kuyukina et al. 2010a).

Trehalose-6,-6'dimycolate (TDM) produced by *Mycobacterium* sp. expressed anti-tumor property, immunomodulating functions, priming of murine macrophages to produce nitric oxide, enhancement of angiogenic activity in mice and induction of the production of cytokines (Franzetti et al. 2010). Though the TLs of *Mycobacteria* sp. possess excellent biological activities, the pathogenicity and relatively high toxicity of their source hinders their potential use.

## *Applications of CBLs*

### *Antimicrobial activity*

Among other glycolipid BSs, CBLs are known for their fungicidal property, and their effective concentrations leading to cell death are between 0.01 to 0.15 mg/mL against *Filobasidiella neoformans* and *Candida tropicalis*, respectively (Kulakovskaya and Kulakovskaya 2013). In another study, the fungicidal activity of CBLs was compared with that of SLs, another class of glycolipid BS since both share the same structural features. MIC of CBLs against *Filobasidiella neoformans* and *Candida tropicalis* were 0.005 and 0.04 mg/mL, respectively when compared to 1 and 15 mg/mL, with SLs respectively, with SLs. The MIC value for CBLs against bacterial strains was higher, ranging from 10–30 mg/mL, when compared to other BSs (Kulakovskaya et al. 2014). Kulakovskaya et al. (2009) studied the fungicidal effects of CBLs produced from *Cryptococcus humicola* and *Pseudozyma fusiformata* against different yeasts including pathogenic *Cryptococcus* and *Candida* species. Mycelial cells of *Filobasidiella neoformans* were completely eradicated in 30 mins when treated with 0.02 mg/mL CBLs while pathogenic yeasts, including *Candida* species needed a concentration of 0.1–0.3 mg/mL for their eradication. Table 3 summarizes the MIC concentrations of CBLs for different classes of fungi. In addition to fungicidal activity, CBLs also exhibit antibacterial property. Memee et al. (2009) reported the antibacterial effect of CBL (flocculosin) produced from

**Table 3.** Antimicrobial activity of cellobiose lipids produced by *C. humicola, P. fusiformata* and *P. flocculosa* against fungal and bacterial cultures.

| Test Culture | Concentration (mg/mL) | | Bacterial Cultures | Concentration (mg/mL) |
| --- | --- | --- | --- | --- |
| | CBLs of *Cr. humicola* | CBLs of *Ps. fusiformata* | | CBLs of *P. flocculosa* |
| *Cryptococcus terreus* | 0.02 | 0.02 | *Pediococcus acidilactici* | 0.32 |
| *Filobasidiella neoformans* | 0.02 | 0.02 | *Bacillus cereus* *Clostridium clostridiiforme* | 0.01 0.08 |
| *Clavispora lusitaniae* | 0.15 | 0.1 | *Staphylococcus aureus* | 0.01 |
| *Candida viswanathii* | 0.15 | 0.1 | *S. aureus* MRSA | 0.013 |
| *Candida albicans* | 0.15 | 0.15 | *Bacteroides ovatus* | 0.16 |
| *Candida glabrata* | 0.3 | 0.3 | | |
| *Saccharomyces cerevisiae* | 0.3 | 0.3 | | |

*Pseudozyma flocculosa.* Gram positive bacteria were more susceptible to CBLs than Gram negative ones and the MIC varied from as minimum as 0.01 to 0.32 mg/mL (Table 3).

The fungicidal property of CBLs is due to their ability to cause membrane damage (Puchkov et al. 2002, Kulakovskaya et al. 2013). The CBLs treated target fungal cells were tested for intracellular ATP content. The amount of the ATP in the control cells was stable after 15 min of incubation while it reduced by 2 and 4 folds in the cells treated with 0.04 mg/mL of CBLS and 4 mg/mL SLs, respectively (Kulakovskaya et al. 2014). The membrane damage causing ability was further confirmed through phase contrast and florescence microscopic studies using ethidium bromide dye (Puchkov et al. 2002). The potential of CBLs to kill many pathogenic fungi when compared to other BSs make them promising bio-based product for preventing spoilage of food by fungi (Imura et al. 2014).

## Conclusion

BSs are becoming promising bio-based chemical entities which can replace the synthetic surfactants. Among BSs, the glycolipids are widely accepted because of their biodegradability, low toxicity, excellent surface activity and interfacial properties, biochemical interactions and self-assembling properties. Three glycolipids discussed here can be potent candidates in biomedicine and therapeutics for their antimicrobial, hemolytic, antiviral, anticarcinogenic and immunemodulating activities. Since CBLs exhibit a very low MIC against diverse fungal pathogens, they can be an efficient pharmaceutical ingredient as well as food preserving agent. Among the three, MELs are studied widely and exploited for use in large number of applications. Their excellent moisturizing activity, skin care, hair repair properties after damage, ease of production (higher yield), and mild production conditions make them the best candidates for cosmetic applications. Further research involving increasing the production of these essential glycolipids using cost-effective renewable resources, scaling up to large sized fermenters, and finding newer applications in various sectors can effectively make these bio-based products to replace their chemical counterparts.

## Acknowledgement

Chandraprasad Madihalli acknowledges IIT Madras for providing opportunity to pursue my PhD in department of biotechnology and Management, BMS College of engineering, Bangalore for sponsorship.

## References

Arutchelvi, J., S. Bhaduri, P.V. Uppara and M. Doble. 2008. Mannosylerythritol lipids: a review. J. Ind. Microbiol. Biot. 35: 1559–1570.

Arutchelvi and M. Doble. 2011. Mannosylerythritol lipids: microbial production and their applications. pp. 145–177. *In*: G. Soberon Chavez (ed.). Biosurfactants: From Genes to Applications. Springer, Berlin, Heidelberg.

Banat, I.M., R.S. Makkar and S.S. Cameotra. 2000. Potential commercial applications of microbial surfactants. Appl. Microbiol. Biot. 53: 495–508.

Banat, I.M., A. Franzetti, I. Gandolfi, G. Bestetti, M.G. Martinotti, L. Fracchia, T., J. Smyth and R. Marchant. 2010. Microbial biosurfactants production, applications and future potential. Appl. Microbiol. Biot. 87: 427–444.

Banat, I.M., M.A.D. De Rienzo and G.A. Quinn. 2014. Microbial biofilms: biosurfactants as antibiofilm agents. Appl. Microbiol. Biot. 98: 9915–9929.

Beal, R. and W.B. Betts. 2000. Role of rhamnolipid biosurfactants in the uptake and mineralization of hexadecane in *Pseudomonas aeruginosa*. J. Appl. Microbiol. 89: 158–168.

Boyd, J.V., C. Parkinson and P. Sherman. 1972. Factors affecting emulsion stability, and the HLB concept. J. Colloid Interf. Sci. 41: 359–370.

Christova, N., S. Lang, V. Wray, K. Kaloyanov, S. Konstantinov and I. Stoineva. 2015. Production, structural elucidation and *in vitro* antitumor activity of trehalose lipid biosurfactant from *Nocardia farcinica* strain. J. Microbiol. Biotechnol. 24: 439–447.

De Almeida, D.G., F. Rita de Cassia, J.M.L. Silva, R.D. Rufino, V.A. Santos, I.M. Banat and L.A. Sarubbo. 2016. Biosurfactants: promising molecules for petroleum biotechnology advances. Front. Microbiol. 7: 1718.

De Rienzo, M.A.D., B. Dolman, F. Guzman, C. Kaisermann, J. Winterburn, I.M. Banat and P. Martin. 2014. Antimicrobial properties of sophorolipids produced by *Candida bombicola* ATCC 22214 against gram positive and Gram-negative bacteria. New Biotechnol. S66–S67.

De Rienzo, M.A.D., I.M. Banat, B. Dolman, J. Winterburn and P.J. Martin. 2015. Sophorolipid biosurfactants: possible uses as antibacterial and antibiofilm agent. New Biotechnol. 32: 720–726.

De Rienzo, D., A. Mayri, P. Stevenson, R. Marchant and I.M. Banat. 2016. Antibacterial properties of biosurfactants against selected Gram-positive and -negative bacteria. FEMS Microbiol. Lett. 363.

Desai, J.D. and I.M. Banat. 1997. Microbial production of surfactants and their commercial potential. Microbiol. Mol. Biol. R. 61: 47–64.

Dziegielewska, E. and M. Adamczak. 2013. Evaluation of waste products in the synthesis of surfactants by yeasts. Chem. Pap. 67: 1113–1122.

Faria, N.T., M. Santos, C. Ferreira, S. Marques, F.C. Ferreira and C. Fonseca. 2014. Conversion of cellulosic materials into glycolipid biosurfactants, mannosylerythritol lipids, by *Pseudozyma* sp. under SHF and SSF processes. Microb. Cell Fact. 13: 155.

Faria, N.T., S. Marques, C. Fonseca and F.C. Ferreira. 2015. Direct xylan conversion into glycolipid biosurfactants, mannosylerythritol lipids, by *Pseudozyma antarctica* PYCC 5048 T. Enzyme Microb. Tech. 71: 58–65.

Fracchia, L., M. Cavallo, M.G. Martinotti and I.M. Banat. 2012. Biosurfactants and bioemulsifiers biomedical and related applications—Present status and future potentials. pp. 325–373. *In*: D.N. Ghista (ed.). Biomedical Science, Engineering and Technology. InTech, London, UK.

Fracchia, L., C. Ceresa, A. Franzetti, M. Cavallo, I. Gandolfi, J. Van Hamme, P. Gkorezis, R. Marchant and I. Banat. 2014. Industrial applications of biosurfactants. pp. 245–260. *In*: F.V.S. Naim Kosaric (ed.). Biosurfactants: Production and Utilization—Processes Technologies and Economics. CRC Press, Boca Raton, Florida.

Fracchia, L., J.J. Banat, M. Cavallo, C. Ceresa and I. Banat. 2015. Potential therapeutic applications of microbial surface-active compounds. AIMS Bioeng. 2: 144–162.

Franzetti, A., P. Caredda, P. La Colla, M. Pintus, E. Tamburini, M. Papacchini and G. Bestetti. 2009. Cultural factors affecting biosurfactant production by *Gordonia* sp. BS29. Int. Biodeterior. Biodegrad. 63: 943–947.

Franzetti, A., I. Gandolfi, G. Bestetti, T.J.P. Smyth and I.M. Banat. 2010. Production and applications of trehalose lipid biosurfactants. Eur. J. Lipid Sci. Technol. 112: 617–627.

Frautz, B., S. Lang and F. Wagner. 1986. Formation of cellobiose lipids by growing and resting cells of *Ustilago maydis*. Biotechnol. Lett. 8: 757–762.

Fukuoka, T., T. Morita, M. Konishi, T. Imura, H. Sakai and D. Kitamoto. 2007. Structural characterization and surface-active properties of a new glycolipid biosurfactant, mono-acylated mannosylerythritol lipid, produced from glucose by *Pseudozyma antarctica*. Appl. Microbiol. Biotechnol. 76: 801–810.

Fukuoka, T., M. Kawamura, T. Morita, T. Imura, H. Sakai, M. Abe and D. Kitamoto. 2008. A basidiomycetous yeast, *Pseudozyma crassa*, produces novel diastereomers of conventional mannosylerythritol lipids as glycolipid biosurfactants. Carbohyd. Res. 343: 2947–2955.

Haddadin, M.S.Y., A.A.A. Arqoub, I.A. Reesh and J. Haddadin. 2009. Kinetics of hydrocarbon extraction from oil shale using biosurfactant producing bacteria. Energy Convers. Manage. 50: 983–990.

Hua, Z., Y. Chen, G. Du and J. Chen. 2004. Effects of biosurfactants produced by *Candida antarctica* on the biodegradation of petroleum compounds. World J. Microbiol. Biotechnol. 20: 25–29.

Im, J.H., T. Nakane, H. Yanagishita, T. Ikegami and D. Kitamoto. 2001. Mannosylerythritol lipid, a yeast extracellular glycolipid, shows high binding affinity towards human immunoglobulin G. BMC Biotechnol. 1: 5.

Imokawa, G., S. Akasaki, Y. Minematsu and M. Kawai. 1989. Importance of intercellular lipids in water-retention properties of the stratum corneum: induction and recovery study of surfactant dry skin. Arch. Dermatol. Res. 281: 45–51.

Imura, T., S. Yamamoto, C. Yamashita, T. Taira, H. Minamikawa, T. Morita and D. Kitamoto. 2014. Aqueous gel formation from sodium salts of cellobiose lipids. J. Oleo Sci. 63: 1005–1010.

Ines, M. and G. Dhouha. 2015. Glycolipid biosurfactants: potential related biomedical and biotechnological applications. Carbohyd. Res. 416: 59–69.

Inoh, Y., D. Kitamoto, N. Hirashima and M. Nakanishi. 2004. Biosurfactant MEL-A dramatically increases gene transfection via membrane fusion. J. Control. Release 94: 423–431.

Isoda, H., D. Kitamoto, H. Shinmoto, M. Matsumura and T. Nakahara. 1997. Microbial extracellular glycolipid induction of differentiation and inhibition of the protein kinase C activity of human promyelocytic leukemia cell line HL60. Biosci. Biotech. Bioch. 61: 609–614.

Isoda, H. and T. Nakahara. 1997. Mannosylerythritol lipid induces granulocytic differentiation and inhibits the tyrosine phosphorylation of human myelogenous leukemia cell line K562. Cytotechnology 25: 191–195.

Ivshina, I.B., M.S. Kuyukina, J.C. Philp and N. Christofi. 1998. Oil desorption from mineral and organic materials using biosurfactant complexes produced by *Rhodococcus* sp. World J. Microb. Biot. 14: 711–717.

Kim, H.-S., J.-W. Jeon, B.-H. Kim, C.-Y. Ahn, H.-M. Oh and B.-D. Yoon. 2006. Extracellular production of a glycolipid biosurfactant, mannosylerythritol lipid, by Candida sp. SY16 using fed-batch fermentation. Appl. Microbiol. Biotechnol. 70: 391–396.

Kim, M.K., E.S. Jeong, K.N. Kim, S.H. Park and J.W. Kim. 2014. Nanoemulsification of pseudo-ceramide by molecular association with mannosylerythritol lipid. Colloid. Surface. B. 116: 597–602.

Kitamoto, D., T. Fuzishiro, H. Yanagishita, T. Nakane and T. Nakahara. 1992. Production of mannosylerythritol lipids as biosurfactants by resting cells of *Candida antarctica*. Biotechnol. Lett. 14: 305–310.

Kitamoto, D., H. Yanagishita, T. Shinbo, T. Nakane, C. Kamisawa and T. Nakahara. 1993. Surface active properties and antimicrobial activities of mannosylerythritol lipids as biosurfactants produced by *Candida antarctica*. J. Biotechnol. 29: 91–96.

Kitamoto, D., T. Ikegami, G.T. Suzuki, A. Sasaki, Y.-i. Takeyama, Y. Idemoto, N. Koura and H. Yanagishita. 2001. Microbial conversion of n-alkanes into glycolipid biosurfactants, mannosylerythritol lipids, by *Pseudozyma (Candida antarctica)*. Biotechnol. Lett. 23: 1709–1714.

Kitamoto, D., H. Isoda and T. Nakahara. 2002. Functions and potential applications of glycolipid biosurfactants-from energy-saving materials to gene delivery carriers. J. Biosci. Bioeng. 94: 187–201.

Kitamoto, D., T. Morita, T. Fukuoka, M.-a. Konishi and T. Imura. 2009. Self-assembling properties of glycolipid biosurfactants and their potential applications. Curr. Opin. Colloid In. 14: 315–328.

Konishi, M., T. Morita, T. Fukuoka, T. Imura, K. Kakugawa and D. Kitamoto. 2007. Production of different types of mannosylerythritol lipids as biosurfactants by the newly isolated yeast strains belonging to the genus *Pseudozyma*. Appl. Microbiol. Biotechnol. 75: 521.

Konishi, M., T. Morita, T. Fukuoka, T. Imura, K. Kakugawa and D. Kitamoto. 2008. Efficient production of mannosylerythritol lipids with high hydrophilicity by *Pseudozyma hubeiensis* KM-59. Appl. Microbiol. Biotechnol. 78: 37–46.

Kulakovskaya, W., I. Golubev, M.A. Tomashevskaya, E.V. Kulakovskaya, A.S. Shashkov, A.A. Grachev, A.S. Chizhov and N.E. Nifantiev. 2010. Production of antifungal cellobiose lipids by *Trichosporon porosum*. Mycopathologia 169: 117–123.

Kulakovskaya, E. and T. Kulakovskaya. 2013. Extracellular glycolipids of yeasts: biodiversity, biochemistry, and prospects. Academic Press, Massachusetts.

Kulakovskaya, E., B. Baskunov and A. Zvonarev. 2014. The antibiotic and membrane-damaging activities of cellobiose lipids and sophorose lipids. J. Oleo Sci. 63: 701–707.

Kuyukina, M.S., I.B. Ivshina, S.O. Makarov, L.V. Litvinenko, C.J. Cunningham and J.C. Philp. 2005. Effect of biosurfactants on crude oil desorption and mobilization in a soil system. Environ. Int. 31: 155–161.

Kuyukina, M.S. and I.B. Ivshina. 2010a. Application of *Rhodococcus* in bioremediation of contaminated environments. pp. 231–262. *In*: Hector, M. Alvarez (ed.). Biology of *Rhodococcus*. Microbiology Monographs Springer, Berlin Heidelberg.

Kuyukina, M.S. and I.B. Ivshina. 2010b. *Rhodococcus* biosurfactants: biosynthesis, properties, and potential applications. pp. 291–313. *In*: Hector, M. Alvarez (ed.). Biology of *Rhodococcus*. Springer, Berlin, Heidelberg.

Lang, S. and J.C. Philp. 1998. Surface-active lipids in *Rhodococci*. A. Van Leeuw. J. Microb. 74: 59–70.

Liu, Z.F., G.M. Zeng, H. Zhong, X.Z. Yuan, H.Y. Fu, M.F. Zhou, X.L. Ma, H. Li and J.B. Li. 2012a. Effect of dirhamnolipid on the removal of phenol catalyzed by laccase in aqueous solution. World J. Microb. Biot. 28: 175–181.

Liu, Z.F., G. Zeng, J. Li, H. Zhong, X. Yuan, Y. Liu, J. Zhang, M. Chen and Y. Liu. 2012b. Influence of rhamnolipids and Triton X-100 on adsorption of phenol by *Penicillium simplicissimum*. Bioresource Technol. 110: 468–473.

Madihalli, C., H. Sudhakar and M. Doble. 2016. Mannosylerythritol lipid-a as a pour point depressant for enhancing the low-temperature fluidity of biodiesel and hydrocarbon fuels. Energ. Fuel. 30: 4118–4125.

Marchant, R. and I.M. Banat. 2012a. Biosurfactants: a sustainable replacement for chemical surfactants? Biotechnol. Lett. 34: 1597–1605.

Marchant, R. and I.M. Banat. 2012b. Microbial biosurfactants: challenges and opportunities for future exploitation. Trends Biotechnol. 30: 558–565.

Marques, A.M., A. Pinazo, M. Farfan, F.J. Aranda, J.A. Teruel, A. Ortiz, A. Manresa and M.J. Espuny. 2009. The physicochemical properties and chemical composition of trehalose lipids produced by *Rhodococcus erythropolis* 51T7. Chem. Phys. Lipids 158: 110–117.

Mimee, B., R. Pelletier and R.R. Belanger. 2009. *In vitro* antibacterial activity and antifungal mode of action of flocculosin, a membrane-active cellobiose lipid. J. Appl. Microbiol. 107: 989–996.

Morita, T., M. Konishi, T. Fukuoka, T. Imura, S. Yamamoto, M. Kitagawa, A. Sogabe and D. Kitamoto. 2008. Identification of *Pseudozyma graminicola* CBS 10092 as a producer of glycolipid biosurfactants, mannosylerythritol lipids. J. Oleo Sci. 57: 123–131.

Morita, T., Y. Ishibashi, T. Fukuoka, T. Imura, H. Sakai, M. Abe and D. Kitamoto. 2009a. Production of glycolipid biosurfactants, mannosylerythritol lipids, using sucrose by fungal and yeast strains, and their interfacial properties. Biosci. Biotech. Bioch. 73: 2352–2355.

Morita, T., M. Kitagawa, M. Suzuki, S. Yamamoto, A. Sogabe, S. Yanagidani, T. Imura, T. Fukuoka, and D. Kitamoto. 2009b. A yeast glycolipid biosurfactant, mannosylerythritol lipid, shows potential moisturizing activity toward cultured human skin cells: the recovery effect of MEL-A on the SDS-damaged human skin cells. J. Oleo Sci. 58: 639–642.

Morita, T., M. Kitagawa, S. Yamamoto, M. Suzuki, A. Sogabe, T. Imura, T. fukuoka and D. Kitamoto. 2010a. Activation of fibroblast and papilla cells by glycolipid biosurfactants, mannosylerythritol lipids. J. Oleo Sci. 59: 451–455.

Morita, T., M. Kitagawa, S. Yamamoto, A. Sogabe, T. Imura, T. Fukuoka and D. Kitamoto. 2010b. Glycolipid biosurfactants, mannosylerythritol lipids, repair the damaged hair. J. Oleo Sci. 59: 267–272.

Morita, T., Y. Ishibashi, T. Fukuoka, T. Imura, H. Sakai, M. Abe and D. Kitamoto. 2011. Production of glycolipid biosurfactants, cellobiose lipids, by *Cryptococcus humicola* JCM 1461 and their interfacial properties. Biosci. Biotech. Bioch. 75: 1597–1599.

Morita, T., T. Fukuoka, T. Imura and D. Kitamoto. 2013a. Accumulation of cellobiose lipids under nitrogen-limiting conditions by two ustilaginomycetous yeasts, *Pseudozyma aphidis* and *Pseudozyma hubeiensis*. FEMS Yeast Res. 13: 44–49.

Morita, T., T. Fukuoka, T. Imura and D. Kitamoto. 2013b. Production of mannosylerythritol lipids and their application in cosmetics. Appl. Microbiol. Biotechnol. 97: 4691–4700.

Morita, T., D. Kawamura, N. Morita, T. Fukuoka, T. Imura, H. Sakai, M. Abe and D. Kitamoto. 2013c. Characterization of mannosylerythritol lipids containing hexadecatetraenoic acid produced from cuttlefish oil by *Pseudozyma churashimaensis* OK96. J. Oleo Sci. 62: 319–327.

Morita, T., T. Fukuoka, T. Imura and D. Kitamoto. 2015. Mannosylerythritol lipids: production and applications. J. Oleo Sci. 64: 133–141.

Mutalik, S.R., B.K. Vaidya, R.M. Joshi, K.M. Desai and S.N. Nene. 2008. Use of response surface optimization for the production of biosurfactant from *Rhodococcus* sp. MTCC 2574. Bioresource Technol. 99: 7875–7880.

Pacheco, G.J., E.M.P. Ciapina, E.d.B. Gomes and N. Pereira Junior. 2010. Biosurfactant production by *Rhodococcus erythropolis* and its application to oil removal. Braz. J. Microbiol. 41: 685–693.

Page, C.A., J.S. Bonner, S.A. Kanga, M.A. Mills and R.L. Autenrieth. 1999. Biosurfactant solubilization of PAHs. Environ. Eng. Sci. 16: 465–474.

Perfumo, A., M. Rudden, R. Marchant and I.M. Banat. 2016. Biodiversity of biosurfactants and roles in enhancing the (Bio) availability of hydrophobic substrates. Cellular Ecophysiology of Microbe: 1–29.

Philp, J., M. Kuyukina, I. Ivshina, S. Dunbar, N. Christofi, S. Lang and V. Wray. 2002. Alkanotrophic *Rhodococcus ruber* as a biosurfactant producer. Appl. Microbiol. Biotechnol. 59: 318–324.

Puchkov, E.O., U. Zahringer, B. Lindner, T.V. Kulakovskaya, U. Seydel and A. Wiese. 2002. The mycocidal, membrane-active complex of *Cryptococcus humicola* is a new type of cellobiose lipid with detergent features. BBA-Biomembranes 1558: 161–170.

Rapp, P., H. Bock, V. Wray and F. Wagner. 1979. Formation, isolation and characterization of trehalose dimycolates from *Rhodococcus erythropolis* grown on n-alkanes. Microbiology 115: 491–503.

Rau, U., L.A. Nguyen, H. Roeper, H. Koch and S. Lang. 2005. Fed-batch bioreactor production of mannosylerythritol lipids secreted by *Pseudozyma aphidis*. Appl. Microbiol. Biotechnol. 68: 607–613.

Ristau, E. and F. Wagner. 1983. Formation of novel anionic trehalosetetraesters from *Rhodococcus erythropolis* under growth limiting conditions. Biotechnol. Lett. 5: 95–100.

Sadouk, Z., H. Hacene and A. Tazerouti. 2008. Biosurfactants production from low cost substrate and degradation of diesel oil by a Rhodococcus strain. Oil Gas Sci. Technol.—Rev. IFP 63: 747–753.

Safdel, M., M.A. Anbaz, A. Daryasafar and M. Jamialahmadi. 2017. Microbial enhanced oil recovery, a critical review on worldwide implemented field trials in different countries. Renew. Sust. Energ. Rev. 74: 159–172.

Saharan, B.S., R.K. Sahu and D. Sharma. 2011. A review on biosurfactants: fermentation, current developments and perspectives. Genet. Eng. Biotechnol. J. 2011: 1–14.

Smyth, T.J.P., A. Perfumo, R. Marchant and I.M. Banat. 2010. Isolation and analysis of low molecular weight microbial glycolipids. pp. 3705–3723. *In*: Kenneth, N. Timmis. (ed.). Handbook of Hydrocarbon and Lipid Microbiology. Springer, Berlin, Heidelberg.

Sugibayashi, K., T. Hayashi, K. Matsumoto and T. Hasegawa. 2004. Utility of a three-dimensional cultured human skin model as a tool to evaluate the simultaneous diffusion and metabolism of ethyl nicotinate in skin. Drug Metab. Pharmacok. 19: 352–362.

Takahashi, M., T. Morita, T. Fukuoka, T. Imura and D. Kitamoto. 2012. Glycolipid biosurfactants, mannosylerythritol lipids, show antioxidant and protective effects against $H_2O_2$-induced oxidative stress in cultured human skin fibroblasts. J. Oleo Sci. 61: 457–464.

Tokumoto, Y., N. Nomura, H. Uchiyama, T. Imura, T. Morita, T. Fukuoka and D. Kitamoto. 2009. Structural characterization and surface-active properties of a succinoyl trehalose lipid produced by Rhodococcus sp. SD-74. J. Oleo Sci. 58: 97–102.

Tuleva, B., N. Christova, R. Cohen, G. Stoev and I. Stoineva. 2008. Production and structural elucidation of trehalose tetraesters (biosurfactants) from a novel alkanothrophic *Rhodococcus wratislaviensis* strain. J. Appl. Microbiol. 104: 1703–1710.

Uchida, Y., S. Misawa, T. Nakahara and T. Tabuchi. 1989. Factors affecting the production of succinoyl trehalose lipids by *Rhodococcus erythropolis* SD-74 grown on n-alkanes. Agr. Biol. Chem. 53: 765–769.

Yu, M., Z. Liu, G. Zeng, H. Zhong, Y. Liu, Y. Jiang, M. Li, X. He and Y. He. 2015. Characteristics of mannosylerythritol lipids and their environmental potential. Carbohyd. Res. 407: 63–72.

Zhang, Y. and R.M. Miller. 1992. Enhanced octadecane dispersion and biodegradation by a *Pseudomonas* rhamnolipid surfactant (biosurfactant). Appl. Environ. Microb. 58: 3276–3282.

Zhang, Y., W.J. Maier and R.M. Miller. 1997. Effect of rhamnolipids on the dissolution, bioavailability, and biodegradation of phenanthrene. Environmental Science & Technology 31: 2211–2217.

Zhao, X., T. Murata, S. Ohno, N. Day, J. Song, N. Nomura, T. Nakahara and K.K. Yokoyama. 2001. Protein kinase $C_\alpha$ plays a critical role in mannosylerythritol lipid-induced differentiation of melanoma B16 cells. J. Biol. Chem. 276: 39903–39910.

# 6

# Microbial Glycoprotein and Lipopeptide Biosurfactants: Production, Properties and Applications

*Ana Belén Moldes,[1,]\* Xanel Vecino,[2] Rodríguez-López L.,[1] Rincón-Fontán M.[1] and José Manuel Cruz[1]*

## Introduction

This chapter makes a comparative study of glycoproteins and lipopeptides with surfactant capacity. It is noticeable that the number of lipopeptides and glycolipids identified at the moment is much superior to the glycoproteins reported with surfactant capacity.

In terms of production, glycoproteins are usually cell-bound to the plasmatic membrane of lactic acid bacteria, whereas lipopeptides are produced by *Bacillus subtilis* strains and are mostly excreted to the fermentation medium, so they are considered extracellular products. Therefore, different extraction protocols can be applied depending on the type of biosurfactant (BSs) (extracellular or cell-bound). Lipopeptides can be obtained from fermentation media by acid precipitation or liquid-liquid extraction, while glycoproteins are extracted from microbial cells using a phosphate bufffered saline (PBS) solution at room temperature (Vecino et al. 2012), or with phosphate buffered (PB) solution, without salt at 65°C (Vecino et al. 2015a, b).

---

[1] University of Vigo, Department of Chemical Engineering, Campus As Lagoas-Marcosende, Pontevedra-Vigo, Spain, 36310.
[2] Polytechnic University of Catalonia, Department of Chemical Engineering, Campus Diagonal-Besòs, Barcelona, Spain, 08930.
\* Corresponding author: amoldes@uvigo.es

Lipopeptides are composed by 6 different types of families: surfactin, iturin, fengycin and more recently, bacitracin, polymyxin and kurstakin, whereas at the moment, there does not exist a classification for glycoproteins with surfactant capacity.

The main applications recommended for lipopeptide BSs include the environmental, pharmaceutical, and cosmetic industries (Gharaei-Fathabad 2011, Gudiña et al. 2016, Vecino et al. 2017a). By the same token, glycoproteins are also involved in the cosmetic and pharmaceutical industries due to its capacity to induce the recovery of the cells from internal and external stress, providing cellular repair (San Miguel et al. 2015).

Other different aspect between lipopeptides and glycoproteins could be established based on their emulsified capacity, and their hydrophilic-lipophilic balance (HLB). Nevertheless, in this sense, more studies would be needed to establish additional differences among them.

However, the major difference between lipopeptides and glycoproteins is their critical micellar concentration (CMC). The CMC is the minimum concentration needed of BS to reduce the surface tension of aqueous solutions to the lowest value. Therefore, BSs with lower CMC are more valuable from an industrial point of view. In general, the CMC of lipopeptides could be even more than 30 times lower than the CMC of glycoproteins. This parameter is important from an industrial point of view, in terms of costs and effectiveness, and would therefore define their further commercialization and application.

## Glycoproteins BSs

Glycoproteins are the lesser-known BSs that are composed of a sugar chain linked to an amino acidic sequence. Like all other BSs, they are amphiphilic substances produced by a heterogeneous group of microorganisms as secondary metabolites, which have a wide variety of applications in different fields, including food, pharmaceutical and cosmetic industries as well as in environmental applications (Faivre and Rosilio 2010, Ara and Mulligan 2015, Gudiña et al. 2016, Vecino et al. 2017b).

The number of publications in indexed international journals related to the production, characterization, and applications of glycoproteins as surfactants is small in comparison with other BSs, with less than 50 publications in the period between 1991 and 2017. However, there has been an increase in the number of publications describing this kind of BSs in the last couple of years (see Figure 1a), mainly from India, China, Japan, Portugal and United Kingdom (see Figure 1b).

Most of these glycoproteins were obtained by biotechnological processes using lactic acid bacteria, in which BSs are extracted through solid-liquid extractions using phosphate buffered saline (Bustos et al. 2007, Moldes et al. 2007, 2011, 2013, Rivera et al. 2007, Portilla-Rivera et al. 2008, 2009, Portilla et al. 2008, Gudiña et al. 2010, 2015, Vecino et al. 2012, 2017b). During the extraction process, biomass is centrifuged, washed twice in demineralized water and suspended in PBS buffer (10 mM $KH_2PO_4/K_2HPO_4$ with 150 mM NaCl, adjusted to pH 7.0). Following this, cell suspensions are left at room temperature (20–25°C) for 2 h, with gentle stirring,

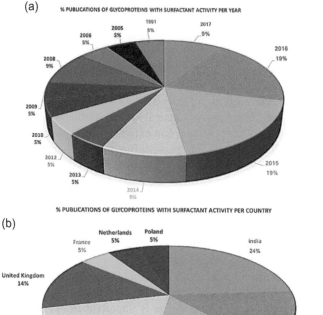

**Figure 1.** Representation of the published articles of glycoproteins with BS activity, regarding the year of publication (a) or the country of origin (b), respectively.

for BS release. After that, the bacteria cells are removed by centrifugation, and the remaining supernatant containing the BS is filtered through a 0.22 μm pore-size filter (to remove the microbial biomass) and dialyzed against demineralized water at 4°C. Finally, the dialysate is freeze-dried and stored. Gudiña et al. (2010) developed this methodology by adding a purification step adjusting the pH of the BS containing PBS solution with 1 M HCl, to pH = 2.0 and keeping it at 4°C for 2 h before precipitation and recovery of BSs by filtration.

Although this is the regular methodology used by different authors to extract glycoproteins, recently, Vecino and coworkers (2015a, b) have also proposed the extraction of cell-bound BSs, in absence of salt, using phosphate buffered at higher temperatures. These authors observed that high yields could be obtained by subjecting the microbial biomass of *Lactobacillus pentosus,* at 65°C during 1.5 h. In this case, the BS was identified as a glycolipopeptide. The advantage of not using phosphate buffered saline during the extraction process is that it provides a better BS, in terms of their direct use in the bioremediation of contaminated sites, avoiding the phytotoxic effect produced by salt contained in the BS extract. Figure 2 shows the two protocols established at the moment for the extraction of cell-bound BSs.

GLYCOPROTEIN BIOSURFACTANTS: EXTRACTION PROCESS

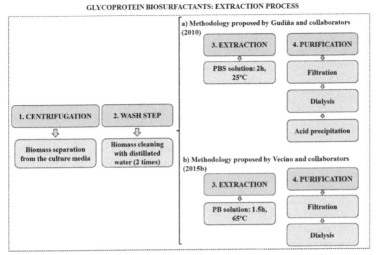

**Figure 2.** Extraction process for cell-bound glycoproteins BSs: (a) with phosphate buffered saline (PBS) or (b) phosphate buffered (PB), respectively.

BSs can be produced in different phases of the microbial growth. Therefore, Velraeds et al. (1996) reported the production of cell-bound glycoproteins by different, *Lactobacilli* strains, at different growing stages. These authors observed that BSs obtained at the stationary phase were richest in protein, in comparison with those produced in the exponential phase of growth. In this study, it was also observed that the BSs produced by *Lactobacillus acidophilus* RC14 and *Lactobacillus fermentum* B54 were more effective in reducing the surface tension (ST) of aqueous solutions than the BSs produced by other *Lactobacilli* strains (like *Lactobacillus casei* subsp. *rhamnosus*). The minimum surface tensions achieved by the BSs studied by Velraeds et al. (1996) were between 30–40 mN/m with CMC values about 1–3 g/L at the stationary phase.

Golek et al. (2009) also have studied other *Lactobacilli* strains (*L. casei, L. fermentum, Lactobacillus plantarum* and *Lactobacillus brevis*), observing that they produce cell-bound BSs that reduced the ST of water between 19–27 units. These BSs were obtained using the same protocol established by Velraeds and collaborators (1996), although the extractions were carried out only at the stationary phase.

Brzozowski et al. (2011) isolated and evaluated the functional properties of two cell-bound glycoproteins produced by *Lactobacillus fermenti* 126 and *L. rhamnosus* CCM 1825, with CMC values of 9.0 and 6.0 g/L which reduce the surface tension of media up to 45 and 36 mN/m, respectively. Additionally, these authors observed that BS from *L. rhamnosus* CCM 1825 has important emulsifying and froth-forming activities, whereas *L. fermenti* 126 presents notable anti-adhesive capacities.

Madhu and Prapulla (2014) have produced a BS composed by protein and polysaccharide fractions using *L. plantarum* CFR 2194, isolated from Kanjika, a rice-based fermented substrate. In this case, maximum surfactant production occurs at 72 h, under stationary conditions, being the CMC of this BS 6.0 g/L, the same as that obtained with *L. rhamnosus* (Brzozowski et al. 2011).

The formation of glycoproteins could be related to the high concentration of proteins floating in or around the cell membrane, which can react with sugars producing glycoproteins. Therefore, most of the glycoprotein BSs reported at present are cell-bound, which is reasonable because glycoproteins are integrated in the cell membranes where they play an important role. The proteins of cell membrane are amphiphilic macromolecules incorporated in their lipid domain. A part of their surface, which is in contact with the lipid bilayer core, is hydrophobic while other parts, which are exposed to the aqueous environment on either site of the membrane, are hydrophilic (Aivaliotis et al. 2003).

Table 1 shows the maximum surface tension reduction achieved in presence of glycoproteins and their CMC. In general, the CMC values for glycoproteins are very high, in the range of g/L, in comparison with the CMC values reported for lipopeptide BSs, described below, that are in the range of mg/L.

**Table 1.** CMC and surface tension reduction of several glycoprotein BSs reported in the literature.

| Microorganism | ST Reduction (mN/m) | CMC (g/L) | Reference |
|---|---|---|---|
| *Lactobacillus plantarum* CFR2194 | 44.3 | 6.0 | Madhu and Prapulla (2014) |
| *Lactobacillus helveticus* | 40.2 | 2.5 | Sharma et al. (2014) |
| *Lactobacillus acidophilus* RC14 | 39.0 | 1.0 | Velraeds et al. (1996) |
| *Lactococcus lactis* 53 | 36.0 | 14.0 | Rodrigues et al. (2006a) |
| *Streptococcus thermophilus* A | 36.0 | 20.0 | Rodrigues et al. (2006b) |
| *Lactobacillus casei* 8/4 | 38.3 | ND* | Golek et al. (2009) |
| *Lactobacillus fermentum* 11/b | 42.1 | ND* | Golek et al. (2009) |
| *Lactobacillus paracasei* A20 | 41.8 | 2.5 | Gudiña et al. (2010) |
| *Lactobacillus fermenti* 126 | 45.1 | 9.0 | Brzozowski et al. (2011) |
| *Lactobacillus rhamnosus* CCM1825 | 43.6 | 6.0 | Brzozowski et al. (2011) |
| *Lactobacillus agilis* CCUG31450 | 42.5 | 7.5 | Gudiña et al. (2015) |

*not determined

## Lipopeptides BSs

The most studied lipopeptides are those produced by *Bacillus* species which have low molecular weights (around 1–1.5 KDa) in comparison with other polymeric surfactants, like Emulsan and Alasan, isolated from *Acinetobacter* species, which have a molecular weight of around 1,000 kDa. In general, lipopeptides lower the surface and interfacial tension more efficiently than other BSs like glycolipids. Lipopeptides are mainly composed of C12–C18 fatty acids chains linked to a fraction of amino acids. Moreover, they have good heterogeneity in accordance with the type and sequence of amino acid moiety and nature, as well as with the length and branching of fatty acid chains and their moieties. Additionally, they proved to be ecofriendly, biodegradable, less toxic and biocompatible compared to chemical surfactants which promotes potential use in pharmaceutical, food and cosmetic industries to greater extent (Chen et al. 2015, Walia and Cameotra 2015).

The number of publications related with the production, characterization and application of lipopeptides reveals a noticeable interest by the research community in this kind of BSs. More than 300 publications were noted between year 2001 to 2017 (see Figure 3a) where India and China have the most number of publications (see Figure 3b).

The first classifications carried out between 1949 and 1986 divided lipopeptides into three different families: surfactins, iturins, and fengycins. In 2000, a new family of lipopeptides named kurstakins, produced by *Bacillus thuringiensis*, was reported and considered as a biomarker of this species. Other lesser known lipopeptides are polymyxin and bacitracin.

Surfactins are characterized as bacterial cyclic lipopeptides. They are produced by various strains of *B. subtilis* and are regarded as one of the most effective BSs. They have the capacity to reduce the surface tension of water from 72 to 27 mN/m at a concentration around 0.05 g/L. Surfactin is composed of a peptide chain formed by seven α-amino acids bonded to a hydroxyl fatty acid (C14) by lactone bond to

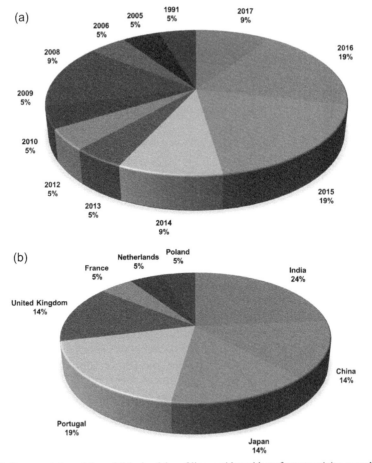

**Figure 3.** Representation of the published articles of lipopeptides with surfactant activity, regarding the year of publication (a) or the country of origin (b), respectively.

form a cyclic lipopeptide. The typical sequence of amino acids in the peptide ring is: L-Glu1-L-Leu2-D-Leu3-L-Val4-L-Asp5-D-Leu6-L-Leu7 (Liu et al. 2004).

Iturins are cyclic lipoheptapeptides compounds which contain a β-amino fatty acid as lipophilic components (Vater et al. 2002). They are used as antibiotics with potent antimicrobial activities. It is important to note that they can be co-produced with surfactin by various *B. subtilis* strains (Asaka and Shoda 1996, Ahimou et al. 2000).

Fengycin is an antifungal lipopeptide complex produced by *B. subtilis* strain F-29-3. It inhibits filamentous fungi but it is ineffective against yeast and bacteria. The inhibition is antagonized by sterols, phospholipids and oleic acid, whereas unsaturated fatty acids increase their antifungal effect. It is characterized by having a hemolytic activity 40-fold lower than surfactin. The structure is composed of a β-hydroxy fatty acid linked to a peptide part comprising 10 amino acids, where 8 of them are organized in a cyclic structure (Deleu et al. 2008). Fengycin A is comprised of 1 D-Ala, 1 L-Ile, 1 L-Pro, 1 D-allo-Thr, 3 L-Glx, 1 D-Tyr, 1 L-Tyr, 1 D-Orn, whereas in fengycin B, the D-Ala is replaced by D-Val. The lipid moiety of both analogs is variable, as fatty acids have been identified as anteiso-pentadecanoic acid (ai-$C_{15}$), iso-hexadecanoic acid (i-$C_{16}$), and n-hexadecanoic acid (n-$C_{16}$) (Vanittanakom et al. 1986).

Bacitracins, produced by *B. subtilis* and *Bacillus licheniformis*, are mixtures of structurally related cyclic polypeptides with antibiotic properties. They act by interfering with the biosynthesis of the bacterial cell wall, showing great potential in the pharmaceutical industry as an emerging drug against resistant pathogens (Lee et al. 1996, Han et al. 2015).

Polymyxins are defined as lipopeptides isolated from a strain of *Bacillus polymyxa* (Stansly et al. 1947). They contain a mixture of D- and L-amino acids and are usually characterized by a heptapeptide ring, a high percentage of 2,4-diaminobutyric acid and a fatty acid attached to the peptide through an amide bond. Polymyxins have molecular weights in the range of 1.2–1.3 KDa (Yang et al. 1967). Two fatty acids can be found attached to the peptide chain, 6-methyl-octanoic acid or 6-methyl-heptanoic acid. The amphipathic character of these peptides might suggest that their disruptive influence on membrane structure is completely analogous to simple cationic detergents.

The new lipopeptide, kurstakin, produced by *B. thuringiensis* subsp. *kurstaki*, was discovered recently, which revealed the presence of different isoforms composed of C11, C12 and C13 fatty acid chains (Béchet et al. 2012). Kurstakins were also detected in other species belonging to *Bacillus genus* such as *Bacillus cereus* (Béchet et al. 2012). Amino acid analysis in kurstakins homologs revealed the same residues composed mainly of Thr, His, Ala, Gly, Ser, and Glu or Gln with molar ratios of 1:1:1:1:1:2, respectively.

In terms of analytical characterization of lipopeptides, Price and coworkers (2007) have analyzed BSs from 54 *Bacillus* strains and species from seven different geographical locations and reported that these BSs fall into three mass ranges: 850–950 m/z, which includes kurstakins; 1000–1100 m/z, which includes surfactins and iturins; and 1450–1550 m/z, which includes fengycins, polymyxins, and bacitracins. These results were also corroborated by Mandal and collaborators

(2013), who reported that lipopeptides' molecular weight ranges between 607.2 to 1,536 Da.

Regarding the methodology used for the separation and purification of lipopeptides, there are two different methods used at present involving precipitation or extraction with organic solvents. Some authors (Kim et al. 2000, Noah et al. 2005, Kim et al. 2009) proposed the separation of BSs by precipitation, followed by liquid-liquid extraction. For instance, Noah et al. (2005) proposed the extraction of lipopeptides from culture media by precipitation with ammonium sulfate 40% (w/v) after incubation overnight at room temperature, followed by extraction of the precipitate with chilled acetone to remove most proteins. However, Satpute et al. (2010) proposed the precipitation of lipopeptides by adding 3 N HCl to pH of 2.0, and then store it at 4°C for 30 min. The acid precipitate is collected by centrifugation and dissolved in a 2:1 (v/v) chloroform:methanol solvent system, followed by evaporation of the organic phase and dissolving the precipitate in methanol and filtering through a 0.22 μm no pyrogenic hydrophilic membrane.

In regards to the use of organic solvents, Vecino et al. (2014) have assayed the liquid-liquid extraction of lipopeptides by using chloroform, ethyl acetate, methyl tert-butyl ether, trichloroethylene, hexane, heptane, or dichloromethane, reporting that among the solvents used, chloroform and ethyl acetate provide a better performance.

Thavasi et al. (2011) also carried out the separation of lipopeptides produced by *Pseudomonas aeruginosa*, by extracting it from the supernatant with a mixture of 2:1 (v/v) chloroform:methanol, followed by a purification step in a silica gel column eluted with chloroform:methanol, ranging from 20:1 to 2:1 (v/v) in a gradient manner. The fractions are pooled and solvents are evaporated. The resulting BS extract is dialyzed against demineralized water and lyophilized. Figure 4 shows some of the protocols established at the moment for the extraction of lipopeptide BSs from culture media.

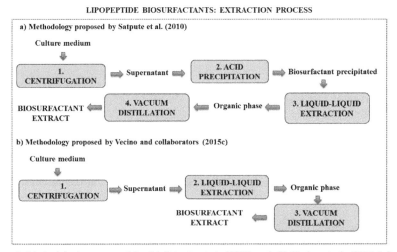

LIPOPEPTIDE BIOSURFACTANTS: EXTRACTION PROCESS

**Figure 4.** Methodologies proposed by (a) Satpute et al. (2010) and (b) Vecino et al. (2015c) for the extraction and purification of lipopeptide BSs from fermentation media.

The extraction process plays an important role in the commercialization of BSs because it is going to determine their industrial applications. For example, BSs applied in the environmental industry should be obtained using the cheapest methods. Probably, the liquid-liquid extraction with organic solvents (chloroform or methanol, among others) provides a suitable method for the separation of BSs destined for the environmental industry, taking into account that the organic solvent can be recovered after the extraction process. On the other hand, when BSs are applied to the cosmetic or pharmaceutical industries, several purification steps including precipitation, liquid-liquid extraction and/or solid-liquid extraction could be applied, due to the higher benefit margins and due to the toxicological requirements established for this kind of industries, in comparison with the environmental sector.

## *Ionic charge of the lipopeptide BSs*

Unlike chemically synthesized surfactants which are usually classified according to the nature of their polar groups (cationic, anionic, zwitterionic and non-ionic), BSs are usually categorized mainly by their chemical structure and microbial origin. However, from an industrial point of view, it would be interesting to know the ionic charge of BSs and classify them on the basis of this characteristic.

Saranya et al. (2015) have found that different *Bacillus* species can produce different lipopeptides with different ionic charge; this aspect is important to determine their application at industrial scale, for example, in the cosmetic industry. Therefore, *B. subtilis* growth on palm oil and *B. cereus* growth on coconut oil, isolated from tannery wastewater-contaminated soil, produces anionic and cationic BSs that reduce the surface tension of water up to 28 mN/m and 23 mN/m, respectively. The anionic lipoprotein BS contained about 61.5% polar amino acids and 38.5% non-polar amino acids, whereas the cationic lipoproteins were 53.0% polar amino acids and 47.0% non-polar amino acids, respectively.

Lipids included in the BSs are a group of compounds negatively charged at the end of the carboxyl groups, although once esterified they do not have charge. Therefore, when a lipopeptide BS possess an ionic charge, this should be provided by the peptide moiety. In nature, there are at least 20 different amino acids, the majority of them being non-ionic; the ionic amino acids are lysine (+), arginine (+), aspartate (–) and glutamate (–). Charged amino acids are polar and, therefore, hydrophilic.

On the other hand, Kowall et al. (1998) have observed a huge variation in the amino acid chain between lipopeptide BSs produced by the same microorganism. For instance, *B. subtilis* is able to produce 44 different types of BSs, named surfactins, depending on the amino acids' composition in their hydrophilic chain. Other authors, Singh and coworkers (2014) and Coronel-León et al. (2015), have also reported differences in the surfactin composition.

Saranya and collaborators (2015) have demonstrated that the anionic BS produced by *B. subtilis* is composed of a hydrophilic fraction of amino acids with mainly amino acids arginine (34.5%), followed by glutamic acid (23.0%) and leucine (11.3%). In comparison, the peptide chain of the cationic BS produced by *B. cereus* is composed of glutamic acid (45.0%), followed by tyrosine (15.5%), arginine (13.8%) and aspartic acid (10.9%).

Rufino et al. (2014) have characterized an anionic BS produced by *Candida lipolytica* UCP 0988 as a glycolipopeptide, composed of 50% proteins, 20% lipids and 8% carbohydrates, and the BS reduced the surface tension of fermentation medium by up to 25 mN/m. Table 2 shows some of the lipopeptide BSs and their ionic charge identified in the literature.

**Table 2.** Ionic charge of different lipopeptide BSs obtained in microbial production.

| Microorganisms | Biosurfactant | Ionic Charge | Reference |
|---|---|---|---|
| *Candida lipolytica* | glycolipopeptide* | anionic | Rufino et al. (2014) |
| *Bacillus subtilis* | lipopeptide | anionic | Saranya et al. (2015) |
| *Bacillus cereus* | lipopeptide | cationic | Saranya et al. (2015) |
| *Serratia marcescens* | lipopeptide | non-ionic | Thies et al. (2014) |
| *Bacillus licheniformis* | lipopeptide | non-ionic | Sineriz et al. (2001) |
| *Bacillus subtilis* | lipopeptide | non-ionic | Isogai et al. (1982) |
| *Lactobacillus* or *Bacillus* species | lipopeptide | amphoteric | Rodríguez-López et al. (2017) |

*glycolipopeptide BSs were included in this table because of their very low content of carbohydrates and their similar behavior to lipopeptides.

## *Agro-industrial streams as source of lipopeptide BSs*

The major challenge for industrial production and application of BSs is their high production cost. They are produced in lower quantity than chemical surfactants, although they have a high value, especially due to their exceptional biological properties. In terms of operation costs, the extraction of BSs from agro-industrial streams, as well as the use of renewable sources, may provide important improvements for the industrial commercialization of microbial surfactants (Banat et al. 2014a, Vecino et al. 2014, 2015c, Delbeke et al. 2016).

Recently, Vecino and collaborators (2014, 2015c), discovered the presence of a BS in corn steep liquor (CSL), an agro-industrial stream obtained from corn wet-milling industry, with similar characteristics to other lipopeptide BSs reported in the literature. They established a liquid-liquid extraction protocol for the recovery of this BS, which is included in Figure 4. The extraction method involves the utilization of CSL:chloroform ratio of 1:2 (v/v) at 56°C during 1 h. Figure 5 shows a picture of the BS extracted from CSL by liquid-liquid extraction, following the methodology proposed by Vecino and collaborators (2015c).

It is likely that the presence of BSs in CSL is due to the spontaneous growth of microorganisms like *Lactobacillus* or *Bacillus* species during the corn wet-milling process. In fact, it is known that CSL has important concentrations of lactic acid that indicate the presence of lactic acid bacteria in this industrial stream (Hull et al. 1996).

The BS, extracted from CSL, is composed of 50.0–55.2% linolelaidic acid, 15.7–22.2% oleic and/or elaidic acid, 5.9–14.6% stearic acid, and 14.9–19.6% palmitic acid (Vecino et al. 2014). The Fourier transform infrared (FTIR) spectroscopy of the BS extract, obtained from CSL, is similar to the FTIR spectrum of surfactin, produced by *B. subtilis* (Vecino et al. 2015c).

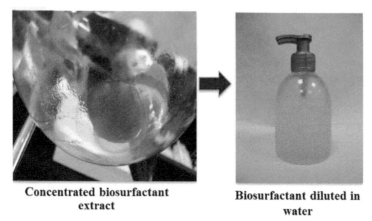

|  |  |
|---|---|
| **Concentrated biosurfactant extract** | **Biosurfactant diluted in water** |

**Figure 5.** Aspect of the crude BS extract obtained from corn steep liquor (left image) as well as the BS extract dissolved in demineralized water (right image).

This BS is able to reduce the surface tension of water between 30–35 units with a CMC about 100–400 mg/L. However, it is necessary to take into account that small differences in the fermented stream of CSL can induce small changes in the concentration and composition of the BS extract. Those differences can also produce alterations in its surface-active capacity as well as in its CMC. The BS extracted from CSL is produced in a non-controlled fermentation process. Consequently, the differences between different lots can be more accentuated than with other families of BSs, produced in a controlled fermentative process.

BS, extracted from CSL, is more soluble in demineralized water than surfactin and can be dissolved at concentrations about 900 mg/L water at pH = 4–5, whereas surfactin is soluble in organic solvents or in water a pH > 10. The highest solubility of BS from CSL can be due to the presence of polar amino acids. This BS was defined by Rodríguez-López et al. (2017) as amphoteric, being entrapped by cationic and anionic resins, probably due to the presence of charged amino acids like lysine (+) and/or arginine (+), aspartate (–) and/or glutamate (–), although its amino acid moiety remains unknown.

## Properties and Industrial Applications of Lipopeptides versus Glycoproteins

It is necessary to remark the huge number of lipopeptides with BS capacity identified at present in comparison with the number of glycoproteins (see Figure 6). In 2017, less than 5 publications related with glycoproteins were found, whereas the number of lipopeptides with surfactant activity was stated around 44.

As mentioned earlier, most of the BSs reported in the literature are mainly composed of fatty acids linked to a small hydrophilic polymeric fraction composed of sugars or proteins. The lipid fraction plays an important role in the BS formation (Youssef et al. 2005), and it can be easily characterized by GC-MS (Rodríguez-López et al. 2016).

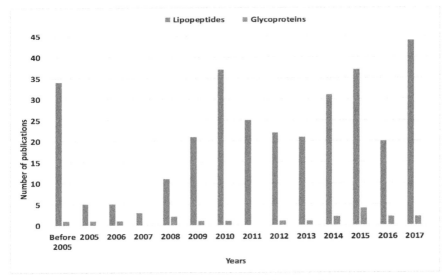

**Figure 6.** Number of publications of lipopeptides versus glycoproteins.

On the other hand, proteins are an important part of glycoproteins BSs, although their determination is quite difficult. The most accurate method of determining proteins is probably acid hydrolysis followed by amino acid analysis. Most colorimetric methods for determining percentages of proteins have significant interferences and the procedure proposed by Lowry et al. (1951) is not an exception; however, they have been widely used for protein estimations as an alternative to more complex methods involving acid or basic hydrolysis (Waterborg et al. 2002).

Some studies have showed interferences, especially working with detergents; therefore, mixtures of Triton X100 with sodium dodecyl sulfate (SDS) or deoxycholic acid at given concentrations interfere with serum albumin bovine determination by the method of Lowry (Bruscalupi et al. 1979).

Other methods, such as Ninhydrin (2,2-dihydroxyindane-1,3-dione) test, could be more useful for confirming the presence of the peptide chain in BSs. This compound is used to detect ammonia or primary and secondary amines, which produces a deep blue or purple color after they react. Most of amino acids, except proline, are hydrolyzed and react with ninhydrin, but certain amino acid chains may also be degraded. Therefore, separated analyses are required for identifying amino acids.

Regarding applications, glycoproteins could play an interesting role in cosmetic industry. They can help different body systems to recover their cells from internal and external stresses and they also are involved in cellular repair, among other functions (San Miguel et al. 2015). However, there is no evidence that they can affect wrinkles when they are applied topically.

Other studies suggest the potential use of glycoproteins against tumor cells. Duarte et al. (2014) have evaluated the inhibitory effect of a glycoprotein (BioEG) produced by *Lactobacillus paracasei* subsp. *paracasei* A20 on proliferation of human breast cancer cells. This glycoprotein BS was tested against two breast cancer cell lines, T47D and MDA-MB-231, and a non-tumor fibroblast cell line (MC-3 T3-E1). Exposure of cells to 0.15 g of this BS, for 48 h, decreased cancer cells viability, without affecting normal fibroblasts. This glycoprotein showed more activity against breast cancer cells than surfactin.

Moreover, some glycoproteins, produced by *L. rhamnosus* CCM 1825, showed emulsifying and froth-forming activities, whereas glycoproteins produced by *L. fermenti* 126 exhibited good anti-adhesive properties against *Escherichia coli*, *Klebsiella pneumoniae* and *P. aeruginosa* (Brzozowski et al. 2011).

Kawahara et al. (2013) reported that the glycoprotein produced from the red yeast, *Rhodotorula mucilaginosa* KUGPP-1, originating in the Antarctic, with a molecular mass about 730 KDa, has dispersive power against astaxanthin. This BS was purified by ultrafiltration followed by precipitation with acetone. It is composed by sugars and proteins in an approximate molar ratio of 9:1, respectively. The sugars involve mannose, galactose and glucose.

Gudiña et al. (2010) also demonstrated that the glycoprotein produced by *L. paracasei* ssp. *paracasei* A20 has antimicrobial properties against *E. coli*, *Streptococcus agalactiae* and *Streptococcus pyogenes*. Additionally, Gudiña et al. (2015) studied the anti-adhesive and antimicrobial activities of the glycoprotein BS, produced by *L. agilis*, observing a considerable anti-adhesive activity against *Staphylococcus aureus*, as well as an important antimicrobial activity against *S. aureus, S. agalactiae* and *P. aeruginosa*. Therefore, 5 g/L of this BS inhibited around 20% the growth of *S. aureus*; 13.5% the growth of *P. aeruginosa* and 10.7% the growth of *S. agalactiae*, whereas no antimicrobial activity for this BS was detected against *E. coli* and *Candida albicans*. The anti-adhesive capacity of this glycoprotein BS was around 64.6%–50.3% at concentrations between 10–1 g/L.

Most of the applications proposed for glycoprotein BSs are related with their antimicrobial, anti-adhesive and anti-tumoral properties, but possible applications in environmental issues have also been demonstrated (Kiran et al. 2016). Some authors have suggested the use of glycoproteins for environmental applications. A novel BS, produced by *Geobacillus stearothermophilus* A-2, with emulsifier properties useful for environmental protection, was discovered recently by Zhou et al. (2016). This BS showed potential applications in the environmental industry enhancing oil recovery from contaminated sites. It was composed by 71.4% of carbohydrates and 27.8% of proteins. Monosaccharides in this BS were identified as mannose (33.5%), glucose (30.9%), galactose (29.7%), and glucuronic acid (5.9%), whereas the peptide chain contained 17 types of amino acids.

Some authors have also observed a good emulsifying capacity of the glycolipopeptide BS produced by *L. pentosus,* in different oil-in-water emulsions systems, like gasoline/water (Vecino et al. 2012), fluorene/water (Portilla-Rivera et al. 2008, Vecino et al. 2015b) and rosemary/water (Vecino et al. 2015a). Additionally, Vecino et al. (2013) studied the adsorption properties of the BS produced by *L. pentosus* onto river sediments, showing an important potential as foaming agent in froth flotation process.

Moldes et al. (2011, 2013) reported that the BS produced by *L. pentosus* could also be employed for the remediation of hydrocarbon-contaminated sites. It was observed that the presence of BS from *L. pentosus* accelerated the biodegradation of octane in comparison with the use of synthetic surfactants, like SDS.

Recently, Vecino and collaborators (2017b) studied the effect of the carbon source on the BS production from *Lactobacilli* strains. The results obtained showed that when glucose from vineyard pruning waste (a lignocellulosic residue from winery industry) was used, the BS from *L. paracasei* was a glycolipopeptide, whereas when the carbon source was replaced by lactose the BS produced was a glycoprotein. Additionally, it was found that the extraction process, either with PB or PBS, influenced the BS chemical structure and emulsion capacity of BSs.

This glycolipopeptide BS produced by *L. paracasei* showed good properties as a stabilizing agent of emulsions, formulated with natural components based on essential oils and/or natural antioxidant extracts (Ferreira et al. 2017), as well as important antimicrobial and anti-adhesive activities against skin pathogens such as *P. aeruginosa*, *S. agalactiae*, *S. aureus*, *E. coli*, *S. pyogenes* and *C. albicans* (Vecino et al. 2017c).

Regarding the applications of lipopeptides, these can be used in the cosmetic, pharmaceutical and environmental industries. In general, lipopeptides have anti-inflammatory properties, interacting with cell membranes and macromolecules (Zhang et al. 2015); they possess antibiotic and anti-adhesive properties avoiding biofilm production and the adhesion of bacteria on different surfaces (Banat et al. 2014b, Gudiña et al. 2016). Additionally, they have been proposed for the control of phytopathogen, even with those organisms that have developed resistance to specific antimicrobial agents. For instance, kurstakins have antifungal activities against *Stachybotrys chartarum* (Pérez-García et al. 2011, Quinn et al. 2013) with a strong biostatic or biocidal activity, at concentrations much lower than the common cationic detergents (Rosenthal et al. 1976). Kurstakins exhibit activity against gram-positive and gram-negative bacteria, yeasts, and protozoa (Schwartz et al. 1959).

In general, surfactin possesses a wide variety of biological activities such as the ability to lyse erythrocytes, bacterial spheroplasts and protoplasts, among others. Therefore, surfactin family of iturins is recommended as biopesticides for plant protection. For instance, iturin A, produced by *B. subtilis* RB14, is effective for the control of damping-off, caused by *Rhizoctonia solani* in tomato plants (Asaka and Shoda 1996).

Some assays have also showed the antifungal activity of *B. subtilis* 20B BSs (surfactins and iturins) against mycelia growth of *Chaetoderma indicum*, *Alternaria burnsii*, *Fusarium oxysporum*, *Fusarium udum*, *Trichoderma harzianum* and *Rhizoctonia bataticola* (Joshi et al. 2008, Ceresa et al. 2016).

On the other hand, it has been revealed that the lipopeptide BS obtained from corn steep liquor not only has surfactant properties but also possess antioxidant properties that could be useful in the cosmetic and food industries, among others (Rodríguez-López et al. 2016). It has been suggested that this lipopeptide is amphoteric and it can be adsorbed onto human hair with a maximum capacity of 3679 μg/g, demonstrating a huge potential in the formulations of hair cosmetic products (Rincón-Fontán et al. 2016, Rodríguez-López et al. 2017).

Lately, Rincón-Fontán and coworkers (2018) evaluated the potential of the BS extract obtained from CSL to act as a sunscreen agent. In this way, the BS can exert this effect of absorbing UV light from sun, by itself or in combination with mica minerals.

Moreover, the BS extracted from CSL was included in the synthesis of silver (Ag) and gold (Au) nanoparticles in a single-step process. The dual functionality of the BS, acting as reducing agent as well as stabilizer, makes it particularly useful for green synthesis approaches. Furthermore, the biocompatible nanoparticles of Ag with BS were used in antimicrobial tests, exhibiting a high antimicrobial activity, against gram-negative bacteria like *E. coli*, at a very low concentration of silver (Gómez-Graña et al. 2017).

The use of lipopeptides has also been proposed by some authors for environmental protection. For instance, Pérez-Ameneiro et al. (2015) used the lipopeptide BS, extracted from CSL, for favoring the formulation of biocomposites used in the treatment of wastewater. Moreover, Vecino et al. (2015d) reported the elimination of polycyclic aromatic hydrocarbons (PAHs) from sewage sludge by using the lipopeptide BS extracted from CSL. The results showed that this BS not only was able to solubilize PAHs but also allowed their biodegradation. Figure 7 includes information about the microbial production and applications of glycoproteins and lipopeptide BSs, respectively.

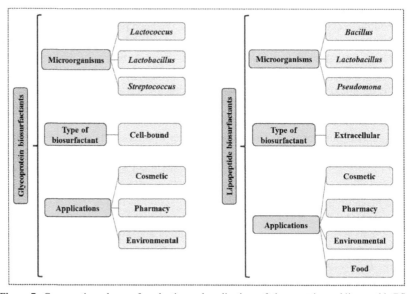

**Figure 7.** Comparative scheme of production and applications of glycoproteins and lipopeptide BSs.

### *Hydrophilic-Lipophilic Balance (HLB) as a challenge to delimit the applications of lipopeptide and glycoprotein BSs*

An important characteristic of BSs is their HLB, which is going to determine their capacity to stabilize oil-in-water (O/W) or water-in-oil (W/O) emulsions, as well as their detergent and humectant properties, establishing their industrial applications. The HLB depends on the number of hydrophobic and hydrophilic groups of the molecule, so it provides information about the balance of the size and strength of the hydrophilic and the lipophilic domains, contained in BSs. For instance, the capacity of lipopeptides to act as emulsifiers should be different from glycoproteins, influenced by the number and nature of hydrophobic and lipophilic groups. However, for the moment, this aspect has not being taken into much consideration by the research community, probably because most of BSs are not completely pure.

The HLB determines the solubility of BSs; therefore, compounds with low HLB (below 9.0) should be soluble in oil and they should have higher amount of lipophilic groups, producing W/O emulsions. In comparison, BSs with high HLB (above 11.0) should be more soluble in water, forming O/W emulsions, because of the predominance of hydrophilic groups. Considering this aspect a rough approximation of the HLB for a specific BS, could be obtained based on their solubility in water. For example, surfactin has a low solubility in water, thus its HLB should be under 9, whereas glycoproteins, extracted from *Lactobacillus* strains, are more soluble in water, therefore they should have high HLB values. BSs, with HLB ranges between 9 and 11, should be considered intermediated, and they should stabilize either O/W or W/O emulsions (Atlas 1963).

Some authors have proposed a method to measure the HLB of a BS based on the HLB values and behavior of known surfactants. For instance, two surfactants, Span 80 (HLB = 4–3) and Tween 80 (HLB = 15), were used by Dehghan-Noudeh et al. (2005) to establish the HLB of a lipopeptide BS produced by *B. subtilis* ATCC 6633. A series of emulsions were prepared with both surfactants and the BS subjected to evaluation in graduated tubes. After emulsion formation, the tubes were centrifuged at 1500 rpm, measuring the amount of water separated. Following this method, the net HLB value of this BS was 21.27, this being appropriate as solubilizing agent and detergent. Table 3 includes different methods established by different authors to calculate the HLB of synthetic surfactants that could be applied to BSs.

**Table 3.** Methods proposed by various authors for calculating the HLB of synthetic surfactants.

| Formula | Variables | Reference |
|---|---|---|
| $HLB = \left(\frac{M_p}{M}\right) \times 20$ | $M_p$ = molecular weight of molecule polar part, $M$ = total molecular weight | Griffin (1949) |
| $HLB = 7 + \sum(HG) - \sum(LG)$ | HG = hydrophilic groups, LG = lipophilic groups | Davies (1957) |
| $HLB = 7 + \sum(SG)$ | SG = shared groups | Apostoluk and Szymanowski (1998), Zieliński (2013) |
| $HLB = 20\left(1 - \frac{S}{A}\right)$ | S = saponification number, A = acid number | Ogonowski and Tomaszkiewicz-Potępa (2004) |
| $HLB = \frac{A \times H}{0.05}(A \times H + B \times L)$ | A = 15 (constant), B = 10 (constant), H = height (mm) of the curve of integration for protons from hydrophilic groups, L = height (mm) of the curve of integration for protons from hydrophobic groups in the 1H NMR spectra | Ogonowski and Tomaszkiewicz-Potępa (2004) |

Various non-ionic, anionic and amphoteric surfactants, including the bile acid salts, saponin and surfactin, were added by Shinichiro et al. (1981) to an insulin preparation, observing that HLB values between 8 and 14 produce the highest promoting effect on the nasal absorption of insulin.

The HLB of glycolipids have been studied more than the HLB of lipopeptides or glycoproteins. Glycolipids have intermediate HLB values. For example, Li et al. (2015) reported that the rhamnolipid BS, produced by *Pseudomonas* sp. LKY-5, has a HLB of 12.3, while the oleil-sophorolipid obtained by Brown et al. (1991) has a HLB value about 7–8, this being an excellent humectant for skin in cosmetic formulations. Additionally, Kim et al. (2002) established that the HLB of the mannosylerythritol lipid (MEL-SY16) produced by *Candida* sp. SY16 was 8.8, which is in the range of the HLB values established by other authors for glycoproteins. Marqués et al. (2009) also reported that the glycolipid synthesized by *Rhodococcus* sp. 5117 has a HLB about 11.

## Biocompatibility of lipopeptide and glycoprotein BSs

The biocompatibility of BSs is directly related with their structure. In this way, BSs have an advantage in comparison with the synthetic surfactants as they are composed of biomolecules, which are present in human cells and natural environments.

In the last years, some industrial sectors, including cosmetic and personal care companies, are eager to include more natural components in their formulations due to the demand of consumers for products with less irritant and cytotoxic effects. Hence, the replacement of chemical surfactants for bio-based surfactants or BSs is of much interest in these industrial sectors (Vecino et al. 2017a).

Some studies have indicated that lipopeptides have good properties for their inclusion in cosmetic and pharmaceutical formulations because of the similarity of their lipids and amino acids with those present in animal cells. However, compared to chemical surfactants, BSs have high production cost, derived from their biotechnological production process and downstream processing including extraction and purification operations. Therefore, the application of these natural surfactants has to be focused towards industrial applications where the final product has important margin of benefits like cosmetic and pharmaceutical industries.

Table 4 shows the content of lipids and proteins of the most studied lipopeptide BSs, highlighting differences among them. For example, the percentages of proteins and lipids contained in the BS produced by *P. aeruginosa* and *Azotobacter chroococcum* are higher than those reported by other authors for different lipopeptides with surfactant capacity. These differences can be due to the different methods used for their separation and purification. Vecino et al. (2015a) reported that *L. pentosus* produces a glycolipopeptide, composed of 10% carbohydrates, 30% proteins and 60% lipids, with a hydrophobic chain composed by linoelaidic acid (C18:2), oleic or elaidic acid (C18:1), palmitic acid (C16) and stearic acid (C18). As it can be observed, the major fractions of this BS comprise lipids and proteins, thus it can be expected to have a similar behavior between this BS and lipopeptide BSs.

Additionally, Rufino et al. (2014) observed that *C. lipolytica* produces a BS composed of 50% protein, 20% lipids and a small percentage of carbohydrates (8%),

Table 4. BS characterization in terms of lipid and protein content.

| Microorganisms | Lipid (%) | Protein (%) | Reference |
|---|---|---|---|
| *Azotobacter chroococcum* | 31.3 | 68.7 | Thavasi et al. (2009) |
| *Bacillus subtilis* | 18.0 | 15.2 | Pemmaraju et al. (2012) |
| *Pseudomonas aeruginosa* | 50.2 | 49.8 | Thavasi et al. (2011) |
| *Bacillus subtilis* | 17.0 | 13.2 | Makkar and Cameotra (1999) |
| CSL microbial biomass* | 64.2 | 21.9 | Vecino et al. (2014) |
| *Candida lipolytica* | 20.0 | 50.0 | Rufino et al. (2014) |
| *Lactobacillus pentosus* | 60.0 | 30.0 | Vecino et al. (2015a) |
| *Lactobacillus paracasei* | 13.7–24.4 | 21.2–58.2 | Vecino et al. (2017b) |

*Spontaneous microbial biomass growth in CSL like *Lactobacillus* or *Bacillus* species.

with a CMC of 300 mg/L. This biochemical composition is similar to that obtained by Vecino et al. (2015a) for the BS produced by *L. pentosus*.

## Final Thoughts

Nowadays, there is an increasing interest in the incorporation of more biocompatible, biodegradable and renewable BSs, like lipopeptides and glycoproteins, in the cosmetic, pharmaceutical and environmental industries. However, before incorporating BSs in these industrial sectors, it would be necessary to carry out detailed investigation into their HLB and ionic character, as well as potential applications and permeation on human cells. Additionally, it would be interesting to increase the number of investigations about glycoproteins, which are scarce in comparison with other BSs like lipopeptides or glycolipids. It can be speculated that the metabolic production of glycoproteins is different than the other BSs. In fact, the CMC of glycoproteins is much higher than the CMC of lipopeptides. Hence, glycoproteins are cell-bound to the plasmatic membrane of microorganisms, whereas lipopeptides and glycolipids are excreted to the fermentation medium. Contrary to lipopeptides and glycolipids, most of the glycoproteins are produced by lactic acid bacteria, which are Generally Recognized As Safe, "GRAS", by the American Food and Drug Administration, providing them a prebiotic connotation.

## Acknowledgments

This study was supported by the Spanish Ministry of Economy and Competitiveness (MINECO) under the project CTM2015-68904 (FEDER funds) and by the Xunta de Galicia under project ED431B 2017/77. X. Vecino acknowledges MINECO for funding through a Juan de la Cierva contract (FJCI-2014-19732). L. Rodríguez-López acknowledges the Spanish Ministry of Education, Culture and Sport for her pre-doctoral fellowship (FPU15/00205). M. Rincón-Fontán expresses her gratitude to University of Vigo for her pre-doctoral scholarship.

# References

Ahimou, F., P. Jacques and M. Deleu. 2000. Surfactin and iturin A effects on *Bacillus subtilis* surface hydrophobicity. Enzyme Microb. Technol. 27: 749–754.

Aivaliotis, M., P. Samolis, E. Neofotistou, H. Remigy, A.K. Rizos and G. Tsiotis. 2003. Molecular size determination of a membrane protein in surfactants by light scattering. Biochim. Biophys. Acta. 1615: 69–76.

Apostoluk, W. and J. Szymanowski. 1998. Estimation of alcohol adsorption parameters at hydrocarbon/ water interfaces in systems containing various hydrocarbons. Colloid. Surface A 132: 137–143.

Ara, I. and C.N. Mulligan. 2015. Reduction of chromium in water and soil using a rhamnolipid biosurfactant. Geotech. Engin. 46: 24–31.

Asaka, O. and M. Shoda. 1996. Biocontrol of *Rhizoctonia solani* damping-off of tomato with *Bacillus subtilis* RB14. Appl. Environ. Microbiol. 62: 4081–4085.

Atlas HLB System. 1963. A Time-saving Guide to Emulsifier Selection, third ed., Atlas Chemical Industries Inc., Wilmington, DE 1: 7–13.

Banat, I.M., S.K. Satpute, S.S. Cameotra, R. Patil and N.V. Nyayanit. 2014a. Cost effective technologies and renewable substrates for biosurfactants' production. Front. Microbiol. 5: 697.

Banat, I.M., M.A. Díaz De Rienzo and G.A. Quinn. 2014b. Microbial biofilms: biosurfactants as antibiofilm agents. Appl. Microbiol. Biot. 98: 9915–9929.

Béchet, M., T. Caradec, W. Hussein, A. Abderrahmani, M. Chollet, V. Leclére, T. Dubois, D. Lereclus, M. Pupin and P. Jacques. 2012. Structure, biosynthesis, and properties of kurstakins, nonribosomal lipopeptides from *Bacillus* spp. Appl. Microbiol. Biot. 95: 593–600.

Brown, M.J. 1991. Biosurfactants for cosmetic applications. Int. J. Cosmetic Sci. 13: 61–64.

Bruscalupi, G., S. Leoni, A.F. Panzali, S. Spagnuolo and G. Vullo. 1979. Interference of some detergent micelles with the titration of protein. Boll. Soc. Ital. Biol. Sper. 55: 450–454.

Brzozowski, B., W. Bednarski and P. Gołek. 2011. The adhesive capability of two *Lactobacillus* strains and physicochemical properties of their synthesized biosurfactants. Food Technol. Biotech. 49: 177–186.

Bustos, G., N. de la Torre, A.B. Moldes, J.M. Cruz and J.M. Domínguez. 2007. Revalorization of hemicellulosic trimming vine shoots hydrolyzates trough continuous production of lactic acid and biosurfactants by *L. pentosus*. J. Food Eng. 78: 405–412.

Ceresa, C., M. Rinaldi, V. Chiono, I. Carmagnola, G. Allegrone and L. Fracchia. 2016. Lipopeptides from *Bacillus subtilis* AC7 inhibit adhesion and biofilm formation of Candida albicans on silicone. Anton. Leeuw. Int. J. G. 109: 1375–1388.

Chen, W.C., R.S. Juang and Y.H. Wei. 2015. Applications of a lipopeptide biosurfactant, surfactin, produced by microorganisms. Biochem. Eng. J. 103: 158–169.

Coronel-León, J., G. de Grau, A. Grau-Campistany, M. Farfan, F. Rabanal, A. Manresa and A.M. Marqués. 2015. Biosurfactant production by AL 1.1, a *Bacillus licheniformis* strain isolated from Antarctica: production, chemical characterization and properties. Annal. Microbiol. 65: 2065–2078.

Davies, J.T. 1957. A quantitative kinetic theory of emulsion type, I. Physical chemistry of the emulsifying agent, gas/liquid and liquid/liquid interface. *In*: Proceedings of the 2nd International Congress of Surface Activity, 426–438, Butterworths, London.

Dehghan-Noudeh, G., M. Housaindokht and B.S.F. Bazzaz. 2005. Isolation, characterization, and investigation of surface and hemolytic activities of a lipopeptide biosurfactant produced by *Bacillus subtilis* ATCC 6633. J. Microbiol. Seoul. 43: 272–276.

Delbeke, E.I.P., J. Everaert, E. Uitterhaegen, S. Verweire, A. Verlee, T. Talou, W. Soetaert, I.N.A. Van Bogaert and C.V. Stevens. 2016. Petroselinic acid purification and its use for the fermentation of new sophorolipids. AMB Express 6: 28–36.

Deleu, M., M. Paquot and T. Nylander. 2008. Effect of fengycin, a lipopeptide produced by *Bacillus subtilis*, on model biomembranes. Biophys. J. 94: 2667–2679.

Duarte, C., E.J. Gudiña, C.F. Lima and L.R. Rodrigues. 2014. Effects of biosurfactants on the viability and proliferation of human breast cancer cells. AMB Express 4: 1–12.

Faivre, V. and V. Rosilio. 2010. Interest of glycolipids in drug delivery: From physicochemical properties to drug targeting. Expert Opin. Drug. Deliv. 7: 1031–1048.

Ferreira, A., X. Vecino, D. Ferreira, J.M. Cruz, A.B. Moldes and L.R. Rodrigues. 2017. Novel cosmetic formulations containing a biosurfactant from *Lactobacillus paracasei*. Colloids Surface B 155: 522−529.

Gharaei-Fathabad, E. 2011. Biosurfactants in pharmaceutical industry: a mini-review. Amer. J. Drug Disc. Devel. 1: 58−69.

Golek, P., W. Bednarski, B. Brzozowski and B. Dziuba. 2009. The obtaining and properties of biosurfactants synthesized by bacteria of the genus Lactobacillus. Annal. Microbiol. 59: 119−126.

Gómez-Graña, S., M. Perez-Ameneiro, X. Vecino, I. Pastoriza-Santos, J. Perez-Juste, J.M. Cruz and A.B. Moldes. 2017. Biogenic synthesis of metal nanoparticles using a biosurfactant extracted from corn and their antimicrobial properties. Nanomaterials 139: 1−14.

Griffin, W.C. 1949. Classification of surface-active agents by HLB. J. Soc. Cosmet. Chem. 1: 311–326.

Gudiña, E.J., J.A. Teixeira and L.R. Rodrigues. 2010. Isolation and functional characterization of a biosurfactant produced by *Lactobacillus paracasei*. Colloid. Surface B 76: 298−304.

Gudiña, E.J., E.C. Fernandes, J.A. Teixeira and L.R. Rodrigues. 2015. Antimicrobial and anti-adhesive activities of cell-bound biosurfactant from *Lactobacillus agilis* CCUG31450. RSC Adv. 5: 90960−90968.

Gudiña, E.J., J.A. Teixeira and L.R. Rodrigues. 2016. Biosurfactants produced by marine microorganisms with therapeutic applications. Mar. Drugs 14: 38−52.

Han, X., X.D. Du, L. Southey, D.M. Bulach, T. Seemann, X.X. Yan et al. 2015. Functional analysis of a bacitracin resistance determinant located on ICECp1, a novel tn916-like element from a conjugative plasmid in clostridium perfringens. Antimicrob. Agents Chemother. 59: 6855−6865.

Hull, S.R., B.Y. Yang, D. Venzke, K. Kulhavy and R. Montgomery. 1996. Composition of corn steep water during steeping. J. Agr. Food Chem. 44: 1857–1863.

Isogai, A., S. Takayama, S. Murakoshi and A. Suzuki. 1982. Structure of β-amino acids in antibiotics iturin A. Tetrahedron Lett. 23: 3065−3068.

Joshi, S., C. Bharucha and A.J. Desai. 2008. Production of biosurfactant and antifungal compound by fermented food isolate *Bacillus subtilis* 20B. Bioresour. Technol. 99: 4603−4608.

Kawahara, H., A. Hirai, T. Minabe and H. Obata. 2013. Stabilization of astaxanthin by a novel biosurfactant produced by *Rhodotorula mucilaginosa* KUGPP-1. Biocontrol. Sci. 18: 21−28.

Kim, H.S., J.W. Jeon, S.B. Kim, H.M. Oh, T.J. Kwon and B.D. Yoon. 2002. Surface and physico-chemical properties of a glycolipid biosurfactant, mannosylerythritol lipid, from *Candida Antarctica*. Biotechnol. Lett. 24: 1637−1641.

Kim, K.M., J.Y. Lee, C.K. Kim and J.S. Kang. 2009. Isolation and characterization of surfactin produced by *Bacillus polyfermenticus* KJS-2. Arch. Pharm. Res. 32: 711−715.

Kim, S.H., E.J. Lim, S.O. Lee, J.D. Lee and T.H. Lee. 2000. Purification and characterization of biosurfactants from Nocardia sp. L-417. Biotechnol. Appl. Biochem. 31: 249−253.

Kiran, G.S., A.S. Ninawe, A.N. Lipton, V. Pandian and J. Selvin. 2016. Rhamnolipid biosurfactants: Evolutionary implications, applications and future prospects from untapped marine resource. Crit. Rev. Biotechnol. 36: 399−415.

Kowall, M., J. Vater, B. Kluge, T. Stein, P. Franke and D. Ziessow. 1998. Separation and characterization of surfactin isoforms produced by *Bacillus subtilis* OKB 105. J. Colloid Interf. Sci. 204: 1−8.

Lee, J., J.H. Griffin and T.I. Nicas. 1996. Solid-phase total synthesis of bacitracin A. J. Org. Chem. 61: 3983−3986.

Li, L., C. Zhao, Q. Liu and C. Liu. 2015. Study on biosurfactant produced by *Pseudomonas* SP. LKY-5 and its stability. Pet. Process. Pet. 46: 18−21.

Liu, X., S. Yang and B. Mou. 2004. Molecular structures of microbial lipopeptides. Biotechnol. Bull. 4: 18−26.

Lowry, O.H., N.J. Rosebrough, A.L. Farr and R.J. Randall. 1951. Protein measurement with the Folin phenol reagent. J. Biol. Chem. 193: 265–275.

Madhu, A.N. and S.G. Prapulla. 2014. Evaluation and functional characterization of a biosurfactant produced by *Lactobacillus plantarum* CFR 2194. Appl. Biochem. Biotech. 172: 1777−1789.

Makkar, R.S. and S.S. Cameotra. 1999. Biosurfactant production by microorganisms on unconventional carbon sources. J. Surfactants Deterg. 2: 237−241.

Mandal, S.M., S. Sharma, A.K. Pinnaka, A. Kumari and S. Korpole. 2013. Isolation and characterization of diverse antimicrobial lipopeptides produced by *Citrobacter* and *Enterobacter*. BMC Microbiol. 13: 152−160.

Marqués, A.M., A. Pinazo, M. Farfan, F.J. Aranda, J.A. Teruel, A. Ortiz and M.J. Espuny. 2009. The physicochemical properties and chemical composition of trehalose lipids produced by *Rhodococcus erythropolis* 51T7. Chem. Phys. Lipids 158: 110−117.

Moldes, A.B., A.M. Torrado, M.T. Barral and J.M. Domínguez. 2007. Evaluation of biosurfactant production from various agricultural residues by *Lactobacillus pentosus*. J. Agr. Food Chem. 55: 4481−4486.

Moldes, A.B., R. Paradelo, D. Rubinos, R. Devesa-Rey, J.M. Cruz and M.T. Barral. 2011. *Ex situ* treatment of hydrocarbon-contaminated soil using biosurfactants from *Lactobacillus pentosus*. J. Agr. Food Chem. 59: 9443−9447.

Moldes, A.B., R. Paradelo, X. Vecino, J.M. Cruz, E. Gudiña, L. Rodrigues, J.A. Teixeira, J.M. Domínguez and M.T. Barral. 2013. Partial characterization of biosurfactant from *Lactobacillus pentosus* and comparison with sodium dodecyl sulphate for the bioremediation of hydrocarbon contaminated soil. BioMed. Res. Int. Article number 961842.

Noah, K.S., D.F. Bruhn and G.A. Bala. 2005. Surfactin production from potato process effluent by *Bacillus subtilis* in a chemostat. In Twenty-Sixth Symposium on Biotechnology for Fuels and Chemicals. Humana Press 122: 465−473.

Ogonowski, J. and A. Tomaszkiewicz-Potępa. 2004. The Analysis of Surfactants, Wydawnictwo IGSMiE PAN, Kraków.

Pemmaraju, S.C., D. Sharma, N. Singh, R. Panwar, S.S. Cameotra and V. Pruthi. 2012. Production of microbial surfactants from oily sludge-contaminated soil by *Bacillus subtilis* DSVP23. Appl. Biochem. Biotechnol. 167: 1119−1131.

Pérez-Ameneiro, M., X. Vecino, J.M. Cruz and A.B. Moldes. 2015. Wastewater treatment enhancement by applying a lipopeptide biosurfactant to a lignocellulosic biocomposite. Carbohydr. Polym. 131: 186−196.

Pérez-García, A., D. Romero and A. de Vicente. 2011. Plant protection and growth stimulation by microorganisms: biotechnological applications of *Bacilli* in agriculture. Curr. Opin. Biotech. 22: 87−193.

Portilla-Rivera, O., A. Torrado, J.M. Domínguez and A.B. Moldes. 2008. Stability and emulsifying capacity of biosurfactants obtained from lignocellulosic sources using *Lactobacillus pentosus*. J. Agr. Food Chem. 56: 8074−8080.

Portilla-Rivera, O.M., A. Torrado-Agrasar, J. Carballo, J.M. Domínguez and A.B. Moldes. 2009. Development of a factorial design to study the effect of the major hemicellulosic sugars on the production of surface-active compounds by *L. pentosus*. J. Agr. Food Chem. 57: 9057−9062.

Portilla, O.M., B. Rivas, A. Torrado, A.B. Moldes and J.M. Domínguez. 2008. Revalorisation of vine trimming wastes using *Lactobacillus acidophilus* and *Debaryomyces hansenii*. J. Sci. Food Agr. 88: 2298−2308.

Price, N.P.J., A.P. Rooney, J.L. Swezey, E. Perry and F.M. Cohan. 2007. Mass spectrometric analysis of lipopeptides from Bacillus strains isolated from diverse geographical locations. FEMS Microbiol. Lett. 271: 83−89.

Quinn, G.A., A.P. Maloy, M.M. Banat and I.M. Banat. 2013. A comparison of effects of broad-spectrum antibiotics and biosurfactants on established bacterial biofilms. Curr. Microbiol. 67: 614−623.

Rincón-Fontán, M., L. Rodríguez-López, X. Vecino, J.M. Cruz and A.B. Moldes. 2016. Adsorption of natural surface active compounds obtained from corn on human hair. RSC Adv. 6: 63064−63070.

Rincón-Fontán, M., L. Rodríguez-López, X. Vecino, J.M. Cruz and A.B. Moldes. 2018. Design and characterization of greener sunscreen formulations based on mica powder and a biosurfactant extract. Powder Technol. 327: 442−448.

Rivera, O.M.P., A.B. Moldes, A.M. Torrado and J.M. Domínguez. 2007. Lactic acid and biosurfactants production from hydrolyzed distilled grape marc. Process Biochem. 42: 1010−1020.

Rodrigues, L.R., J.A. Teixeira, H.C. Van der Mei and R. Oliveira. 2006a. Physicochemical and functional characterization of a biosurfactant produced by *Lactococcus lactis* 53. Colloids Surface B 49: 79−86.

Rodrigues, L.R., J.A. Teixeira, H.C. Van der Mei and R. Oliveira. 2006b. Isolation and partial characterization of a biosurfactant produced by *Streptococcus thermophilus A.* Colloids Surface B 53: 105–112.

Rodríguez-López, L., X. Vecino, L. Barbosa-Pereira, A.B. Moldes and J.M. Cruz. 2016. A multifunctional extract from corn steep liquor: Antioxidant and surfactant activities. Food and Funct. 7: 3724–3732.

Rodríguez-López, L., M. Rincón-Fontán, X. Vecino, J.M. Cruz and A.B. Moldes. 2017. Ionic behavior assessment of surface-active compounds from corn steep liquor by exchange resins. J. Surfactants Deterg. 20: 207–217.

Rosenthal, K.S., P.E. Swanson and D.R. Storm. 1976. Disruption of *Escherichia coli* outer membranes by EM 49. A new membrane active peptide antibiotic. Biochem. 15: 5783–5792.

Rufino, R.D., J.M. de Luna, G.M. Takaki and L.A. Sarubbo. 2014. Characterization and properties of the biosurfactant produced by *Candida lipolytica* UCP 0988. Electron. J. Biotechn. 17: 34–38.

San Miguel, G.S., J.M.S. Nuñez and A.T. Tomas. 2015. Cosmetic product having DNA repair properties. U.S. Patent # 20160256375.

Saranya, P., P. Bhavani, S. Swarnalatha and G. Sekaran. 2015. Biosequestration of chromium (III) in an aqueous solution using cationic and anionic biosurfactants produced from two different *Bacillus* sp.—a comparative study. RSC Adv. 5: 80596–80611.

Satpute, S.K., A.G. Banpurkar, P.K. Dhakephalkar, I.M. Banat and B.A. Chopade. 2010. Methods for investigating biosurfactants and bioemulsifiers: a review. Crit. Rev. Biotechnol. 30: 127–144.

Schwartz, B.S., M.R. Warren, F.A. Barkley and L. Landis. 1959. Microbiological and pharmacological studies of colistin sulfate and sodium colistinmethanesulfonate. Antibiot. Annu. 7: 41–60.

Sharma, D., B.S. Saharan, N. Chauhan, A. Bansal and S. Procha. 2014. Production and structural characterization of *Lactobacillus helveticus* derived biosurfactant. Sci. World J. 2014: 493548–493557.

Shinichiro, H., Y. Takatsuka and M. Hiroyuki. 1981. Effect of surfactants on the nasal absorption of insulin in rats. Int. J. Pharm. 9: 165–172.

Sineriz, F., R.K. Hommel and H.P. Kleber. 2001. Production of biosurfactants. Encyclopedia of Life Support Systems. Eolls Publishers, Oxford.

Singh, A.K., R. Rautela and S.S. Cameotra. 2014. Substrate dependent *in vitro* antifungal activity of *Bacillus* sp. strain AR2. Microb. Cell Fact. 13: 2–11.

Stansly, P.G., R.G. Shepherd and H.J. White. 1947. Polymyxin: a new chemotherapeutic agent. Bull Johns Hopkins Hosp. 81: 43–54.

Thavasi, R., V.R.M.S. Nambaru, S. Jayalakshmi, T. Balasubramanian and I.M. Banat. 2009. Biosurfactant production by *Azotobacter chroococcum* isolated from the marine environment. Mar. Biotechnol. 11: 551–556.

Thavasi, R., V.R.M.S. Nambaru, S. Jayalakshmi, T. Balasubramanian and I.M. Banat. 2011. Biosurfactant production by *Pseudomonas aeruginosa* from renewable resources. Indian J. Microbiol. 51: 30–36.

Thies, S., B. Santiago-Schübel, F. Kovačić, F. Rosenau, R. Hausmann and K.E. Jaeger. 2014. Heterologous production of the lipopeptide biosurfactant serrawettin W1 in *Escherichia coli*. J. Biotechnol. 181: 27–30.

Vanittanakom, N., W. Loeffler, U. Koch and G. Jung. 1986. Fengycin-A novel antifungal lipopeptide antibiotic produced by *Bacillus subtilis* F-29-3. J. Antibiot. 39: 888–901.

Vater, J., B. Kablitz, C. Wilde, P. Franke, N. Mehta and S.S. Cameotra. 2002. Matrix-assisted laser desorption ionization-time of flight mass spectrometry of lipopeptide biosurfactants in whole cells and culture filtrates of *Bacillus subtilis* C-1 isolated from petroleum sludge. Appl. Environ. Microbiol. 68: 6210–6219.

Vecino, X., R. Devesa-Rey, J.M. Cruz and A.B. Moldes. 2012. Study of the synergistic effects of salinity, pH, and temperature on the surface-active properties of biosurfactants produced by *Lactobacillus pentosus*. J. Agr. Food Chem. 60: 1258–1265.

Vecino, X., R. Devesa-Rey, J.M. Cruz and A.B. Moldes. 2013. Evaluation of biosurfactant obtained from *Lactobacillus pentosus* as foaming agent in froth flotation. J. Environ. Manage. 128: 655–660.

Vecino, X., L. Barbosa-Pereira, R. Devesa-Rey, J.M. Cruz and A.B. Moldes. 2014. Study of the surfactant properties of aqueous stream from the corn milling industry. J. Agr. Food Chem. 62: 5451–5457.

Vecino, X., L. Barbosa-Pereira, R. Devesa-Rey, J.M. Cruz and A.B. Moldes. 2015a. Optimization of extraction conditions and fatty acid characterization of *Lactobacillus pentosus* cell-bound biosurfactant/bioemulsifier. J. Sci. Food Agr. 95: 313−320.

Vecino, X., G. Bustos, R. Devesa-Rey, J.M. Cruz and A.B. Moldes. 2015b. Salt-free aqueous extraction of a cell-bound biosurfactant: A kinetic study. J. Surfactants Deterg. 18: 267−274.

Vecino, X., L. Barbosa-Pereira, R. Devesa-Rey, J.M. Cruz and A.B. Moldes. 2015c. Optimization of liquid-liquid extraction of biosurfactants from corn steep liquor. Bioproc. Biosyst. Eng. 38: 1629−1637.

Vecino, X., L. Rodríguez-López, J.M. Cruz and A.B. Moldes. 2015d. Sewage sludge polycyclic aromatic hydrocarbon (PAH) decontamination technique based on the utilization of a lipopeptide biosurfactant extracted from corn steep liquor. J. Agric. Food Chem. 63: 7143−7150.

Vecino, X., J.M. Cruz, A.B. Moldes and L.R. Rodrigues. 2017a. Biosurfactants in cosmetic formulations: Trends and challenges. Crit. Rev. Biotechnol. 37: 911−923.

Vecino, X., L. Rodríguez-López, E.J. Gudiña, J.M. Cruz, A.B. Moldes and L.R. Rodrigues. 2017b. Vineyard pruning waste as an alternative carbon source to produce novel biosurfactants by *Lactobacillus paracasei*. J. Ind. Eng. Chem. 55: 40−49.

Vecino, X., L. Rodríguez-López, D. Ferreira, J.M. Cruz, A.B. Moldes and L.R. Rodrigues. 2017c. Bioactivity of glycolipopeptide cell-bound biosurfactants against skin pathogens. Int. J. Biol. Macromol. Doi.org/10.1016/j.ijbiomac.2017.11.088.

Velraeds, M.M.C., H.C. Van Der Mei, G. Reid and H.J. Busscher. 1996. Physicochemical and biochemical characterization of biosurfactants released by Lactobacillus strains. Colloid Surface B 8: 51−61.

Walia, N.K. and S.S. Cameotra. 2015. Lipopeptides: Biosynthesis and applications. J. Microb. Biochem. Technol. 7: 103−107.

Waterborg, J.H. 2002. The Protein Protocols Handbook. Humana Press Inc., Totowa, NJ.

Yang, C.C., C.C. Chang, K. Hamaguchi, K. Ikeda, K. Hayashi and T. Suzuki. 1967. Optical rotatory dispersion of cobrotoxin. J. Biochem. 61: 272−274.

Youssef, N.H., K.E. Duncan and M.J. McInerney. 2005. Importance of 3-hydroxy fatty acid composition of lipopeptides for biosurfactant activity. Appl. Environ. Microbiol. 71: 7690−7695.

Zhang, Y., C. Liu, B. Dong, X. Ma, L. Hou, X. Cao and C. Wang. 2015. Anti-inflammatory activity and mechanism of surfactin in lipopolysaccharide-activated macrophages. Inflammation 38: 756−764.

Zhou, J.F., G.Q. Li, J.J. Xie, X.Y. Cui, X.H. Dai, H.M. Tian, P.K. Gao, M.M. Wu and T. Ma. 2016. A novel bioemulsifier from *Geobacillus stearothermophilus* A-2 and its potential application in microbial enhanced oil recovery. RSC Adv. 6: 96347−96354.

Zieliński, R. 2013. Surfactants Structure, Properties, Applications. 2nd ed. Wydawnictwo Uniwersytetu Ekonomicznego w Poznaniu.

# 7

# Biosurfactants from Lactic Acid Bacteria

*Surekha K. Satpute,*[1,*] *Parijat Das,*[1] *Karishma R. Pardesi,*[1]
*Nishigandha S. Mone,*[1] *Deepansh Sharma*[2] and
*Ibrahim M. Banat*[3]

## Introduction

### *Lactic acid bacteria (LAB)-General outlook*

Louis Pasteur was the first to discover the relationship between food fermentation process and lactic acid bacteria (LAB) which has since achieved a significant status in our life, becoming a vital part of our food industries (Coeuret et al. 2003). Some of the frequently reported genera of LAB used in food include *Lactobacillus, Streptococcus, Lactococcus, Carnobacterium, Enterococcus Tetragenococcus, Pediococcus, Leuconostoc, Weissella, Oenococcus*, and *Vagococcus*. Apart from lactic acid production, other compounds are produced by LAB with roles in the food industry including starter culture, providing: aroma, flavoring, bioactive peptides, texture improvement, exopolysaccharides and bacteriocins. LABS are also used in food processing because they lower the carbohydrate content of the foods they ferment. Lactobacilli are used for the production of cheese, yoghurt, fermented plant foods, fermented meats, beer and sourdough bread (Stiles and Holzapfel 1997). In addition, they are commensals on the urogenital tract of healthy pre-menopausal women (Barrons and Tassone 2008). When Lactobacilli in the urinary tract are overwhelmed by uropathogens, urinary tract infections (UTI) are often detected. Lactobacilli can interfere with pathogens through several mechanisms: (i) their cells (viable and non-viable) and cell wall fragments can competitively inhibit uropathogens

[1] Department of Microbiology, Savitribai Phule Pune University, Pune-411007, Maharashtra, India.
[2] Amity Institute of Microbial Technology, Amity University, Jaipur, Rajasthan, India.
[3] School of Biomedical Sciences, University of Ulster, Coleraine, N. Ireland, UK.
* Corresponding author: drsurekhasatpute@gmail.com

adherence to uroepithelial cells, polymers and catheter surfaces, (ii) Lactobacilli can co-aggregate with uropathogens to eliminate them or (iii) they may produce a variety of metabolic by-products with antimicrobial activity, like hydrogen peroxide, lactic acid, bacteriocins and biosurfactants (BSs) (Rivera et al. 2007, Reid et al. 2000, 1998, Velraeds et al. 1997, 1996a, b). Goudarzi et al. (2016) have demonstrated anti-swarming effects as well as antimicrobial bacteriocins and BS from probiotics bacteria against *Proteus* spp. These species belonging to *Enterobacteriaceae* family represent the third most, UTI infection causing organisms. They also observed that bacteriocin successfully inhibits *Proteus* spp. interfering with their swarming migration abilities whereas BS in a solvent form does not exhibit considerable effect on the same organism.

LAB represents first and foremost indigenous normal microflora on various surfaces of skin. It is also found frequently on surfaces of vegetables and fruits. During surface cleaning procedures, there is a high chance that these indigenous organisms are washed away or eliminated. Fracchia et al. (2010) isolated fifteen LAB from fresh fruits (apples and pears) and vegetables (cucumbers, lettuce and cabbage) from a rural area of Piedmont, Italy. Twelve isolates of these were genotypically characterized under genus *Lactobacillus*. This indicates the dominance of *Lactobacillus* spp. on fruits and vegetable surfaces. BS producing LAB isolated from brined cucumbers exhibits antifungal activities against some pathogenic fungi belonging to the *Penicillium* and *Aspergillus* spp. Researchers have not only used dairy, fruits, vegetable and human system but chicken ceca is also used for isolation of BS producing *Lactobacillus* strain. BS obtained from *L. curvatus* DN317 demonstrated antimicrobial activity against various strains like *Campylobacter jejuni*, *Listeria monocytogenes* and *Bacillus subtilis* (Zommiti et al. 2017). LAB are therefore useful for biotechnological purposes due to their broad spectrum antimicrobial activities.

Various fermented products like cheese, yogurt, sauerkraut, cream, sausages are routine part of our life throughout many years. Currently, consumers are mostly interested in nutritionally rich and low carbohydrate content in foods—a requirement which is readily met by fermented food products which ideally exhibit such properties. Proliferation of LAB in fermented food enhances their palatability along with increasing their nutritional values. However, this outcome is not guaranteed every time, since some of the microbes may exhibit spoilage activities in all food products including dairy, meat and beverages. In addition, some of the other antimicrobial metabolites have other beneficial roles in the human intestine including:

i. Co-aggregation molecules (formation of normal flora of helpful organisms which impedes the growth of pathogens).
ii. Production of various acids (Lactic, formic, acetic acids).
iii. Hydrogen peroxide (Harmful to pathogens).
iv. Bacteriocins (Antagonistic effect to pathogens).
v. Antiadhesion factor (Impeding the adhesion of microorganism).

The above mentioned metabolites released by LAB can be termed as "antibiotic assistants" since they can assist to solve the problems associated with antibiotic resistance (Sgibnev and Kremleva 2017). Additionally, LAB also release a wide variety of antibiotics, enzymes and anti-carcinogenic substances. Their major roles are contributed by several acids, particularly, lactic acid which provides ideal preservation capabilities. LAB are "generally recognized as safe" (GRAS) microorganisms; therefore, these organisms can be exploited in the best possible ways for the benefit of mankind.

## The Benefit of Biosurfactant (BS) Production by Lactic Acid Bacteria (LAB)

Several researchers have proven the hypothesis that LAB and their BS products play a vital role in controlling infections of pathogenic organisms (Gómez et al. 2016, Kanmani et al. 2013). The attraction towards such role has been increasing due to the escalating antimicrobial resistance developing for many microorganisms (Diaz De Rienzo et al. 2016).

Literature suggests that LAB plays a critical role in reducing the occurrence or period of antibiotic use related conditions such as diarrhea (Issa and Moucari 2014). LAB works like an obstacle to reduce the incidence of pathogenic diseases such as vaginal candidiasis and bacterial vaginosis. Urinary tract infections (UTI) are often treated or prevented through ingestion of LAB (Borchert et al. 2008). Additionally, LAB improve the immunological defense system and also diminishes the activity of some toxic metabolites. All these approaches sound highly effective, feasible and safe. Pathogenic organisms dominate surroundings through biofilm formation where an extracellular matrix of polysaccharides protects them from harsh conditions. Biofilms represent the aggregations that facilitate pathogens to act as potent competitors over planktonic or free-floating organisms (Satpute et al. 2016a, b, 2018). Such biofilm environment is extremely suitable for antibiotic resistance development and antibiotic gene transfer, proliferation and quorum sensing. Surrounding planktonic microbes also try to occupy the available space and access the nutrients through production of some antagonistic effect. Basically, it is an 'Attack and Defense' mechanism carried out to determine existence and predominance.

Biofilm formers can be successfully reduced or eliminated through microbial systems that can secrete inhibitory compounds (Donlan 2002, Valle et al. 2006). In this regard, BSs produced by LAB exhibit antimicrobial, antibiofilm and antiadhesive antagonistic properties against many pathogens (Servin 2004). One of the recent works documented by Morais et al. (2017) characterized BSs produced from two strains viz., *L. jensenii* $P_{6A}$ and *L. gasseri* $P_{65}$. They also reported antimicrobial properties of both BSs against *E. coli* and *C. albicans*. Both BSs also exhibited anti-adhesive potential against *S. saprophyticus, E. coli* and *E. aerogenes*. Their extensive studies on stability at different pH (2 to 10) and temperatures (100°C) has importance for medical related applications. Sharma and Saharan (2016) worked on *L. helveticus* derived BS for its biomedical potential and demonstrated antimicrobial and antiadhesive properties and inhibition of biofilm forming pathogenic organisms on silicone surfaces. The stability of BSs over different pH ranges (4.0–12.0) and high temperatures (125°C/15 min) were interesting characteristics. Table 1 provides detailed summary of LAB that mainly produce cell associated (CA) and cell free (CF) type BS.

**Table 1.** Detailed summary of lactic acid bacteria (LAB) producing diverse types of biosurfactant (BS) and their potential applications.

| Organism and its Source | CFBS/CABS | Type of Biosurfactant (BS) | Potential Application | Reference |
|---|---|---|---|---|
| *Lactobacillus jensenii* P$_{6A}$ | CABS | Glycolipoproteins | Antimicrobial against *E. coli* and *C. albicans* | Morais et al. 2017 |
| *Lactobacillus gasseri* P$_{65}$ | CFBS | Glycolipoproteins | Antiadhesive activity against *E. coli*, *E. aerogenes*, *S. saprophyticus* | |
| *Lactobacillus casei* | CABS | – | Antioxidant and antiproliferative potential, antiadhesive and anti-biofilm | Merghni et al. 2017 |
| *Lactobacillus lactis* CECT-4434 | CABS | – | Inhibition of *L. sakei* and *S. aureus* | Souza et al. 2017 |
| *Lactobacillus paracasei* | CABS | Glycolipopeptide Glycoprotein | Different types of BS can be produced from same strain with alteration in carbon sources | Vecino et al. 2017 |
| *Lactobacillus* | CABS | – | Anti Swarming Effects and antimicrobial Bacteriocins against *Proteus* spp. No considerable activity of BS | Goudarzi et al. 2016 |
| *Lactobacillus paracasei* | CABS | Glycolipopeptide Glycoprotein | Low-cost carbon source for BS production Same strain but changing the carbon source | Vecino et al. 2016 |
| *Lactobacillus helveticus* MRTL91 | CABS | – | Antimicrobial properties—*B. cereus*, *S. aureus*, *S. epidermidis*, *P. aeruginosa*, *S. typhi*, *S. flexneri* | Sharma and Saharan 2016 |
| *Lactobacillus brevis* (Fresh cabbage) | CFBS | Mixture of components including sugars | Antifungal, Antiadhesion, antibiofilm assay | Ceresa et al. 2015 |
| *Enterococcus faecium* MRTL9 | CABS | Glycolipids (xylolipid) Hexadecanoic acid | Antiadhesive and antimicrobial activity against biofilm forming bacteria and yeast | Sharma et al. 2015 |
| *Lactobacillus agilis* CCUG31450 | CABS | Glycoprotein | Antiadhesive | Gudiña et al. 2015 |
| *Lactobacillus pentosus* | CABS | Glycolipopeptide | Different extraction conditions utilized to produce BS with emulsifier property | Vecino et al. 2015 |
| *Lactobacillus acidophilus* ATCC 4356 | CABS | Surlactin-Protein rich | Antiadhesive properties against *S. marcescens* strains | Shokouhfard et al. 2015 |

| Organism | | BS Type | Activity | Reference |
|---|---|---|---|---|
| *Lactobacillus plantarum* CFR2194 (Kanjika–rice—fermented product) | CABS | Glycoprotein | Antimicrobial and antiadhesive | Madhu and Prapulla 2014 |
| *Lactobacillus casei* MRTL3 | CABS | Glycolipid | Produces both BS and bacteriocin | Sharma and Saharan 2014 |
| *Lactobacillus helveticus* MRTL91 | CABS | Glycolipid closely similar to xylolipids. Hexadecanoic acid | Phytotoxicity Assay—seed germination and root elongation of the *Brassica nigra* and *Triticum aestivum* Cytotoxicity Assessment—mouse fibroblast cell line | Sharma et al. 2014 |
| *Lactobacillus jensenii* *Lactobacillus rhamnosus* American type culture collection (ATCC) | CABS | — | Biofilm dispersal and antimicrobial | Sambanthamoorthy et al. 2014 |
| *Lactobacillus reuteri* (DSM) | CABS | — | Effect of BS on gene expression of essential adhesion genes (gtfB, gtfC and ftf) of *Str. mutans* | Salehi et al. 2014 |
| *Lactobacillus rhamnosus* (Vagina of Iraqi healthy woman) | CFBS | — | Antibacterial, Antibiofilm, Antiadhesive | Salman and Alimer 2014 |
| *Lactobacillus pentosus* and *Lactobacillus plantarum* co-culture (CECT 4023 and CECT 221) | CABS | — | Antimicrobial | Rodriguez-Pazo et al. 2013 |
| *Lactobacillus pentosus* | CABS | Surlactin-Protein rich | Structural properties foaming agents; sediments treated with BS | Vecino et al. 2013 |
| *Lactobacillus* spp. (Egyptian dairy product) | CABS and CFBS | — | Antimicrobial | Gomaa et al. 2013 |
| *Lactobacillus pentosus* | CABS | Glycolipopeptide Glycoprotein | Bioremediation of octane-contaminated soil by improving the solubilization of octane in the water phase of soil | Moldes et al. 2013 |
| *Lactobacillus plantarum* KSBT 5 (tradiitional food product of India) | CFBS | — | Inhibition of growth and pathogenicity of *Salmonella enterica serovar* | Das et al. 2013 |
| *Lactobacillus* spp. (Yogurt, Cheese and Silage) | CFBS | — | Antagonistic activity; bacteriocin and BS production | Kermanshahi et al. 2012 |

*Table 1 contd. ...*

*...Table 1 contd.*

| Organism and its Source | CFBS/CABS | Type of Biosurfactant (BS) | Potential Application | Reference |
|---|---|---|---|---|
| *Lactobacillus plantarum* | CABS | Glycolipid | Structural characterization | Sauvageau et al. 2012 |
| *Lactobacillus* spp. (Pendidam) | CFBS | – | Antibacterial | Mbawala et al. 2012 |
| *Lactobacillus acidophilus* | CABS | – | Antibiofilm activity against *Pro. mirabilis* | Ali 2012 |
| *Lactobacillus delbrueckii* | CFBS | Glycolipid | BS production and structural characterization | Thavasi et al. 2011 |
| *Lactobacillus acidophilus* (DSM) | CABS | Surlactin-protein rich Glycoprotein | Effect on GTFB and GTFC expression level | Tahmourespour et al. 2011 |
| *Lactobacillus fermentii* and *Lactobacillus rhamnosus* (CCCIFM, Polnad) | CABS | Surlactin type protein rich | Antiadhesive properties against the *K. pneumonia, P. aeruginosa* and *E. coli* | Brzozowski et al. 2011 |
| *Lactococcus lactis* (Curd sample) | CFBS | Xylolipid, Octadecanoic acid | Antibacterial activity against the multi-drug resistant pathogens | Sarvanakumari and Mani 2010 |
| *Lactobacillus paracasei* (Portuguese dairy industry) | CABS | – | Antimicrobial and antiadhesive | Gudiña et al. 2010a |
| *Lactococcus paracasei* subsp. *paracasei* A20 (Dairy industry) | CABS | – | Antimicrobial and antiadhesive | Gudiña et al. 2010b |
| *Lactococcus lactis* (CECT 4434) | CABS | – | Simultaneous extraction of BS and bacteriocin | Rodriguez et al. 2010 |
| *Lactobacillus delbrueckii* sp. *delbruckii* (DSMZ) | CFBS | Mixture-components containing sugars | Inhibition against *C. albicans* | Fracchia et al. 2010 |
| *Lactobacillus acidophilus* | CABS | Surlactin-Protein rich | Presence of MgSO$_4$ and MnSO$_4$ is important bacterial growth and surlactin production | Fouad et al. 2010 |
| *Lactobacillus casei* 8/4 | CABS | Glycoproteins or glycoproteins with phosphoric groups | Antimicrobial properties gains pathogenic microflora | Gołek et al. 2009 |

| Organism | CABS | Compound | Application/Property | Reference |
|---|---|---|---|---|
| *Lactobacillus acidophilus* | CABS | – | Antiadhesive activity against *S. aureus* | Walencka et al. 2008 |
| *Lactobacillus plantarum* | CABS | – | Strain utilizes glucose from cellulose films hydrolysis No need for any liqueurs clarification procedures | Portilla-Rivera et al. 2007b |
| *Lactobacillus pentosus* (CECT 4023T) | CABS | – | – | Moldes et al. 2007 |
| *Lactobacillus* strains | CABS | – | Adhesive properties | Gołek et al. 2007 |
| *Streptococcus thermophiles* A | CABS | Glycolipid | Antimicrobial and antiadhesive | Rodrigues et al. 2006d |
| *Lactobacillus casei* CECT525, *Lactobacillus rhamnosus* CECT288, *Lactobacillus pentosus* C ECT4023 and *Lactobacillus coryniformis* subsp. *Torquens* (CECT 25600 Spanish culture collection center) | CABS | Fraction rich in glycoproteins | Kinetics of BSs production | Rodrigues et al. 2006c |
| *Lactococcus lactis* 53 *Streptococcus thermophilus* A | CABS | – | Antibiofilm | Rodrigues et al. 2004 |
| *Lactobacillus fermentum* RC-14 | CABS | – | Inhibit *Staphylococcus aureus* infection of surfaces of surgical implants in rat | Gan et al. 2002 |
| *Streptococci mitis* (Human oral cavity) | CABS | Rhamnolipid like compound | Inhibition of *Streptococci mutans* | van Hoogmoed et al. 2004 |
| *Lactobacillus fermentum* RC-14, *L. rhamnosus* GR-1 and 36, and *Lactobacillus casei* | CABS | Surlactin-Protein rich | Utilized surface-enhanced laser desorption/ionization (SELDI) to characterize clinically important BS solutions | Howard et al. 2000 |
| *Lactobacillus fermentum* RC-14 (Urogenital isolate–healthy woman) | CABS | Surlactin-Protein rich | Inhibits adhesion of *Enterococcus faecalis* | Heinemann et al. 2000 |
| *Lactobacillus fermentum* *Lactobacillus casei rhamnosus* ATCC 7469 | CABS | – | Antiadhesive; Inhibition of biofilms - uropathogens on silicone rubber | Velraeds et al. 2000 |
| *Lactobacillus* | CABS | Surlactin | Inhibit the adhesion of uropathogen and enterococci on polymer surfaces | Reid et al. 1999 |

*Table 1 contd....*

*...Table 1 contd.*

| Organism and its Source | CFBS/ CABS | Type of Biosurfactant (BS) | Potential Application | Reference |
|---|---|---|---|---|
| *Lactobacillus acidophilus* | CABS | Surlactin-Protein rich | Antiadhesion effect on uropathogenic bacteria and yeasts on silicone rubber surface | Velraeds et al. 1998 |
| *Streptococcus thermophilus* (Heat exchanger plate of pasteurizer) | CABS | mixtures having glycolipid-like moiety-surface active property | Antiadhesive activity against *Candida* sp. | Busscher et al. 1997a |
| *Lactobacillus acidophilus* | CABS | Surlactin-Protein rich | Antiadhesive | Munira et al. 1997 |
| *Lactobacillus acidophilus* | CABS | Surlactin-Protein rich | Antiadhesive | Velraeds et al. 1997 |
| *Lactobacillus* isolates | CABS | Surlactin-Protein rich | Antiadhesive | Velraeds et al. 1996a |
| *Lactobacillus* spp. (Urogenital tract of healthy women) | CABS | Surlactin-Protein rich | Antiadhesive against the uropathogenic *E. faecalis* | Velraeds et al. 1996b |
| *Lactobacillus casei* subsp., *rhamnosus* 36; *Lactobacillus casei* subsp., *rhamnosus* ATCC 7469; *Lactobacillus fermentum* B54; *Lactobacillus acidophilus* RC14 | CABS | Surlactin-Protein rich | Inhibit the adhesion of uropathogenic *E. faecalis*, of UTI isolate | Velraeds et al. 1996b |

Microbial BSs therefore provide huge challenges and great opportunities for future exploitation due to their favorable properties and potential applications (Hamza et al. 2017, Marchant and Banat 2012a, b, Ghagi et al. 2011). Due to the ability of BSs to partition on the inter-phases of different immiscible liquids through surface tension (SFT) reduction, they have been used as coating material on biomedical devices to reduce microbial colonization associated with infections. Research into the use of BS in biomedical sciences and healthcare industry has therefore been steadily increasing. The extraordinary potential of the probiotics/BSs producing LAB and their other active metabolites offer antagonistic effects and are useful in combating many potential pathogenic micro-organisms. However, the fact cannot be just ignored that using some probiotics and/or their products may not be adequate to hold down or control several pathogens. Instead, at best we may achieve a reduction in the load of pathogens to a greater extent (Falagas and Makris 2009). A wise selection of suitable probiotics with their products would be effective strategies to tip the balance against pathogenic microbial population.

## Widely Explored Lactic Acid Bacteria (LAB) for Biosurfactant Production

Out of the 12 genera reported frequently as LAB, only five genera, namely, *Lactobacillus, Lactococcus, Streptococcus, Enterococcus* are explored thoroughly for BS production (see Figure 1). Highest number of reports have been documented on *Lactobacillus* spp. (87%), followed by *Sterptococcus* (7%), and *Lactococcus* (5%). Just a single report (1%) (Sharma et al. 2015) discusses *Enterococcus* spp. in the literature. The remaining genera are usually known for their probiotic purposes only. Again from those four genera, Lactobacilli spp. has been the favorite candidate among researchers for BS production mainly due to the following reasons:

1. Easy to isolate (from milk, fermented products, uterine sample, oral cavity, gut, etc.).
2. Abilities to grow in diverse culturing conditions (also in presence of oxygen, microaerophilic), growth can be seen easily due to short generation time. Ability to utilize wide number of substrates.
3. Lack of stringency in the choice of substrate and growth conditions (pH, temperature, minerals, salts).
4. Contributing to competitive inhibition of MDR (multi drug resistant) pathogens belonging to *Pseudomonas, Bacillus, Staphylococcus, Klebsiella, Candida, Salmonella* and *Escherichia*.
5. Probiotic nature extensively explored by researchers for the improvement of human and animal health (dairy, poultry industry).
6. Potential applications in bioimplants–silicon/glass based surfaces due to their antimicrobial, antibiofilm and antiadhesion property.
7. 'Multitasking' nature: Resistance to acid and bile salts makes them stronger candidate for study. Production of several metabolites makes it a very potent probiotic species.

8. Both types of BSs (cell bound and cell free) are reported from this genus. Diverse types of BS molecules are produced such as glycolipid, glycolipoprotein, xylolipid, lipoprotein and surlactin.

9. Typically represents the LAB group for production of surface active agent.

In literature, the highest number of papers describes *L. pentosus* (12%), followed by *L. casei* (11%), *L. acidophilus* (9%) and *L. paracasei* (8%). Within the *Lactobacillus* group, ~ 12% are not known up to the species level while being explored for BS production. Few publications (~ 5%), have been documented on *L. plantarum, L. jensenii, L. rhamnosus* while two percent reports are shared on *L. delbrueckii*. For other species viz., *L. agillis, L. brevis, L. fermentii, L. reutri* among others have only a single report contributing ~ 1% of the literature. The overall percentage distribution for various BSs producing *Lactobacillus* spp. is represented in Figure 2.

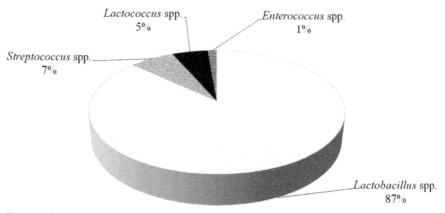

**Figure 1.** Percentage wise distribution of lactic acid bacteria (LAB) reported for biosurfactant production.

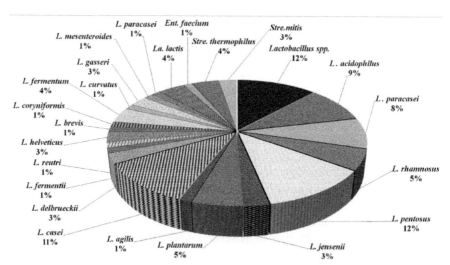

**Figure 2.** Representation for diverse lactic acid bacteria (LAB) reported for biosurfactant production.

# Different Media and Cheap/Renewable Substrates used for Growth and Production of Biosurfactant (BS) from Lactic Acid Bacteria (LAB)

Several research laboratories usually used MRS (DE man Rogosa Sharpe) medium to grow LABs (De Man et al. 1960). One of the earliest report on cultivation of the Lactobacilli was documented by Mclaughlin in 1946 and suggested a 'Trypticase sugar agar' medium containing pancreatic digest of casein (2.0%), lactose (0.5%), glucose (0.5%), sucrose (0.5%), gelatin (0.25%), and agar (1.0%). This is actually very simple and can be easily prepared and supports the growth of Lactobacilli as large colonies that can be easily identified. Trypticase sugar agar' medium had been found successful in various combinations with soybean peptone (0.1%), yeast extract, liver extract, cystine, etc. These alterations resulted in large clear colonies. It is well known that media ingredients certainly create significant impact on the compositions of BS (Banat et al. 2014) produced by different species of LAB. Sugar, lipid, protein and polysaccharide fractions of BS are affected by medium composition. Other environmental parameters such as pH, temperature (Gudiña et al. 2010a, b), incubation conditions, inoculum volume and the growth phase of bacteria (Fouad et al. 2010) have significant impacts on BS production (Satpute et al. 2018, 2016b).

It has been suggested that yeast extract is critical for growth of bacteria, whereas peptone is essential for synthesis of BS. This has been demonstrated by Gudiña et al. (2011) who suggested the use of peptone and meat extract results in a higher production of BS in comparison with de Man, Rogosa, and Sharpe medium (De man et al. 1960). The addition of divalent elements such as magnesium and manganese also supports growth of bacteria and enhances the production of protein rich BS, namely, surlactin (Fracchia et al. 2010). Currently, several reports recommend the use of the MRS media for growth as well as production of various types of BSs from LAB. Alteration in media components affects the compositions of microbial products. This is evident from the work demonstrated by Gołek et al. (2009) on glycoproteins or glycoproteins (with additional phosphoric groups) production by *L. casei* 8/4, where protein and polysaccharide fractions are changed based on the composition of medium. In addition to media components modifications, the growth phase of BS producing microbial system influences composition of BS. Use of minimal medium supplemented with trace element solution and a suitable carbon source is routine practice for BS production from *Pseudomonas, Bacillus* and *Acinetobacter* spps., etc. (Kanna et al. 2014, Al-Wahaibi et al. 2014, Deshpande et al. 2010, Satpute et al. 2018, 2017, 2010a, b, c). However, in case of LAB, ~ 63% of the reports describes the use of MRS broth in BS production technology probably due to their fastidious nutritional requirement.

Hemicellulosic sugars, corn steep liquor (7%) seems to be used for BS from LAB. Fermentation medium supplemented with whey and yeast extract (5%) has been used for few LAB. In other categories, (i) dairy wastes- cheese whey, agriculture waste-molasses, Peanut oil cake (~ 2%), (ii) minimal medium (MM) supplemented with desired carbon source (~ 5%) and (iii) raw/renewable agricultural waste substrates (~ 18%) are used. Many industries generate various wastes such as distillery wastes, molasses, plant oils, and oil wastes. Literature especially highlights use of hemicellulosic sugar hydrolysates (vine shoots), Grape marc, Peanut oil cake, and Vineyard pruning waste for BS production from LAB (Satpute et al. 2018, 2016b) (see Figure 3 and Table 2). Vecino et al. (2017) used Vineyard pruning waste (VPW) for production of novel type of BS from *L. paracasei*. Authors have documented very interesting observation with respect to change in the composition of BS. Glycolipopeptide type BS is produced by *L. paracasei* strain when glucose from VPW is used as carbon source. When glucose is replaced by lactose, the glycoprotein type BS is produced. In conclusion, same strain can produce different types of BS depending upon change in the substrates.

The foremost important requirement of an economically feasible technology is continued reliable supply of renewable substrates. Most of the raw substrates require significant pretreatment or clarification processing procedures. This is often the costly stage from an economical point of view. Moreover, raw substrates may contaminate the final product. Kandler and Weiss (1986) suggested that among LAB, members belonging to *Lactobacillus* genus exhibit complex nutritional requirements which could be supplemented in a culture or production medium. Various components or growth-stimulatory substances can be added as energy sources to support cell growth and cell division. The use of renewable natural resources has played a significant role to develop innovative and smart technologies for BS production (Satpute et al. 2017).

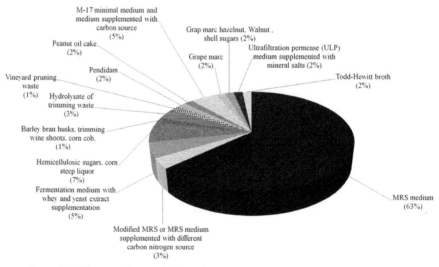

**Figure 3.** Different media used for biosurfactant production from lactic acid bacteria (LAB).

## Diversity and Structural Complexity of BSs Produced by LAB

LAB, particularly *Lactobacilli* spp., has been reported to be involved in the production of diverse types of BS. It is important to highlight that not much reports have disclosed the complete structure of these BSs, most likely due to structural complexity. It has been well established that many BS molecules have a complex composition including

**Table 2.** Summary for surface tension (SFT), critical micelle concentration (*cmc*) of biosurfactant produced by Lactic Acid Bacteria along with media, methods used for their growth and production process.

| Organism | Type of BS and Fatty Acid Reported in BS | Medium used for Production of BS | Methods used for Characterization of BS | Surface Tension (mN/m) | Critical Micelle Concentration (CMC) | Reference |
|---|---|---|---|---|---|---|
| *Lactobacillus jensenii* $P_{6A}$ *Lactobacillus asseri* $P_{65}$ | Glycoproteins | MRS broth | • Centrifugation<br>• Cell lysis<br>• Filtration<br>• Dialysis<br>• Acidification<br>• Precipitation | 43.2<br>42.5 | 7.1 mg/mL<br>8.58 mg/mL | Morais et al. 2017 |
| *Lactobacillus casei* | – | MRS broth | • Centrifugation<br>• Cell lysis<br>• Filtration<br>• Dialysis<br>• Lyophilization<br>• Freeze-drying | – | – | Merghni  et al. 2017 |
| *Lactobacillus lactis* CECT-4434 | – | MRS broth | • Centrifugation<br>• Cell lysis<br>• Filtration<br>• Dialysis<br>• Lyophilization<br>• Freeze-drying | 22.5 and 24.7 | 0.1–4.6 mg/mL | Souza et al. 2017 |
| *Enterococcus faecium* MRTL9 | Xylolipids Hexadecanoic acid | MRS broth | • Centrifugation<br>• Cell lysis<br>• GCMS<br>• FTIR | 40.2 | 2.25 µg/mL | Sharma et al. 2015 |
| *Lactobacillus pentosus* | Surlactin type protein rich | Hemicellulosic sugars, corn steep liquor | • FTIR<br>• GCMS | | 2 µg/mL | Vecino et al. 2015 |

*Table 2 contd. ...*

*...Table 2 contd.*

| Organism | Type of BS and Fatty Acid Reported in BS | Medium used for Production of BS | Methods used for Characterization of BS | Surface Tension (mN/m) | Critical Micelle Concentration (CMC) | Reference |
|---|---|---|---|---|---|---|
| Lactobacillus brevis | Mixture of components along with sugar | MRS broth | • Centrifugation, acidification<br>• Solvent extraction: $CH_3COOC_2H_5$: MeOH Drying under vacuum condition | – | – | Ceresa et al. 2015 |
| Lactobacillus agilis CCUG31450 | Glycoprotein | • MRS broth<br>• Cheese whey as an alternative culture medium | • Centrifugation<br>• Cell lysis<br>• Filtration<br>• Dialysis<br>• Freeze-drying | 42.5 | 7.5 mg/mL | Gudiña et al. 2015 |
| Lactobacillus helveticus MRTL91 | Glycolipid (Xylolipids) Hexadecanoic acid | Fermentation medium with whey and yeast extract | • Centrifugation<br>• Cell lysis<br>• Filtration<br>• Dialysis<br>• Freeze-drying<br>• FTIR<br>• Liquid Chromatography (UPLC) and Mass Spectroscopy<br>• GCMS<br>• Thermal Gravimetric (TG) Analysis | 39.5 | 2.5 mg/mL | Sharma et al. 2014 |
| Lactobacillus jensenii Lactobacillus rhamnosus | – | MRS broth | • Centrifugation<br>• Cell lysis<br>• Filtration<br>• Dialysis<br>• Freeze-drying | – | – | Sambanthamoorthy et al. 2014 |

| | | | | | | |
|---|---|---|---|---|---|---|
| *Lactobacillus plantarum* CFR2194 | Glycoprotein | MRS broth | 44.3 | 6 mg/mL | • Centrifugation<br>• Cell lysis<br>• Filtration<br>• Dialysis<br>• Freeze-drying<br>• FTIR | Madhu and Prapulla 2014 |
| *Lactobacillus reutri* | – | MRS broth | – | – | • Centrifugation<br>• Filtration<br>• Dialysis<br>• Freeze-drying | Salehi et al. 2014 |
| *Lactobacillus rhamnosus* | – | MRS broth | – | 40 mg/mL | • Centrifugation<br>• Solvent extraction: $CHCl_3$:MeOH<br>• Dialysis | Salman and Alimer 2014 |
| *Lactobacillus casei* MRTL3 | Glycolipid | MRS Broth | 40.8 | – | • FTIR<br>• GCMS | Sharma and Saharan 2014 |
| *Lactobacillus pentosus* | Glycoprotein (lipid content determination req) | Hemicellulosic sugars | 53.0 | 10 mg/mL | • Centrifugation<br>• Filtration Dialysis<br>• Freeze-drying<br>• FTIR | Vecino et al. 2013 |
| *Lactobacillus* spp. | – | MRS broth | – | – | • Centrifugation<br>• Acidification<br>• Solvent extraction: $CH_3COOC_2H_5$:MeOH<br>• Drying under vacuum condition | Gomma et al. 2013 |

*Table 2 contd....*

*...Table 2 contd.*

| Organism | Type of BS and Fatty Acid Reported in BS | Medium used for Production of BS | Methods used for Characterization of BS | Surface Tension (mN/m) | Critical Micelle Concentration (CMC) | Reference |
|---|---|---|---|---|---|---|
| *Lactobacillus* spp. | – | MRS broth | • Centrifugation<br>• Acidification<br>• Solvent extraction: $CH_3COOC_2H_5$:MeOH<br>• Drying under vacuum condition<br>• Recovery with acetone | 45 | 20 mg/mL | Mbawala et al. 2013 |
| *Lactobacillus pentosus* | Glycoprotein or Glycolipopeptide | vineyard pruning waste–Hemicellulosic sugars: 18 g/L xylose, 10.6 g/L glucose, 3.9 g/L arabinose | • Centrifugation<br>• Cell lysis<br>• Filtration<br>• Dialysis | 54 | 2.65 μg/mL | Moldes et al. 2013 |
| 15 *Lactobacillus* spp. | – | Pendidam, Heating for denaturation of proteins | • Centrifugation<br>• Cell lysis<br>• Filtration | 54–29 | – | Mbawala and Hippolyte 2012 |
| *Lactobacillus plantarum* | Glycolipid Palmitic acid, Oleic acid and Octadecanoic acid | Carbohydrate-free modified MRS | • Extraction of dried cell-mass $CHCl_3$/MeOH/ water<br>• Filtration<br>• Evaporation up to dryness<br>• Suspended $CHCl_3$/ MeOH/water<br>• Evaporation of organic fraction<br>• TLC | – | – | Sauvageau et al. 2012 |

| Strain | Composition | Medium | Method | | | Reference |
|---|---|---|---|---|---|---|
| *Lactobacillus* spp. | — | MRS broth | • Analysis of cell free supernatant | 23.3 | — | Kermanshahi et al. 2012 |
| *Lactobacillus coryniformis* ssp. *torquens* CECT 25600, *Lactobacillus paracasei* ssp. *paracasei* A20 *Lactobacillus plantarum* A14 | — | MRS-Lac medium (Glucose replaced by lactose) | • Centrifugation<br>• Cell lysis<br>• Filtration | 47.5<br>45.6<br>49.9<br>49.9 | — | Gudina et al. 2011 |
| *Lactobacillus acidophilus* DSM 20079 | Glycoprotein Surlactin type protein rich | MRS broth | • Centrifugation<br>• Cell lysis<br>• Filtration<br>• Dialysis<br>• Freeze-drying | — | | Tahmourspour 2011 |
| *Lactobacillus delbrueckii* | Glycolipid | Peanut oil cake | • Centrifugation<br>• Acidification<br>• Solvent extraction: CHCl₃:MeOH<br>• Drying under vacuum condition<br>• Column chromatography<br>• Solvent extraction<br>• Dialysis<br>• Freeze drying | — | | Thavasi et al. 2011 |
| *Lactobacillus fermentum* *Lactobacillus rhamnosus* | Protein, polysaccharide, phosphate Surlactin type protein rich | MRS broth | • Centrifugation<br>• Cell lysis<br>• Filtration<br>• Dialysis<br>• Freeze-drying | 43–46<br>43–46 | 9.0–6.0 mg/mL<br>6.0–4.5 mg/mL | Brzozowski et al. 2011 |

*Table 2 contd. ...*

*...Table 2 contd.*

| Organism | Type of BS and Fatty Acid Reported in BS | Medium used for Production of BS | Methods used for Characterization of BS | Surface Tension (mN/m) | Critical Micelle Concentration (CMC) | Reference |
|---|---|---|---|---|---|---|
| *Lactobacillus paracasei* | – | MRS broth | • Centrifugation<br>• Cell lysis<br>• Filtration<br>• Dialysis<br>• Freeze-drying | 41.9 | 2.5 mg/mL | Gudiña et al. 2010b |
| *Lactobacillus paracasei* spp. *paracasei* A20 | – | MRS broth | • Centrifugation<br>• Cell lysis<br>• Filtration<br>• Dialysis<br>• Freeze-drying | – | – | Gudiña et al. 2010a |
| *Lactococcus lactis* | – | Hemicellulosic hydrolyzates vinasses containing the solid waste | • Centrifugation<br>• Cell lysis | 14.4 | 1.2–2.4 µg/mL | Rodriguez et al. 2010 |
| *Lactobacillus* spp. | Mixture of compounds including sugars | MRS broth | • Centrifugation, acidification<br>• Solvent extraction: $CH_3COOC_2H_5$:MeOH Drying under vacuum<br>• TLC | 45.4 | 106 µg/mL | Fracchia et al. 2010 |
| *Streptococcus thermophilus* A | Glycolipid | M17 broth | • Centrifugation<br>• Cell lysis<br>• Filtration<br>• Dialysis<br>• Freeze-drying | 35 | 0.5 mg/mL | Busscher et al. 1997a |

| | | | | | | |
|---|---|---|---|---|---|---|
| *Lactococcus lactis* | Xylolipids Octadecanoic acid | Minimal media (MM) with 2% paraffin | • Centrifugation<br>• Acidification<br>• Extraction with acetone<br>• Air-dried precipitate | 40.5 | — | Sarvanakumari and Mani 2010 |
| *Lactobacillus casei* 1825<br>*Lactobacillus casei* 8/4<br>*Lactobacillus fermentum* 126<br>*Lactobacillus plantarum* 91<br>*Lactobacillus brevis* T16<br>*Lactobacillus brevis* 37 | Glycoproteins or glycoproteins with additional phosphoric groups | MRS broth<br>Whey ultrafiltration permease (WUP) medium supplemented with mineral salts | • Centrifugation<br>• Cell lysis<br>• Filtration<br>• Dialysis<br>• Freeze-drying<br>• Electrophoretic separation–NuPAGE method | 53.9<br>27.3<br>53.7<br>57.3<br>58.5<br>55.2 | — | Gołek et al. 2009 |
| *Lactobacillus pentosus* | — | • Grape marc hazelnut<br>• Walnut<br>• Shell sugars | • Centrifugation<br>• Cell lysis | 53.9 | $F_{CMC}$ 0.33<br>$F_{CMC}$ 0.42<br>$F_{CMC}$ 0.28 | Portilla et al. 2008a |
| *Lactobacillus acidophilus* | — | MRS broth | • Centrifugation<br>• Cell lysis<br>• Filtration<br>• Dialysis<br>• Freeze-drying | — | — | Walencka et al. 2008 |
| *Lactobacillus acidophilus* CECT-4179 (ATCC 832) | — | Vine trimming waste | • Centrifugation<br>• Cell lysis<br>• Filtration<br>• Dialysis<br>• Freeze-drying | 24.5<br>–1<br>24.5<br>–1<br>24.5 | $F_{CMC}$ | Portilla et al. 2008b |

*Table 2 contd. ...*

...Table 2 contd.

| Organism | Type of BS and Fatty Acid Reported in BS | Medium used for Production of BS | Methods used for Characterization of BS | Surface Tension (mN/m) | Critical Micelle Concentration (CMC) | Reference |
|---|---|---|---|---|---|---|
| *Lactobacillus pentosus* | – | • Barley bran husks<br>• Trimming vine shoots<br>• Corn cob<br>• *E. globulus* chips | • Centrifugation<br>• Cell lysis<br>• Filtration<br>• Dialysis<br>• Freeze-drying | 56<br>51<br>54<br>55 | $F_{CMC}$ 1.6<br>$F_{CMC}$ 2.8<br>$F_{CMC}$ 2.3<br>$F_{CMC}$ 2.0 | Moldes et al. 2007 |
| *Lactobacillus pentosus* | – | Grape marc | • Centrifugation<br>• Cell lysis<br>• Filtration | 56.6 | 4.8 µg/mL | Rivera et al. 2007a |
| *L. casei*<br>*L. rhamnosus*<br>*L. pentosus*<br>*L. corniformis* | – | MRS broth<br>Whey broth | • Centrifugation<br>• Cell lysis<br>• Filtration<br>• Dialysis<br>• Freeze-drying | 53.0<br>51.5<br>50.5<br>55.0 | – | Rodrigues et al. 2006c |
| *Streptococcus thermophilus* A | Glycolipid | M-17 broth | • Centrifugation<br>• Cell lysis<br>• Filtration<br>• Dialysis<br>• Freeze-drying | 37.0 | 20 mg/mL | Rodrigues et al. 2006b |
| *Lactococcus lactis* 53 | Fraction rich in glycoproteins | MRS broth | • Centrifugation<br>• Cell lysis<br>• Filtration<br>• Dialysis<br>• Freeze-drying | 36 | 14 mg/mL | Rodrigues et al. 2006a |
| *Lactococcus lactis* 53 *Streptococcus thermophiles* A | – | MRS broth | • Centrifugation<br>• Cell lysis<br>• Filtration<br>• Dialysis<br>• Freeze-drying | – | – | Rodrigues et al. 2004 |

| | | | | 30 to 40 | Crude BS | |
|---|---|---|---|---|---|---|
| *Streptococcus mitis* BS *Streptococcus mitis* BMS | Rhamnolipid like | Todd-Hewitt broth (THB) + 0.5% sucrose | • Centrifugation<br>• Cell lysis<br>• Filtration<br>• Dialysis<br>• Freeze-drying | 30 to 40 | Crude BS | van Hoogmoed et al. 2004 |
| *Lactobacillus fermentum* RC14 | Protein | MRS broth | • Centrifugation<br>• Cell lysis<br>• Suspended in buffer<br>• Filtration<br>• Dialysis<br>• Size exclusion chromatography<br>• N-terminal sequencing | — | — | Heinemann et al. 2000 |
| *Lactobacillus fermentum* RC-14, *Lactobacillus rhamnosus* GR 11 *Lactobacillus rhamnosus* GR-36 *Lactobacillus casei* | Surlactin type Protein rich | MRS broth | • Centrifugation<br>• Cell lysis<br>• Filtration<br>• Dialysis<br>• Lyophilization | — | — | Howard et al. 2000 |
| *Lactobacillus acidophilus* | Surlactin | MRS broth | • Centrifugation<br>• Cell lysis<br>• Suspended in buffer<br>• Filtration<br>• Dialysis | — | — | Valreads et al. 1998 |
| *Lactobacillus acidophilus* | Surlactin | MRS broth | • Centrifugation<br>• Cell lysis<br>• Suspended in buffer<br>• Filtration<br>• Dialysis | — | — | Valreads et al. 1997 |

*Table 2 contd. ...*

...*Table 2 contd.*

| Organism | Type of BS and Fatty Acid Reported in BS | Medium used for Production of BS | Methods used for Characterization of BS | Surface Tension (mN/m) | Critical Micelle Concentration (CMC) | Reference |
|---|---|---|---|---|---|---|
| *Lactobacillus casei* subsp., *rhamnosus* 36 *Lactobacillus casei* subsp., *rhamnosus* ATCC 7469 *Lactobacillus fermentum* B54 *Lactobacillus acidophilus* RC14 | Protein rich with high amount of polysaccharides and phosphate content | MRS broth | • Centrifugation • Cell lysis • Filtration • Dialysis • Freeze-drying | 52.0 12.0 39.0 39.0 | – 0.5 mg/mL 1.0 mg/mL | Valreads et al. 1996c |
| *Lactobacillus fermentum* B54 Other *Lactobacillus* strain | Surlactin | MRS broth | • Centrifugation • Cell lysis • Filtration • Dialysis • Freeze-drying | 29–40 | 2.0 µg/mL | Valreads et al. 1996a |
| *Lactobacillus acidophilus* | Surlactin | MRS broth | • Centrifugation • Cell lysis • Filtration • Dialysis • Freeze-drying | 39–40 | 1.0 µg/mL | Valreads et al. 1996b |

proteins, lipids and carbohydrates mixtures, collectively known as glycolipopeptide, glycolipids or glycoprotein. Some BS may contain additional phosphoric groups. Researchers are characterizing various compositions with the help of several analytical techniques including gas chromatography-mass spectrometry (GC-MS), liquid chromatography-mass spectrometry (LC-MS), Fourier transform infrared (FTIR) spectroscopy, high performance liquid chromatography (HPLC), and nuclear magnetic resonance (1H and 13C). One of the important observations to be noted in the literature is that LAB generally produces protein-based BSs (Satpute et al. 2016b). In these amphiphilic molecules, the hydrophilic moiety is mostly composed of proteins or sugar component and the hydrophobic moiety is composed of lipids. Due to the combinations of such hydrophilic and hydrophobic moieties, BS molecules confer distinctive properties. Such combination, however, makes it difficult to elucidate the complete structural details of BS. Table 1 represents the summary for different types of BS produced by different LAB. Figures 4 and 5 represent the structural diversity of LAB involved in production of BS.

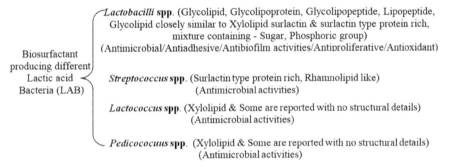

**Figure 4.** Summary for different lactic acid bacteria (LAB) reported for production of biosurfactant along with their potential applications.

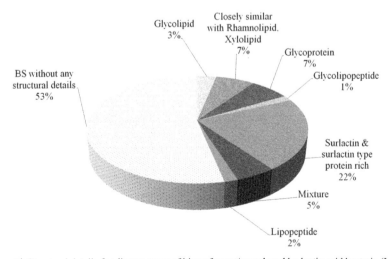

**Figure 5.** Structural details for diverse types of biosurfactant/s produced by lactic acid bacteria (LAB).

***Lactobacillus* spp.:** BSs produced by *Lactobacillus* species can be categorized under two categories. 1. CABS obtained from cell lysis. 2. CFBS obtained from the supernatant. Around 89% reports available describe CABS whereas only 11% reports CFBS type BS. The complete protocol used in production, purification, characterization and application of CABS and CFBS is represented in Figure 6. Depending upon the type of BS, different methods have been used to isolate and characterize BS produced by various genera of LAB described in Table 2. Our literature search indicates that Lactobacillus strains have been extensively explored for BSs production. At a commercial scale, Lactobacilli spp. has been used in combination with *Streptococcus* for several dairy products. Their combination results in acid production and is responsible for flavoring properties. Both species are equally imperative to reduce the growth and adherence of uropathogenic bacteria. However, *Lactobacillus* spp. has gained much popularity among consumers and is generally assumed as a safe bacterial system; however, the fact is even this microbial system could be responsible for pathogenicity related issues in food and health.

It is well known that Lactobacilli have been exploited comprehensively for lactic acid production and other uses in several applications in food industries (Campos et al. 2013). We would like to emphasize that some of the strains have been detected in patients suffering from human immunodeficiency virus HIV. Therefore, Lactobacilli strains, even though known to be harmless and favorable for good immune system, can play a critical role in infective procedures in immunocompromised individuals. The wide range properties of BS/BE offers several applications in various areas including

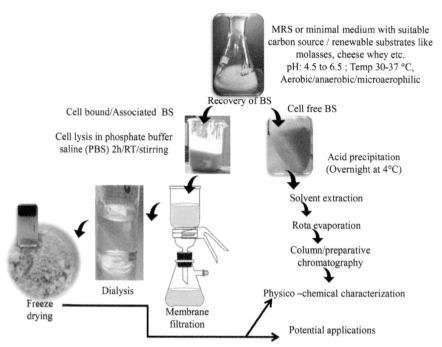

**Figure 6.** Overall procedure to isolate and characterize cell associated (CABS) and cell free (CFBS) biosurfactant from Lactobacilli and other lactic acid bacteria (LAB).

biodegradation, oil recovery, bioremediation (Banat et al. 2010, Satpute et al. 2005). Some of the BS produced by *Lactobacillus* spp. facilitates the bioremediation process in comparison with well exploited synthetic anionic natural surfactant, namely, sodium dodecyl sulfate (SDS) (Moldes et al. 2013). For degradation of any hydrocarbons like octane, n-hexane, n-heptane, n-hexadecane, etc., they need to get solubilized in water bodies. Under such circumstances, some mediator molecules like BS expedites the solubilization of hydrocarbons in water and improves the degradation of those contaminants with the help of microbial biomass present in that particular niche. In fact, it is advantageous to use surfactant/emulsifier molecules of microbial origin rather than chemical products. CABS potentially works in such application due to the presence of protein fractions. For example, literature reported by Moldes et al. (2013) suggests that *L. pentosus* produces cell bound BS which can reduce ≈ 65.1% hydrocarbon concentration, in comparison with SDS (37.2%) in bioremediation experiments proving that *Lactobacillus* derived BS are applicable for bioremediation purposes.

## Glycoprotein, Glycolipoprotein and Glycolipopeptide

Moldes et al. (2013) showed cell bound BS production from *L. pentosus* and elucidated the chemical structure proposing the presence of NH, OH groups, C=O stretching of carbonyl groups. In addition, other observations like NH bending (peptide linkage), $CH_2–CH_3$ and C–O stretching matching with FTIR spectra of earlier reported BS from other LAB were also documented. Some of the other LAB viz., *L. paracasei, La. Lactis, Str. thermophilus* A, *Str. thermophilus* B (Busscher et al. 1997a, b, Rodrigues et al. 2006a, b, d) also showed the presence of such functional groups giving the evidence for surface active agents. Glycoproteins type BS shows existence of a 3200–3500 $cm^{-1}$ peak in the BS spectrum signifying the presence of NH and OH groups. These observations have been previously reported from glycoproteins type BS produced by *La. lactis* (Rodrigues et al. 2006a) and *L. paracasei* (Gudiña et al. 2010a, b). Another type of BS viz. glycolipid-like compounds produced by *Str. thermophilus* also showed the presence of distinguishing peptide/protein groups (Busscher et al. 1997a, Rodrigues et al. 2006a). From the literature, it appears that FTIR techniques are the most popular characterization tool for LAB derived BS. Infrared (IR) transmission spectra represent peaks characteristic of any specific chemical bonds of the compound under investigation. Therefore, this technique has been exploited by several researchers to determine the components of BS mixtures. However, it is essential to determine total soluble protein as well as reducing sugar's content of BS which can be analyzed by Lowry (1951) and the phenol sulfuric method (Dubois et al. 1956), respectively.

## Glycolipids

The term glycolipid suggests the most common rhamnolipids or sophorolipids type BS produced. However, depending upon the molecular weight, BSs are categorized as low molecular weight (LMW) and high molecular weight (HMW) substances. Under LMW types, glycolipids and short chain containing lipopetides have often been reported. Rhamnolipids (RHL), sophorolipids (SPL) and surfactin (SRF) are the most comprehensively characterized LMWBS. We would like to highlight that glycolipid type LMWBS are not restricted to only *Pseudomonas* spp.; few *Lactobacillus* spp. also produced BS containing sugar and lipid moieties together. Under HMWBS, researchers are familiar with bioemulsifier based polymeric and lipopeptides types. BS produced by *Pseudomonas* spp. is popularly known as RHL. It is extremely effective surface active agent which has been explored thoroughly – in terms of their structural and functional properties. The first report appeared in 1946 indicating the linking of rhamnose moiety to fatty acid chains which was produced by *Pseudomonas pyocyanea* (presently recognized as *P. aeruginosa*) (Bergström et al. 1946). Since then, several species of *Pseudomonas* have been explored for production and their applications in several fields. Similarly, SRF exhibit strong surfactant potential in comparison with chemical based sodium lauryl sulfate (SDS). Exploring molecules bearing both antimicrobial and surfactant related properties from *Lactobacillus* and other LAB have attracted researchers around to explore such molecules.

Literature survey clearly indicates that *Lactobacillus* spp. are known for the production of protein rich CABS. From total 75 reports, only five reports are documented on glycolipid type BS production from *Lactobacillus* spp. *L. casei* MRTL91 produces glycolipid BS along with bacteriocin (Sharma and Saharan 2014). The same group of researchers (Sharma et al. 2014) documented glycolipid type BS which were closely similar to xylolipids. Their contribution indicated CABS production from *L. helveticus* MRTL91. Sauvageau et al. (2012) have also characterized glycolipid type CABS from *L. plantarum*. All these three reports are on CABS with glycolipid nature. *Lactobacillus* spp. also produces glycolipid type BS as CFBS. This is apparent from Thavasi et al. (2011) experimental evidences suggesting that *L. delbrueckii* produces glycolipid type CFBS composed of 30% of carbohydrate and 70% of lipid components. This data was supported by authors through FTIR, MS studies. BS derived from the *L. lactis* was identified structurally as glycolipids containing Methyl-2-*O*-methyl-*β*-d-xylopyranoside with octadecanoic acid as major fatty acid with a cumulative molecular weight of about 476 g/mol after GC–MS and NMR interpretation Saravanakumari and Mani (2010). Report published by Ceresa et al. (2015) which illustrated that CFBS produced by *L. brevis* contains a mixture of sugar (no structural details) are capable of preventing the adhesion of *C. albicans* on surfaces of medical-grade silicone elastomeric disks. Ceresa et al. (2015) recommended that 2000 μg/ml concentration can effectively decrease biofilm formation (~ 90%) in *C. albicans*. Similarly, Frachhia et al. (2010) reported the production of CFBS from *L. delbrueckii* spp. delbrueckii CV8LAC possessing antibiofilm activity against *Candida* cultures. Authors also claimed that CFBS possess a mixture of components along with sugar.

## Surlactin—A protein rich BS produced by *Lactobacillus* species

Velraeds et al. (1996a) investigated BS production from various Lactobacilli strains and reported that *L. acidophilus* produce protein rich CABS with small fractions of phosphates and polysaccharides and coined the term 'Surlactin' for the surface active product (Velraeds et al. 1996b). The term Surlactin is now widely accepted for protein rich CABS derived from *Lactobacillus* spp. (Satpute et al. 2018, 2016b). Soon after, Velraeds et al. (1998) also reported the ability of surlactin to inhibit the adhesion of uropathogens on silicone-based surface. It is evident from the literature that surlactin efficiently inhibits growth of bacterial as well as fungal growth (Munira et al. 2013, Vecino et al. 2013, Fouad et al. 2010, Velraeds et al. 2000, 1998, 1997). Fouad et al. (2010) demonstrated that surlactin derived from *L. acidophilus* is glycolipoprotein type. Their studies were also supported by gel filtration proving the molecular weight of glycolipoprotein as 60–80 kDa. Munira et al. (2013) also reported surlactin production from several cultures of *L. acidophilus.* They also highlighted the role of surlactin in inhibition of adhesion of biofilm formation by members of a *Pseudomoas* strain on contact lenses' surface. Supporting applications were also demonstrated by Munira et al. (2013) using surlactin for treating *P. aeruginosa* rabbits' infected eye. Nevertheless, they reported that surlactin was ineffective against *S. aureus.*

Currently, researchers are using several analytical techniques for Surlactin characterization. However, it is important to mention here that like several other biological surface active molecules (RHL, SPL and SRF), the structure of surlactin has not been fully determined. FTIR studies supported by Vecino et al. (2013) indicated presence of protein rich fractions. In general, *Lactobacillus* spp. produce protein rich (glycoprotein or glycolipoprotein complex) BS. In addition to protein moieties, CABS are also associated with sugar or lipid fractions. Several researchers have not used the term "surlactin" for BS produced by *Lactobacillus* spp. explored by them and have simply described them as protein rich BS (Shokouhfard et al. 2015, Brzozowski et al. 2011, Tahmourespour et al. 2011).

## Lactobacilli Surface Active Compounds without any Structural Details

Lactobacilli species isolated from fruits and vegetable surfaces produce BS in the mid-exponential phase (Fracchia et al. 2010). BS produced by one of the strains *Lactobacillus* spp. CV8LAC, reduces SFT readings from 70.92 to 47.68 mN/m and with a *cmc* of 106 µg/ml. No structural details had been elucidated by Fracchia et al. (2010). The CV8LAC BS considerably inhibits the adhesion of biofilm forming pathogenic strain *C. albicans* strains CA-2894 and DSMZ 11225. Generally, pre-incubation and co-incubation assays were carried out to evaluate the reduction in adhesion and the inhibition of pathogenic microorganisms. Biofilms formation by CA-2894 was effectively reduced by 82% (at concentration of 312.5 µg/ml) and for DSMZ 11225 by 81% (at concentration 625 µg/ml). In co-incubation assays, biofilm formation of strain CA-2894 and DSMZ 11225 were reduced by 70% at a concentration 160.5 µg/well and by 81% at 19.95 µg/well respectively. The cell bound BS failed to inhibit planktonic cells growth for both *Candida* strains. This indicated that CABS possess antibiofilm activities rather than general antimicrobial properties. On similar lines, Ceresa et al. (2015) also reported CFBS production from *L. brevis* (CV8LAC) inhibiting adhesion and biofilms of *C. albicans* on medical-grade silicone based elastomeric disks (SEDs). No structural components of BS-CV8LAC were revealed. BS CV8LAC effectively interfered with the early deposition and adherence of *C. albicans* to silicone surfaces and was extremely effective against *C. albicans* biofilm. Wang et al. (2017) have shown secretion of antimicrobial compounds produced by vaginal isolate *L. crispatus* exhibiting powerful inhibition of *C. albicans*. Such studies are valuable for defining the antiadhesive properties of BS when designing and developing coating formulations to prevent fungal infection on silicone based medical devices.

During organ or bioimplant transplantation procedures, patients are under high risk of exposure to harmful pathogens. Microorganisms present on medical implants or in the surrounding environments like intensive care unit (ICU) are possible sources for pathogenic organism including fungal related infections. Some of the representative organisms like *Enterococcus* spp., *S. aureus*, *E. coli*, *Klebsiella* spp. and *Acinetobacter* spp. may continue to exist for extended periods on inanimate surfaces. Usually members of the *Candida* spp. are responsible for approximately 10% of infections on cardiac devices prosthetic valves and intravenous catheters and 21% of those on urinary catheters. It is also proposed that high mortality rate (20–40%) is also associated with these infections. Treatment issues therefore become expensive (Kojic and Darouiche 2004) and *C. albicans* is one of the most frequently occurring fungal infections on such medical devices (Crump and Collignon 2000), particularly for immunosuppressed patients. Conventionally, few antimicrobial therapies can provide some kind of protection against these dangerous infections and the use of BS coating of bioimplant material may be a promising application for the prevention or reduction of biofilm forming pathogenic organisms. This is one of the possible ways as well to deal with antibiotic resistance ever increasing. Table 3 represents LAB derived BS for their biomedical potentials. Some researchers have also tested the potential application of BS on silicone tubing, catheter tubing and glass surfaces.

Mbawala and Hippolyte (2012) investigated the production of CFBS by *Lactobacillus* spp. which was isolated from a Pendidam—a local fermented milk of Ngaoundere (Cameroon). Initial screening procedures including drop collapse (diameter 7.30 mm), emulsification abilities (56.80%) and interfacial tension (IFT) values (reduction up to 45.09 mN/m) indicated BS production. No structural details were revealed for the BS produced by *Lactobacillus* spp. Authors demonstrated broad spectrum of antimicrobial activity of CFBS against bacterial strains like *B. cereus, E. coli, E. faecalis,* and *Salmonella* spp. The BSs were quite stable at various pH and temperature ranges and therefore offer several opportunities to use as food preserving molecules. Another report by Mbawala et al. (2013) showed production of CFBS (without composition detail) from *Lactobacillus* spp. with good stability at wide range of salinity (5.0–15.0%) and pH values (6.0–12.0) and potential antibacterial activities. In addition to physiological parameters, purity of BS is also one of the major criterions which need to be considered for various applications. Both crude as well as purified form of BS has been tested. Salman and Alimer (2014) compared CFBS in its crude and partially purified form to evaluate inhibitory effects against UTI pathogens viz., *B. cepacia, K. pneumonia* and *S. aureus*. They recommended that BS in their crude form can display good surface activity and also inhibits the adherence of biofilm forming pathogens. The *in vitro* studies conducted by Das et al.

**Table 3.** Use of biosurfactant (BS) produced by different lactic acid bacterial strains to demonstrate their respective applications on bioimplants.

| Microorganism used for Biosurfactant (BS) Production | Potential Applications | Bioimplant used to Demonstrate the Application | Concentration of BS Recommended for the Activity | Reference |
|---|---|---|---|---|
| *Lactobacillus jensenii* P$_{6A}$ *Lactobacillus gasseri* P$_{65}$ | Antimicrobial activity on *E. coli* Antimicrobial effects on *K. pneumoniae*, *E. aerogenes* and *S. saprophyticus* Disruption of the biofilms | – | 16 µg/mL 128 µg/mL 180 µg/mL | Morais et al. 2017 |
| *Lactobacillus casei* | Antioxidant and antiproliferative potential, antiadhesive and antibiofilm | – | 5.0 mg/mL IC50 and its value ranged from 109.1 ± 0.84 mg/mL to 129.7 ± 0.52 mg/mL 1.56 mg/mL | Merghni et al. 2017 |
| *Lactobacillus pentosus* | Bioactivity antimicrobial and anti-adhesive against *Stre. agalactiae*, *P. aeruginosa*, *S. aureus*, *Stre. Pyogenes* | – | 50 mg/mL | Vecino et al. 2017 |
| *Lactococcus lactis* subsp. *lactis* CECT-4434 | Antimicrobial activity against *L. sakei*, *S. aureus* | – | – | Souza et al. 2017 |
| *L. helveticus* MRTL91 | Antimicrobial properties against *B. cereus*, *S. aureus*, *S. epidermidis*, *P. aeruginosa*, *S. typhi*, *S. Flexner* Antiadhesive activity Biofilm hindrance on silicone tubes | Silicone tubes | 25 to 50 mg/mL 25 mg/mL 312.5 µg/ml | Sharma and Saharan 2016 |
| *Lactobacillus brevis* CV8LAC | Antifungal, Antiadhesion and antibiofilm activity against *C. albicans* | Medical grade silicone elastomeric disks (SEDs) | 2000 µg/mL | Ceresa et al. 2015 |
| *Lactobacillus agilis* CCUG31450 | Antiadhesive activity against *S. aureus*, Antimicrobial activity against *S. aureus*, *Stre. agalactiae*, *P. aeruginosa* | – | – | Gudiña et al. 2015 |
| *Enterococcus faecium* MRTL 9 | *E. coli* and *S. aureus*, *P. aeruginosa*, *B. cereus*, *L. monocytogenes*, *C. albicans* | Silicone rubber tubes | 25 mg/ml | Sharma et al. 2015 |

*Table 3 contd. ...*

*...Table 3 contd.*

| Microorganism used for Biosurfactant (BS) Production | Potential Applications | Bioimplant used to Demonstrate the Application | Concentration of BS Recommended for the Activity | Reference |
|---|---|---|---|---|
| *Lactobacillus casei* MRTL3 | Antimicrobial activity against *S. aureus, S. epidermidis, B. cereus, Listeria monocytogenes* and *L. innocua, Shigella flexneri, S. typhi* | – | – | Sharma and Saharan 2014 |
| *Lactobacillus plantarum* CFR2194 | Antimicrobial activity against *E. coli, S. typhi, Y. enterocolitica, S. aureus* | – | 25 mg/mL | Madhu and Prapulla 2014 |
| *Lactobacillus jensenii* *Lactobacillus rhamnosus* | Antimicrobial, antiadhesive and antimicrobial activities MDR *A. baumannii, E. coli, S. aureus* | – | 25–100 mg/mL 25–50 mg/mL | Sambanthamoorthy et al. 2014 |
| *Lactobacillus rhamnosus* | Antibacterial, antibiofilm and antiadhesive properties: UTI- *K. pneumonia, B. cepacia, E. coli, S. aureus* | – | 32 mg/mL 64 mg/mL | Salman and Alimer 2014 |
| *Lactobacillus paracasei* | Inhibition of *E. coli, S. agalactiae, S. pyogenes* | – | 25 mg/mL | Gudiña et al. 2010b |
| *Lactobacillus paracasei* A20 | Antimicrobial activity against various Gram positive and Gram negative bacteria | – | 3.12 mg/mL to 50 mg/mL | Gudiña et al. 2010a |
| Ten Lactobacilli species | Antimicrobial and antibiofilm activity: *B. subtilis, B. subtilis, B. cereus, S. aureus, P. aeruginosa, E. coli, S. typhi, P. vulgaris, S. marcescens, Candida albicans, Bacillus* sp. | – | Crude biosurfactant | Gomaa 2010 |
| *Lactococcus lactis* | Antimicrobial activity of BS (Xylolipids) against the multi-drug resistant (MDR) *S. aureus, E. coli* | – | – | Sarvanakumari and Mani 2010 |
| *Lactobacillus casei* | Antimicrobial activity against *S. aureus, B. subtilis* and *M. roseus*; Antiadhesive activity against *K. pneumonia, P. aeruginosa, E. coli* | – | 4% wt/v | Gołek et al. 2009 |
| *Lactococcus lactis* 53 *Streptococcus thermophiles* A | Antimicrobial activity against *S. epidermidis, Str. salivarius, S. aureus, C. tropicalis* | Silicone-rubber voice prosthesis treatment-BS | 100 mg/mL 40 g/L | Rodrigues et al. 2006d |

| Strain | Target | Substratum/method | Concentration | Reference |
|---|---|---|---|---|
| *Streptococcus thermophilus* A | Anti-adhesive and antimicrobial activity against bacteria, yeast strains (voice prostheses) | – | 40 g/L | Rodrigues et al. 2006b |
| *Streptococcus mitis* strains: BA and BMS | *Stre. mutans* | Glass with and without a salivary conditioning film/presence of probiotic or coating with its BS only | 20 mg/mL | van Hoogmoed et al. 2000 |
| *Lactobacillus* spp. | Uropathogens | Silicon rubber discs in non-pooled specimens of male and female urine samples | *Lactobacillus* suspension $3 \times 10^8$ cells per mL | Velraeds et al. 2000 |
| *Lactobacillus acidophilus* RC 14 | *E. faecalis* | Glass or silicone rubber substratum/pre-conditioning with surlactin BS | 1 mg/mL | Velraeds et al. 1997 |
| *Streptococcus thermophilus* B | *C. albicans* and two *C. tropicalis* strains | Silicone rubber with and without a salivary conditioning film/presence of probiotic or coating with its BS only | 2% surfactant solution | Busscher et al. 1997a |
| *Lactobacillus* spp. | *E. faecalis* | Glass substratum/adsorption of BS | 2% surfactant solution | Velraeds et al. 1996a |

(2013) demonstrated that cell free supernatant derived BS from *L. plantarum* (KSBT 56) prevents the growth and pathogenicity of *S. enterica*. Studies are significant to facilitate the use of those strains as prospective probiotics strains with having substantial valuable effect for human system.

## BSs from *Lactococcus* Species

*Lactococcus* is Gram positive, homo-fermenter LAB, formerly a member of genus *Streptococcus* group 1. *Lactococcus* is used as a dairy starter culture for various dairy fermented foods, particularly cheese, because of their rapid acidification potential. The major species within the genus are *La. lactis*, *La. lactis* subspp. *Cremoris*, *La. plantarum* and *La. fujiensi*. Out of them, *La. lactis* is extensively used for the production of buttermilk and cheese and is known for high lactic acid production (Madigan and Martinko 2005).

La. lactis has been given generally recognized as safe (GRAS) status and have been used for dairy fermented food production (Wessels et al. 2004). During the past decade there have been many reports on BS production by *La. lactis* (Rodrigues et al. 2008, 2006a, b, c, d, e, f, 2004, Saravanakumari and Mani 2010, Rodriguez-Pazo et al. 2010a, b). The very first report was published by Rodrigues et al. (2004) demonstrating the role of *La. lactis* 53 derived BS in preventing the adhesion of human pathogens on silicone rubber in a parallel plate flow chamber. Silicone rubber with and without BS pre-adsorption were characterized on the basis of contact angle measurement. It is a well-known fact that application of BS on any solid surface decreases the contact angle and reduces the microbial adhesion to the surface (Nitschke et al. 2009). The results confirmed that the ability of *La. lactis* derived BS surface active agent was effective in the reduction of the initial deposition of the pathogenic strains. The outcome of the study constitutes an additional method suitable in developing approaches for the prevention of microbial colonization on biomedical surfaces.

This BS was structurally characterized as a glycoprotein rich fraction after hydrophobic interaction chromatography and X-ray photoelectron spectroscopy (Rodrigues et al. 2006a). It was also found stable to a wide range of pH ranging from 5.0–9.0 with similar functional properties. As the production of a cost effective BS is vital for its functional application, structural attributes and large scale production, potential use of alternative nutrient source for formulation of growth medium for *La. lactis* 53 was also investigated (Rodrigues et al. 2006a). Different sole carbon sources were studied viz., glucose, lactose and sucrose in synthetic media MRS and M17 and cell growth and BS production were monitored. When synthetic media was replaced with low cost whey and molasses based nutrients, the fermentation yield had higher productivities of BS. An upsurge of ~ 1.2–1.5 times in the mass of produced BS per gram of cell dry weight (CDW) and reduction of 60–80% upstream cost were reported.

The major outcome of the study showed that low cost cheese whey and molasses supplementation can be used as a comparatively reasonable and economical substitute to synthetic media for BS production by *La. lactis*. The structural composition of the *Lactococcus* is quiet important for further food and biomedical properties. Structurally characterized BS from genus *Lactococcus* was found effective in controlling growth of multidrug resistant pathogens (Saravanakumari and Mani 2010). The BS was reported to be a glycolipid in nature; more specifically, it was structurally drawn as xylolipids containing octadecanoic acid as the major fatty acid with no toxicity. BS obtained from *La. lactis* displayed broad range of antibacterial potential and were safe to administer orally and intradermal route after acute toxicity evaluation. BS could be used as an effective therapeutic material in bio-pharmaceutical or in food industries as a future food additive or preservative. LABs are not only known for the production of BS, lactic acid and starter culture but were also reported widely for the production of the bacteriocins, i.e., Nisin (Beasley et al. 2004, Zhang et al. 2016). The majority of the LAB producing CABS, however, had bacteriocins production as an extracellular activity. *La. lactis* CECT 4434 was used in simultaneous production of BS and bacteriocins for industrial applications using the opportunity to downstream separately (Rodriguez et al. 2010). In conclusion, strain *La. lactis* CECT-4434 produces both a cell-bound BS and bacteriocins in fermentation without pH control. The protocol was developed and validated for optimal extraction of bacteriocins and BS in extracts. The protocol is established as two steps reaction viz., initial acidification of the fermentation medium containing biomass, and subsequent incubation of the biomass with phosphate buffer saline. Further development in *Lactococcus* genus derived BS will depend upon the utilization of the facts reported earlier.

## BS Producing *Streptococcus* Species

*Streptococcus* is a Gram positive LAB which is further separated into two different genera, *Enterococcus* and *Lactococcus* with more than 50 recognized spp. *Streptococcis* have been found to be part of oral microbiome of humans and animals and are often involved in dental caries, upper respiratory tract and urinary tract infections. They, however, have been reported to have various health benefits in the production of probiotics, bacteriocins, vitamins, exopolysacchrides and BSs (Iyer et al. 2010). *Str. thermophilus* has been extensively regarded as GRAS used in various foods and beverages processing like yogurt production (*Str. thermophilus* in consortium with *L. dlbrueckii*), probiotic beverages (*Str. salivarius* and *Str. thermophilus*) and cheddar cheese (Iyer et al. 2010). In addition, it also produces BS with functional attributes like antiadhesive and antimicrobial to various bacterial and fungal pathogens associated with human health and food processing (Rodrigues et al. 2006a, b, c, d, Busscher et al. 1997a, b).

BS of *Streptococci* origin have been structurally characterized as a mixture of components mainly composed of polysaccharides, amino acids and lipids and characterized as glycolipids (Busscher et al. 1997a). The report advocates the utilization of BS obtained as coating material for voice prostheses and surgical implants to inhibit biofilm formation. Rodrigues et al. (2006a) reported similar findings while working with the silicone rubber tubes. The BS was found to be an antibiofilm agent which prevents the formation of biofilm on silicone rubber typically used for voice prostheses and urinary catheters. They also characterized the BS as a glycolipid rich fraction after hydrophobic interaction chromatography and elemental composition determined by X-ray photoelectron. The glycolipid rich BS were also reported to be an antiadhesive and antimicrobial agent of various pathogenic bacteria and yeast strains. The yield of BS produced, however, was very low and the optimization of the cultural conditions and nutrient composition was recommended (Rodrigues et al. 2006b). In one approach of BS production from *Str. thermophilus*, optimization of medium was carried out using response surface methodology (Rodrigues et al. 2006f). The selected factors based on MRS medium were meat extract, peptone, lactose, ammonium citrate and $KH_2PO_4$. The product yield was doubled after the optimization of the medium which resulted in enhanced BS production.

## BS from *Enterococcus* Species

*Enterococci* are commensal in gastrointestinal tract of humans and animals. Some species of *E. faecium* are used as probiotics for both animals and humans. But certain reports of vancomycin resistant enterococci as multidrug resistant pathogen have also been emerging. The role of Enterococci in BS production is not well explored. *E. faecium* MRTL9 was the only available report on BS production (Sharma et al. 2015). The strain was isolated from a fermented food (buttermilk). Furthermore, gas and column chromatography characterization of the BS showed glycolipid (Xylolipid) with hexadecanoic acid as major fatty acid composition. The BS product displayed significant antiadhesive potential with non-cytotoxic nature. This indicates potential uses as a cleaning agent for bio-medical and surgical equipment.

## Toxicity Studies on LAB BSs

The exploration of microbial surfactants depends on their functional properties and compatibilities. The lack of toxicity of BS is of essential significance for their applications in the food and biomedical sector (Fletcher 1991). At the present time, there is a vast apprehension concerning the toxicity and safety of BS used for food and biomedical purpose. Noteworthy potential properties like foaming, wetting, emulsification, stabilization BS have been reported for food industries (Campos et al. 2013). The toxicity assessment of BS are generally carried out using seed germination test (Yerushalmi et al. 2003, da Souza Sobrinho et al. 2013), toxicity assessment against *Artemia salina* growth or survival (da Souza Sobrinho et al. 2013), acute oral toxicity (Saravanakumari and Mani 2010) and cytotoxicity effects on human cell lines (Sharma et al. 2015, Cochis et al. 2012). The key explanation that limits LAB derived BS is the lack of toxicity data, so as to use them appropriately in food and pharmaceuticals. The toxicity assessment of the BS derived from the LAB is not a well explored segment. Only a handful of reports are available which assessed the toxicity fate of the BS of LAB origin (Table 4) (Sharma et al. 2014, 2015, Saravanakumari and Mani 2010).

Initial work of toxicity assessment has been carried out on BS obtained from *La. lactis* (Saravanakumari and Mani 2010). The acute toxicity of the BS obtained from *La. lactis* was evaluated at a concentration of the 5000 mg/kg administered to a set of male and female rats orally and intra-dermally and monitored for two months

**Table 4.** Data available on toxicity assessment for BS obtained from LAB in literature.

| Strain | Toxicity Assessment | Reference |
|---|---|---|
| *Lactococccus lactis* | Oral and acute toxicity | Sarvanakumari and Mani 2010 |
| *Lactobacillus helveticus* MRTL91 | Phytotoxicity and cytotoxicity | Sharma et al. 2014 |
| *Enterococcus faecium* MRTL91 | Cytotoxicity | Sharma et al. 2015 |

for weight, fluctuations in eating food, excretion and general wellbeing. No visible or quantifiable ill effects or death occurred throughout the toxicity assessment period. For this reason, the BS was positioned in the toxicity group IV and is regarded as safe for oral consumptions. Intra-dermal dosages were also carefully observed for any allergic reactions like reddening, inflammation and irritation in the tail region. No apparent deaths were observed throughout the dermal toxicity assessment. Therefore, the BS can be grouped in the toxicity category IV (safe to use for dermal applications) as per the regulations of FDA (Food Additives Amendment 1958).

Furthermore, the BS obtained from *L. helveticus* was assayed for phytotoxicity and cytotoxicity at their *cmc* and double *cmc* values (Sharma et al. 2014). The phytotoxicity of the BS was determined by static seed germination and root elongation experiment on *Brassica nigra* and *Triticum aestivum* seeds. They reported 100% seed germination for both types of seeds. In comparison, seed germination declined in the treatment with SDS (amount equal to *cmc*). All the tested parameters such as root length and seed germination were enhanced in the BS treatment. Cytotoxicity test also showed highest viability, about 43.3% at 6.25 mg/ml, in BS derived from *L. helveticus*, while RHL (Jeneil) and SDS showed 35.3% and 35.99%, respectively (Sharma et al. 2015).

In conclusion, the toxicity assessment of BS obtained from the LAB is still questionable or not fully demonstrated in the light of FDA food additives amendment (1958). LAB is generally regarded as safe (GRAS) which meets FDA approval for microbial origin food additives and applications. But the data related to the animal testing and extended testing outside of the laboratory are still a long way to complete. There is therefore a need to evaluate the BS of LAB origin for their toxicity fate as per regulatory bodies' and agencies' guidelines to popularize and commercialize the microbial BS for food and pharmaceutical applications. The research strategies related to the BS toxicity assessment and outcomes should be mapped according the guidelines of FDA and other regulatory agencies.

# Possible Limitations to Explore the Biosurfactant Production from Lactic Acid Bacteria

The Divers range of BS produced by the various LAB possesses significant potential in biomedical field. This sector is always in search of appropriate surface active agents which can be utilized effectively for various application purposes. A need to explore innovative approaches in designing and development of novel BS based formulations remain and efforts in this regard faces many challenges including:

a) Difficulty to mimic the exact *in vivo* conditions (intestine, vagina, skin, oral cavity, etc.) suitable for production and functionality of exact type of BS molecules from LAB.

b) Accessibility and continuous supply of cheap/renewable substrates rich in sugar, amino acids and other essentially required nutrients.

c) Lack of successful trials for LAB derived BS production technology.

d) Inadequate availability of innovative strains proficient in consuming cheap/renewable substrates to yield good quality and quantity of the products.

e) Difficulty in revealing the complete structural details of BSs due to their constituent's complexity.

f) Insufficient knowledge on the exact mechanisms/interaction of different types of BS against pathogens.

g) Alteration in structural and BS production capacity on sub-culturing procedures.

h) Insufficient yield of BS produced from LAB.

i) Limited number of LAB has been tested for BS production technology.

j) Deficiency in research/knowledge on some BS extraction and downstream processing procedures.

k) Production of high lactic acid during fermentation which leads to end product inhibition to their own growth. Low lactic acid LAB should be screened for future applications.

l) Lack of *in vitro* and *in vivo* data and facts related to the cellular toxicity of BS.

m) Lack of molecular level studies—Gene responsible for production and regulation of BS.

n) Production of lactic acid as a major end product during production which leads to end product inhibition and control the growth of LAB in late exponential phase. The low lactic acid producer or homofermentative LAB could be an efficient strain of choice.

We should endeavour to overcome the above mentioned challenges to be able to utilize the diverse BS products produced by these probiotic bacteria towards therapeutic purposes. Around 12 different screening methods are available to detect BS/BE producers in literature. Those methods can be vigorously utilized with some alteration procedures (Satpute et al. 2008) to identify potent LAB. Bakhshi et al. (2017) suggested that an ideal surfactant should decrease dynamic SFT in addition to low equilibrium surface and rapid reduction of SFT with time.

## Conclusions

Despite an availability of enormous literature on BS/BE production from varied microorganisms, limited number of reports discloses LAB originated BS. Each habitat provides a unique environment to construct particular metabolic machinery in flora and fauna. BS/BE produced by LAB do exhibit unique structural and functional diversity. High as well as low molecular weight BS/BE are produced by LAB, thus making available a large number of potent BS producing microorganisms. A concentration of BS ranging between 1 to 130 mg/ml is generally recommended for antimicrobial activities. Nevertheless, concentration between 20–25 mg/ml is used more frequently. BS reduces SFT ranging between 30–55 mN/m and *cmc* values ranging between 1 µg/ml to 20 mg/ml. LAB are used in food related applications. LAB derived BS are used in antimicrobial, antibiofilm and antiadhesive activities. With the help of innovative approaches, newer BS based formulations can be designed. Therefore, we need to widen our screening methodologies and improve further purification techniques to accomplish better qualities of surface active molecules and we need to overcome the challenges facing the exploration of innovative approaches in designing and development of novel BS based formulations.

## Future Prospects

Rather than just adhering to antimicrobial, antiadhesive, antibiofilm of LAB derived BS, other challenging characteristics like antioxidant, antiproliferative properties and abilities for synthesis of nanoparticles (NPs) can be investigated in detail. LAB produces high and low molecular weight BSs which can perform extremely well for biomedical purposes. Researchers working particularly in the area of BS production technology explore novel strains proficient in utilizing raw, renewable substrates raising high yield and fine toning the downstream separation/purification processing technology (Fracchia et al. 2012, 2015). Commercial robust investigation for advanced piece of applications is being carried out consistently by considerable number of industries. Significant promising prospects are currently obtainable in developing/designing original formulations based on BS. Exciting applications reported by Ferreira et al. (2017) suggests a novel cosmetic formulation based on BS derived from *L. paracasei*. The development of this new green cosmetic preparation is particularly attractive compared to using synthetic chemicals. Our personal and cosmetic care products like creams, shampoos, lotions, etc. are generally based on petroleum derived surfactants. "Green" products are appealing options to take care of all different products used by customers. Ferreira et al. (2017) used BS from *L. paracasei* in combination with several essential oils and a grape seeds derived natural antioxidant extract to develop formulations. Vecino et al. (2017) demonstrated bioactivity (antimicrobial and antiadhesive) of *L. pentosus* CECT-4023T derived glycolipopeptide type CABS against skin pathogens and, therefore, are useful for cosmetic and pharmaceutical formulations. The ideal comparison between BS and SDS is extremely useful for recommending the use of BS in designing newer formulations. The authors evaluated different emulsions formed during formulations and data was also supported through cytotoxicity studies supporting the use of BS in formulation designing. Synergistic effect of bioactive molecules can be enhanced by using them in combination with NP. Similarly, biogenic synthesis of metal NPs is also carried out with the help of BS from corn steep liquor. On this aspect, work has been contributed by Gómez-Graña et al. (2017) to develop green technologies and reduce the wastage in the environment. We can reduce our dependence on antibiotics and synthetic antimicrobials through the combination of different approaches (Elshikh et al. 2017). The broad spectrum of antimicrobial potential of Lactobacilli derived BS is recognized widely. One of the first reports documented by Foschi et al. (2017) suggests the effective role of Lactobacilli strains and their BS to prevent the occurrence of *Neisseria gonorrhoeae*. Abruzzo et al. (2017) have introduced innovative promising vaginal delivery systems to eradicate chronic infections of *Candida* spp. They prepared novel formulation for vaginal delivery containing econazolenitrate and phosphatidylcholine—mixed vesicles with BS from *L. gasseri* BC9. The formulation appears to effectively tackle fungal infections. The extensive market demand is encouraging the identification and developments in this particular area. One of the mysterious challenges faced by researchers is the structural complexity of diverse variety of BS and determination of their complete chemical composition. To some extent, achievements have been reported on structure related aspects. Many valuable properties bared by BS offers their superior competence in the direction of therapeutic approaches. Utilization of BSs belonging to LAB groups in precluding hospital acquired infections can be undertaken. In addition, the uses of BSs can be further extended more widely for the reduction or removal of microbial biofilm formation on various surfaces.

## Acknowledgements

Surekha K. Satpute is highly grateful to Department of Science and Technology (DST), Government of India, Ministry of Science and Technology, New Delhi, India for financial support {SR/WOS-A/LS-1076/2014(G)}. Without DST's financial support, it would not have been possible to carry out this work. We also acknowledges Department Research and Development Programs (DRDP), Savitribai Phule Pune University (SPPU) for financial support.

## References

Abruzzo, A., B. Giordani, C. Parolin, B. Vitali, M. Protti, L. Mercolini, M. Cappelletti, S. Fedi, F. Bigucci, T. Cerchiara and B. Luppi. 2017. Novel mixed vesicles containing lactobacilli biosurfactant for vaginal delivery of an anti-Candida agent. Eur. J. Pharm. Sci. 112: 95–101.

Ali, O.A. 2012. Prevention of *Proteus mirabilis* biofilm by surfactant solution. Egypt. Acad. J. Biology Sci. 4(1): 1–8.

Al-Wahaibi, Y., S. Joshi, S. Al. Bahry, A. Elshafie and A. Al-Bemani. 2014. Biosurfactant production by *Bacillus subtilis* B30 and its application in enhancing oil recovery. Colloids and Surfaces B: Biointerfaces 114: 324–333.

Bakhshi, N., S. Soleimanian-Zad and M. Sheikh-Zeinoddin. 2017. Dynamic surface tension measurement for the screening of biosurfactants produced by *Lactobacillus plantarum* subsp. *plantarum* PTCC 1896. Enzyme Microb. Technol. 101: 1–8.

Banat, I.M., A. Franzetti, I. Gandolfi, G. Bestetti, M.G. Martinotti, L. Fracchia, T.J. Smyth and R. Marchant. 2010. Microbial biosurfactants production, applications and future potential. Appl. Microbiol. Biotechnol. 87: 427–444.

Banat, I.M., S.K. Satpute, S.S. Cameotra, R. Patil and N.V. Nyayanit. 2014. Cost effective technologies and renewable substrates for biosurfactants' production. Front. Microbiol. 5: 697.

Barrons, R. and D. Tassone. 2008. Use of *Lactobacillus* probiotics for bacterial genitourinary infections in women: a review. Clin Ther. 30: 453–468.

Beasley, S.S., T.M. Takala, j. Reunanen, J.N.D. Apajalahti and P.E.J. Saris. 2004. Characterization and electro transformation of *Lactobacillus crispatus* isolated from chicken crop and intestine. Poultry Sci. 83(1): 45–48.

Bergström, S., H. Theorell and H. Davide. 1946. On a metabolic product of *Ps pyocyania*. pyolipic acid active against *M. tuberculosis*. Arkiv. Kemi. Mineral. Geol. 23A: 1–12.

Blomberg, L., A. Hendriksson and P.L. Conway. 1993. Inhibition of adhesion of *Escherichia coli* K88 to piglet Ileal mucus by *Lactobacillus* spp. Appl. Environ. Microbiol. 59: 34–39.

Borchert, D., L. Sheridan, A. Papatsoris, Z. Faruquz, J.M. Barua, I. Junaid, Y. Pati, F. Chinegwundoh and N. Buchholz. 2008. Prevention and treatment of urinary tract infection with probiotics: Review and research perspective. Indian J. Urol. 24(2): 139–144.

Brzozowski, B., W. Bednarski and P. Gołek. 2011. The adhesive capability of two *Lactobacillus* strains and physicochemical properties of their synthesized biosurfactants. Food Technol. Biotechnol. 49: 177–186.

Busscher, H.J., C.G van Hoogmoed, G.I. Geertsema-Doornbusch, M. van der Kuijl-Booij and H.C. van der Mei. 1997a. *Streptococcus thermophilus* and its biosurfactants inhibit adhesion by Candida spp. on silicone rubber. Appl. Environ. Microbiol. 63(10): 3810–3817.

Busscher, H.J. and H.C. Van der Mei. 1997b. Physico-chemical interactions in initial microbial adhesion and relevance for biofilm formation. Adv. Dent. Res. 11(1): 24–32.

Campos, J.M., T.L. Stamford, L.A. Sarubbo, J.M. de Luna, R.D. Rufino and I.M. Banat. 2013. Microbial biosurfactants as additives for food industries. Biotechnol. Prog. 29: 1097–1108.

Ceresa, C., F. Tessarolo, I. Caola, G. Nollo, M. Cavallo, M. Rinaldi and L. Fracchia. 2015. Inhibition of *Candida albicans* adhesion on medical-grade silicone by a *Lactobacillus*-derived biosurfactant. J. Appl. Microbiol. 118: 1116–1125.

Cochis, A., L. Fracchia, M.G. Martinotti and L. Rimondini. 2012. Biosurfactants prevent *in vitro Candida albicans* biofilm formation on resins and silicon materials for prosthetic devices. Oral Surgery, Oral Medicine, Oral Pathology and Oral Radiology 113(6): 755–761.

Coeuret, V., S. Dubernet, M. Bernardeau, M. Gueguen and J. Vernoux. 2003. Isolation, characterization and identification of Lactobacilli focusing mainly on cheeses and other dairy products. Lait. 83: 269–306.

Crump, J.A. and P.J. Collignon. 2000. Intravascular catheter-associated infections. Eur. J. Clin. Microbiol. Infect. Dis. 19: 1–8.

Da Souza Sobrinho, H.B., J.M. da Luna, R.D. Rufino, A.L.F. Porto and L.A. Sarubbo. 2013. Application of biosurfactant from *Candida sphaerica* UCP 0995 in removal of petroleum derivative from soil and sea water. J. Life Sciences 7(6): 559.

Das, J.K., D. Mishra, P. Ray, P. Tripathy, T.K. Beuria, N. Singh and M. Suar. 2013. *In vitro* evaluation of anti-infective activity of a *Lactobacillus plantarum* strain against *Salmonella enterica* serovar Enteritidis. Gut Pathog. 5: 11 10.1186/1757-4749-5-11.

De Man, J.C., M. Rogosa and M.E. Sharpe. 1960. A medium for cultivation of Lactobacilli. J. Appl. Bacteriol. 23: 130–135.

Deshpande, S.A., S. Pawar, N.V. Nyayanit and S.K. Satpute. 2010. Glycolipid biosurfactant production from *Pseudomonas aeruginosa* SCOS 46 isolated from oil contaminated sites of Pune city. J. Adva. Sci. Technol. 13(3): 15–19.

Diaz De Rienzo, M.A., P. Stevenson, R. Marchant and I.M. Banat. 2016. Antibacterial properties of biosurfactants against selected Gram-positive and -negative bacteria. FEMS Microbiol. Lett. 363(2): 224: 1–8.

Donlan, R.M. 2002. Biofilms: microbial life on surfaces. Emerg. Infect. Dis. 8: 881–890.

Dubois, M., K.A. Gilles, J.K. Hamilton, P.A. Rebers and F. Smith. 1956. Colorimetric method for determination of sugars and related substances. Ana. Chem. 28: 350–356.

Elshikh, M., I. Moya-Ramírez, H. Moens, S. Roelants, W. Soetaert, R. Marchant and I.M. Banat. 2017. Rhamnolipids and lactonic sophorolipids: natural antimicrobial surfactants for oral hygiene. J. Appl. Microbiol. 123: 1111–1123.

Falagas, M.E. and G.C. Makris. 2009. Probiotic bacteria and biosurfactants for nosocomial infection control: a hypothesis. J. Hosp. Infec. 71: 301e306.

Ferreira, A., X. Vecino, D. Ferreira, J.M. Cruz, A.B. Moldes and L.R. Rodrigues. 2017. Novel cosmetic formulations containing a biosurfactant from *Lactobacillus paracasei*. Colloids and Surfaces B: Biointerfaces 155: 522–529.

Fletcher, J. 1991. Keynote speech: a brief overview of plant toxicity testing. In Plants for toxicity assessment: Second volume. ASTM International.

Food Additives Amendments of 1958. Public Law No. 85–529, Ch. 4.72 Stat. 1785 [codified at 21 USC § 348 (1981)].

Foschi, C., M. Salvo, R. Cevenini, C. Parolin, B. Vitali and A. Marangoni. 2017. Vaginal lactobacilli reduce *Neisseria gonorrhoeae* viability through multiple strategies: An *in vitro* study. Front. Cell. Infect. Microbiol. 7: 502.

Fouad, H.K., H.H. Khanaqa and Ch.I. Munira. 2010. Purification and characterization of surlactin produced by *Lactobacillus acidophilus*. IRAQI Acad. Sci. J. 8: 34–39.

Fracchia, L., M. Cavallo, G. Allegrone and M.G. Martinotti. 2010. A *Lactobacillus*-derived biosurfactant inhibits biofilm formation of human pathogenic *Candida albicans* biofilm producers. pp. 827–837. *In*: A. Méndez-Vilas (ed.). Current Research, Technology and Education Topics in Applied Microbiology and Microbial Biotechnology (Microbiology Book Series, No. 2, Vol. 2), FORMATEX, Spain. ISBN-13:978-84-614-6195-0.

Fracchia, L., M. Cavallo, M.G. Martinotti and I.M. Banat. 2012. Chapter 14, Biosurfactants and bioemulsifiers biomedical and related applications—present status and future potentials. pp. 325–370. *In*: D.N. Ghista (ed.). Biomedical Science, Engineering and Technology, InTech, Rijeka, Croatia, Europe, ISBN: 978-953-307-471-9, DOI 10.5772/23821.

Fracchia, L., J.J. Banat, M. Cavallo, C. Ceresa and I.M. Banat. 2015. Potential therapeutic applications of microbial surface active compounds. AIMS Bioeng. 2: 144–162.

Gan, B.S., J. Kim, G. Reid, P. Cadieux and J.C. Howard. 2002. *Lactobacillus fermentum* RC-14 inhibits *Staphylococcus aureus* infection of surgical implants in rats. J. Infect. Dis. 185: 1369–1372.

Ghagi, R., S.K. Satpute, B.A. Chopade and A.G. Banpurkar. 2011. Isolation and properties of crude biosurfactant from *Sapindus mukorossi* (Ritha). Indian J. Sci. Technol. 4(5): 500–533.

Gołek, P., W. Bednarski and M. Lewandowska. 2007. Characteristics of adhesive properties of *Lactobacillus* strains synthesising biosurfactants. Polish J. Natural Sci. 22: 333–342.

Gołek, P., W. Bednarski, B. Brzozowski and B. Dziuba. 2009. The obtaining and properties of biosurfactants synthesized by bacteria of the genus *Lactobacillus*. Ann. Microbiol. 59: 119–126.

Gomaa, E.Z. 2013. Antimicrobial and anti-adhesive properties of biosurfactant produced by Lactobacilli isolates, biofilm formation and aggregation ability. J. Gen. Appl. Microbiol. 59: 425–436.

Gómez, N.C., J.M.P. Ramiro, B.X.V. Quecan, D.G. Bernadette and F. de Melo Franco. 2016. Use of potential probiotic lactic acid bacteria (lab) biofilms for the control of *Listeria monocytogenes, Salmonella typhimurium*, and *Escherichia coli* O157:H7 Biofilms Formation. Front Microbiol. 7: 863.

Gómez-Graña, S., M. Perez-Ameneiro, X. Vecino, I. Pastoriza-Santos, J. Perez-Juste, José Manuel Cruz and A.B. Moldes. 2017. Biogenic synthesis of metal nanoparticles using a biosurfactant extracted from corn and their antimicrobial properties. Nanomaterials 7(6): 139.

Goudarzi, L., R.K. Kermanshahi, Z. Mousavinezha and M.M.S. Dallal. 2016. Antimicrobial and anti-swarming effects of bacteriocins and biosurfactants from probiotic bacterial strains against *Proteus* spp. J. Medical Bacteriol. 5(5-6): 1–12.

Gudiña, E.J., V. Rocha, J.A. Teixeira and L.R. Rodrigues. 2010a. Antimicrobial and antiadhesive properties of a biosurfactant isolated from *Lactobacillus paracasei* ssp. *paracasei* A20. Lett. Appl. Microbiol. 50: 419–424.

Gudiña, E.J., J.A. Teixeira and L.R. Rodrigues. 2010b. Isolation and functional characterization of a biosurfactant produced by *Lactobacillus paracasei*. Colloids Surf. B 76: 298–304.

Gudiña, E.J., T.A. Teixeira and L.R. Rodrigues. 2011. Biosurfactant-producing lactobacilli: screening, production profiles, and effect of medium composition. Appl. Environ. Soil Sci. 1–9.

Gudiña, E.J., E.C. Fernandes, T.A. Teixeira and L.R. Rodrigues. 2015. Antimicrobial and anti-adhesive activities of cell-bound biosurfactant from *Lactobacillus agilis* CCUG31450. RSC Adv. 5: 909–960.

Hamza, F., S.K. Satpute, A. Banpurkar, A. Ravi Kumar and S. Zinjarde. 2017. Biosurfactant from the tropical marine bacterium *Staphylococcus lentus* SZ2: Disruption of aquaculture associated biofilms and efficacy in protecting *Artemia salina* against infections. FEMS Microbiol. Ecol. (93): 11.

Heinemann, C., J.E. van Hylckama Vlieg, D.B. Janssen, H.J. Busscher, H.C. van der Mei and G. Reid. 2000. Purification and characterization of a surface-binding protein from *Lactobacillus fermentum* RC-14 that inhibits adhesion of *Enterococcus faecalis* 1131. FEMS Microbiol. Lett. 190(1): 177–180.

Howard, J.C., C. Heinemann, B.J. Thatcher, B. Martin, B.S. Gan and G. Reid. 2000. Identification of collagen-binding proteins in *Lactobacillus* spp. with surface-enhanced laser desorption/ionization-time of flight protein chip technology. Appl. Environ. Microbiol. 66: 4396–4400.

Issa, I. and R. Moucari. 2014. Probiotics for antibiotic-associated diarrhea: Do we have a verdict? World J. Gastroenterol. 20(47): 17788–17795.

Iyer, R., S.K. Tomar, T.U. Maheswari and R. Singh. 2010. *Streptococcus thermophilus* strains: Multifunctional lactic acid bacteria. International Dairy J. 20(3): 133–141.

Kandler, O. and N. Weiss. 1986. Genus *Lactobacillus* Beijerinck 1901, 212AL. pp. 1209–1234. *In*: P.H.A. Sneath, N.S. Mair, M.E. Sharpe and J.G. Holt (eds.). Bergey's Manual of Systematic Bacteriology. Baltimore: Williams and Wilkins.

Kanmani, P., R. Satish Kumar, N. Yuvaraj, K. Paari, V. Pattukumar and V. Arul. 2013. Probiotics and its functionally valuable products—A review. Crit. Rev. Food Sci. 53: 641–658.

Kanna, R., N.S. Gummadi and G. Suresh Kumar. 2014. Production and characterization of biosurfactant by *Pseudomonas putida* MTCC 2467. J. Biological Sciences 14: 436–445.

Kermanshahi, R.K. and S.H. Peymanfar. 2012. Isolation and identification of Lactobacilli from cheese, yoghurt and silage by 16S rDNA gene and study of bacteriocin and biosurfactant production. Judishapur J. Microbiol. 5: 528–532.

Kojic, E.M. and R.O. Darouiche. 2004. *Candida* infections of medical devices. Clin. Microbiol. Rev. 17: 255–267.

Lowry, O.H., N.J. Rosebrough, A.L. Farr and R.J. Randall. 1951. Protein measurement with the Folin phenol reagent. J. Biol. Chem. 193: 265–275.

Madhu, A.N. and S.G. Prapulla. 2014. Evaluation and functional characterization of a biosurfactant produced by *Lactobacillus plantarum* CFR 2194. Appl. Biochem. Biotechnol. 172: 1777–1789.

Madigan, M. and J. Martinko. 2005. Brock Biology of Microorganisms, 11th Edition. Upper Saddle River, NJ: Pearson Prentice Hall, 2006.

Marchant, R. and I.M. Banat. 2012a. Biosurfactants: a sustainable replacement for chemical surfactants? Biotechnol. Lett. 34: 1597–1605.

Marchant, R. and I.M. Banat. 2012b. Microbial biosurfactants: challenges and opportunities for future exploitation. Trends Biotechnol. 11: 558–565.

Matei, G.M., S. Matei, A. Matei and E. Draghici. 2017. Antifungal activity of a biosurfactant-producing lactic acid bacteria strain. The Euro. Biotech. J. 1(3): 212–216.

Mbawala, A. and M.T. Hippolyte. 2012. Screening of biosurfactants properties of cell-free supernatants of cultures of *Lactobacillus* spp. isolated from a local fermented milk (Pendidam) of Ngaoundere (Cameroon). Int. J. Eng. Res. Appl. 2: 974–985.

Mbawala, A., T.H. Mouafo and R.R. Kom. 2013. Antibacterial activity of *Lactobacillus*' biosurfactants against *Pseudomonas* spp. isolated from fresh beef. Novus Int'l. J. Biotechnol. Biosci. 2(1): 7–22.

McLaughlin, B. 1946. A readily prepared medium for the cultivation of the lactobacilli. J. Bact. 51: 560–561.

Merghni, A., I. Dallel, E. Noumi, Y.H. Hentati, S. Tobji, A. Ben Amor and M. Mastouri. 2017. Antioxidant and antiproliferative potential of biosurfactants isolated from *Lactobacillus casei* and their anti-biofilm effect in oral *Staphylococcus aureus* strains. Microb. Pathog. 104: 84–89.

Moldes, A.B., A.M. Torrado, M.T. Barral and J.M. Domianguez. 2007. Evaluation of biosurfactant production from various agricultural residues by *Lactobacillus pentosus*. J. Agric. Food Chem. 55: 4481–4486.

Moldes, A.B., R. Paradelo, X. Vecino, J.M. Cruz, E. Gudiña, L. Rodrigues, J.A. Teixeira, J.M. Domínguez and M.T. Barral. 2013. Partial characterization of biosurfactant from *Lactobacillus pentosus* and comparison with sodium dodecyl sulfate for the bioremediation of hydrocarbon contaminated soil. BioMed. Res. Int. 1–9.

Morais, I.M.C., A.L. Cordeiro, G.S. Teixeira, V.S. Domingues, R.M.D. Nardi, A.S. Monteiro, R.J. Alves, E.P. Siqueira and V.L. Santos. 2017. Biological and physicochemical properties of biosurfactants produced by *Lactobacillus jensenii* $P_{6A}$ and *Lactobacillus gasseri* $P_{65}$. Microbial Cell Factories 16: 155.

Munira, Ch.I., M.I. Kadhim and Kh. Al-M Mayasaa. 2013. The effect of surlactin produced by *Lactobacillus acidophilus* on eye infectious bacteria in rabbits. J. Baghdad Sci. 10: 133–142.

Nitschke, M., L.V. Araújo, S.G. Costa, R.C. Pires, A.E. Zeraik, A.C. Fernandes, D.M. Freire and J. Contiero. 2009. Surfactin reduces the adhesion of food-borne pathogenic bacteria to solid surfaces. Lett. Appl. Microbiol. 49(2): 241–247.

Portilla, O.M., B. Rivas, A. Torrado, A.B. Moldes and J.M. Domíngue. 2008b. Revalorisation of vine trimming wastes using *Lactobacillus acidophilus* and *Debaryomyces hansenii*. J. Sci. of Food Agri. 88(13): 2298–2308.

Portilla-Rivera, O.M., M.A.B. Moldes, A.M. Torrado Agrasar and J.M. Dom Inguez Gonz Alez. 2007b. Biosurfactants from grape marc: stability study. J. Biotechnol. 131S: S136.

Portilla-Rivera, O., A. Torrado, J.M. Dominguez and A.B. Moldes. 2008a. Stability and emulsifying capacity of biosurfactants obtained from lignocellulosic sources using *Lactobacillus pentosus*. J. Agric. Food Chem. 56: 8074–8080.

Reid, G., A. Bruce and V. Smeianov. 1998. The role of Lactobacilli in preventing urogenital and intestinal infections. Int. Dairy. J. 8: 555–562.

Reid, G., C. Heinemann, M. Velraeds, H.C. van der Mei and H.J. Busscher. 1999. Biosurfactants produced by *Lactobacillus*. Methods Enzymol. 310: 426–433.

Reid, G., A.W. Bruce, H.J. Busscher and H.C. van der Mei. 2000. *Lactobacillus* therapies. United State Patent. US006051552A, Patent no. 6,051,552.

Rivera, O.M., A.B. Moldes, A.M. Torrado and J.M. Dominguez. 2007. Lactic acid and biosurfactants production from hydrolyzed distilled grape marc. Proc. Biochem. 42: 1010–1020.

Rodrigues, L.R., H.C. van der Mei, J.A. Teixeira and R. Oliveira. 2004. Influence of biosurfactants from probiotic bacteria on formation of biofilms on voice prosthesis. Appl. Environ. Microbiol. 70: 4408–4410.

Rodrigues, L.R., J.A. Teixeira, H.C. van der Mei and R. Oliveira. 2006a. Physicochemical and functional characterization of a biosurfactant produced by *Lactococcus lactis* 53. Colloids and Surfaces B 49: 79–86.

Rodrigues, L.R., J.A. Teixeira, H.C. van der Mei and R. Oliveira. 2006b. Isolation and partial characterization of a biosurfactant produced by *Streptococcus thermophilus* A. Colloids and surfaces B: Biointerfaces 53(1): 105–112.

Rodrigues, L., A. Moldes, J. Teixeira and R. Oliveira. 2006c. Kinetic study of fermentative biosurfactant production by *Lactobacillus* strains. Biochem. Eng. J. 28: 109–116.

Rodrigues, L., H. van Der Mei, I.M. Banat, J. Teixeira and R. Oliveira. 2006d. Inhibition of microbial adhesion to silicone rubber treated with biosurfactant from *Streptococcus thermophilus* A. Pathog. Dis. 46(1): 107–112.

Rodrigues, L., I.M. Banat, J. Teixeira and R. Oliveira. 2006e. Biosurfactants: potential applications in medicine. J. Antimicrob. Chemother. 57: 609–618.

Rodrigues, L., J. Teixeira, R. Oliveira and H.C. Van Der Mei. 2006f. Response surface optimization of the medium components for the production of biosurfactants by probiotic bacteria. Process Biochem. 41(1): 1–10.

Rodrigues, L.R. and J.A. Teixeira. 2008. Biosurfactants production from cheese whey. pp. 81–104. *In*: Ma. E. Cerdán, Ma I. González-Siso and M. Becerra (eds.). Advances in Cheese Whey Utilization: ISBN: 978-81-7895-359-5.

Rodríguez-Pazo, N., J.M. Salgado, S. Cortés and J.M. Domínguez. 2010. Alternatives for biosurfactants and bacteriocins extraction from *Lactococcus lactis* cultures produced under different pH conditions. Lett. Appl. Microbiol. 51(2): 226.

Rodríguez-Pazo, N., J.M. Salgado, S. Cortés-Diéguez and J.M. Domínguez. 2013. Biotechnological production of phenyllactic acid and biosurfactants from trimming vine shoot hydrolyzates by microbial co-culture fermentation. Appl. Biochem. Biotechnol. 169(7): 2175–2188.

Rodríguez-Pazo, N., L. Vázquez-Araújo, N. Pérez-Rodríguez, S. Cortés-Diéguez and J.M. Domínguez. 2013. Cell-free supernatants obtained from fermentation of cheese whey hydrolyzates and phenylpyruvic acid by *Lactobacillus plantarum* as a source of antimicrobial compounds, bacteriocins, and natural aromas. Appl. Biochem. Biotechnol. 171(4): 1042–1060.

Saguir, F.N. and M.M. de Nadra. 2007. Improvement of a chemically defined medium for the sustained growth of *Lactobacillus plantarum*: Nutritional requirements. Curr. Microbiol. 54: 414–418.

Salehi, R., O. Savabi, M. Kazemi, S. Kamali, A.R. Salehi, G. Eslami and A. Tahmourespour. 2014. Effects of *Lactobacillus reuteri*-derived biosurfactant on the gene expression profile of essential adhesion genes (gtfB, gtfC and ftf) of *Streptococcus mutans*. Adv. Biomed. Res. 3: 169.

Salman, J.A.S. and A.D. Alimer. 2014. *Lactobacillus rhamnosus* against some bacteria causing UTI in Iraqi women. Int. J. Curr. Res. 6: 5368–5374.

Sambanthamoorthy, K., X. Feng, R. Patel, S. Patel and S.C. Paranavitana. 2014. Antimicrobial and antibiofilm potential of biosurfactants isolated from Lactobacilli against multi-drug resistant pathogens. BMC Microbiol. 14: 197.

Saravanakumari, P. and K. Mani. 2010. Structural characterization of a novel xylolipid biosurfactant from *Lactococcus lactis* and analysis of antibacterial activity against multi-drug resistant pathogens. Bioresource Technol. 101(22): 8851–8854.

Satpute, S.K., P.K. Dhakephalkar and B.A. Chopade. 2005. Biosurfactants and bioemulsifiers in hydrocarbon biodegradation and spilled oil bioremediation. In Technology transfer for industrial applications of novel methods and materials for environmental problems. Bharti Vidyapeeth, Dept. of Chemical Engineering Pune and Italy. 5th–6th December.

Satpute, S.K., B.D. Bhawsar, P.K. Dhakephalkar and B.A. Chopade. 2008. Assessment of different screening methods for selecting biosurfactant producing marine bacteria. Indian J. Marine Sci. 37(3): 243–250.

Satpute, S.K., S.S. Bhuyan, K.R. Pardesi, S.S. Mujumdar, P.K. Dhakephalkar, A.M. Shete and B.A. Chopade. 2010a. Molecular genetics of biosurfactant synthesis in microorganisms. pp. 14–41. *In*: R.K. Sen (ed.). Biosurfactants—Advances in Experimental Medicine and Biology. Landes Bioscience, Intelligence Unit, Springer.

Satpute, S.K., A.G. Banpurkar, P.K. Dhakephalkar, I.M. Banat and B.A. Chopade. 2010b. Methods for investigating biosurfactants and bioemulsifiers: A review. Crit. Rev. Biotechnol. 30: 127–144.

Satpute, S.K., I.M. Banat, P.K. Dhakephalkar, A.G. Banpurkar and B.A. Chopade. 2010c. Biosurfactants, bioemulsifiers and exopolysaccharides from marine microorganisms. Biotechnol. Adv. 28: 436–450.

Satpute, S.K., A.G. Banpurkar, I.M. Banat, J.N. Sangshetti, R.H. Patil and W.N. Gade. 2016a. Multiple roles of biosurfactants in biofilms. Curr. Pharm. Des. 22: 429–448.

Satpute, S.K., G.R. Kulkarni, A.G. Banpurkar, I.M. Banat, N.S. Mone, R.H. Patil and S.S. Cameotra. 2016b. Biosurfactant/s from Lactobacilli species: Properties, challenges and potential biomedical applications. J. Basic Microbiol. 56(11): 1140–1158.

Satpute, S.K., G.A. Płaza and A.G. Banpurkar. 2017. Biosurfactants' production from renewable natural resources: example of innovative and smart technology in circular bioeconomy. Management Sys. Prod. Eng. 25(1): 46–54.

Satpute, S., S. Zinjarde and I. Banat. 2018. Recent updates on biosurfactant/s in Food industry. *In*: D. Sharma and B.D. Saharan (eds.). Microbial Cell Factories. CRC Press, Taylor and Francis ISSN No: 978-1-13-806138-5.

Sauvageau, J., J. Ryan, K. Lagutin, I.M. Sims, B.L. Stocker and M.S. Timmer. 2012. Isolation and structural characterisation of the major glycolipids from *Lactobacillus plantarum*. Carbohydr. Res. 357: 151–156.

Servin, A.L. 2004. Antagonistic activities of lactobacilli and bifido bacteria against microbial pathogens. FEMS Microbiol. Rev. 28: 405e440.

Sgibnev, A. and E. Kremleva. 2017. Influence of hydrogen peroxide, lactic acid, and surfactants from vaginal lactobacilli on the antibiotic sensitivity of opportunistic bacteria. Probiotics and Antimicrob. Proteins. 9(2): 131–141.

Sharma, D. and B. Singh Saharan. 2014. Simultaneous production of biosurfactants and bacteriocins by probiotic *Lactobacillus casei* MRTL3. Inter. J. Microbiol. 1–7.

Sharma, D., B. Singh Saharan, N. Chauhan, A. Bansal and S. Procha. 2014. Production and structural characterization of *Lactobacillus helveticus* derived biosurfactant. The Scientific World J. 1–9.

Sharma, D., B.S. Saharan, N. Chauhan, S. Procha and S. Lal. 2015. Isolation and functional characterization of novel biosurfactant produced by *Enterococcus faecium*. Springer Plus. 4(1): 4.

Sharma, D., B. Singh Saharan and K. Shailly. 2016. Biosurfactants of lactic acid bacteria. eBook ISBN. 978-3-319-26215-4. Springer Briefs in Microbiology. Springer International Publishing AG Switzerland. DOI 10.1007/978-3-319-26215-4.

Shokouhfard, M., R.K. Kermanshahi, R.V. Shahandashti, M.M. Feizabadi and S. Teimourian. 2015. The inhibitory effect of a *Lactobacillus acidophilus* derived biosurfactant on biofilm producer *Serratia marcescens*. Iran J. Basic Med. Sci. 18: 1001–1007.

Souza, E.C.P.O. de Souza de Azevedo, J.M. Domínguez, A. Converti and R.P. de Souza Oliveira. 2017. Influence of temperature and pH on the production of biosurfactant, bacteriocin and lactic acid by *Lactococcus lactis* CECT-4434, CyTA. J. Food. 15(4): 525–530.

Sperti, G.S. 1971. Probiotics. AVI Publishing Company, Westport, Connecticut, U.S.A. pp. 1–120.

Stiles, M.E. and W.H. Holzapfel. 1997. Lactic acid bacteria of foods and their current taxonomy. Int. J. Food Microbiol. 36: 1–29.

Tahmourespour, A., R. Salehi, R.K. Kermanshahi and G. Eslami. 2011. The anti-biofouling effect of *Lactobacillus fermentum*-derived biosurfactant against *Streptococcus mutans*. Biofouling 27: 385–392.

Thavasi, R., S. Jayalakshmi and I.M. Banat. 2011. Application of biosurfactant produced from peanut oil cake by *Lactobacillus delbrueckii* in biodegradation of crude oil. Bioresour. Technol. 102: 3372–3666.

US Environmental Protection Agency Office of Pesticide Programs (USEPAOPP). Biopesticides Registration action document rhamnolipid biosurfactant (PC Code 110029).

Valle, J., S. Da Re, N. Henry, T. Fontaine, D. Balestrino, P. Latour-Lambert and Jean-Marc Ghigo. 2006. Broad-spectrum biofilm inhibition by a secreted bacterial polysaccharide. Proc Natl. Acad. Sci. USA 103: 12558e12563.

van Hoogmoed, C.G., H.C. van der Mei and H.J. Busscher. 2004. The influence of biosurfactants released by *S. mitis* BMS on the adhesion of pioneer strains and cariogenic bacteria. Biofouling 20(6): 261–267.

Vecino, X., R. Devesa-Rey, J.M. Cruz and A.B. Moldes. 2013. Evaluation of biosurfactant obtained from *Lactobacillus pentosus* as foaming agent in froth flotation. J. Environ. Manag. 128: 655–660.

Vecino, X., L. Barbosa-Pereira, R. Devesa-Rey, J.M. Cruz and A.B. Moldes. 2015. Optimization of extraction conditions and fatty acid characterization of *Lactobacillus pentosus* cell-bound biosurfactant/bioemulsifier. J. Sci. Food Agric. 95: 313–320.

Vecino, X., L. Rodriguez-Lopez, B. Cruz, J.M., Moldes, A.B. and L.R. Rodrigues. 2016. The influence of the medium composition on the biosurfactants produced by *Lactobacillus paracasei*. Nano formulation 2016. Formulate your innovation. Innovate your formulation. Conference held at Barcelona July 4th to 7th July. Barcelona. Poster P 49.

Vecino, X., L. Rodríguez-López, D. Ferreira, J.M. Cruz, A.B. Moldes and L.R. Rodrigues. 2017. Bioactivity of glycolipopeptide cell-bound biosurfactants against skin pathogens. Int. J. Biol. Macromol. S0141–813.

Vecinoa, X., L. Rodríguez-López, E.J. Gudiña, J.M. Cruz, A.B. Moldes and L.R. Rodrigues. 2017. Vineyard pruning waste as an alternative carbon source to produce novel biosurfactants by *Lactobacillus paracasei*. J. Indus. and Eng. Chem. 55: 40–49.

Velraeds, M.M.C., H.C. van der Mei, G. Reid and H. Busscher. 1996a. Inhibition of initial adhesion of uropathogenic *Enterococcus faecalis* by biosurfactants from *Lactobacillus* isolates. Appl. Environ. Microbiol. 62: 1958–1963.

Velraeds, M.M.C., H.C. van der Mei, G. Reid and H.J. Busscher. 1996b. Physicochemical and biochemical characterization of biosurfactants released by *Lactobacillus* strains. Coll. Surf. B 8: 51–61.

Velraeds, M.M.C., H.C. van der Mei, G. Reid and H.C. Busscher. 1997. Inhibition of initial adhesion of uropathogenic *Enterococcus faecalis* to solid substrata by an adsorbed biosurfactant layer from *Lactobacillus acidophilus*. Urology 49: 790–794.

Velraeds, M.M.C., B. van de Belt-Gritter, H.C. van der Mei, G. Reid and H.J. Busscher. 1998. Interference in initial adhesion of uropathogenic bacteria and yeasts to silicone rubber by a *Lactobacillus acidophilus* biosurfactant. J. Med. Microbiol. 47: 1081–1085.

Velraeds, M.M.C., B. van de Belt-Gritter, H.J. Busscher, G. Reid and H.C. van der Mei. 2000. Inhibition of uropathogenic biofilm growth on silicone rubber in human urine by Lactobacilli—a teleologic approach. World. J. Urol. 18: 422–426.

Walencka, E., R.O. Zalska, B. Sadowska and B. lska. 2008. The influence of *Lactobacillus acidophilus* derived surfactants on staphylococcal adhesion and biofilm formation. Folia Microbiol. 53: 61–66.

Wang, S., Q. Wang, E. Yang, L. Yan, T. Li and H. Zhuang. 2017. Antimicrobial compounds produced by vaginal *Lactobacillus crispatus* are able to strongly inhibit *Candida albicans* growth, hyphal formation and regulate virulence-related gene expressions. Frontiers in Microbiol. 8: 564.

Wesselsa, S., L. Axelssonb, E.B. Hansenc, L.D. Vuystd, S. Laulunde, L. La¨hteenma¨kif, S. Lindgreng, B. Molleth, S. Salmineni and A. von Wright. 2004. The lactic acid bacteria, the food chain, and their regulation. Trends Food Sci. Technol. 15: 498–505.

Yerushalmi, L., S. Rocheleau, R. Cimpoia, M. Sarrazin, G. Sunahara, A. Peisajovich and S.R. Guiot. 2003. Enhanced biodegradation of petroleum hydrocarbons in contaminated soil. Bioremediation J. 7(1): 37–51.

Zhang, J., Q. Caiyin, W. Feng, X. Zhao, B. Qiao, G. Zhao and J. Qiao. 2016. Enhance nisin yield via improving acid-tolerant capability of *Lactococcus lactis* F44. Scientific Reports. 6.

Zommiti, M., N. Connil, J.B. Hamida and M. Ferchichi. 2017. Probiotic characteristics of *Lactobacillus curvatus* DN317, a strain isolated from chicken ceca probiotics. Antimicrob. Prot. 9: 415–424.

# 8

# Biosurfactants from Actinobacteria: State of the Art and Future Perspectives

*Marilize Le Roes-Hill,*[1,]* *Kim A. Durrell*[1] *and*
*Johannes H. Kügler*[2]

## Introduction

With the ever increasing human population and its demands on various industries, the need for chemical substances have increased dramatically. Surfactants are one such commodity that are utilised by a range of industries. It is therefore not surprising that it is estimated that the need for surfactants will increase to approximately 24 million tonnes by 2020 (Otzen 2016). At the moment, the surfactant market is still largely dominated by chemical surfactants, with only a few companies trading in surfactants of biological origin (biosurfactants) (Sekhon and Rahman 2014, Sajna et al. 2015). However, chemical surfactants, being derived from fossil fuels, are not sustainable, hard to biodegrade (resulting in bioaccumulation in sensitive ecosystems), and toxic to the environment, fauna and flora. The development of global bio-economy approaches has resulted in an increased need for a shift towards the development of sustainable bio-based processes, and therefore a need for a 'green' alternative to chemical surfactants (Jackson et al. 2015, Otzen 2016).

[1] Cape Peninsula University of Technology, Biocatalysis and Technical Biology Research Group, Institute of Biomedical and Microbial Biotechnology, Symphony Way, Bellville, Western Cape Province, South Africa, 7530.
[2] Karlsruhe Institute of Technology (KIT), Institute of Process Engineering in Life Sciences Section II: Technical Biology, Engler-Bunte Ring 1 Karlsruhe, Germany, 76131.
Emails: durrellka@gmail.com; j_kuegler@gmx.net
* Corresponding author: leroesm@cput.ac.za

Biosurfactants (BSs) are predominantly produced by microorganisms, but also include oleochemicals (vegetable oils, animal fats, and their derivatives), and saponins, a natural soap found in plants (Sajna et al. 2015). The way in which these BSs are produced varies, but it is typically highly regulated, involving polyketide synthase (PKS) and/or non-ribosomal peptide synthase (NRPS) biosynthetic pathways. In contrast to chemical surfactants, these natural products are easily biodegradable, exhibit improved emulsification and foaming properties, are less/non toxic and hypo-allergenic, and stable at extremes of pH, temperature and salinity, making them suitable replacements for chemical surfactants (Jackson et al. 2015). However, various challenges such as expensive production costs, low yields, and complicated downstream processing, have precluded their widespread application (Marchant and Banat 2012a).

BSs play an important role in numerous biological processes whenever a microbe comes in contact with an interface. These processes include: bacterial cell signalling, motility, cellular differentiation, biofilm formation, accessing substrates, and bacterial pathogenesis (Cameotra and Makkar 2004, Van Hamme et al. 2006). Among BS-producing microorganisms, bacteria such as *Pseudomonas* spp. and *Bacillus* spp. have received the greatest attention, mainly because of the type of BSs they produce (Hausmann and Syldakt 2014). In addition, organisms belonging to the Class *Actinobacteria*, notably the Orders *Actinomycetales* and *Micrococcales* have been exploited for their ability to produce structurally diverse BSs. This chapter therefore aims to introduce the readers to the different types of BSs produced by actinobacterial strains, their biotechnological potential, the challenges in producing BSs from actinobacteria, and future perspectives.

## BSs Produced by Actinobacteria

The '*Actinobacteria*' is notably one of the largest bacterial phyla (Figure 1). All members of this phylum are typically referred to as actinobacteria, while those that are grouped into the Order *Actinomycetales* and exhibit filamentous growth are referred to as actinomycetes (Goodfellow and Fiedler 2010). Members of the Class *Actinobacteria* are Gram positive to Gram variable bacteria, and typically contain a high guanine:cytosine ratio in their DNA. They are found in diverse environments and play important roles in the turnover of organic matter and the overall carbon cycle (Goodfellow and Fiedler 2010). Members of the Order *Actinomycetales* are well-known for their ability to produce bio-active compounds: since the discovery of the anti-tuberculosis compound, streptomycin, more than 13,000 bio-active metabolites have been isolated from actinobacterial strains, with more than 10,000 being produced by members of the genus *Streptomyces* (Bérdy 2012), notably one of the biggest bacterial genera (> 600 validly published species; Parte 2017).

The type of compounds that are produced by the actinobacteria is strongly influenced by their natural environment. Since actinobacteria are found in diverse environments (terrestrial and marine), it is not surprising that the BSs they produce have diverse chemical structures, and include various types of glycolipids, glycoside BSs, lipopeptides, and polymeric BSs (Figure 2). The production of BSs by actinobacterial strains have recently been reviewed by Kügler et al. (2015b). It is

**Figure 1.** The phylum '*Actinobacteria*' and the six classes within the phylum. BSs production is notably limited to the Order *Actinomycetales* and the Order *Micrococcales*. Groupings are as defined on the list of prokaryotic names with standing in nomenclature website (Parte 2017).

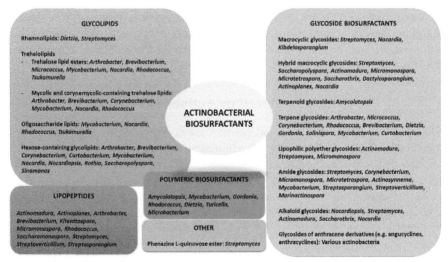

**Figure 2.** Different types of BSs produced by actinobacterial strains (representative genera are indicated). Grey blocks represent BSs typically evaluated for surface activity and application in a wide range of industries, while the black blocks represent BSs typically exploited for biomedical and pharmaceutical applications. The polymeric BSs are of interest due to their role in pathogenesis.

clear from the review article that there is still a great scope for the discovery of BSs from the *Actinobacteria* and that we do not have a clear idea of the true diversity of the BSs produced by actinobacterial strains, with many BSs still requiring structural elucidation. In the following sections, different types of BSs produced by the actinobacteria will be highlighted.

## Glycolipids

Glycolipids (GLs) have been the focus of various research studies and are considered to be ideal for industrial application (Marchant and Banat 2012b). This is mainly due to their good surface activity and chemical properties. The polar component of GLs consists of a mono- or, di- or, oligosaccharide, or sugar alcohol, while the non-polar component consists of long chain aliphatic or hydroxyaliphatic acid (Sajna et al. 2015). The main types of GLs that have been produced by actinobacterial strains include trehalose lipid mycolic acid esters, trehalose lipid esters, trehalose-based oligosaccharide lipids, hexose-containing glycolipids, and various other glycoside BSs (Kügler et al. 2015b). More recently, rhamnolipids (RLs) have also been detected in actinobacterial cultures (see RLs section below).

## Rhamnolipids

Rhamnolipid (RL) production by *Pseudomonas pyocyanea* (now *P. aeruginosa*) was first described in 1947. RLs are often considered to be "smart molecules" mainly due to their involvement in various processes such as hydrocarbon degradation, quorum sensing, biofilm formation or disruption, and as nanoparticle stabilising agents (Kiran et al. 2016b). To date, there have been various reports on the abilities of actinobacterial strains to produce RLs. However, since the majority of these reports are not substantiated by structural information (Kügler et al. 2015b), the validity of these claims should be questioned (Irorere et al. 2017).

To date, RL production (confirmed by structural information) has been reported for *Cellulomonas cellulans* (*Oerskovia xanthineolytica*; based on acid hydrolysis and gas chromatography; Arino et al. 1998), *Renibacterium salmoninarum* (based on thin layer chromatography and infrared spectroscopy; Christova et al. 2004), a *Nocardiopsis* sp. (based on acid hydrolysis of the BS and thin layer chromatography; Vasileva-Tonkova and Gesheva 2005), and *Rhodococcus fascians* (same analyses as for the previous strain; Gesheva et al. 2010). More recently, Kalyani et al. (2014a, b) and Yan et al. (2014), for the first time, reported the production of RLs by *Streptomyces* spp. The structure of the RLs produced by *S. coelicoflavus* NBRC 15399[T] (Kalyani et al. 2014a) and *S. matensis* NBRC 12889[T] (Kalyani et al. 2014b) was not characterized, but the RL produced by *Streptomyces* sp. ISP2-49E, was determined to consist of two rhamnose units conjugated to two $C_{10}$ fatty acid chains (Yan et al. 2014). A study by Allada et al. (2015) confirmed that the RLs produced by *S. coelicoflavus* NBRC 15399[T], are mono- and di-RL (Rha-$C_{10}$-$C_{10}$ and Rha$_2$-$C_{10}$-$C_{10}$). The production of a di-RL (Rha$_2$-$C_{10}$-$C_{10}$) was also reported for the deep sea isolate, *Dietzia maris* (Wang et al. 2014). The discovery of RL production outside the proteobacteria emphasises the need to extend the screening widely among other untapped microbial groups.

## Trehalose lipids

Trehalose lipids (TLs) were first discovered in fat extracts from tubercle bacilli in 1933, but were only later purified from *Mycobacterium tuberculosis* in 1956. The purified compound was identified as 6,6'-dimycoloyl-α',α'-D-trehalose, best known as the "cord factor" which is an important virulence factor in *Mycobacterium* spp. After this discovery, several groups of TLs have been isolated, mostly from *Corynebacterium*, *Mycobacterium* and *Nocardia* spp. (Christova and Stoineva 2014). TLs possess a large spectrum of biological activities as well as other favourable characteristics such as unique surface properties, low critical micelle concentration (CMC; mg/L), low toxicity, and biodegradability (Franzetti et al. 2010). TLs can be divided into three general subclasses: Subclass I—6,6 substituted trehalose esters, i.e., trehalose dimycolates (TDMs, cord factor), trehalose dicorynomycolates (TDCMs), and fatty acid trehalose diesters (TDEs); subclass II—2,3-diesters of trehalose sulphates; and subclass III—succinoyl di- and tetra-esters. Since these subclasses have recently been reviewed by Christova and Stoineva (2014), it will not be reviewed in this chapter.

## Oligosaccharide lipids

The oligosaccharide-containing lipids (OLs) are complex mixtures of different compounds which generally are di-, tri- and tetra-saccharides, such as trehalose bound to molecules of galactose and glucose, which are bound to acetyl groups and fatty acids with 4 to 18 carbon chain length (Konishi et al. 2014). OLs can be divided into two main groups, namely (1) mycosides—which occurs when oligosaccharide moieties are linked to an unusual lipidic residue (specific to mycobacteria); and (2) lipooligosaccharides, acylated by long chain fatty acids.

Phenolglycolipids (mycosides A, B, and G) are used to detect infections by *M. tuberculosis*. The structure of these compounds comprises of a lipidic aglycone moiety called phenol phthiocerol (Gastambide-Odier et al. 1970). Some mycobacterial strains only bear one methyl rhamnoside residue (*M. bovis* and *M. marinum*), while others, as in the case of *M. kansasii*, contain oligosaccharides in which one of them comprises of a labile dideoxyhexose at its non-reducing end (Fournié et al. 1987). Non-polar glycopeptidolipids (mycoside C), mainly found in *M. avium*, comprises of three amino acids positioned in a D series. These amino acids are linked by the C-terminal group to alaninol and by the N-terminal group to a 3-hydroxy-$C_{28}$ fatty acid in which the fatty acid may be methylated or free (Lanéelle and Asselineau 1968).

*Tsukamurella* sp. DSM 44370 produced a mixture of di-, tri- and tetra-saccharide lipids when cultivated on medium containing sunflower oil; the production increased by 60% when marigold oil was used as a sole carbon source (Vollbrecht et al. 1999). A trisaccharide succinic tetraester was isolated from producing organism *R. fascians* NBRC 12155 (Konishi et al. 2014), while oligosaccharide lipids were detected within the growth media of *T. spumae* and *T. pseudospumae*. It was suggested by Kügler et al. (2015a) that compounds produced by these organisms are more than likely to contain hydrophobic units and aromatic moieties which differ from typical fatty acid residues produced.

Other GLs, partially or not structurally elucidated, have been reported for a range of actinobacterial genera, including *Brachybacterium paraconglomeratum* MSA21 (Kiran et al. 2014b), *Brevibacterium* spp. (Graziano et al. 2016), *Corynebacterium lepus* (Cooper et al. 1979), *Frankia* sp. Cp11 (Tunlid et al. 1989), *Gordonia* sp.

BS29 (Franzetti et al. 2010), *Nesterenkonia* spp. (Mohanram et al. 2016), *Nocardia erythropolis* ATCC 4277 (Macdonald et al. 1981), *Nocardioides* sp. A-8 (Vasileva-Tonkova and Gesheva 2005), *Oerskovia xanthineolytica* CIP 104849 (Arino et al. 1998), and *R. salmoninarum* 27BN (Christova et al. 2004).

## Other GLs

In their review on BSs produced by actinobacterial strains, Kügler et al. (2015b) further identified other types of GLs produced by the actinobacteria. These include the hexose-containing GLs: sucrose- or fructose-containing GLs which were produced by *Arthrobacter*, *Brevibacterium*, *Corynebacterium*, *Mycobacterium*, and *Nocardia* spp. in response to replacing *n*-alkanes with sucrose or fructose as sole carbon source, in effect replacing the trehalose component; di-mannose GLs produced by *Arthrobacter*, *Curtobacterium*, *Micrococcus*, *Nocardiopsis*, *Rothia*, and *Saccharopolyspora* spp., as well as *Sinomonas astrocyaneus*; and the galactosyl diglycerides produced by *Arthrobacter scleromae* and *A. globiformis* (Kügler et al. 2015b). In addition, Vollbrecht et al. (1999) reported the production of a TL as well as tri- and tetra-saccharide lipids when *Tsukamurella* sp. DSM 44370 was cultivated in the presence of sunflower oil ($C_{18:1}$). The tri- and tetra-saccharide lipids were able to reduce the surface tension of water from 72 mN/m to 23 and 24 mN/m. Structural elucidation showed that the trisaccharide consisted of three glucose units, while the tetrasaccahride consisted of two glucose units and one galactose unit (Vollbrecht et al. 1999).

Kügler et al. (2015b) and Dembitsky (2005a, b, c, d, e) also identified a range of glycoside BSs that are produced by the actinobacteria. These glycoside BSs are typically exploited by the biomedical and pharmaceutical industries, mainly because they exhibit a range of bio-activities, such as antibacterial, antifungal, antiviral, antitumour, and immunosuppressant activity, or may serve as a basis for new drug design. These glycoside BSs are classified based on their structural properties: macrocyclic glycosides, hybrid macrocyclic glycosides, terpenoid glycosides, terpene glycosides, alkaloid glycosides, fatty acid amide glycosides, and lipophilic polyether glycosides. Examples of some of these compounds described from actinobacteria are summarised in Tables 1–5 and their great structural diversity can be seen in Figure 3. Glycosides of anthracene derivatives, such as angucyclines and anthracyclines are also considered to be BSs (Dembitsky 2005d). However, these bio-active compounds represent a vast group of structurally diverse compounds and have been extensively reviewed (e.g., Elshahawi et al. 2015) and will not be included in this chapter.

## Macrocyclic glycosides

Macrocyclic glycosides or macrolides are a large group of structurally diverse compounds. They consist of an aglycone (highly substituted monocyclic lactone) attached to one or more saccharides glycosidically linked. The aglycone is synthesised via polyketide biosynthesis pathways and vary in the number of carbons making up the cyclic structure (Dembitsky 2005a). The dimeric macrocyclic glycosides are a rare group of compounds isolated from various microorganisms and sea sponges.

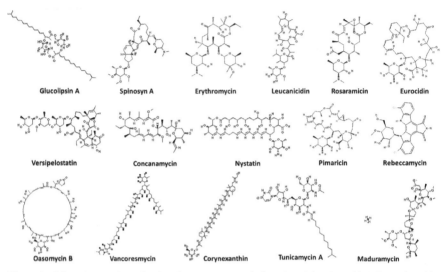

**Figure 3.** Selected examples reflecting the great structural diversity of the glycoside BSs produced by various actinobacterial strains.

They exhibit a range of bio-activities such as antibacterial, antitumour, and antiviral activities (Dembitsky 2005a). The hybrid macrocyclic glycosides consist of fatty acid macrolactones (or derivatives) with one or more rare sugar. This group consists of a vast number of compounds ranging from compounds with a 12-membered ring system (e.g., the spinosyns produced by *Saccharopolyspora spinosa*) to compounds with a 42-membered ring system (e.g., oasomycin B produced by *Streptoverticillium baldacci* subsp. *netropse*). Other examples of this group are summarised in Table 1.

## *Terpenoid and terpene glycosides*

Terpenoid and terpene glycosides are isoprenoid based compounds that are linked to a hydrophilic sugar component. Vancoresmycin, a potent antibacterial agent, is a C65 terpenoid glycoside produced by an *Amycolatopsis* sp. (Hopmann et al. 2002). Notably, it exhibited antibacterial activity against vancomycin-resistant pathogenic strains. Its complex structure consists of a tetrameric acid unit glycosidically linked to a methylated carbohydrate. In addition to this terpenoid glycoside, various terpene glycosides have been isolated from actinobacterial strains. These compounds are natural pigments that play a key role in the survival of the producing organism. These pigments are believed to be involved in the dissipation of excess energy, exhibit an antioxidant effect (scavenges free radicals), and modulate membrane fluidity and proton permeability. These are high value compounds and can be applied as natural pigments and nutraceuticals (Klassen 2010). Even though the occurrence of these compounds among the actinobacteria appears to be limited, a prokaryotic carotenoid database compiled by Nupur et al. (2016; www.bioinfo.imtech.res.in/servers/ procardb), shows that the actinobacteria produces > 500 carotenoid compounds and are produced by various genera. The monocyclic and bicyclic carotenoid glycosides produced by actinobacterial strains are summarised in Table 2.

**Table 1.** Selected examples of macrocyclic glycoside BSs produced by actinobacterial strains (as reviewed by Dembitsky 2005a, Elshahawi et al. 2015, Kügler et al. 2015b).

| Type of Biosurfactant | Actinobacterium | Biosurfactant |
|---|---|---|
| Macrocyclic glycoside | *Streptomyces* spp. HKI-0113 and HKI-0114; *S. melanosporus; S. hygroscopicus* var *azalomyceticus; S. violaceoniger* Tu 905 | Elaiophylin |
| | *S. hygroscopicus* | Halichoblelide |
| | *S. purpurogeniscleroticus* WC71634 and *Nocardia vaccinia* WC65712 | Glucolipsin A and B |
| | *S. microflavus* 2445 | Fattiviracin FV-8 |
| | *Kibdelosporangium albatum* R761-7 (ATCC 55061) | Cycloviracins B1 and B2 |
| | *Streptomyces* sp. BA-2836.1 | Macroviracins |
| Hybrid macrocyclic glycoside: 12-membered ring system | *Streptomyces* spp. | Methymycin, neomethymycin |
| | *S. venezuelae* | Novamethymycin |
| | *S. spinosa* | Spinosyns, lepicidin A |
| | *Actinomadura* sp. MK73-NF4 | Decatromicins A and B |
| | *Micromonospora* sp. C39217-R109-7 | Pyrrolosporin A |
| Hybrid macrocyclic glycoside: 14-membered ring system | *Streptomyces erythreus* and *Nocardia* spp. | Erythromycins A-E |
| | *S. antibioticus* | Oleandomycin |
| | *S. felleus* and *S. narbonensis* | Pikromycin, narbomycin, 5-*O*-mycaminosyl-narbonolide |
| | *Micromonospora megalomicea* and *M. inositola* | Megalomicin A, B, C1, C2, and XK-41-B2 |
| | *S. violaceoniger* and *S. spinichromogenes* | Lankamycin and its derivatives |
| | *Actinomadura vulgaris* subsp. *lanata* | Fluviricin B1 and B2 |
| | *Microtetraspora tyrrhenii, M. pusilla, Saccharothrix mutabilis,* and *Streptomyces* sp. MJ677-72F5 | Variations of fluviricin |
| Hybrid macrocyclic glycoside: 15-membered ring system | *Streptomyces* spp. A233, A239, and A240 | Leucanicidin and derivatives |
| Hybrid macrocyclic glycoside: 16-membered ring system | *Streptomyces avermitilis* | Avermectins |
| | *S. ambofaciens, S. fradiae,* and *S. hygroscopicus* | Spiramicins and related compounds |
| | *S. kitasatoensis* and *S. josamyceticus* | Various leucomycins |
| | *S. hygroscopicus* | Maridomycins and platenomycins |
| | *S. platensis* | Carbomycin B, platenomycin W1 and W2, niddamycin, midecamycin A3 and A4 |
| | *Micromonospora rosaria* | Rosaramicin |
| | *M. capillata* | Izenamicin A3, juvenimicin A2 and A4 |

*Table 1 contd. ...*

*...Table 1 contd.*

| Type of Biosurfactant | Actinobacterium | Biosurfactant |
|---|---|---|
| Hybrid macrocyclic glycoside: 17-membered ring system | *Streptomyces versipellis* 4083-SVS6 | Versipelostatin |
| Hybrid macrocyclic glycoside: 18-membered ring system | *S. subflavus* subsp. *irumaensis* | Irunamycin |
| | *S. diastatochromogenes* S-45 | Concanamycin A-C |
| | *Dactylosporangium aurantiacum* subsp. *hamdensis* | Tiacumycins |
| | *Actinoplanes deccanensis* | Lipiarmycin |
| | *M. echinospora* | Clostomicin B1 |
| Hybrid macrocyclic glycoside: 20-membered ring system | *Streptomyces* sp. HC34 | Vicenistatin |
| Hybrid macrocyclic glycoside: 24-membered ring system | *Streptomyces* sp. GK9244 | Tetrin, tetramycins |
| | *S. natalensis* | Pimaricin |
| | *Streptoverticillium aeurocidicum* IFO 13491 | Eurocidins C-E |
| Hybrid macrocyclic glycoside: 30-membered ring system | *Nocardia brasiliensis* | Brasilinolide A |
| | *Streptomyces* sp. LL-C13122 | Colubricidin A |
| | *S. noursei* ATCC 11455 | Nystatin |
| | *S. griseus* IMRU 3570 | Candicidin |
| Hybrid macrocyclic glycoside: 42-membered ring system | *S. baldacci* subsp. *netropse* | Oasomycin B and desertomycin |

**Table 2.** Selected examples of terpenoid and terpene glycoside BSs produced by actinobacterial strains (as reviewed by Dembitsky 2005b, Elshahawi et al. 2015, Kügler et al. 2015b).

| Type of Biosurfactant | Actinobacterium | Biosurfactant |
|---|---|---|
| Terpenoid glycoside | *Amycolatopsis* sp. DSM 12216 | Vancoresmycin (C65 terpenoid) |
| Terpene glycoside | *Arthrobacter* sp. M3 and *Curtobacterium flaccumfaciens* pvar *poinsettiae* | Corynexanthin mono- and diglycosides (C50 terpene) |
| | *Corynebacterium* sp. CMB 8 | Corynexanthin (C50 terpene) |
| | *Micrococcus yunnanensis* AOY-1 *M. luteus* | Sarcinaxanthin, sarcinaxanthin mono- and diglycosides (C50 terpene) |
| | *Rhodococcus rhodochrous* RNMS1 | Carotenoid (C40 terpene) glycoside (C36-C50 mycolic) |
| | *Nocardia kirovani* | Fatty acid ester of pheixanthophyll |
| | *Corynebacterium autrophicum* and *Rhodobacter sphaeroides* | Zeaxanthin mono- and diglucosides and rhamnosides |

## Fatty acid amide glycosides

Fatty acid amide glycosides include the well-known class of antibiotics, the aminoglycosides. These broad spectrum antibiotics exhibit activity against both Gram positive and Gram negative bacteria, making then of great interest to the pharmaceutical and biomedical industries. They typically contain an aminocyclitol group that is glycosidically linked to two or more amino sugars (Dembitsky 2005c). In addition to the aminoglycosides, this group also includes other antimicrobial, antitumour, antiviral, antimycoplasma, and bio-insecticide compounds, including the potent group of nucleoside antibiotics (Dembitsky 2005c, Hiratsuka et al. 2014, Chen et al. 2016, Price et al. 2016) as presented in Table 3.

**Table 3.** Selected examples of fatty acid amide glycoside BSs produced by actinobacterial strains (as reviewed by Dembitsky 2005c, Elshahawi et al. 2015, Kügler et al. 2015b).

| Type of Biosurfactant | Actinobacterium | Biosurfactant |
|---|---|---|
| Fatty acid amide glycosides | *Micromonospora* sp. MK-7 | Fortimicins A and B |
| | *Streptomyces tenjimariensis* | Istamycin A and B |
| | *M. tuberculosis* and *M. bovis* | Mycothiol |
| | *S. lysosuperficus* | Tunicamycins I-X |
| | *S. griseoflavus* | Streptovirudins |
| | *Corynebacterium rathayi* | Corynetoxins |
| | *S. noursei* and *S. vinaasdrappus* SANK-62394 | Streptothricins and derivatives |
| | *S. noursei* | Nourseothricins |
| | *S. griseosporus* | Liposidomycins A-C |
| | *S. antibioticus* Tü 6040 | Simocyclinones |
| | *Streptoverticillium fervens* HP-891 | Jawsamycin |
| | *Streptomyces niger* NRRL B-3857 | Quinovosamycins |

## Alkaloid glycosides

Alkaloid glycosides represent another large group of compounds with potential application in the pharmaceutical and biomedical industries as antioxidants, anticancer agents, and antimicrobial compounds (Table 4). Various types of alkaloid glycosides have been reported, notably indole alkaloid glycosides (an indole ring forms a part of the structure), enediyne alkaloid glycosides (contains an enediyne core unit), and the piperidine, pyridine, pyrrolidine, and pyrrolizidine alkaloid glycosides (Dembitsky 2005e, Elshehawi et al. 2015).

## Lipophilic polyether glycosides

Lipophilic polyether glycosides typically consist of tetrahydrofuran or tetrahydropyrane rings that are connected via aliphatic bridges, and contain one or more sugars (Dembitsky 2005a). The presence of free carboxyl groups, alkyl groups, and oxygen functional groups, typically allows for the formation of cyclic structures. The production of these compounds occurs via polyketide synthases for the production of the polyketide backbone. The well-known antibiotics, maduramycin (produced by various *Actinomadura* spp.), and the everninomicins produced by *Micromonospora carbonacea*, are but two examples of this structurally diverse group of compounds (see Dembitsky 2005a for a more in-depth review; Table 5).

**Table 4.** Selected examples of alkaloid glycoside BSs produced by actinobacterial strains (as reviewed by Dembitsky 2005e, Elshahawi et al. 2015, Kügler et al. 2015b).

| Type of Biosurfactant | Actinobacterium | Biosurfactant |
|---|---|---|
| Alkaloid glycosides | *Nocardiopsis dassonvillei* | Kahakamides A and B |
| | *Streptomyces hygroscopicus* | Neosidomycin |
| | *S. rugosporus* LL-42D005 | Pyrroindomycins |
| | *Streptomyces* spp., *Nocardia aerocoligenes*, *Saccharothrix aerocolonigenes* ATCC 39243 | Rebeccamycin |
| | *Actinomadura madurae* H710-49 | Madurapeptin |
| | *Streptomyces caeruleus* | Caerulomycin |

**Table 5.** Selected examples of lipophilic polyether glycoside BSs produced by actinobacterial strains (as reviewed by Dembitsky 2005a, Elshahawi et al. 2015, Kügler et al. 2015b).

| Type of Biosurfactant | Actinobacterium | Biosurfactant |
|---|---|---|
| Lipophilic polyether glycoside | *Actinomadura yumaensis* | Maduramycin |
| | *Actinomadura* sp. ATCC 53708 | Maderamycin |
| | *Streptomyces hygroscopicus* X-14540 | Lenoremycin |
| | *M. carbonacea* | Everninomicins B, C, D |
| | *S. hygroscopicus* T-42082 | Carriomycin |
| | *S. endus* subsp. *aureus* ATCC 39574 | Endusamycin |
| | *S. hygroscopicus* TM-581 and *Streptomyces* sp. X-14931 | Dianemycin |
| | *Streptomyces* sp. C20-12 FERM P-2736 | Etheromycin |
| | *S. nanchangensis* | Nanchangmycin |

## *Lipopeptides*

Of the different types of BSs produced by actinobacterial strains, the lipopeptides are probably more well-known for their antibiotic properties rather than their BS properties. Indeed, it is their BS nature that has made them such potent antimicrobial and antitumour agents. Lipopeptides consist of a peptide component and a fatty acid component. Variations in the lipopeptide structure are typified by variations in the composition of the peptide, the length of the fatty acid, and the way the two components are linked to each other (Mnif and Ghribi 2015b). Lipopeptide

BSs (LPBSs) are classified into two main types, namely: cyclic and short linear lipopeptides (see Figure 4 for examples) of which cyclic lipopeptides (CLPs) are the most common. CLPs, because of their complex structures, serve various functions such as promotion of bacterial swarming, and antagonistic interactions with other organisms (e.g., antibacterial, antifungal, antiviral, plant pathogenicity and cytotoxic activity) (Mnif and Ghribi 2015b).

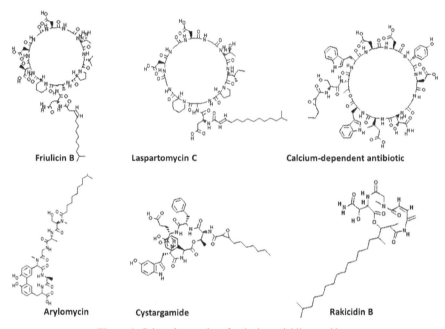

**Figure 4.** Selected examples of actinobacterial lipopeptides.

## *Cyclic lipopeptides*

A number of CLPs have been reported within the Class *Actinobacteria* (Table 6). Lipopeptide antibiotics have significant clinical potential due to their potency against multidrug-resistant strains. Most antimicrobial CLPs are non-ribosomally synthesized and produced by NRPSs with variation of the long tail fatty acid chain (Strieker and Marahiel 2009, Roongsawang et al. 2011). Their structures consist of 11 to 13 amino acid residues rigidified by a ten-membered ring structure and an exocyclic Asp1 or Asn1 residue linked with an acyl group (Strieker and Marahiel 2009). Confusion in the class amphomycin, where identical or similar antibiotics were reported, but under different names, was clarified by Baltz et al. (2005). Structural determination of laspartomycin C led to the identification of glycinocins. Glycinocin A was shown to have an identical structure to laspartomycin C, while glycinocin B and C contain the same peptide core, but differ in the length of the fatty acid tail, while glycinocin D contain the same lipid part (Kong and Carter 2003).

**Table 6.** Selected examples of cyclic lipopeptide BSs produced by actinobacterial strains.

| Actinobacterium | Biosurfactant | Reference |
| --- | --- | --- |
| *Actinoplanes friuliensis* | Friulimicin | Aretz et al. 2000 |
| *A. friuliensis* | A-1437 | Vertesy et al. 2000 |
| *Actinoplanes* sp. ATCC 30076 | Ramoplanin (A-16686) | Ciabatti et al. 1989 |
| *Kitasatospora cystarginea* | Cystargamide | Gill et al. 2014 |
| *S. griseoflavus* | Tsushimycin | Shoji et al. 1968 |
| *S. griseus* var. *spiralis* and *Streptomyces violaceus* var. *aspartocinius* | Aspartocin | Shay et al. 1958 |
| *S. roseosporus* | Daptomycin | Debono et al. 1987 |
| *S. parvulus* var. *parvuli* | Parvuline | Toth-Sarudy et al. 1974 |
| *Streptomyces amritsarensis* | Lipopeptide | Sharma et al. 2014 |
| *Streptomyces canus* | Amphomycins | Heinemann et al. 1953 |
| *Streptomyces coelicolor* A3(2) | Calcium-dependent antibiotics | Lakey et al. 1983 |
| *Streptomyces fradiae* and *S. refuineus* | A54145 | Farnet et al. 2002 |
| *Streptomyces fungicides* no. B5477 | Enduracidin | Yin et al. 2012 |
| *Streptomyces viridochromogenes* | Laspartomycins | Wang et al. 2011 |

## *Linear lipopeptides*

New lipopeptides, arylomycins, a type I peptidase inhibitor, was isolated from *Streptomyces* sp. HCCB10043. The biosynthetic gene clusters (BGC) of arylomycin in this strain is identical to that in the daptomycin producer *S. roseosporus*. The gene cluster *ary*C encoding for a P450 enzyme is responsible for biaryl bond formation; inactivation of this gene led to the generation of two linear lipopentapeptides and abolishment of aryloycin production (Jin et al. 2012). *Streptosporangium amethystogenes* also produces a linear lipopeptide for which various structures have been described. All structures share a 4'-thio C7 fatty acid chain with one amide linked C13–C15 fatty acid chain and two ester linked C16–C19 fatty acid chains (Takizawa et al. 1995). Other LPBSs (partially or not structurally elucidated) are also produced by *Actinopolyspora* sp. A18 (Doshi et al. 2010), *C. lepus* (Cooper et al. 1979), *Corynebacterium xerosis* (Margaritis et al. 1979), *Dietzia* sp. S-JS-1 (Liu et al. 2009), *Nocardiopsis alba* MSA10 (Gandhimathi et al. 2009), *Brevibacterium aureum* MSA13 (Kiran et al. 2010c), *Nocardiopsis* sp. B4 (Khopade et al. 2012a), *Leucobacter komagatae* 183 (Saimmai et al. 2012), *Kocuria marina* BS-15 (Sarafin et al. 2014) and *Brevibacterium luteolum* (Vilela et al. 2014). Streptofactin produced by *Streptomyces tendae* Tü901/8c also shares similarities with microbial lipopeptides (Richter et al. 1998).

## *Polymeric BSs*

One specific type of polymeric BS that has received great attention is the lipoglycan, lipoarabinomannan. Lipoglycans are typically found in the cell envelope of *Mycobacterium* spp. and related organisms (Gibson et al. 2003), and include phosphatidyl-*myo*-inositol mannosides (PIMs), lipomannan (LM), and lipoarabinomannan (LAM). LAM consists of three components: a carbohydrate backbone, a mannosyl-phosphatidyl-*myo*-inositol anchor, and various capping motifs. The carbohydrate backbone typically consists of two homopolysaccharides of D-mannan and D-arabinan, while comparative studies have shown that there is a great variation in the acylation state of the anchor as well as in the composition of the capping motifs. It is believed that it is the capping motif that plays a key role in the virulence and persistence of pathogenic mycobacteria (Sutcliffe 2005). Initially it was thought that LAMs are limited to pathogenic, mycolic acid containing actinobacteria such as *Corynebacterium matruchotii*, *Gordonia bronchialis*, *G. rubripertincta*, *Mycobacterium chelonae*, *Rhodococcus rhodnii*, and *R. equi* (Gibson et al. 2003). However, it soon became clear that LAMs can be found in the cell envelopes of non-pathogenic actinobacteria (e.g., the deep sea isolate, *Dietzia maris*), as well as those that do not contain mycolic acids in their cell walls (e.g., *Amycolatopsis sulphurea*, and *Turicella otitidis*; Gilleron et al. 2005, Gibson et al. 2003), highlighting the fact that the natural role of LAM must therefore be independent of the pathogenic potential of the producing strain. It is clear from literature that there is still a need for further research in order to understand the role of LAM in actinobacterial strains (Sutcliffe 2005).

More recently, a polymeric BS produced by *Microbacterium* sp. BS-2 (MTCC 5822) was patented for its potential application in various industries. Microsan, or β-D-glucoronyl-(1,2)-D-mannosyl-(1,4)-D-glucose exopolysaccharide, exhibited surface activity, antibacterial activity, antioxidant properties, anti-inflammatory, as well as immunomodulatory activities (Kumar and Pombala 2015).

### *Other types of BSs from actinobacteria*

Another type of BS reported to be produced by an actinobacterial strain is the rare phenazine L-quinovose ester produced by a marine *Streptomyces* sp. CNB-253 (Kügler et al. 2015b). This antibacterial compound was only detected when the strain was cultivated in the presence of seawater, highlighting the importance of mimicking the natural environment for induction of BS production. Structural variants of the compound were produced, with variations in the hydroxylation and acetylation of the desoxyglucose unit (Kügler et al. 2015b).

## Challenges in the Production of Actinobacterial BSs

Throughout literature, it is clear that even though there has been an increased research interest in BSs, one of the main challenges is the large scale production of BSs, especially those produced by actinobacterial strains, preventing full scale industrial application. This is linked to the high costs of the production media, the high costs involved in the downstream processing, and the low yields (Otzen 2016). To make the production of BSs more effective, inexpensive growth substrates, optimised bioprocesses for higher yields, and effective product recovery and purification would be required (Jackson et al. 2015). In addition, the development of overproducing strains would allow for increased production of the BS of interest (Franzetti et al. 2010).

## The importance of media optimisation

Whole genome sequencing has shown that natural product producers have the ability to produce more compounds than was first believed. Standard laboratory cultivation conditions are not a direct reflection of what the producing organism is exposed to and/ or capable of in the natural environments, and we are therefore not activating a range of biosynthetic gene clusters (Bode et al. 2002). Various studies on actinobacteria have shown that very small changes in cultivation conditions can have a dramatic effect, either inducing or suppressing the production of secondary metabolites (Bode et al. 2002). In the production of BSs, the media components and precursors (e.g., phosphate, fatty acids, etc.) provided play a key role in the types of BSs produced, the amount produced, as well as the quality (Kiran et al. 2010a). A recent review by Banat et al. (2014a) highlighted the potential use of inexpensive, renewable carbon sources for BS production. They further elaborated that in some cases, the use of only pure carbon sources may not result in a high yield of BS. In addition, nitrogen limitation is essential in the production and composition of BSs. The substrate composition, concentration in the media, along with the carbon/nitrogen ratio and nature of the nitrogen source may also affect BS production (e.g., Ristau and Wagner 1983). A number of studies have reported the use of various nitrogen sources for BS production. These include, but are not limited to, yeast extract (Khopade et al. 2012a, Manivasagan et al. 2014), amino acids, urea and ammonium salts (Khopade et al. 2012b).

Environmental factors such as temperature, pH, agitation speed, aeration, $CO_2$ levels and salinity play a significant role in the characteristics and yield of the BSs produced. Notably, there are no clear guidelines for actinobacteria: (1) In the majority of literature reports, BS production takes place at temperatures ranging from 25 to 30°C. However, improved production of trehalose-2,2',3,4-tetraesters by *R. erythropolis* DSM 43215 was observed when the temperature was shifted from 30 to 22°C (Ristau and Wagner 1983). (2) Even though the production of BSs by actinobacteria are typically in the neutral to weak acidic pH range, the production of the di- and penta-saccharide lipids by *Nocardia corynebacteroides* remained unaffected at pH ranging from 6.5 to 8 (Powalla et al. 1989). (3) High agitation speeds resulted in the shearing of *N. erythropolis* which in turn reduced BS yields (Kretschmer et al. 1982). (4) It was found that the BS produced by the marine *Nocardiopsis* sp. B4 was stable up to 8% (w/v) NaCl, whereas chemical surfactants are deactivated by 2–3% (w/v) salt concentration (Khopade et al. 2012a).

The use of statistical experimental design processes is therefore essential for BS production from actinobacterial strains. Various examples can be found on the successful application of response surface methodology (RSM). The Plackett-Burman design was used by Kalyani et al. (2014a) for the production of a RL by *S. coelicoflavus* NBRC 15399[T]; Korayem et al. (2015) used the Plackett-Burman design for the optimal production of a BS from *Streptomyces* sp. 5S, and Manivasagan et al. (2014) followed a similar approach in the optimisation of glycolipid production by *Streptomyces* sp. MAB36. Colin et al. (2013b) made use of a univariate analysis using a dose-response approach for determining the optimal conditions for the production of a glycolipid by *Streptomyces* sp. MC1, while Kiran et al. (2010b) used RSM (Box-Behnken design) to optimise the production of a glycolipid. In all these studies, parameters such as the carbon source, nitrogen source, pH, temperature, salt concentration, metals, amino acids, inoculum size, and incubation time were evaluated. These studies clearly highlight the unique conditions required for the production of different types of BSs from different actinobacterial strains (Kiran et al. 2010c).

## Downstream processing of actinobacterial BSs

Kügler et al. (2015b) highlighted the main problems associated with the production, extraction, and purification of BSs produced by the actinobacteria. In particular, the actinobacteria produce a wide range of secondary metabolites, thereby complicating the purification process, especially when a specific compound is targeted. In addition, many actinobacterial strains either produce extracellular BSs, and/or membrane-bound BSs, greatly influencing the extraction processes applied (Kügler et al. 2015b). Various approaches have been evaluated for the extraction of BSs from actinobacterial culture media. From the various reports in literature, it is clear that the extraction method applied is not only dictated by the type of BS, but also the producing strain, e.g., Nakano et al. (2011) explored the ability of *D. maris* WR-3 to form foam under cultivation conditions. The foam was harvested and extracted twice with chloroform:methanol (2:1, v/v) to access the wax ester-like BS. Similarly, the foam layer of the *R. erythropolis* 3C-9 culture was extracted with methyl tertiary-butyl ether to access a mixture of free fatty acids, glucolipids and trehalolipids (Peng et al. 2007). For the extraction of various types of BSs, the culture filtrate is typically either extracted directly with a solvent system (e.g., Gesheva et al. 2010, Kiran et al. 2010a, b, Yan et al. 2014), or the culture filtrate is first subjected to a precipitation method (acid or acetone) followed by solvent extraction (e.g., Kiran et al. 2010c, Bajaj et al. 2014, Kalyani et al. 2014a). Other approaches also include salt precipitation followed by solvent extraction (Vasileva-Tonkova and Gasheva 2005), acid precipitation only (Khopade et al. 2012a, b), acetone precipitation only (Colin et al. 2013a, b, Chakraborty et al. 2015), and the use of ultrasonication to release cell-bound BSs (Ivshina et al. 2016).

According to Kügler et al. (2015b), various purification approaches have been applied to access actinobacterial BSs. Crude extracts containing glycolipids are typically further purified via the use of silica gel chromatography. Compounds are eluted through the use of solvent gradients of non-polar to polar solvent systems. Preparative high performance liquid chromatography is also applied, with C18 reverse phase chromatography or cellulose-based ionic interaction chromatography being the most common approaches taken. Another approach is the use of gel filtration as a purification step for nucleoside antibiotics, but this varies from one BS type to the next and has only been explored by a few researchers (e.g., Eckardt 1983). Various staining techniques are used in thin layer chromatography analyses to determine the presence of sugar or carbohydrate moieties, amino acids, peptides, etc., but in the majority of the reports on glycolipid production, there is a general lack of structural information (Kügler et al. 2015b). Quite a few lipopeptides have been structurally elucidated, while the majority of structural elucidations have been observed for the vast group of glycosidic BSs.

## Biotechnological Potential of Actinobacterial BSs

With the great structural diversity of the BSs produced by the actinobacteria, it is not surprising that they have either found application in various industries, or have shown great promise for application. In some instances, one type of BS could be applied in various industries, e.g., *Rhodococcus* trehalolipids have pore forming capabilities and membrane permeabilization properties. This allows for widespread application: biocontrol agents in agriculture, preservatives in the food industry, antibacterial and antiviral agents in medicine and pharmaceutics, as well as inhibitors of fibrin clot formation (Mnif and Ghribi 2015a). In the following sections, potential applications of actinobacterial BSs are briefly discussed. It is difficult to cover all the different BSs that have found application and only selected examples are provided. Figure 5 is a graphical overview of selected applications/potential applications of actinobacterial BSs reported in literature.

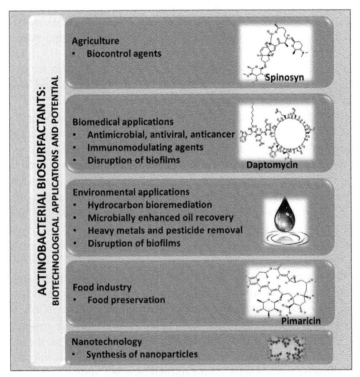

**Figure 5.** Examples of biotechnological applications and potential applications of BSs produced by actinobacterial strains.

## *Agriculture*

BSs may find various applications in the agricultural sector: the bioremediation of contaminated soil (see *Environmental applications*), as antimicrobial agents against phytopathogens, to assist microbe-plant interactions, and as natural pesticides (Sachdev and Cameotra 2013). Chemical surfactants are typically used in pesticide formulations and their large-scale application has resulted in their accumulation in soil, affecting soil quality and plant productivity. Various BSs have been evaluated for their application as natural pesticides/biocontrol agents. Among these, the macrocyclic glycosides spinosyns A and D that are produced by *S. spinosa* have been combined and marketed as Spinosad, which is used as an insect control agent in the agricultural sector (Kirst 2010). Since its initial development as an insecticide, its application has been extended to the treatment of external parasites of livestock, and spinosad-containing products have been approved in various countries for organic agriculture (Kirst 2010).

## *Biomedical applications*

As indicated previously, there is potential for application of actinobacterial BSs in the biomedical field, mostly because of their biological activities: antimicrobial, antiviral, antiadhesive, anticancer, and immunomodulatory activities. In general, they are considered to be safer than synthetic pharmaceuticals, e.g., the effect of the trehalolipid (TL) produced by *R. erythropolis* on a skin irritation mouse model showed that the TL is less irritating than the surfactant, sodium dodecyl sulfate, which is typically used in cosmetics (Franzetti et al. 2010).

Trehalose-6,6'-dimycolate (TDM; cord factor) has been extensively studied for its role in pathogenesis. Despite this, TDM has been shown to exhibit various bio-activities: (1) antitumour activity, (2) augmentation of non-specific immunity to microbial infections, (3) immunomodulating activity, e.g., granuloma-forming activity, (4) priming murine macrophages to produce nitric oxide, (5) induction of production of cytokines, and (6) enhancement of angiogenic activity in mice. The use of TDM from mycobacteria is not recommended due to its toxicity. However, the TDM from *Rhodococcus* spp. have been shown to exhibit similar properties, but are not as toxic as the mycobacterial TDM and it is produced by a non-pathogenic bacterium (Franzetti et al. 2010). TLs have also been shown to exhibit antiviral activities, e.g., the TDM produced by *Rhodococcus* spp. inferred resistance to intranasal infection by the influenza virus in mice (Franzetti et al. 2010).

There has been various reports on the antimicrobial activity of actinobacterial BSs, e.g., by TLs produced by *Tsukamurella* sp. DSM 44370 (Franzetti et al. 2010), glycolipids produced by *Streptomyces* sp. B3 and *Streptomyces* sp. MAB36 (Khopade et al. 2012b, Manivasagan et al. 2014), and the glycolipids produced by *R. erythropolis* (Adel-Megeed et al. 2011). Some of the BSs produced by the actinobacteria have found application in industry: daptomycin has been commercialised as Cubicin® (Raja et al. 2003), and the glycoside BS, amphotericin B (macrolide polyene produced by *Streptomyces nodosus*), was approved in 1959 by the FDA for clinical use as an antifungal agent (Hamill 2013).

The mono- and di-RLs produced by *S. coelicoflavus* NBRC 15399[T] have been shown to exhibit antitumour activity against MCF-7 (breast cancer) cell line and therefore have potential as an anticancer drug (Allada et al. 2015). The lipopeptides, rakicidin A-E produced by a marine *Micromonospora* sp. exhibit various activities. Rakicidin A has been shown to be hypoxia-selective cytotoxic against cancer cell lines (HCT-8 and PANC-1); rakicidin B is active against oesophageal squamus carcinoma cells (EC109), lung cancer cells (A549 and 95D), gastric cancer cells (SGC7901), uterine cervix cancer cells (HeLa), and hepato-cellular carcinoma cells (HepH2); rakicidins C and D are non-cytotoxic; and rakicidin E was found to interfere with invasiveness of aggressive breast cancer cells. The main mode of action of rakicidin B was found to involve the induction of apoptosis via the activation of caspase-3, -7 and -9. It was also found to block the MARK and JNK/P38 signalling pathways (Gudiña et al. 2016).

Kiran et al. (2010d) evaluated the potential use of the glycolipid produced by *Brevibacterium casei* MSA19 as an anti-biofilm agent. Biofilms may consist of microbial consortia or out of a single species and are typically difficult to remove. Kiran et al. (2010d) noted that marine sponges have been shown to prevent biofilm formation on their outer surfaces—this phenomenon may be due to the sponge's ability to produce specific compounds and/or the BS activity of microorganisms found in association with marine sponges. The glycolipid produced by *B. casei* MSA19 was shown to act as an antiadhesive, preventing biofilm formation by single and mixed cultures (Kiran et al. 2010d). Kuyukina et al. (2016) also showed that the TL produced by *Rhodococcus ruber* IEGM 231 exhibit antiadhesive effects, but did not inhibit bacterial growth. Similarly, the analysis of the glycolipid produced by *Nocardiopsis* sp. MSA13A, showed the ability to disrupt a pre-formed biofilm consisting of *Vibrio* spp. (Kiran et al. 2014a). For further details on the potential role of BSs as antibiofilm agents, see Banat et al. (2014b).

## *Environmental applications*

Actinobacteria have great potential for application in the bioremediation of various contaminated environments. This can be attributed to their ability to utilize a wide range of compounds as sole carbon sources and their ability to degrade complex polymers (Pizzul et al. 2006). Polyaromatic hydrocarbons (PAHs) persist in the natural environment and are not readily degradable: most PAHs have low water solubility, adsorb to soil particles and can be trapped in soil micropores and organic matter, resulting in a very low bioavailability to PAH-degrading microorganisms. BSs have been shown to assist with the removal of PAHs from the environment, making these compounds more readily available for biodegradation. Members of the genera *Mycobacterium, Gordonia, Rhodococcus, Corynebacterium, Streptomyces, Micrococcus*, and *Nocardioides* have previously been implemented in removing PAHs from the environment and a more recent study have shown the ability of *Streptomyces rochei* PAH-13 to degrade PAH (Pizzul et al. 2006, Zhang et al. 2006, Ferradji et al. 2014, Sharma et al. 2016).

4-nitrotoluene (4-NT) is used in the production of various products. The extensive use of 4-NT has resulted in the contamination of various ecosystems. Kundu et al. (2013) therefore analysed the use of a *Rhodococcus pyridinivorans* strain and its BS in the emulsification of 4-NT. The strain was able to degrade 400 mg/L of 4-NT, making it a great contender as a bioaugmentor for 4-NT contaminated environments.

Similar to PAH, hydrocarbons are difficult to remove from the environment, mainly because these compounds tend to bind to soil particles, effectively reducing their bio-availability. The addition of glycolipids to contaminated soil has shown to assist bioremediation, increasing oil degradation and the number of oil-degrading bacteria. *Mycobacterium flavescens* EX-91 has been used for the development of a commercial product, Ekoil. The product was developed for the decontamination of oil-polluted water. It was successfully applied in the treatment of engine oil-contaminated wastewater of a nuclear power plant (Franzetti et al. 2010). Similarly, *Nocardiopsis* sp. VITSISB isolated from seawater samples have been shown to have the ability to produce a RL. The strain was able to degrade engine oil, highlighting the potential use of the strain in bioremediation of marine environments impacted by oil spills (Roy et al. 2015). Brevifactin, the BS produced by and isolated from *B. aurum* MSA13, showed great potential for application in microbially enhance oil recovery (MEOR). MEOR can be achieved using three approaches: (1) the production of a BS in a bioreactor and adding the BS to the oil reservoir, (2) inject the BS-producing organisms into the oil reservoir, or (3) add nutrients to the oil reservoir to stimulate BS-producing indigenous organisms (Franzetti et al. 2010). In the case of brevifactin, oil was applied to a column of sand and allowed to settle. Upon addition of the brevifactin to the soil column, oil release took place (Kiran et al. 2010c). The potential application of TLs produced by *Rhodococcus* spp. in MEOR has also been reported (Franzetti et al. 2010). Unique approaches in the treatment of oil-contaminated environments have also been described: Roy et al. (2015) showed effective bioremediation when the BS and the strain was immobilised in alginate beads; Laorrattanasak et al. (2016) reported on the immobilisation of *Gordonia westfalica* GY40 onto chitosan flakes for oil spill remediation, and Ivshina et al. (2013) reported the use of BSs produced by *R. ruber* as a hydrophobic agent used as a coating on sawdust for the effective immobilization of *R. ruber* cells used in the transformation and degradation of the hydrocarbon, *n*-hexadecane.

In a recent review by Alvarez et al. (2017), the importance of actinobacterial strains that produce BSs was highlighted, especially within the context of the bioremediation of soil contaminated with pesticides and heavy metals. BSs increase the solubility and therefore bioavailability of pesticides and heavy metals, facilitating their easy removal from the environment. For example, Bajaj et al. (2014) reported the ability of a TL-producing *Rhodococcus* sp. IITR03 to degrade 1,1,1-trichloro-2,2-bis(4-chlorophenyl)ethane (DDT) and Colin et al. (2013a) showed that bioemulsifiers produced by *Amycolatopsis tucumanensis* DSM 45259 are effective in the bioremediation of soil contaminated with chromium. The authors showed the potential for either a bioaugmentation approach (addition of the BS-producing strain to the soil) or a bioremediation approach (addition of the BS to the contaminated soil).

## Food industry

The macrocyclic glycoside, pimaricin (also known as natamycin), is produced by *S. natalensis* and various other *Streptomyces* spp. This antifungal agent has been used in the food industry for more than 40 years, particularly the dairy industry, to prevent fungal contamination of cheeses, yoghurts, and other non-sterile food products (e.g., sausages) (Aparicio et al. 2016). Its application in the food industry has also been extended to include other food products such as salads, sauces, fish, beverages, and many more. In addition, it has also been incorporated into packaging materials for the preservation of food. Besides its application in the food industry, it was also approved by the FDA in 1978 as the first-line drug for the treatment of fungal keratitis, and was approved in 2012 by the U.S. Environmental Protection Agency as a biopesticide for the use in the cultivation of mushrooms (Aparicio et al. 2016). An overview of the application of BSs in the food industry can be found in the reviews by Campos et al. (2016) and Campos et al. (2013).

## Nanotechnology

Nanoparticles form the basis for many technologies. Commercially available nanoparticles are typically synthesised via physical and chemical processes. During these processes, it is difficult to control the size of the nanoparticles synthesised, and since the size and shape of nanoparticles determine their bioactivity and application, it is essential to find an alternative means for nanoparticle synthesis (Kiran et al. 2010a). The potential of BSs to act as stabilisers was exploited by Kiran et al. (2010a). The glycolipid produced by the actinobacterium, *B. casei* MSA19, was successfully applied in the synthesis and stabilisation of silver nanoparticles (Kiran et al. 2010a). The glycolipid was produced under solid state fermentation conditions using inexpensive industrial and agri-industrial waste. The nanoparticles were uniform in size and stable for two months. The glycolipid produced by *B. casei* therefore has great potential to replace synthetic nano-emulsions, highlighting the potential of BSs as a green alternative for the controlled synthesis of nanoparticles (Plaza et al. 2014, Kiran et al. 2016b).

# Future Perspectives

In a recent review by Katz and Baltz (2016), the authors highlighted the importance of actinobacteria for the discovery of novel natural products. Many actinobacterial strains have large genomes and it is believed that to date < 10% of the secondary metabolite gene clusters have been expressed at sufficient levels to allow for detection. The actinobacteria therefore represent a vast potential resource of novel natural products. Even though there have been great advances in the area of recombinant technology and genome sequencing, our ability to manipulate the biosynthetic gene clusters involved in BS production (PKS and NRPS) is limited, mainly due to our limited understanding of how these systems work. As highlighted by Katz and Baltz (2016), more research is required in the following areas: genome sequencing to access novel BGCs, understanding the systems involved in the synthesis of natural products (e.g., via combinatorial biosynthesis), and new isolation programs, so that new isolates can become the focus of genome sequencing programs.

## Genome mining, metagenomics, and genetic manipulation

BSs described in this chapter have been accessed via studies on isolated organisms. The production of actinobacterial BSs via heterologous expression remains limited to certain types of BSs (e.g., lipopeptides). According to Jackson et al. (2015), this may be linked to the high degree of regulation required for the production of these compounds and the general lack of information on the BGCs involved in their production. Various approaches are currently being explored in accessing these compounds and their BGCs. These approaches are briefly discussed below.

### Genome mining

The past few years have seen a dramatic increase in the number of sequenced actinobacterial genomes. This has mainly been part of a drive to use genome sequence information as taxonomic indicators. The sequencing of the genomes of two industrial strains, *S. coelicolor* and *S. avermitilis*, highlighted the value of genome sequencing in identifying BGCs involved in the production of natural products (Yamanaka et al. 2014). It soon became clear that not all BGCs are expressed under laboratory conditions. These so-called 'silent'/'cryptic'/'orphan' BGCs may hold the key to the discovery of unique natural products. In an attempt to access one such BGC, Yamanaka et al. (2014) developed a transformation-associated recombination (TAR) cloning platform that allows the cloning of large genomic loci. *Saccharomonospora* sp. CNQ-490 was isolated from a marine sediment sample and its genome sequence analysed for unique BGCs. Using the TAR system, Yamanaka et al. (2014) was able to express a 67 kb NRPS BGC in *S. coelicolor*. This resulted in the production of the dechlorinated lipopeptide antibiotic, taromycin A. The authors described the TAR method as a 'plug-and-play' approach that can be used in the discovery of many more 'silent'/'cryptic'/'orphan' biosynthetic pathways.

To date, there have only been a few reports focused on identifying the biosynthetic pathways involved in the production of actinobacterial glycolipids. The overexpression of a putative acyl coenzyme A and the use of an alkane monooxygenase promoter resulted in a two-fold increase in the production of the succinoyl trehalose lipid produced by *R. erythropolis* SD-74. In the second study, the insertion of the *Vitreoscilla* haemoglobin gene (vgb) into the genome of *Gordonia amarae* NRRL B-8176 resulted in a 2–4 fold increase in the production of a trehalose lipid (Kuyukina et al. 2015). Further studies by Kiran et al. (2016a) focused on the marine BS producers previously isolated and described by them. The organisms studied included *N. alba* MSA10, *B. aureum* MSA13, *B. paraconglomeratum* MSA21, *Streptomyces* sp. MAD01, and *S. dendra* MS1051, all isolated from marine sponges. Kiran et al. (2016a) determined that BS production by the marine actinobacteria may be linked to type II iterative PKS activity, but concluded that further studies are required.

Peptide natural products have diverse biological functions and can be produced either via ribosomal or non-ribosomal biosynthetic pathways (Kersten et al. 2011).

In a 'chemotype-to-genotype' genome mining approach, which combines mass spectrometry and genomics for the discovery of new peptide natural products, Kersten et al. (2011) detected a lipopeptide produced by *S. hygroscopicus* ATCC 53653. The process starts with the analysis of culture media using MALDI-TOF mass spectrometry (MS), followed by MSn analysis and sequence tagging via genome mining. The chemotype prediction is then linked to structure elucidation as determined via NMR (Kersten et al. 2011). This process, called 'natural product peptidogenomics', allowed for the identification of the stendomycin type lipo-tetradecapeptide production by *S. hygroscopicus*. Stendomycin I was first described from *S. endus*, but this approach showed that *S. hygroscopicus* has the ability to produce five or more stendomycin analogues, all encoded via an NPRS biosynthetic pathway (Kersten et al. 2011).

## *Metagenomics*

Over the past few decades, metagenomics has provided researchers access to the genetic material of various environments. Through sequence-based and function-based screening of clone libraries, an array of novel natural products have been discovered (Jackson et al. 2015). Sequence-based screening tend to be limited to our current knowledge, which is reflected by the information captured in sequence databases and may result in overlooking truly unique compounds. However, numerous online tools and databases such as antiSMASH, NaPDoS, PRISM, Bagel3 and Bactibase, to name a few, have greatly simplified the identification of potentially novel sequence information. The Secondary Metabolite Bioinformatics Portal (SMBP) provides invaluable assistance to those interested in exploring sequence information, providing links to the abovementioned online tools and many more (http://www.secondarymetabolites.org/; Weber and Kim 2016). The websites of the various databases and tools are provided on the webpage as well as in their recent publication, making it a 'one-stop-shop' for beginners.

The alternative approach of function-based screening allows researchers to access unique natural products often missed by sequence-based screening. Even though there has been great advances in this approach, such as codon optimisation, development of suitable expression hosts, etc., there are still quite a few problems that need to be overcome: (1) an entire operon or BGC needs to be present; (2) if the BGC or operon is present in low abundance, it may be out-diluted during the screening process; (3) more suitable expression hosts are required; (4) suitable screening methods are required; (5) gene expression is influenced by various factors such as gene promoter recognition, transcription initiation factors, protein folding and the export of gene products; and (6) expressed products may be toxic to the expression host (Jackson et al. 2015).

Walter et al. (2010) summarised the different screening techniques currently in use for the detection of BS activity. Interestingly, most of the screening methods used for the detection of BS production are suited for application in high throughput screening to some degree, a key aspect in function-based screening. The drop collapse assay, oil spreading assay, blue agar assay, microtiter plate assay, emulsification assay, and the haemolytic assay are all suitable for large scale screening programs

(Jackson et al. 2015), but may be limited in the type of BS detected. In addition, some of these methods would also require the cultivation of the clones in liquid media prior to performing the assay, which is time-consuming. The atomized oil assay developed by Burch et al. (2010) may be more suited for the screening of metagenomic libraries. It has been reported to be more sensitive than the drop collapse assay and allows for the detection of various types of surfactants (chemical surfactants and BSs). In this assay, a fine mist of oil is sprayed with an airbrush over the surface of an agar plate containing the colonies or transformants of interest. The presence of BSs can be visualised from the halos that form around positive colonies (Burch et al. 2010).

Recently, the first successful metagenomic discovery of BSs was reported. Thies et al. (2016) constructed a metagenomic library of a bacterial community involved in biofilm formation in a slaughterhouse drain. Function-based screening employing the haemolytic assay, atomized oil assay, and the microtiter plate assay, confirmed the presence of two transformants with the ability to produce BSs. The successful expression of the genes encoding for the two BSs by the *Escherichia coli* DH10b host allowed for structure determination via NMR. The two BSs were found to be *N*-acyltyrosines, with *N*-myristoyltyrosine being the predominant species (Thies et al. 2016). Even though 16S rRNA gene sequence analysis showed the predominant bacterial genera to belong to the *Flavobacteriaceae*, this first report holds a promise that the metagenomic discovery of BSs from other bacterial groups (such as the actinobacteria), may be within reach.

## Genetic engineering

The majority of reports on genetic engineering of actinobacterial BS BGCs are focused on lipopeptides. The use of genetic engineering has allowed for the production of peptide cores of daptomycin-like lipopeptides. The coupling of the peptide cores with various acyl chains resulted in a wide range of bioactive lipopeptides (Patel et al. 2015). In addition, the use of engineered enzymes and the expression of biosynthetic pathways allowed for the development of drug analogs. Similarly, the use of combinatorial biosynthesis via domain swapping, clustering and recombination, has allowed for the production of novel lipopeptides with increased efficiency compared to the parent compound, daptomycin (Patel et al. 2015). Nguyen et al. (2010) reported constructing hybrid molecules of daptomycin via genetic engineering. The new constructs showed a change in amino acid composition, which resulted in a decreased toxicity, but increased activity against *Streptococcus pneumoniae* in mice.

Du et al. (2014) performed various mutation studies to determine the biosynthetic pathway involved in the production of the lipopeptide, polyoxypeptin A. It was determined that the BGC consists of 37 open reading frames. With the elucidation of the biosynthetic pathway, it will allow for a greater insight into the biosynthesis of this type of BS and would allow for the application of combinatorial biosynthesis. This compound is part of the so-called azinothricin family of compounds. This group has shown to exhibit diverse bio-activities: antibacterial, antitumour, anti-inflammatory, and accelerate wound healing (Du et al. 2014). Structurally similar compounds that have been produced by actinobacteria, include A8356C, IC101,

L-156,602, pipalamycin, aurantimycins, azinothricin, citropeptin, diperamycin, kettapeptin, variapeptin, and verucopeptin (Du et al. 2014). Furthermore, the identification of the BGC involved in the production of the lipopeptide antibiotic, telomycin, allowed Fu et al. (2015) to perform genetic manipulation, resulting in the generation of various derivatives. The derivatives showed effective and rapid killing of bacteria, specifically multi-drug resistant Gram positive pathogens.

### *Lowering the costs of production: Waste products in the production of BSs*

In an attempt to lower the production costs and develop more economically viable processes for the production of BSs, various researchers have been focusing on the use of alternative, inexpensive substrates, including industrial wastes, oily by-products, and municipal wastes rich in organic pollutants (Makkar et al. 2011). This approach would allow for the reduction in cost of BS production as well as the need to treat waste (Khopade et al. 2012a). Even though various researchers have shown the potential viability of using inexpensive substrates for BS production, in some instances, product yield remained low due to the requirement of gene regulation, requiring the supplementation of the waste material with a suitable carbon and/or nitrogen source (Jackson et al. 2015). Product recovery and purification also adds to the complexity: various industries would require BSs of variable purity, e.g., in the cosmetics industry, BSs of high purity would be required, while in bioremediation processes, a crude preparation may be sufficient (Jackson et al. 2015). This was particularly highlighted for members of the genus *Rhodococcus* where it has been shown that production of TLs are dependent on the substrate provided. The use of an undefined feedstock may result in the production of an undefined set of TLs with unknown biological activities (Kuyukina et al. 2015). It is therefore essential to understand the producing organism and to have a clear idea of the application of the final product.

BS production from various actinobacterial strains has been successfully demonstrated in the presence of various hydrocarbons, but the hazardous nature of these hydrocarbons precludes large scale production (Laorrattanasak et al. 2016). However, industrial wastes may serve as suitable alternatives. For the production of BSs from *Nocardiopsis lucentensis* MSA04, Kiran et al. (2010b) tried different types of agri-industrial wastes such as wheat bran, ground nut oil cake, rice bran, oil seed cake, and industrial wastes, including furnished leather powder, diesel contaminated soil, petrol bunk soil, milk processing waste (pre-treated and treated), tannery treated and pre-treated sludge, pre-treated molasses, and treated molasses (distillery waste), in an attempt to explore inexpensive substrates. Wheat bran was identified as the optimal substrate in solid-state fermentation for the production of the glycolipid, but had to be supplemented with kerosene, beef extract and copper sulphate for optimal BS production (Kiran et al. 2010b). In a similar approach, Kiran et al. (2010c) also evaluated the effect of these substrates in the production of brevifactin, a lipopeptide produced by *B. aurum* MSA13. For the lipopeptide production, pre-treated molasses was found to be the optimal substrate, while glucose, acrylamide and ferrous chloride were identified as key factors influencing biosurfactant production (Kiran et al. 2010c). In related studies, as summarised by Kiran et al. (2016b), treated molasses, oil seed cake, and tannery pre-treated sludge, were found to be the optimal substrate for BS production by *Nocardiopsis* sp. MSA13A, *B. casei* MSA19, and *B. paraconglomeratum* MSA21, respectively. Other examples of BS production from waste include the glycolipopeptide produced by *Corynebacterium kutscheri* from waste motor lubricant oil and peanut oil cake (Thavasi et al. 2007); *R. erythropolis* sp. 51T7 produced a 2,3,4,2'-trehalose tetraester when cultivated on waste lubricant oil as a sole carbon source (Espuny et al. 1996); *Kocuria turfanesis* strain-J produced BSs in curd whey waste generated during the preparing of Chakka (an intermediate product obtained by draining of curd/fermented milk) (Dubey et al. 2012); and *M. luteus* MFW1 isolated from milk factory wastewater were able to utilize whey wastewater to produce BSs (Yilmaz et al. 2009).

## Exploring new environments

Even though it was first believed that actinobacteria occur only in soil and freshwater, subsequent studies have shown that actinobacteria can be found in virtually any type of environment (Goodfellow and Fiedler 2010). The environment from which a BS-producing strain has been isolated as well as the producing organism itself strongly influences the biological properties of the BS. Therefore, another approach to bypass the main economic obstacle for the commercialisation of BSs is the search for new BS-producing strains with the ability to produce new types of BSs from inexpensive substrates (Colin et al. 2013a). This approach would not only require the use of intelligent isolation approaches, but would also require new screening processes that would allow for the targeting of strains with the ability to utilise inexpensive substrates for the production of high yields of BSs (Goodfellow and Fiedler 2010, Colin et al. 2013a).

Kügler et al. (2015b) noted that BS-producing actinobacteria have predominantly been isolated from three main environments, specifically environments in which the production of BSs would be to the benefit of the producing organism: hydrocarbon-contaminated soil (e.g., Vollbrecht et al. 1999, Pizzul et al. 2006, Kalyani et al. 2014a), human, animal, and plant infected tissues (e.g., Sutcliffe 2005), and the marine environment (e.g., Roy et al. 2015, Ettoumi et al. 2016), notably marine ascidians (e.g., Kiran et al. 2010a, b, c, 2014a, b). In addition, a vast amount of BS-producing organisms have been isolated from soil samples (e.g., Korayem et al. 2015, Laorrattanasak et al. 2016). Some unique environments have also been explored as a source of BS-producing organisms: copper polluted sediments (Colin et al. 2013a), water and sediment samples from oil-polluted seasonal ponds (Kavyanifard et al. 2016), soil enriched with chlorophenols (Pizzul et al. 2006), ornithogenic soil (Vasileva-Tonkova and Gesheva 2005), sediment from an industrial and coal mine region (Chakraborty et al. 2015), beach sand near an old harbour (Peng et al. 2007), effluent sediment from a pesticide manufacturing facility (Kundu et al. 2013), DDT-contaminated soil (Bajaj et al. 2014), water spring near an oil-extracting enterprise (Ivshina et al. 2016), soil contaminated with poultry waste (Kalyani et al. 2014b), sugar cane (Colin et al. 2013b), saltpan soil from a coastal region (Shubrasekhar et al. 2013), and activated sludge foam (Kügler et al. 2014, 2015a, Gomes et al. 2016).

In the case of pristine environments such as Antarctica, it would be beneficial if the bioremediation of oil spills could be performed via the use of indigenous microorganisms, especially since it is prohibited to introduce foreign organisms into this type of environment (Gesheva et al. 2010, Malavenda et al. 2015). In a study by Malavenda et al. (2015), BS-producing strains were isolated from Arctic and Antarctic shoreline sediments. Of the 199 isolates, 18 showed the ability to grow in the presence of crude oil, producing BSs. Thin layer chromatography analyses showed that the majority of the BSs produced were glycolipids, and one actinobacterial genus, *Rhodococcus*, was represented among the 18 BS-producing strains (Malavenda et al. 2015). Cai et al. (2015) also highlighted the need of exploring various environments. In their study, they collected crude oil, formation water, drilling mud, and treated produced water samples from offshore oil and gas reservoirs. These reservoirs represent unique environments with extreme temperatures, high pressures and low oxygen content, which makes for a unique microbial population. Of the 59 strains isolated, two were *Rhodococcus* strains closely related to *Rhodococcus erythropolis* and *R. phenolicus*. The strains exhibited different BS properties and in the natural environment may act synergistically in dispersing oil spills (Cai et al. 2015).

## Final Thoughts

It is clear from the great diversity of actinobacterial BSs that it is virtually impossible to capture all relevant information in one chapter. There are, however, certain key aspects that became clear from the studies reported in literature.

1) Various environments still need to be explored for potentially novel BS-producing actinobacterial strains. Plant-associated actinobacteria are vastly underexplored, especially rhizosphere-associated actinobacteria. In this environment, BSs may play a key role in the uptake of nutrients and metals required for the symbiotic relationship with the "host" plant. Other possible sources that needs to be evaluated as a source of BS-producing actinobacteria is the waste/wastewater generated in industries involved in the processing of edible oil, animal-based products (tanneries, slaughterhouses), and sugar. These actinobacteria would already be capable of utilising waste from these industries and their cultivation conditions may only require minimal optimisation.

2) Some of the BSs reported in this chapter were accessed from actinobacteria purchased from international culture collections. These culture collections serve as vast resources of novel actinobacterial strains that have previously not been evaluated for their ability to produce BSs.

3) There is a need for the development of reproducible high throughput screening protocols such as the atomised oil spray assay that would allow for the screening of large numbers of cultures and transformants from metagenomic studies.

4) Understanding the physiological role of BSs in actinobacteria (keeping in mind the ecological niche the strains were isolated from) would be required in order to understand the 'triggers' for BS production. This would lead to a better understanding of the optimal growth conditions required and how these can be manipulated for increased BS production.

5) There is a need for simplified downstream processing and more structural elucidation (this will further help us to understand the molecular basis for BS production), which can ultimately lead to genetic manipulation of producer strains and/or the cloning of the BGCs involved in BS production.

6) Understanding the requirements of industry and the development of translatable research and economically-viable bioprocesses for the production of BSs, e.g., the development of protocols for general to directed applications.

7) BS production from waste materials shows great potential for incorporation into a biorefinery approach where waste material can be utilised in the production of BSs, while the remainder of the waste, including cell mass, can be used in energy recovery. Industry symbiosis programs, where the waste generated in one industry can be used by others for the production of other products should also be explored for the production of BSs.

It is clear that in general, we have only explored the surface of the potential of actinobacterial BSs and our understanding of BS production by the *Actinobacteria*. Members of the phylum '*Actinobacteria*' still remains underexplored, serving as a vast resource of novel BSs still to be accessed.

## Acknowledgements

MLR-H and KD wish to acknowledge the funding received from the National Research Foundation (NRF) of South Africa for supporting the research on actinobacterial bio-active compounds currently taking place in their research group. Any opinions, findings and conclusions or recommendations expressed in this material are those of the authors and therefore the NRF do not accept any liability in regard thereto.

## References

Abdel-Megeed, A., A.N. Al-Rahma, A.A. Mostafa and K. Husnu Can Baser. 2011. Biochemical characterisation of anti-microbial activity of glycolipids produced by *Rhodococcus erythropolis*. Pak. J. Bot. 43: 1323–1334.

Allada, L.T.K., G.S. Guntuku, M.K.K. Muthyala, M.K. Duddu and N.S. Golla. 2015. Characterization of bioactive compound obtained from *Streptomyces coelicoflavus* NBRC (13599[T]) and its anticancer activity. Int. J. Chem. Pharm. Anal. 2: eISSN 2348-0726.

Alvarez, A., J.M. Saez, J.S.D. Costa, V.L. Colin, M.S. Fuentes, S.A. Cuozzo, C.S. Benimeli, M.A. Polti and M.J. Amoroso. 2017. Actinobacteria: Current research and perspectives for bioremediation of pesticides and heavy metals. Chemosphere 166: 41–62.

Aparicio, J.F., E.G. Barreales, T.D. Payero, C.M. Vicente, A. de Pedro and J. Santos-Aberturas. 2016. Biotechnological production and application of the antibiotic pimaricin: biosynthesis and its regulation. Appl. Microbiol. Biotechnol. 100: 61–78.

Aretz, W., J. Meiwes, G. Seibert, G. Vobis and J. Wink. 2000. Friulimicins: novel lipopeptide antibiotics with peptidoglycan synthesis inhibiting activity from *Actinoplanes friuliensis* sp. nov. I. Taxonomic studies of the producing microorganism and fermentation. J. Antibiot. (Tokyo) 53: 807–815.

Arino, S., R. Marchal and J.-P. Vandecasteele. 1998. Production of new extracellular glycolipids by a strain of *Cellulomonas cellulans* (*Oerskovia xanthineolytica*) and their structural characterization. Can. J. Microbiol. 44: 238–243.

Bajaj, A., S. Mayilraj, M.K.R. Mudiam, D.K. Patel and N. Manickam. 2014. Isolation and functional analysis of a glycolipid producing *Rhodococcus* sp. IITR03 with potential for degradation of 1,1,1-trichloro-2,2-bis(4-chlorophenyl)ethane (DDT). Bioresour. Technol. 167: 398–406.

Baltz, R.H., V. Miao and S.K. Wrigley. 2005. Natural products to drugs: daptomycin and related lipopeptide antibiotics. Nat. Prod. Rep. 22: 717–741.

Banat, I.M., S.K. Satpute, S.S. Cameotra, R. Patil and N.V. Nyayanit. 2014a. Cost effective technologies and renewable substrates for biosurfactants' production. Front. Microbiol. 5.

Banat, I.M., M.A. Díaz De Rienzo and G.A. Quinn. 2014b. Microbial biofilms: biosurfactants as antibiofilm agents. Appl. Microbiol. Biot. 98: 9915–9929.

Bérdy, J. 2012. Thoughts and facts about antibiotics: where we are now and where we are heading. J. Antibiot. (Tokyo) 65: 385–395.

Bode, H.B., B. Bethe, R. Höfs and A. Zeeck. 2002. Big effects from small changes: Possible ways to explore Nature's chemical diversity. ChemBioChem. 3: 619–627.

Burch, A.Y., B.K. Shimada, P.J. Browne and S.E. Lindow. 2010. Novel high-throughput detection method to assess bacterial surfactant production. Appl. Environ. Microbiol. 76: 5363–5372.

Cai, Q., B. Zhang, B. Chen, X. Song, Z. Zhu and T. Cao. 2015. Screening of biosurfactant-producing bacteria from offshore oil and gas platforms in North Atlantic Canada. Environ. Monit. Assess 187: 284.

Cameotra, S.S. and R.S. Makkar. 2004. Recent applications of biosurfactants as biological and immunological molecules. Curr. Opin. Microbiol. 7: 262–266.

Campos, J.M., T.L. Montenegro Stamford, L.A. Sarubbo, J.M. de Luna, R.D. Rufino and I.M. Banat. 2013. Microbial biosurfactants as additives for food industries; A review. Biotechnol. Progr. 29: 1097–1108.

Campos, J.M., L.A. Sarubbo, J.M. de Luna, R.D. Rufino and I.M. Banat. 2016. Use of (bio)surfactants in foods. pp. 435–459. *In*: J.L. Bicas, M.R. Maróstica and G.M. Pastore (eds.). Biotechnological Production of Natural Ingredients for Food Industry. Bentham Science Publishers. 1 ed. 2016, v. 1. Ch. 11.

Chakraborty, S., M. Ghosh, S. Chakraborti, S. Jana, K.K. Sen, C. Kokare and L. Zhang. 2015. Biosurfactant produced from actinomycetes *Nocardiopsis* A17: Characterization and its biological evaluation. Int. J. Biol. Macromol. 79: 405–412.

Chen, W., J. Qi, P. Wu, D. Wan, J. Liu, X. Feng and Z. Deng. 2016. Natural and engineered biosynthesis of nucleoside antibiotics in actinomycetes. J. Ind. Microbiol. Biotechnol. 43: 401–417.

Christova, N., B. Tuleva, Z. Lalchev, A. Jordanova and B. Jordanov. 2004. Rhamnolipid biosurfactants produced by *Renibacterium salmoninarum* 27BN during growth on *n*-hexadecane. Z Naturforsch C 59: 70–74.

Christova, N. and I. Stoineva. 2014. Trehalose biosurfactants. pp. 197–216. *In*: C.N. Mulligan, S.K. Sharma and A. Mudhoo (eds.). Biosurfactants Research Trends and Applications. CRC Press.

Ciabatti, R., J.K. Kettenring, G. Winters, G. Tuan, L. Zerilli and B. Cavalleri. 1989. Ramoplanin (A-16686), a new glycolipodepsipeptide antibiotic. III. Structure elucidation. J. Antibiot. (Tokyo) 42: 254–267.

Colin, V.L., M.F. Castro, M.J. Amoroso and L.B. Villegas. 2013a. Production of bioemulsifiers by *Amycolatopsis tucumanensis* DSM 45259 and their potential application in remediation technologies for soil contaminated with hexavalent chromium. J. Hazard. Mater. 261: 577–583.

Colin, V.L., C.E. Pereira, L.B. Villegas, M.J. Amoroso and C.M. Abate. 2013b. Production and partial characterization of bioemulsifier from a chromium-resistant actinobacteria. Chemosphere 90: 1372–1378.

Cooper, D.G., J.E. Zajic and D.F. Gerson. 1979. Production of surface-active lipids by *Corynebacterium lepus*. Appl. Environ. Microbiol. 37: 4–10.

Debono, M., M. Barnhart, C.B. Carrell, J.A. Hoffmann, J.L. Occolowitz, B.J. Abbott, D.S. Fukuda, R.L. Hamill, K. Biemann and W.C. Herlihy. 1987. A21978C, a complex of new acidic peptide antibiotics. Isolation, chemistry, and mass spectral structure elucidation. J. Antibiot. (Tokyo) 40: 761–777.

Dembitsky, V.M. 2005a. Astonishing diversity of natural surfactants. 2. Polyether glycoside ionophores and macrocyclic glycosides. Lipids 40: 219–248.

Dembitsky, V.M. 2005b. Astonishing diversity of natural surfactants. 3. Carotenoid glycosides and isoprenoid glycolipids. Lipids 40: 535–557.

Dembitsky, V.M. 2005c. Astonishing diversity of natural surfactants. 4. Fatty acid amide glycosides, their analogs and derivatives. Lipids 40: 641–660.

Dembitsky, V.M. 2005d. Astonishing diversity of natural surfactants. 5. Biologically active glycosides of aromatic metabolites. Lipids 40: 869–900.

Dembitsky, V.M. 2005e. Astonishing diversity of natural surfactants. 6. Biologically active marine and terrestrial alkaloid glycosides. Lipids 40: 1081–1105.

Doshi, D.V., J.P. Maniyar, S.S. Bhuyan and S.S. Mujumdar. 2010. Studies on bioemulsifier production by *Actinopolyspora* sp. A 18 isolated from garden soil. Indian J. Biotechnol. 9: 391–396.

Du, Y., Y. Wang, T. Huang, M. Tao, Z. Deng and S. Lin. 2014. Identification and characterization of the biosynthetic gene cluster of polyoxypeptin A, a potent apoptosis inducer. BMC Microbiol. 14: 30.

Dubey, K.V., P.N. Charde, S.U. Meshram, L.P. Shendre, V.S. Dubey and A.A. Juwarkar. 2012. Surface-active potential of biosurfactants produced in curd whey by *Pseudomonas aeruginosa* strain-PP2 and *Kocuria turfanesis* strain-J at extreme environmental conditions. Bioresour. Technol. 126: 368–374.

Eckardt, K. 1983. Tunicamycins, streptovirudins, and corynetoxins, a special subclass of nucleoside antibiotics. J. Nat. Prod. 46: 544–550.

Elshahawi, S.I., K.A. Shaaban, M.K. Kharel and J.S. Thorson. 2015. A comprehensive review of glycosylated bacterial natural products. Chem. Soc. Rev. 44: 7591–7697.

Espuny, M., S. Egido, I. Rodón, A. Manresa and M. Mercadé. 1996. Nutritional requirements of a biosurfactant producing strain *Rhodococcus* sp. 51T7. Biotechnol. Lett. 18: 521–526.

Ettoumi, B., H. Chouchane, A. Guesmi, M. Mahjoubi, L. Brusetti, M. Neifar, S. Borin, D. Daffonchio and A. Chrif. 2016. Diversity, ecological distribution and biotechnological potential of *Actinobacteria* inhabiting seamounts and non-seamounts in the Tyrrhenian Sea. Microbiol. Res. 186-187: 71–80.

Farnet, C., A. Staffa and E. Zazopoulos. 2002. Compositions, methods and systems for discovery of lipopeptides. U.S. Patent Application 10/329,027.

Ferradji, F.Z., S. Mnif, A. Badis, S. Rebbani, D. Fodil, K. Eddouaouda and S. Saycidi. 2014. Naphthalene and crude oil degradation by biosurfactant producing *Streptomyces* spp. isolated from Mitidja plain soil (North of Algeria). Int. Biodeterior. Biodegradation 86: 300–308.

Fournié, J.J., M. Riviere and G. Puzo. 1987. Structural elucidation of the major phenolic glycolipid from *Mycobacterium kansasii*. I. Evidence for tetrasaccharide structure of the oligosaccharide moiety. J. Biol. Chem. 262: 3174–3179.

Franzetti, A., I. Gandolfi, G. Bestetti, T.J.P. Smyth and I.M. Banat. 2010. Production and application of trehalose lipid biosurfactants. Eur. J. Lipid Sci. Technol. 112: 617–627.

Fu, C., L. Keller, A. Bauer, M. Brönstrup, A. Froidbise, P. Hammann, J. Herrmann, G. Mondesert, M. Kurz, M. Schiell, D. Schummer, L. Toti, J. Wink and R. Müller. 2015. Biosynthetic studies of telomycin reveal new lipopeptides with enhanced activity. J. Am. Chem. Soc. 137: 7692–7705.

Gandhimathi, R., G.S. Kiran, T.A. Hema, J. Selvin, T.R. Raviji and S. Shanmughapriya. 2009. Production and characterization of lipopeptide biosurfactant by a sponge-associated marine actinomycetes *Nocardiopsis alba* MSA10. Bioprocess Biosyst. Eng. 32: 825–835.

Gastambide-Odier, M. and P. Sarda. 1970. Contribution à l'étude de la structure et de la biosynthèse de glycolipides spécifiques isolés de *mycobactéries*: les mycosides a et b. Pneumonologie 142: 241–255.

Gesheva, V., E. Stackebrandt and E. Vasileva-Tonkova. 2010. Biosurfactant production by halotolerant *Rhodococcus fascians* from Casey station, Wilkes Land, Antarctica. Curr. Microbiol. 61: 112–117.

Gibson, K.J.C., M. Gilleron, P. Constant, G. Puzo, J. Nigou and G.S. Besra. 2003. Identification of a novel mannose-capped lipoarabinomannan from *Amycolatopsis sulphurea*. Biochem. J. 372: 821–829.

Gill, K.A., F. Berrue, J.C. Arens and R.G. Kerr. 2014. Isolation and structure elucidation of cystargamide, a lipopeptide from *Kitasatospora cystarginea*. J. Nat. Prod. 77: 1372–1376.

Gilleron, M., N.J. Garton, J. Nigou, T. Brando, G. Puzo and I.C. Sutcliffe. 2005. Characterization of a truncated lipoarabinomannan from the actinomycete *Turicella otitidis*. J. Bacteriol. 187: 854–861.

Gomes, M.B., E.E. Gonzales-Limache, S.T.P. Sousa, B.M. Dellagnezze, A. Sartoratto, L.C.F. Silva, L.M. Grieg, E. Valoni, R.S. Souza, A.P.R. Torres, M.P. Sousa, S.O. De Paula, C.C. Silva and V.M. Oliveira. 2016. Exploring the potential of halophilic bacteria from oil terminal environments for biosurfactant production and hydrocarbon degradation under high-salinity conditions. Int. Biodeterior. Biodegradation (in press).

Goodfellow, M. and H.-P. Fiedler. 2010. A guide to successful bioprospecting: informed by actinobacterial systematics. Antonie van Leeuwenhoek 98: 119–142.

Graziano, M., C. Rizzo, L. Michaud, E.M.D. Porporato, E. De Domenico, N. Spanò and A. Lo Giudice. 2016. Biosurfactant production by hydrocarbon-degrading *Brevibacterium* and *Vibrio* isolates from the sea pen *Pteroeides spinosum* (Ellis, 1764). J. Basic Microbiol. 56: 963–974.

Gudiňa, E.J., J.A. Teixera and L.R. Rodrigues. 2016. Biosurfactants produced by marine microorganisms with therapeutic applications. Mar. Drugs. 14: 38.

Haferburg, D., R. Hommel, R. Claus and H.-P. Kleber. 2005. Extracellular microbial lipids as biosurfactants. Adv. Biochem. Eng. Biotechnol. 33: 53–93.

Hamill, R.J. 2013. Amphotericin B formulations: a comparative review of efficacy and toxicity. Drugs 73: 919–934.

Hausmann, R. and C. Syldakt. 2014. Types and classification of microbial surfactants. p. 1. *In*: N. Kosaric and F. Vader-Sukan (eds.). Biosurfactants: Production and Utilization—Processes, Technologies, and Economics. Vol. 159. CRC Press, Boca Raton, London, New York.

Heinemann, B., M.A. Kaplan, R.D. Muir and I.R. Hooper. 1953. Amphomycin, a new antibiotic. Antibiot. Chemother. (Northfield, Ill.) 3: 1239–1242.

Hiratsuka, T., H. Suzuki, R. Kariya, T. Seo, A. Minami and H. Oikawa. 2014. Biosynthesis of the structurally unique polycyclopropanated polyketide-nucleoside hybrid jawsamycin (FR-900848). Aangew. Chem. Int. Ed. Engl. 53: 5423–5426.

Hopmann, C., M. Kurz, M. Brönstrup, J. Wink and D. Lebeller. 2002. Isolation and structure elucidation of vancoresmycin—a new antibiotic from *Amycolatopsis* sp. ST 101170. Tetrahedron Lett. 43: 435–438.

Irorere, V.U., L. Tripathi, R. Marchant, S. McClean and I.M. Banat. 2017. Microbial rhamnolipid production: A critical re-evaluation of published data and suggested future publication criteria. Appl. Microbiol. Biot. 101: 3941–3951.

Ivshina, I., L. Kostina, A. Krivoruchko, M. Kuyukina, T. Peshkur, P. Anderson and C. Cunningham. 2016. Removal of polycyclic aromatic hydrocarbons in soil spiked with model mixtures of petroleum hydrocarbons and heterocycles using biosurfactants from *Rhodococcus ruber* IEGM 231. J. Hazard. Mater. 312: 8–17.

Ivshina, I.B., M.S. Kuyukina, A.V. Krivoruchko, O.A. Plekhov, O.B. Naimark, E.A. Podorozhko and V.I. Lozinsky. 2013. Biosurfactant-enhanced immobilization of hydrocarbon-oxidizing *Rhodococcus ruber* on sawdust. Appl. Microbiol. Biotechnol. 97: 5315–5327.

Jackson, S.A., E. Borchert, F. O'Gara and A.D.W. Dobson. 2015. Metagenomics for the discovery of novel biosurfactants of environmental interest from marine ecosystems. Curr. Opin. Biotechnol. 33: 176–182.

Jin, X., M. Rao, W. Wei, M. Ge, J. Liu, D. Chen and Y. Liang. 2012. Biosynthesis of new lipopentapeptides by an engineered strain of *Streptomyces* sp. Biotechnol. Lett. 34: 2283–2289.

Kalyani, A.L.T., G. Naga Siresha, A.K.G. Aditya, G. Girija Sankar and T. Prabhakar. 2014a. Production optimization of rhamnolipid biosurfactant by *Streptomyces coelicoflavus* (NBRC 15399[T]) using Plackett-Burman design. Eur. J. Biotechnol. Biosci. 1: 7–13.

Kalyani, A.L.T., G. Girija Sankar and T. Prabhakar. 2014b. Optimization of rhamnolipid biosurfactant production by *Streptomyces matensis* (NBRC 12889[T]) using Plackett-Burman design. J. Biomed. Pharm. Res. 3: 1–7.

Katz, L. and R.H. Baltz. 2016. Natural product discovery: past, present, and future. J. Ind. Microb. Biotechnol. (in press).

Kavyanifard, A., G. Ebrahimipour and A. Ghasempour. 2016. Structure characterization of a methylated ester biosurfactant produced by a newly isolated *Dietzia cinnamea* KA1. Microbiol. 85: 430–435.

Kersten, R.D., Y.-L. Yang, Y. Xu, P. Cimermancic, S.-J. Nam, W. Fenical, M.A. Fischbach, B.S. Moore and P.C. Dorrestein. 2011. A mass spectrometry-guided genome mining approach for natural product peptidogenomics. Nat. Chem. Biol. 7: 794–802.

Khopade, A., R. Biao, X. Liu, K. Mahadik, L. Zhang and C. Kokare. 2012a. Production and stability studies of the biosurfactant isolated from marine *Nocardiopsis* sp. B4. Desalination 285: 198–204.

Khopada, A., B. Ren, X.-Y. Liu, K. Mahadik, L. Zhang and C. Kokare. 2012b. Production and characterisation of biosurfactant from marine *Streptomyces* species B3. J. Colloid Interface Sci. 367: 311–318.

Kiran, G.S., A. Sabu and J. Selvin. 2010a. Synthesis of silver nanoparticles by glycolipid biosurfactant produced from marine *Brevibacterium casei* MSA19. J. Biotechnol. 148: 221–225.

Kiran, G.S., T.A. Thomas and J. Selvin. 2010b. Production of a new glycolipid biosurfactant from marine *Nocardiopsis lucentensis* MSA04 in solid-state cultivation. Colloids Surf. B Biointerfaces 78: 8–16.

Kiran, G.S., T.A. Thomas, J. Selvin, B. Sabarathnam and A.P. Lipton. 2010c. Optimization and characterization of a new lipopeptide biosurfactant produced by marine *Brevibacterium aureum* MSA13 in solid state culture. Bioresour. Technol. 101: 2389–2396.

Kiran, G.S., B. Sabarathnam and J. Selvin. 2010d. Biofilm disruption potential of a glycolipid biosurfactant from marine *Brevibacterium casei*. FEMS Immunol. Med. Microbiol. 59: 432–438.

Kiran, G.S., L.A. Nishanth, S. Priyadharshini, K. Anitha and J. Selvin. 2014a. Effect of Fe nanoparticle on growth and glycolipid biosurfactant production under solid state culture by marine *Nocardiopsis* sp. MSA13A. BMC Biotechnol. 14: 48.

Kiran, G.S., B. Sabarathnam, N. Thajuddin and J. Selvin. 2014b. Production of glycolipid biosurfactant from sponge-associated marine actinobacterium *Brachybacterium paraconglomeratum* MSA21. J. Surfactants Deterg. 17: 531–542.

Kiran, G.S., J. Selvin, S. Ganesan, L.A. Nishanth, N.A. Al-Dhabi and M.V. Arasu. 2016a. Ketide synthase (KS) domain prediction and analysis of iterative type II PKS gene in marine sponge associated actinobacteria producing biosurfactants and antimicrobial agents. Front. Microbiol. 7: 63.

Kiran, G.S., A.S. Ninawe, A.N. Lipton, V. Pandian and J. Selvin. 2016b. Rhamnolipid biosurfactants: evolutionary implications, applications and future prospects from untapped marine resources. Crit. Rev. Biotechnol. 36: 399–415.

Kirst, H.A. 2010. The spinosyn family of insecticides: realizing the potential of natural products research. J. Antibiot. (Tokyo) 63: 101–111.

Klassen, J.L. 2010. Phylogenetic and evolutionary patterns in microbial carotenoid biosynthesis are revealed by comparative genomics. PLoS One 5: e11257.

Kong, F. and G.T. Carter. 2003. Structure determination of glycinocins A to D, further evidence for the cyclic structure of the amphomycin antibiotics. J. Antibiot. (Tokyo) 56: 557–564.

Konishi, M., S. Nishi, T. Fukuoka, D. Kitamoto, T.O. Watsuji, Y. Nagano, A. Yabuki, S. Nakagawa, Y. Hatada and J.I. Horiuchi. 2014. Deep-sea *Rhodococcus* sp. BS-15, lacking the phytopathogenic fas genes, produces a novel glucotriose lipid biosurfactant. Mar. Biotechnol. 16: 484–493.

Korayem, A.S., A.A. Abdelhafez, M.M. Zaki and E.A. Saleh. 2015. Optimization of biosurfactant production by *Streptomyces* isolated from Egyptian arid soil using Plackett-Burman design. Ann. Agric. Sci. 60: 209–217.

Kretschmer, A., H. Bock and F. Wagner. 1982. Chemical and physical characterization of interfacial active lipids from *Rhodococcus erythropolis* grown on normal-alkanes. Appl. Environ. Microbiol. 44: 864–870.

Kügler, J.H., C. Muhle-Goll, B. Kühl, A. Kraft, R. Heinzler, F. Kirschhöfer, M. Henkel, V. Wray, B. Luy, G. Brenner-Weiss, S. Lang, C. Syldakt and R. Hausmann. 2014. Trehalose lipid biosurfactants produced by the actinomycetes *Tsukamurella spumae* and *T. pseudospumae*. Appl. Microbiol. Biotechnol. 98: 8905–8915.

Kügler, J.H., A. Kraft, S. Heißler, C. Muhle-Goll, B. Luy, W. Schwack, C. Syldakt and R. Hausmann. 2015a. Extracellular aromatic biosurfactant produced by *Tsukamurella pseudospumae* and *T. spumae* during growth on *n*-hexadecane. J. Biotechnol. 211: 107–114.

Kügler, J.H., M. Le Roes-Hill, C. Syldakt and R. Hausmann. 2015b. Surfactants tailored by the class *Actinobacteria*. Front. Microbiol. 6: 1–23.

Kumar, C.G. and S. Pombala. 2015. Process for the preparation of polymeric biosurfactants. US patent: US 2015/0322173A1.

Kundu, P., C. Hazra, N. Dandi and A. Chaudhari. 2013. Biodegradation of 4-nitrotoluene with biosurfactant production by *Rhodococcus pyridinivorans* NT2: metabolic pathway, cell surface properties and toxicological characterization. Biodegradation 24: 775–793.

Kuyukina, M.S., I.B. Ivshina, T.A. Baeva, O.A. Kochina, S.V. Gein and V.A. Chereshnev. 2015. Trehalolipid biosurfactants from non-pathogenic *Rhodococcus* actinobacteria with diverse immunomodulatory activities. New Biotechnol. 32: 559–568.

Kuyukina, M.S., I.B. Ivshina, I.O. Korshunova, G.I. Stukova and A.V. Krivoruchka. 2016. Diverse effects of a biosurfactant from *Rhodococcus ruber* IEGM 231 on the adhesion of resting and growing bacteria to polystyrene. AMB Express 6: 14.

Lakey, J.H., E.J.A. Lea, B.A.M. Rudd, H.M. Wright and D.A. Hopwood. 1983. A new channel-forming antibiotic from *Streptomyces coelicolor* A3 (2) which requires calcium for its activity. Microbiol. 129: 3565–3573.

Lanéelle, G. and J. Asselineau. 1968. Structure d'un glycoside de peptidolipide isolé d'une *mycobactérie*. Eur. J. Biochem. 5: 487–491.

Laorrattanasak, S., W. Rongsayamanont, N. Khondee, N. Paorach, S. Soonglerdsongpha, O. Pinyakong and E. Luepromchai. 2016. Production and application of *Gordonia westfalica* GY40 biosurfactant for remediation of fuel oil spill. Water Air Soil Pollut. 227: 325.

Liu, J., X.F. Huang, L.J. Lu, J.C. Xu, Y. Wen, D.H. Yang and Q. Zhou. 2009. Comparison between waste frying oil and paraffin as carbon source in the production of biodemulsifier by *Dietzia* sp. S-JS-1. Bioresour. Technol. 100: 6481–6487.

MacDonald, C.R., D.G. Cooper and J.E. Zajic. 1981. Surface-active lipids from *Nocardia erythropolis* grown on hydrocarbons. Appl. Environ. Microbiol. 41: 117–123.

Makkar, R.S., S.S. Cameotra and I.M. Banat. 2011. Advances in utilization of renewable substrates for biosurfactant production. AMB Express 1: 5.

Malavenda, R., C. Rizzo, L. Michaud, B. Gerçe, V. Bruni, C. Syldakt, R. Hausmann and A. Lo Giudice. 2015. Biosurfactant production by Arctic and Antarctic bacteria growing on hydrocarbons. Polar Biol. 38: 1565–1574.

Manivasagan, P., P. Sivasankar, J. Venkatesan, K. Sivakumar and S.-K. Kim. 2014. Optimization, production and characterization of glycolipid biosurfactant from the marine actinobacterium, *Streptomyces* sp. MAB36. Bioprocess Biosyst. Eng. 37: 783–797.

Marchant, R. and I.M. Banat. 2012a. Microbial biosurfactants: challenges and opportunities for future exploitation. Trends Biotechnol. 30: 558–565.

Marchant, R. and I.M. Banat. 2012b. Biosurfactants: a sustainable replacement for chemical surfactants? Biotechnol. Lett. 34: 1597–1605.

Margaritis, A., J.E. Zajic and D.F. Gerson. 1979. Production and surface-active properties of microbial surfactants. Biotechnol. Bioeng. 21: 1151–1162.

Mnif, I. and D. Ghribi. 2015a. Glycolipid biosurfactants: Potential related biomedical and biotechnological applications. Carbohydr. Res. 416: 59–69.

Mnif, I. and D. Ghribi. 2015b. Review: lipopeptides biosurfactants: mean classes and new insights for industrial, biomedical, and environmental applications. Biopolymers 104: 129–147.

Mohanram, R., C. Jagtap and P. Kumar. 2016. Isolation, screening, and characterization of surface-active agent-producing, oil-degrading marine bacteria of Mumbai Harbor. Mar. Pollut. Bull. 105: 131–138.

Nakano, M., M. Kihara, S. Iehata, R. Tanaka, H. Maedla and T. Yoshikawa. 2011. Wax ester-like compounds as biosurfactants produced by *Dietzia maris* from *n*-alkane as a sole carbon source. J. Basic Microbiol. 51: 490–498.

Nguyen, K.T., X. He, D.C. Alexander, C. Li, J.-Q. Gu, C. Mascio, A. Van Praagh, L. Martin, M. Chu, J.A. Silverman, P. Brian and R.H. Baltz. 2010. Genetically engineered lipopeptide antibiotics related to A54145 and daptomycin with improved properties. Antimicrob. Agents Chemother. 54: 1404–1413.

Nupur, L.N.U., A. Vats, S.K. Dhanda, G.P.S. Raghava, A.K. Pinnaka and A. Kumar. 2016. ProCarDB: a database of bacterial carotenoids. BMC Microbiol. 16: 96.

Ohadi, M., G. Dehghan-Noudeh, M. Shakibaie, I.M. Banat, M. Pournamdari and H. Forootanfar. 2017. Isolation, characterization, and optimization of biosurfactant production by an oil-degrading *Acinetobacter junii* B6 isolated from an Iranian oil excavation site. Biocatal. Agric. Biotechnol. 12: 1–9.

Otzen, D. 2016. Biosurfactants and surfactants interacting with membranes and proteins: same but different? BBA—Biomembranes (in press).

Parte, A.C. 2017. List of prokaryotic names with standing in nomenclature (LPSN). http://www.bacterio.net.

Patel, S., S. Ahmed and J.S. Eswari. 2015. Therapeutic cyclic lipopeptides mining from microbes: latest strides and hurdles. World J. Microbiol. Biotechnol. 31: 1177–1193.

Peng, F., Z. Liu, L. Wang and Z. Shao. 2007. An oil-degrading bacterium: *Rhodococcus erythropolis* 3C-9 and its biosurfactants. J. Appl. Microbiol. 102: 1603–1611.

Pizzul, L., M. del PilarCastillo and J. Stenström. 2006. Characterization of selected actinomycetes degrading polyaromatic hydrocarbons in liquid culture and spiked soil. World J. Microbiol. Biotechnol. 22: 745–752.

Plaza, G.A., J. Chojniak and I.M. Banat. 2014. Biosurfactant mediated biosynthesis of selected metallic nanoparticles. Int. J. Mol. Sci. 15: 13720–13737.

Powalla, M., S. Lang and V. Wray. 1989. Penta- and disaccharide lipid formation by *Nocardia corynebacteroides* grown on n-alkanes. Appl. Microbiol. Biotechnol. 31: 473–479.

Price, N.P.J., D.P. Labeda, T.A. Naumann, K.E. Vermillion, M.J. Bowman, M.A. Berhow, W.W. Metcalf and K.M. Boschoff. 2016. Quinovosamycins: new tunicamycin-type antibiotics in which the α, β-1", 11'-linked *N*-acetylglucosamine residue is replaced by *N*-acetylquinovosamine. J. Antibiot. (Tokyo) 69: 637–646.

Raja, A., J. LaBonte, J. Lebbos and P. Kirkpatrick. 2003. Daptomycin. Nat. Rev. Drug Discov. 2: 943–944.

Richter, M., J.M. Willey, R. Süßmuth, G. Jung and H.P. Fiedler. 1998. Streptofactin, a novel biosurfactant with aerial mycelium inducing activity from *Streptomyces tendae* Tü 901/8c. FEMS Microbiol. Lett. 163: 165–171.

Ristau, E. and F. Wagner. 1983. Formation of novel anionic trehalosetetraesters from *Rhodococcus erythropolis* under growth limiting conditions. Biotechnol. Lett. 5: 95–100.

Roongsawang, N., K. Washio and M. Morikawa. 2011. Diversity of nonribosomal peptide synthetases involved in the biosynthesis of lipopeptide biosurfactants. Int. J. Mol. Sci. 12: 141–172.

Roy, S., S. Chandni, I. Das, L. Karthik, G. Kumar and K.V.B. Rao. 2015. Aquatic model for engine oil degradation by rhamnolipid producing *Nocardiopsis* VITSISB. 3Biotech 5: 153–164.

Sachdev, D.P. and S.S. Cameotra. 2013. Biosurfactants in agriculture. Appl. Microbiol. Biotechnol. 97: 1005–1016.

Saimmai, A., V. Sobhon and S. Maneerat. 2012. Production of biosurfactant from a new and promising strain of *Leucobacter komagatae* 183. Ann. Microbiol. 62: 391–402.

Sajna, K.V., R. Höfer, R.K. Sukumaran, L.D. Gottumukkala and A. Pandey. 2015. White biotechnology in biosurfactants. pp. 499–521. *In*: A. Pandey, R. Höfer, M. Taherzadeh, K.M. Nampoothiri and C. Larroche (eds.). Industrial Biorefineries and White Biotechnology. Elsevier, Netherlands.

Sarafin, Y., M.B.S. Donio, S. Velmurugan, M. Michaelbabu and T. Citarasu. 2014. *Kocuria marina* BS-15 a biosurfactant producing halophilic bacteria isolated from solar salt works in India. Saudi J. Biol. Sci. 21: 511–519.

Sekhon, K.K. and P.K. Rahman. 2014. Rhamnolipid biosurfactants—Past, present and future scenario of global market. Front. Microbiol. 5: 454.

Sharma, A., S.B. Singh, R. Sharma, P. Chaudhary, A. Pandey, R. Ansari, V. Vasudevan, A. Arora, S. Singh, S. Saha and L. Nain. 2016. Enhanced biodegradation of PAHs by microbial consortium with different amendment and their fate in *in-situ* condition. J. Environ. Manage. 181: 728–736.

Sharma, D., S.M. Mandal and R.K. Manhas. 2014. Purification and characterization of a novel lipopeptide from *Streptomyces amritsarensis* sp. nov. active against methicillin-resistant *Staphylococcus aureus*. AMB Express 4: 50.

Shay, A.J., J. Adam, J.H. Martin, W.K. Hausmann, P. Shu and N. Bohonos. 1958. Aspartocin. I. Production, isolation, and characteristics. Antibiot. Ann. 7: 194–198.

Shoji, J.-I., S. Kozuki, S. Okamoto, R. Sakazaki and H. Otsuka. 1968. Studies on tsushimycin. I. Isolation and characterization of an acidic acylpeptide containing a new fatty acid. J. Antibiot. (Tokyo) 21: 439–443.

Shubrasekhar, C., M. Supriya, L. Karthik, K. Gaurev and K.V. Bhaskara Rao. 2013. Isolation, characterization and application of biosurfactant produced by marine actinobacteria isolated from saltpan soil from coastal area of Andhra Pradesh, India. Res. J. Biotechnol. 8: 18–25.

Strieker, M. and M.A. Marahiel. 2009. The structural diversity of acidic Lipopeptide antibiotics. ChemBioChem. 10: 607–616.

Sutcliffe, I. 2005. Lipoarabinomannans—structurally diverse and functionally enigmatic macroamphiphiles of mycobacteria and related actinomycetes. Tuberculosis 85: 205–206.

Takizawa, M., T. Hida, T. Horiguchi, A. Hiramoto, S. Harada and S. Tanida. 1995. TAN-1511 A, B and C, Microbial lipopeptides with G-CSF and GM-CSF inducing activity. J. Antibiot. (Tokyo) 48: 579–588.

Thavasi, R., S. Jayalakshmi, T. Balasubramanian and I.M. Banat. 2007. Biosurfactant production by *Corynebacterium kutscheri* from waste motor lubricant oil and peanut oil cake. Lett. Appl. Microbiol. 45: 686–691.

Thies, S., S.C. Rausch, F. Kovacic, A. Schmidt-Thaler, S. Wilhelm, F. Rosenau, R. Daniel, W. Streit, J. Pietruszka and K.-E. Jaeger. 2016. Metagenomic discovery of novel enzymes and biosurfactants in a slaughterhouse biofilm microbial community. Sci. Rep. 6: 27035.

Toth-Sarudy, E., I. Horvath, J. Gyimesi, I. Ott, L. Alfoldi, J. Berdy, I. Koczka, V. Scholtz, V. Szell and E. Laszio. 1974. US patent 3798129: Process for producing parvulines.

Tunlid, A., N.A. Schultz, D.R. Benson, D.B. Steele and D.C. White. 1989. Differences in fatty acid composition between vegetative cells and N2-fixing vesicles of *Frankia* sp. strain CpI1. Proc. Nat. Acad. Sci. 86: 3399–3403.

Van Hamme, J.D., A. Singh and O.P. Ward. 2006. Physiological aspects. Part I in a series of papers devoted to surfactants in microbiology and biotechnology. Biotechnol. Adv. 24: 604–620.

Vasileva-Tonkova, E. and V. Gasheva. 2005. Glycolipids produced by Antarctic *Nocardioides* sp. during growth on *n*-paraffin. Process Biochem. 40: 2387–2391.

Vertesy, L., E. Ehlers, H. Kogler, M. Kurz, J. Meiwes, G. Seibert, M. Vogel and P. Hammann. 2000. Friulimicins. Novel lipopeptide antibiotics with peptidoglycan synthesis inhibiting activity from *Actinoplanes friuliensis* sp. nov. II. Isolation and structural characterization. J. Antibiot. (Tokyo) 53: 816–827.

Vilela, W.F.D., S.G. Fonseca, F. Fantinatti-Garboggini, V.M. Oliveira and M. Nitschke. 2014. Production and properties of a surface-active lipopeptide produced by a new marine *Brevibacterium luteolum* strain. Appl. Biochem. Biotechnol. 174: 2245–2256.

Vollbrecht, E., U. Rau and S. Lang. 1999. Microbial conversion of vegetable oils into surface-active di-, tri-, and tetrasaccharide lipids (biosurfactants) by the bacterial strain *Tsukamurella* spec. Lipid 101: 389–394.

Walter, V., C. Syldakt and R. Hausmann. 2010. Screening concepts for the isolation of biosurfactant producing microorganisms. pp. 1–13. *In*: R. Sen (ed.). Biosurfactants. Springer, New York.

Wang, W., Z. Shao and B. Cai. 2014. Oil degradation and biosurfactant production by the deep sea bacterium *Dietzia maris* As-13-3. Front. Microbiol. 5: 711.

Wang, Y., Y. Chen, Q. Shen and X. Yin. 2011. Molecular cloning and identification of the laspartomycin biosynthetic gene cluster from *Streptomyces viridochromogenes*. Gene 483: 11–21.

Weber, T. and H.U. Kim. 2016. The secondary metabolite bioinformatics portal: Computational tools to facilitate synthetic biology of secondary metabolite production. Synth. Syst. Biotechnol. 1: 69–79.

Yamanaka, K., K.A. Reynolds, R.D. Kersten, K.S. Ryan, D.J. Gonzalez, V. Nizet, P.C. Dorrestein and B.S. Moore. 2014. Direct cloning and refactoring of a silent lipopeptide biosynthetic gene cluster yields the antibiotic taromycin A. Proc. Natl. Acad. Sci. USA 111: 1957–1962.

Yan, X., J. Sims, B. Wang and M.T. Hamann. 2014. Marine actinomycete *Streptomyces* sp. ISP2-49E, a new source of rhamnolipid. Biochem. Syst. Ecol. 55: 292–295.

Yilmaz, F., A. Ergene, E. Yalçin and S. Tan. 2009. Production and characterization of biosurfactants produced by microorganisms isolated from milk factory wastewaters. Environ. Technol. 30: 1397–1404.

Yin, X. and T.M. Zabriskie. 2012. Enduracidin biosynthetic gene cluster from *Streptomyces fungicidicus*. U.S. Patent 8,188,245.

Zhang, X.X., S.P. Cheng, Z.H.U. Cheng-Jun and S.U.N. Shi-Lei. 2006. Microbial PAH-degradation in soil: degradation pathways and contributing factors. Pedosphere 16: 555–565.

# 9

# Biosurfactants and their Applications in Petroleum Industry

*Nayereh Saborimanesh* and *Catherine N. Mulligan\**

## Introduction

The petroleum industry contains several major sectors including crude oil extraction, production and transportation. Extracted crude oil is delivered to refineries or other consumers where crude oil is processed into petroleum compounds and products (Wong and Hung 2004). Contamination of water and soil with crude oil and petroleum compounds frequently happens at different stages and causes environmental contamination that is either difficult to be removed by traditional remediation techniques or is not economically feasible. Therefore, development of effective and feasible processes and technologies that can lessen the damaging effects of oil contamination and lower treatment costs are of great importance.

Bioemulsifiers (BEs) and biosurfactants (BSs) are produced by microorganisms either as an integral part of microbial cell surface or as extracellular compounds released into the surrounding environments. BSs have amphipathic molecular structures and can interact with both hydrophobic and hydrophilic compounds. They have several distinctive features such as high surface and interfacial activities, temperature and ionic tolerance (Inoue et al. 1980, Hirata et al. 2009a, Daverey and Pakshirajan 2010, Van Bogaert and Soetaert 2010), biodegradability (Hirata et al. 2009b) and low toxicity (Hirata et al. 2009b). The molecular structures of BEs/BSs are less affected by salinity, pH, and temperature thus making them readily applicable

Department of Building, Civil and Environmental Engineering, Concordia University, 1455 de Maisonneuve Blvd., West Montreal, Quebec, Canada, H3G 1M8.
Email: n_sabori@encs.concordia.ca
* Corresponding author: mulligan@civil.concordia.ca

in oil-related activities (Banat et al. 1990, Banat 1995, Daverey and Pakshirajan 2010, Al-Bahry et al. 2013). For example, sophorolipid (SL) BSs can maintain their surface activity over a broad range of pH (e.g., pH 2–10), salinity (e.g., 0–20% NaCl salt), and have thermal stability in boiling water for 120 min (Daverey and Pakshirajan 2010). Similarly, BSs produced by bacterial strains such as Pet 1006, grown on two different carbon sources (e.g., glucose and an immiscible hydrocarbon), can preserve their chemical structures after sterilization procedure with minimum changes in their surface active properties (Banat et al. 1990).

Given the increasing environmental awareness and unlimited applicability of BEs/BSs (Kitamoto et al. 2002, Hirata et al. 2009b, Sajna et al. 2013, Vaughn et al. 2014) and the fact that numerous BS-producing microorganisms have been isolated from the oil-related environments, more attention must be given to the application of BEs/BSs in the oil-related activities (Bognolo 1999). In this regard, petroleum biotechnology has been developed to explore innovative techniques that implement biological processes and reagents for oil extraction, purification, and management and remediation of oil contaminants (Silva et al. 2014). BSs'/BEs' diverse properties as wetting, washing, emulsifying/demulsifying, solubilization and foaming agents, corrosion inhibitors, and viscosity reducers and important roles in increasing the bioavailability of hydrophobic molecules to oil-degrading microorganisms and promoting biodegradation of oil contaminants have been investigated by several researchers in the last decades (Bodour et al. 2004, Malayoglu 2004, Özdemir and Malayoglu 2004, Dagbert et al. 2006, Soudmand-asli et al. 2007, Alavi 2011, Thavasi et al. 2011a, b, Joshi-Navare et al. 2013).

To date, as full-scale applications of BEs/BSs in the oil industry have not yet achieved a widespread recognition, the aim of this chapter is to provide successful examples of BSs application in various aspects of the oil industry and cleanup of petroleum-contaminated environments to encourage the oil industry and decision makers to consider the utilization of BEs/BSs in places where their chemical counter parts are being used. In this chapter, laboratory and field studies and full-scale applications of BEs and BSs in the petroleum industry and oil-contaminated environments will be reviewed.

## BSs in Crude Oil Production and Recovery

Crude oil or bitumen is extracted and recovered from oil sands through two well-established techniques comprising of (i) surface mining (see Figure 1) and (ii) underground oil sands recovery (see Figure 4) (Clark and Pasternack 1932, Mullins et al. 2007, Rao and Liu 2013, Suncor Energy Inc. 2015). In the surface mining technique, hot water is mixed with crushed oil sands to make a slurry (this technique is known as Clark Hot Water Extraction process (CHWE)) (Rao and Liu 2013). Bitumen froth which contains 60 wt % bitumen, 30 wt % water and 10 wt % mineral solids are recovered from the slurry at a separation tank and transported to the froth treatment plant for further purification (Clark and Pasternack 1932, Long et al. 2002, Kirpalani and Matsuoka 2008, Rao and Liu 2013) while in the underground recovery, crude oil or bitumen is recovered from underground oil deposits through injecting hot water, gas or chemicals. In general, two main sources of oily sludge/waste including oil sand extraction tailings and froth cleaning tailings are generated during the crude oil extraction and recovery processes (Gosselin et al. 2010).

The separation of bitumen from the Athabasca oil sand with crude BSs through a cold-water process was examined in the laboratory experiment and their surface and interfacial tensions properties (Table 1) were compared with the synthetic surfactant (e.g., Petrostep-A-50). First, a mixture of oil sands and BS (a 0.02% solution of a whole microbial broth) was shaken at 200 rpm for 48 h. Following the shaking period, four fractions of bitumen including (i) surface oil fraction, (ii) residual tar sand balls, (iii) sand + clay fraction, and (iv) emulsion bitumen were separately analysed for the bitumen content. This study showed that the BSs effectively separated bitumen from the oil sands through two main mechanisms including (i) bitumen flotation and (ii) bitumen release (Zajic and Gerson 1978).

Similarly, the laboratory application of BSs from *Starmerella bombicola* ATCC 22214 (former *Torulopsis bombicola*, anamorph *Candida bombicola*) for the release of bitumen from oil sands with a cold-water process was examined by Cooper and Paddock (1984). A direct relationship between the BS concentration and the bitumen released from the oil sand was observed so that, as shown in Figure 2, higher amounts of bitumen were released from oil sands as the concentration of BS increased.

Moreover, in a U.S. patent, the use of microbial products generated by a mixed microbial community, isolated from indigenous and/or non-indigenous microbial community in the bitumen froth tailings and amended with nutrients, was proposed as an effective method for the biotreatment of bitumen froth tailings. The enriched microbial community was used to degrade asphaltenes content of the froth tailing (Duyvesteyn et al. 1999). This study showed that a lower waste volume can be generated via this method and the production of the by-product mixture (known as "bioliquor") would be suitable for reuse in multiple processes from initial oil sands mining to the final oil recovery process.

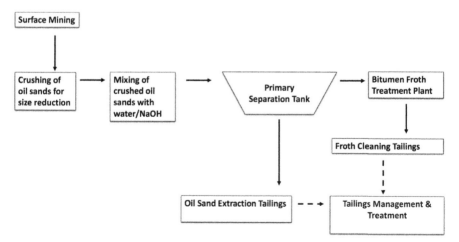

**Figure 1.** Schematic presentation of extraction of bitumen from oil sands by surface mining technique using flotation process (adapted from Rao and Liu 2013).

**Table 1.** Surface tension and Critical Micelle Concentration (CMC) of BSs produced by microorganisms grown on kerosene (adapted from Zajic and Gerson 1978).

| BS-Producing Microorganisms | Strains | Minimum Surface Tension (mN/m) | CMC (% of whole broth) |
|---|---|---|---|
| *Corynebacterium* sp. | OSGB1 | 28 | 0.05 |
| *Pseudomonas* sp. | Aspha 1 | 31 | 6 |
| *Candida lipolytica* | GA | 33 | 90 |
| *Vibrio* sp. | Chry-B | 65 | 100 |
| *Corynebacterium* sp. | CD1 | 32 | 0.5 |

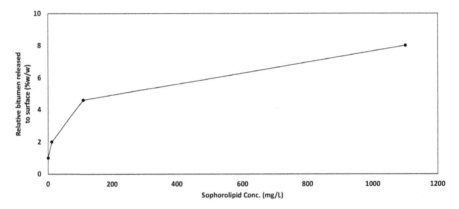

Relative bitumen released to surface = the amount of bitumen released to the surface / corresponding control value

**Figure 2.** Release of bitumen from tar sands by BS produced by *T. bombicola* (adapted from Cooper and Paddock 1984).

A recent study on the microbial pretreatment of bitumen froth with BS produced by *Bacillus subtilis* and *Pseudomonas aeruginosa* strains during the biodegradation of bitumen froth showed that through this technique higher amounts of bitumen can be recovered from oil sands, mainly because the produced BSs alter the solids wettability, degrade the asphaltene content of bitumen, and reduce the bitumen viscosity which leads to bitumen detachment from the sand grains (Ding et al. 2014a, b).

The second common technique for crude oil extraction is recovering crude oil from underground deposits. The crude oil is recovered by primary, secondary, and enhanced/tertiary recovery. In the primary and/or secondary recovery, the energy for bringing the oil-containing fluid to surface is provided by natural energy (e.g., existing gas and water in reservoir), while in the enhanced recovery, artificial reagents and chemicals (e.g., steam, surfactants) are used to bring the fluid to surface (Wong and Hung 2004). As small amounts of crude oil present in the underground oil sands deposits can be recovered through primary (10%) and secondary production (20–40%) processes, the tertiary/advanced techniques (known as Enhanced Oil Recovery (EOR)) have been developed to overcome the capillary, gravitational, and viscous forces in order to increase the amount of extractable crude oil from an oil field by thermal recovery and gas or chemical injections (Table 2) (Bognolo 1999, Energy.Gov 2016). For example, chemical surfactants are injected into deep wells to (i) lower the interfacial tension between the oil-rocks, (ii) overcome the capillary forces that prevent the detachment of oil from rocks, and (iii) increase the displacement of oil (Pornsunthorntawee et al. 2008).

Several biotreatment processes and methods have received increasing attention due to their environmental compatibility, low toxicity, and high effectivity. *In situ* and *ex situ* microbial enhanced oil recovery techniques (MEOR) using BEs/BSs have been practiced due to the stability of these molecules at extreme well/reservoir conditions such as high temperature and salinity (Bognolo 1999, Das and Mukherjee 2007a). In the *ex situ* techniques, BSs are first produced and then introduced/injected into an oil reservoir, while in the *in situ* techniques non-indigenous BS-producing microorganisms are directly injected into the oil reservoir to synthesize BSs in the reservoir (Shennan and Levi 1987) or essential nutrients for microbial growth are injected into the wells to stimulate the growth of indigenous BS-producing microorganisms. Figure 3 represents oil extraction with a MEOR technique.

Table 2. Common enhanced oil recovery methods (adapted from Energy.Gov 2016).

| Recovery Techniques | Injected Solutions | Aims |
| --- | --- | --- |
| Thermal recovery | Hot water/steam | To increase the oil mobility by reducing oil viscosity and interfacial tension |
| Gas injection | $CO_2$, natural gas, $N_2$ | |
| Chemical injection | - Alkaline/caustic solutions react with the existing organic acids in the reservoir and form surface-active compounds<br>- Polymers<br>- Chemical/microbial surfactants | |

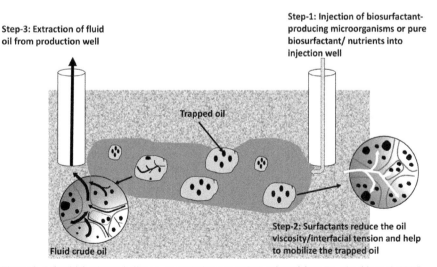

**Step-3: Extraction of fluid oil from production well**

**Step-1: Injection of biosurfactant-producing microorganisms or pure biosurfactant/ nutrients into injection well**

Trapped oil

**Step-2: Surfactants reduce the oil viscosity/interfacial tension and help to mobilize the trapped oil**

Fluid crude oil

**Figure 3.** Microbial enhanced oil recovery techniques (MEOR) (adapted from De Almeida et al. 2016).

According to Lazar et al. (2007), systematic laboratory investigations on MEOR were initiated by ZoBell and his researchers in 1974 and the first field trial was conducted in the Lisbon Oil Field (Union County, Arkansas, USA, in 1954) (Yarbrough and Coty 1983). A comparative study on the applicability of BSs produced by *B. subtilis* PT2 and *P. aeruginosa* SP4 (isolated from oily sludge and petroleum-contaminated soil), and synthetic surfactants including polyoxyethylene sorbitan monooleate (Tween 80), sodium dodecylbenzenesulfonate (SDBS), and sodium alkyl polypropylene oxide sulfate (Alfoterra 145-5PO) for enhanced oil recovery revealed that the BSs were more effective than those synthetic surfactants in the oil recovery (Pornsunthorntawee et al. 2008).

Similarly, the effects of two microbial strains, a BS-producing *B. subtilis* PTCC 1365 and an exopolymer-producing bacterium *Leuconostoc mesenteroides* PTCC 1059 used in enhanced oil recovery in a lab-scale *in situ* study, showed that the recovery of oil increased between 10–40% in the presence of *B. subtilis* PTCC 1365, while the oil recovery was not significantly enhanced with the *L. mesenteroides* PTCC 1059. Oil viscosity, porous media permeability and wettability measurements revealed that the lower efficiency of *L. mesenteroides* PTCC 1059 was due to the production of an exopolymer by PTCC 1059 that plugged the "matrix-fracture interfaces" while the produced BS by *B. subtilis* effectively reduced both the interfacial tension and oil viscosity (Soudmand-asli et al. 2007).

The effect of oil-degrading microorganisms (*B. licheniformis* BNP29), isolated from oil reservoirs at depths of 866 to 1520 m, for *in situ* enhanced oil recovery showed a 9.3% to 22.1% increase in oil recovery due to possible acid production and selective plugging with positive impact on rock porosity and wettability (Yakimov et al. 1997). Das and Mukherjee (2007a) conducted a lab-scale *ex situ* MEOR study with two lipopeptide BSs, produced by strains of *B. subtilis* (e.g., DM-03 and DM-04) grown on potato peels as a carbon source. They found that both BSs successfully enhanced oil recovery (Das and Mukherjee 2007a).

The applicability of rhamnolipids (RLs) (Rha-C10-C10 and Rha2-C10-C10) produced by *P. aeruginosa* MM1011 for MEOR applications was also confirmed by Amani et al. (2013). The produced RL showed strong stability at various salinities, pH, and temperature and lowered the surface and interfacial tensions of water to 26 mN/m and 2 mN/m, respectively, with a critical micelle concentration (CMC) of 120 mg/L at which 27% of the original oil from sand pack by water flooding technique was recovered. Similar oil recovery (27.27%) with SL biosurfactant was also reported by other investigators (Elshafie et al. 2015). C13-, C14- and C15-surfactin BSs produced by *B. subtilis* strains (#309, #311 and #573), isolated from crude oils and grown on different carbon and nitrogen sources, showed stronger potential for MEOR than the chemical surfactants such as Enordet and Petrostep due to effective reduction of interfacial tension (Pereira et al. 2013).

A recent field trial showed the important role of microbial activities of surfactant-producing bacteria in the enhanced oil recovery from a crude oil reservoir in China. Following nutrient injection to the reservoir, heavy crude oil recovery increased by 1872 tons (e.g., an increase in oil recovery of 0.5–4.3 t/d) due to possible emulsifying and surface activities of emulsan (a strong emulsifier) and RL (a strong surface-active BS) produced by the dominant bacteria in the reservoir (Chai et al. 2015).

In another field study, the feasibility of simultaneous bioaugmentation (injection of lipopeptide BS-producing *Bacillus* strains) and biostimulation (addition of a glucose-nutrient mixture) was investigated for recovery of entrapped oil from oil wells in a 60-day period. This study showed that the entrapped oil was successfully recovered (e.g., an additional recovery of 52.5 m$^3$ of oil) with this method via oil mobilization by the BSs produced by the injected microorganisms. The BS production in the wells was measured and nearly 20 and 28 mg/L of lipopeptide BS were produced. Moreover, it was found that acids, alcohols, and carbon dioxide were produced due to the microbial activities which further improved the oil recovery (Youssef et al. 2013). Study of microbial population dynamics by molecular techniques (16S rDNA) in a high pour-point oil reservoir field trial revealed that the indigenous microbial communities (*Petrobacter* and *Alishewanella*) in the production wells were replaced with the BS producing *P. aeruginosa* and *P. pseudoalcaligenes* as a result of biostimulation with nutrient addition during the MEOR process (Zhang et al. 2012).

## Treatment of Oil Waste and Refinery Sludge

As mentioned previously, two sources of wastes are generated during oil extraction and purification. The waste that is generated during bitumen extraction is known as extraction tailings (see Figure 1) and it contains water, sand, clay, salts, metals, residual bitumen, and hydrocarbon diluents (Gosselin et al. 2010) and is usually stored in tailings ponds (Gosselin et al. 2010) to facilitate the gravitational separation of solids from water. Figure 4 illustrates the gravitational separation of water and solids in a tailings pond.

Bitumen and mixed layers of fine particles in the tailing prevent the proper separation and settlement of small particles so that the tailings pond represent a colloidal system (Gosselin et al. 2010). Flocculants are used to separate particles which cannot be settled and colloids from tailing stream mainly for two purposes, first, to improve the quality of recycled water from the tailings pond which results in reduced freshwater consumption in the system and second, to reduce the volume of the tailings pond. Flocculation processes separate fine particles (under gravity) from tailing stream through particle destabilization (e.g., reducing particle charges, compressing the electrical double layers, to allow formation of doublets, triplets, etc., upon collision due to Brownian motion) (Javan Roshtkhari 2016).

Densification of tailing wastes by microbial community activities (*B. subtilis* and strains isolated from weathered oil) and RL biosurfactant (JBR 425) was the focus

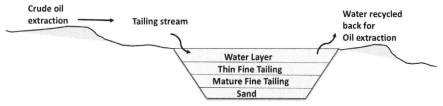

**Figure 4.** Schematic layers of tailing wastes in a tailings pond (adapted from Mamer 2007).

of research conducted by Mulligan and Roshtkhari (2016). Results of comparison of microbial sedimentation treatments of tailing particles with and without RL at $15 \pm 2°C$ showed a substantial increase (e.g., by a factor of 5.1) in particle sedimentation. In addition to increased sedimentation of fine particles, sedimentation of larger particles also increased by a factor of 2.63. Zeta potential measurements were performed to study the effect of RL on the particle charge. The measurements reveal the minor effect of RL on negatively charged particles following the treatment with RL. This implies that the main mechanism in the particle sedimentation was not the electrical charges reduction but was the RL impacts on the hydrophobicity of particles. That is, the particle sedimentation increased because of possible interactions of RL with extracellular polymeric substances produced by the microbial community (Figure 5) (Mulligan and Roshtkhari 2016).

Studies also revealed that in addition to densification, biological treatments play important roles in crude oil sludge treatment. According to Cameotra and Singh (2008), *P. aeruginosa* and *Rhodococcus* (isolated from oily sludge-contaminated soil) biodegraded 90% of crude oil sludge in a six-week field trial. The study showed that time required for biodegradation reduced considerably from six weeks to four weeks when such bacteria were separately amended with both essential nutrients and a laboratory produced RL. For example, maximum of 91–95% of crude oil sludge was degraded in two separate treatments including a treatment with bacteria and nutrients and another treatment with bacteria and RL. Further, 98% biodegradation was achieved in the treatment with the combination of nutrients and RL (Cameotra and Singh 2008).

Biodegradation of aged fine tailings (known as mature fine tailings (MFT) which form due to the long-term storage of tailing streams at tailings ponds) containing n-alkanes (C6, C7, C8, and C10) with indigenous microorganisms in a tailings pond was examined following a 29 week biodegradation period (Siddique et al. 2006). The study revealed that the microorganisms mineralized n-alkanes to methane when they were spiked with 0.2% or 0.5% w/v n-alkanes, respectively, and microorganisms biodegraded nearly 44% of C6 to 100% of C10.

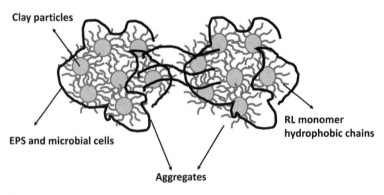

EPS: Extracellular Polymeric Substances
RL: Rhamnolipid Biosurfactant

**Figure 5.** Schematic presentation of aggregation of clay particles with the rhamnolipid-microbial cells (adapted from Mulligan and Roshtkhari 2016).

In another study, the application of BS-producing microorganisms (*P. fluorescens*, *P. putida*, *Acinetobacter lwoffii* and *Microbacterium oxydans*) for biodegradation of tailing waste was investigated. RL and synthetic surfactants (Tween 80, and Triton X-100) concentrations on the removal of total petroleum hydrocarbons (TPHs) from tailing streams were examined under various contact times. Results showed that the addition of chemical surfactants and RL to contaminated tailing considerably (i) reduced surface tension, (ii) increased the solubility and bioavailability of hydrocarbons and (iii) stimulated hydrocarbon biodegradation. Moreover, results of treatment with the RL revealed better effectiveness of RL on TPH removal than those chemical surfactants. For example, a 200 mg/L of RL, Triton X-100, and Tween 80 removed 62.1%, 41.9%, and 38.9% of TPHs, respectively. This study indicated the excellent performance of RL in oil mobilization from the contaminated soil and bioremediation of oil-polluted tailings (Huang et al. 2017).

Oil recovery from oily sludge with surfactin, lichenysin, RL, and emulsan BSs were the focus of the study conducted by Zheng et al. (2012). Batch washing tests were conducted with BSs at a broad range of pH, salinity, temperature, and n-butanol solvent concentrations and showed that 76.81% of the oil was recovered under the optimal conditions (e.g., 2 g/L of RL, pH 12, 10 g/L of NaCl, and 5.0 g/L of n-butanol).

In another laboratory study, the bioremediation of refinery oily sludge containing n-alkenes (C10-C16, C16-C34 and C34-C50) and total petroleum hydrocarbons (TPHs) by a *Luteibacter* sp. strain isolated from a sludge land farming site and a RL biosurfactant (JBR 425) were examined. Results showed that nearly 51% of TPHs in the oily sludge were successfully degraded by the isolated strain after 40 days of bioremediation (Ju et al. 2011). Moreover, a laboratory and field pilot-scale study showed a 91.5% recovery of oil from refinery oily sludge deposited at a separating tank at the Dalian Petrochemical Branch Company, PetroChina, in Liaoning Province, China by a RL biosurfactant, synthesized by *P. aeruginosa* F-2 (Yan et al. 2012).

Drill cuttings are another major source of oily waste that is generated during oil extraction. Drill cuttings (also known as muds) are composed of (i) drilling fluids, used as lubricants and pumped down to the well, (ii) coolants used during oil drilling and (iii) ground rock (Leonard and Stegemann 2010, Ball et al. 2012). Drill cutting wastes (Table 3) need proper management because high volumes of drill cutting waste are generated and the waste contains organic and inorganic contaminants, heavy metals, and polychlorinated biphenyls (Leonard and Stegemann 2010).

Conventional drill cutting treatments include non-biological methods such as burial pits, landfills, reinjection, chemical stabilization, solidification, and thermal treatments (e.g., incineration and thermal desorption) (Leonard and Stegemann 2010, Ball et al. 2012). However, recent efforts have focused on the application and development of biological technologies and processes due to the successful applications of alternative biological methods for drill cuttings treatment and the known environmental risks associated with the traditional methods. The *ex situ* bioremediation technologies and processes include (i) composting and biopile remediation, (ii) land application including land farming and land spreading, (iii) bioreactors, and (iv) vermiculture, which is the addition of warms to drilling waste to enhance the biodegradation of waste (Ball et al. 2012). Although the application

**Table 3.** Specifications of a typical drilling cutting (adapted from Ball et al. 2012, Patin 1999).

| Major Components | Examples |
|---|---|
| Base | Fresh water, salt water, paraffin, diesel, polyemulsion, air, foam, synthetic base |
| Oil | Gasoline, diesel, lubricant oil, crude oil, synthetic oil |
| Clay mixture | Bentonite, kaolinites, organophilic clays |
| Organic polymers | Cellulose, starch, gum, tannins |
| Weighting agents | $BaSO_4$, calcite, carbonates |
| Heavy metals | Chromium, nickel, lead, cadmium, zinc |
| Biocides | Carbamate, sodium, sulfide, aldehyde, chlorinated phenols |
| Chlorides | Potassium chloride, magnesium chloride, sodium chloride, calcium chloride |
| Other additives | Deformers, corrosion and scale inhibitors |

of BSs for enhancing bioremediation of drill cutting waste is not well studied, a recent study performed by Yan et al. (2011) showed that a prewashing step of oil-based drill cuttings by RL biosurfactant before biodegradation can effectively remove solids and increase the biodegradation of prewashed cuttings by bacterial culture containing *Pseudomonas*, *Acinetobacter*, *Alcaligenes*, *Agrobacterium*, and *Comamonas*. A prewashing step by RL led to the removal of 83% of organics under optimal conditions (e.g., liquid/solid ratio: 3:1; washing time: 20 min; stirring speed: 200 rpm; RL concentration: 360 mg/L; and temperature: 60°C). Moreover, the total petroleum hydrocarbon concentrations were reduced from 85,000 to 12,600 mg/kg and the saturated and aromatic content of drill cuttings considerably decreased during the 120-day biodegradation period.

## Clean-up of Oil Tanker/Storage Tank/Pipeline Wastes

Wastewater produced during the storage of crude oil and petroleum products contain free and emulsified oil and suspended solids (Wong and Hung 2004). Heavy fractions of crude oil settle at storage tanks and due to their high viscosity, they cannot be easily removed. Mechanical or chemical (e.g., solvent washing) cleaning methods are applied to remove the deposits. For example, reagents such as emulsifiers, surfactants, and BSs have been utilized to form less viscous oil-in-water emulsions to help mobilize the deposited fractions and to increase the extraction of recoverable compounds from the sludge through breaking or destabilizing the oil-in-water emulsion formed by the reagents (Banat et al. 1990, Bognolo 1999).

In a microbiological cleaning method, oil-degrading microorganisms were used and optimal conditions for production of extracellular emulsifying agents (named Emulsan) by the microorganisms were provided to examine its potential application for cleaning of the oily sludge deposited at the bottom of an oil tanker under controlled conditions. Following four days of biodegradation, it was observed that the oily sludge was completely removed by the emulsifying action of Emulsan (Bourquin et al. 1974, Rosenberg et al. 1975, Gutnick and Rosenberg 1977).

In another study, a BS (named Pet 1006) was used to cleanup deposited oily sludge from a crude oil storage tank. In this field trial, nearly 1.5 tons of the Pet

1006 biosurfactant were added to an 850 m³ oily sludge containing 750 m³ and 2000 m³ of crude oil and saline water, respectively, and the mixture was mechanically mixed for five days. The formation of a stable emulsion was observed following the treatment period which was subsequently destabilized with the addition of an emulsion breaker. This led to the recovery of nearly 90% of a high-quality crude oil from the destabilized emulsion (Banat et al. 1990).

Viable transportation of extracted crude oil through pipelines could be challenging because high contents of paraffin and asphaltenes in the heavy oils not only reduce the fluidity of crude oil but also precipitate on the inner walls of pipelines (Perfumo et al. 2010, Cerón-Camacho et al. 2013). Conventionally, heat or solvents (e.g., xylene, toluene) are applied to lower the viscosity of the oil and dissolve precipitated paraffin and asphaltenes (Perfumo et al. 2010). However, a recent study showed the successful utilization of high molecular weight emulsifying BSs with emulsifying property for pipeline cleaning. These BSs have several reactive groups that enable them to attach to the oil constituents and reduced oil droplets coalescence (Perfumo et al. 2010). Moreover, the emulsifying effect of mixtures of two BSs (long alkyl chain glucosides and tetradecanoyl-O-cellobioside) on the formation of stable oil-in-water emulsions for the easier transport of heavy crude oil by pipelines showed the positive effect of mixtures of both BSs in the formation of stable emulsions (Cerón-Camacho et al. 2013).

Although, field trials on the application of BEs for crude oil transportation are limited, in a field study, an emulsan BE showed promising results for oil transportation. The application of a 70% w/w of oil-in-water emulsion with a viscosity of 70 mPa.s and an emulsifier to oil ratio of 1:500 showed significant viscosity reduction of a viscous heavy crude oil (e.g., Boscan, viscosity of 200,000 mPa.s) when applied and pumped 610 km over 64 hours (Hayes et al. 1987).

Another BS that showed a viscosity reduction property was RL produced by *P. aeruginosa* strain USB-CS1. The emulsion produced by rhamnolipid-crude oil not only was stable for 14 days but also reduced the viscosity of the tested crude oil to less than 500 cP (Rocha et al. 2000). Likewise, the Lazar et al. (1999) study showed that *Pseudomonas* and *Bacillus* species and a mixed consortium were able to effectively reduce the oil viscosity at 7°C–9°C and degrade paraffin up to 90%. Table 4 summarizes the proposed mechanisms that led to paraffin deposition and biodegradation by the isolated bacteria.

**Table 4.** Proposed mechanisms for preventive effects of bacteria on pipeline paraffin deposition (adapted from Lazar et al. 1999).

| Proposed Action | Mechanisms of Actions |
| --- | --- |
| Uptake of paraffin by non-mobile bacteria | Metabolism of the existing paraffin deposits |
| Partial digestion of paraffin | Breaking the chemical bonds between carbon atoms in the paraffin chain by bacteria to the point that the solid paraffin becomes a liquid oil |
| Solvent or BS production by bacteria | Further aids in paraffin breakdown |
| Uptake of paraffin by mobile/swimming bacteria | Bacteria moved toward paraffin deposits |

## Role of BSs in Treatment of Petroleum Wastewater

Wastewater (known as process water) is generated as a result of consumption of water by various water-based oil activities (e.g., refineries) that have been in direct contact with hydrocarbons. The common types of process water include desalter effluent, sour water, tank bottom draws (e.g., crude tanks, gasoline tanks, and slop tanks), and spent caustic. Treatment of wastewater poses a major challenge in petroleum refinery because the effluent must be effectively treated before release into the environment due to the presence of several toxic compounds in the wastewater (IPIECA 2010). For example, polycyclic aromatic hydrocarbons (PAHs) with a general formula of $C_{4n} + 2H_{2n+4}$ ($n$ represents the number of rings) are produced during petroleum refining (Park et al. 1990) and are some of the challenging compounds in the wastewater to be completely removed due to resistance to biodegradation. This is because as the number of rings increases, the volatility and solubility of these compounds decreases but their sorption increases (Mulligan 2005). However, it has been known that some microorganisms uptake hydrocarbons as carbon and energy sources and thus several batch and continuous biologically-based technologies have been developed to treat oily effluents (Tong et al. 2013). Successful treatment of oily wastewater by activated sludge reactors (Tellez et al. 2002), activated sludge reactors coupled with immobilized biological aerated filters (Tong et al. 2013), continuously stirred tank bioreactors (Gargouri et al. 2011), rotating biological contactors (RBC) (Tyagi et al. 1993), continuous upflow anoxic sludge-blanket/fixed-film hybrid bioreactors (UAnSFB) (Moussavi and Ghorbanian 2015), upflow anoxic fixed-bed bioreactors (UAnFB), sequencing anoxic batch reactors (SAnBR) (Ghorbanian et al. 2014), membrane bioreactors (Mannina et al. 2016) and hybrid membrane-aerated biofilm reactors (Li et al. 2015) have been reported.

Moreover, the positive effect of BSs for effective removal of oily wastewater has been reported by several investigators. For example, Sponza and Gok (2011) studied the effect of RL, emulsan, and surfactin BSs on the removal of PAHs with low and high numbers of benzene rings including acenaphthene (ACT), fluorene (FLN) and phenanthrene (PHE), benzo[b]fluoranthene (BbF), benzo[k]fluoranthene (BkF), benzo[a]pyrene (BaP), indeno[1,2,3-cd]pyrene, dibenz[a,h]anthracene (DahA), benzo[g,h,i]perylene (BghiP) with a laboratory-scale aerobic activated sludge reactor (AASR) system. This study revealed that the positive effect of all BSs on PAH removal was by stimulating the biomass growth. However, PAHs and chemical oxygen demand (COD) removals were higher with the RL than those of other BSs. For example, 95% of total PAHs and dissolved COD, 75% of COD associated with the inert organics (CODinert), and 96% of COD associated with the inert soluble microbial products (CODimp) were removed with the RL at 15 mg/L concentration by aerobic treatment for 25 days. Further assessment showed that nearly 88%, 4%, 3%, and 5% of PAHs, respectively, were (i) biodegraded, (ii) accumulated in the system, (iii) released in the effluent, and (iv) remained in the waste sludge with 15 mg/L of RL using a 25 day settling retention time (SRT) (Sponza and Gok 2011).

The capability of bacterial communities for the treatment of PAHs at 60°C was studied by Congiu and Ortega-Calvo (2014). The study showed that bacteria effectively (between 80–90%) degraded low molecular weight compounds including naphthalene, phenanthrene, fluorene and anthracene and high molecular weight compounds including pyrene, benzo(e)pyrene and benzo(k)fluoranthene, respectively, within 10 days. Surface tension measurements and metabolite characterization suggested the important role of BSs produced by the thermophilic polycyclic aromatic hydrocarbon degraders such as *P. aeruginosa* strain CEES1 (KU664514) and *B. thermosaudia* (KU664515) strain CEES2. Moreover, the assessment of bacteria for the treatment of petroleum wastewater in a continuous stirred tank reactor revealed that the bacteria successfully removed nearly $96 \pm 2\%$ of COD and PAHs in 24 days (Congiu and Ortega-Calvo 2014).

In another study, RL application in a bench scale sequential batch reactor (SBR) was examined to determine its effectiveness on the treatment of oil refinery wastewater effluent. Results revealed that a minimum of 50 mg RL/L was required to (i) reduce sludge disposal up to 52%, (ii) remove 81–97% of COD and (iii) provide optimal sludge settling properties (e.g., Sludge Volume Index (SVI) 120 mL/g). RL application resulted in sludge reduction and a subsequent smaller (e.g., 39% to 52%) secondary clarifier (Alexandre et al. 2016).

Likewise, a study was conducted to determine the accelerated biodegradation of TPHs by peroxidase and BS-producing bacteria in a sequencing continuous-inflow reactor (SCR) under various operating conditions. Results showed that the bacterial degradation of TPHs was maximal when the ratio of $H_2O_2$/TPH (4 kg TPH/m$^3$.d) mass was 0.35 and nearly 96.7% of COD was removed in this condition. The results revealed that a total of 38 mg/L of RL and 93 mg/L of surfactin BSs were synthesized during the SCR operation at TPH loading rate of 6 kg/m$^3$.d. The study suggested that accelerated biodegradation and COD removal of petroleum hydrocarbons can be successfully achieved with an enhanced enzymatic biodegradation technique (Moussavi et al. 2017).

## Environmental Contamination by Petroleum Hydrocarbons

Crude oil, petroleum products, and biofuels contain various levels of organics (e.g., long-chain alkyl esters), volatile aromatic compounds (e.g., toluene, styrene), aromatics (e.g., benzene, xylene, polycyclic aromatic compounds) and heavy metals (e.g., zinc, lead) (CSIC 2003, Prego and Cobelo-García 2004, Pérez-Cadahía et al. 2007, Kirkeleit et al. 2008, Smith 2010, Solomon and Janssen 2010). Table 5 summarizes fresh biodiesel, diesel, and light crude oil properties. It is estimated that nearly 50% of produced oil ends up in the environment (Bognolo 1999) and directly or indirectly impacts human health (e.g., through drinking oil-contaminated water, skin contact, and consumption of contaminated seafood) and aquatic ecosystems (e.g., primarily by forming an oil layer that blocks the oxygen from entering the water (Kontogiannis and Barnett 1973, Kanicky et al. 2002, Pérez-Cadahía et al. 2007, Solomon and Janssen 2010, Gross et al. 2013)). In the following section, the application of BSs for cleanup and remediation of oil-contaminated aquatic and terrestrial environments will be reviewed.

**Table 5.** Characteristics of fresh light crude oil, diesel fuel, and biodiesel (adapted from US National Renewable Energy Laboratory 2009, Fingas 2011).

| Parameter | Biodiesel (B100)[4] | | Diesel Fuel[5] | Light Crude Oil[5] |
|---|---|---|---|---|
| Physical state | Liquid | Liquid | Liquid | Liquid |
| Formula | $C_{18}H_{34}O_2$ | $C_{19}H_{36}O_2$ | – | – |
| Molecular weight | 282.5 | 296.5 | – | – |
| Cetane number | – | 47.2–55 | – | – |
| Melting point (°C) | 16 | –20 | – | – |
| Density (kg/m³) | 860–900 | | 840 at 15°C | 780 to 880 at 15°C |
| Kinematic viscosity | 1.9 to 6 at 40°C (mm²/s) | | 2 mPa.s at 15°C 1.3–4.1 at 40°C | 5 to 50 mPa.s at 15°C |
| Saturates | – | | 65 to 95 | 55 to 90 |
| Alkanes | – | | 35 to 45 | – |
| Cyclo-alkanes | – | | 30 to 50 | – |
| Waxes | – | | 0 to 1 | 0 to 20 |
| Olefins | – | | 0 to 10 | – |
| Aromatics | – | | 5 to 25 | 10 to 35 |
| [1]BTEX | – | | 0.5 to 2 | 0.1 to 2.5 |
| [2]PAHs | – | | 0 to 5 | 10 to 35 |
| Polar compounds | – | | 0 to 2 | 1 to 15 |
| Resins | – | | 0 to 2 | 0 to 10 |
| Asphaltenes | – | | – | 0 to 10 |
| Solubility in water (ppm) | – | | 40 | 10–50 |
| [3]IFT (mN/m) at 15°C | – | | 27 | 10 to 30 |

[1] Benzene, toluene, ethylbenzene, and xylenes
[2] Polycyclic aromatic hydrocarbons
[3] Interfacial tension
[4] (US National Renewable Energy Laboratory 2009)
[5] (Fingas 2011)

# Role of BSs in Oil Degradation and Remediation of Oceans and Seas

Oil pollutants enter the aquatic environments (McKew et al. 2007) as a result of activities such as oil extraction, transportation, and accidental spills. The spilled oil is subjected to several natural transformation processes (known as weathering) by underwater mixing, and wave and wind actions (Hollebone 2011). As a result of these processes, the lighter fractions of oil evaporate to the atmosphere, while the heavier fractions may disperse, sink, and biodegrade by oil-degrading microorganisms (Clayton et al. 1993, Hollebone 2011).

The effectiveness of natural oil biodegradation is generally influenced by (i) presence/absence of indigenous oil-degrading microorganisms (e.g., unpolluted environments contain only 0.1–1% and oil-polluted environments contain 1–10%

of oil-degrading microorganisms), (ii) the solubility properties of hydrocarbons in the aqueous environments and (iii) environmental factors such as presence/lack of dissolved oxygen and nutrients, pH, temperature, and water salinity, and (iv) toxicity of oil compounds to microorganisms (Zhang and Miller 1992, Whang et al. 2009, Ward 2010, Okafor 2011).

In general, oil-degrading microorganisms can uptake the oil droplets through direct attachment/interaction to oil droplets by modifying their cell surface (Bouchez-Naïtali et al. 1999, Ward 2010, Saborimanesh and Mulligan 2015). Moreover, they interact with the solubilized parts of oil in the aqueous phase or the oil droplets that encapsulated in the micellar aggregates (Ward 2010, Saborimanesh and Mulligan 2015). As most hydrocarbons are not readily water-soluble, their uptake by the microbial cells can be challenging (Zhang and Miller 1992, Ward 2010). BS production is one way that helps some microorganisms such as *Pseudomonas* species overcome such challenges (Bouchez-Naïtali et al. 1999, Ward 2010). For example, a study of the mechanisms involved in long-chain alkanes uptake by microorganisms (e.g., *Corynebacterium, Mycobacterium, Nocardia*) isolated from oil-polluted and unpolluted soil revealed that direct attachment to oil and micellar transfer of oil to microbial cells through emulsification or solubilization processes by extracellular BSs produced by the microorganisms were the main ways of oil uptake. The cellular hydrophobicity tests suggested two main alkane transfer mechanisms. Micellar transfer was adapted by 11% of the isolated strains while 42% of the isolated strains adapted BS enhanced interfacial uptake (Bouchez-Naïtali et al. 1999). Figure 6 shows the uptake of hydrocarbons by a BS-enhanced interfacial mechanism.

Following the hydrocarbon uptake, their degradation is initiated with oxygenase enzyme by introducing an oxygen atom (derived from molecular oxygen) into the alkane substrate (Kohno et al. 2002, Heiss-Blanquet et al. 2005, Kloos et al. 2006, Mehdi and Giti 2008). The role of plasmids in oil degradation was investigated by several researchers (Singer and Finnerty 1990, David et al. 1995, Thavasi et al. 2006). It was found that plasmids play significant roles in oil degradation by *P. putida*,

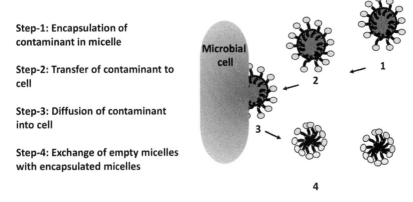

**Step-1: Encapsulation of contaminant in micelle**

**Step-2: Transfer of contaminant to cell**

**Step-3: Diffusion of contaminant into cell**

**Step-4: Exchange of empty micelles with encapsulated micelles**

**Empty micelles leave the cell**

**Figure 6.** Bioavailability enhancement by contaminant micellization (adapted from Christofi and Ivshina 2016).

*P. diminuta, B. coagulans, Brevibacterium linesepidermidis* (David et al. 1995) and *Rhodococcus* sp. (Singer and Finnerty 1990) because plasmids encode the oxygenase enzymes with the ability to initiate the oil degradation in the bacteria. For example, the influence of two plasmids with molecular weights of 4788 and 2400 base pairs in oil degradation was confirmed by Thavasi et al. (2006). Chromosomal genes were also found to encode enzymes other than oxygenase (Rosenberg et al. 1996). Table 6 summarizes the importance of some known enzymes in oil degradation.

As mentioned previously, natural degradation of oil can be a slow process. Therefore, chemical dispersants (with surfactants or surface-active compounds as their main reactive agents) are commonly applied to stimulate the natural biodegradation rate. Dispersants are amphipathic molecules with hydrophilic (water-liking) and hydrophobic (water-repellent) moieties in their structures and can simultaneously interact with water molecules and oils/hydrocarbons and lower the surface and interfacial tensions to the point that break the oil slicks (with the action of mixing) and form small oil droplets which then can be dispersed into the water column for longer periods and facilitate the availability of hydrocarbons to oil-degrading microorganisms. Figure 7 presents oil solubilization/emulsification through micellar encapsulation by dispersants.

**Table 6.** Some of the important enzymes involved in oil biodegradation (adapted from Kohno et al. 2002, Heiss-Blanquet et al. 2005, Kloos et al. 2006, Mehdi and Giti 2008).

| Alkane Hydroxylase Group | Encoding Gene | Impacts |
|---|---|---|
| I | Alk-B | Catalyzes medium chain length (C6-C12) n-alkanes |
| II | Alk-M | Catalyze long chain alkanes > C12 |
| III | Alk-B | Unknown substrate specificity, alkane oxidation pathway and oxidation system |

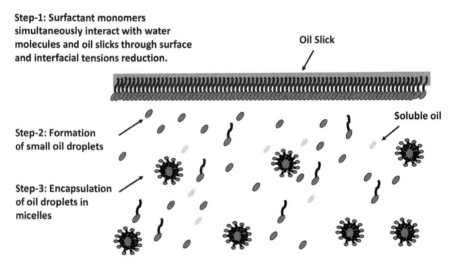

**Step-1: Surfactant monomers simultaneously interact with water molecules and oil slicks through surface and interfacial tensions reduction.**

Oil Slick

Soluble oil

**Step-2: Formation of small oil droplets**

**Step-3: Encapsulation of oil droplets in micelles**

**Figure 7.** Schematic of oil solubilization/emulsification through micellar encapsulation by dispersants (adapted from Saborimanesh 2016).

Studies showed that substrate mass transfer and contact between microorganisms and hydrocarbons increased by the addition of BSs including surfactin, RLs, and SLs (Zhang and Miller 1992, Whang et al. 2009, Chen et al. 2013, Saborimanesh and Mulligan 2015) as a result of (i) modifications of the microbial cell surface properties (Zhang and Miller 1994, Kaczorek 2012, Kaczorek et al. 2008, 2012, Saborimanesh and Mulligan 2015), (ii) encapsulation of oil droplets in micelles and (iii) increases in surface area (Rosenberg 1993, Ron and Rosenberg 2002, Franzetti et al. 2010). For example, potential application of SLs for enhanced dispersion of weathered biodiesel, diesel, and light crude oil-contaminated water under salinities (0, 10, 20, 30 ppt), temperatures (8°C, 22°C, 35°C) and pHs (6–8) using bench-scale experiments showed that SLs reduced surface tension of water with various salinities to a minimum of 34 mN/m at the critical micelle concentration (CMC) of 38 mg/L and increased the weathered oil dispersion as the SL concentrations increased. The dispersion of weathered biodiesel, diesel, and light crude oil in seawater with a salinity of 30 ppt reached 27%, 16%, and 12%, respectively, by 80 mg/L of SL. Further assessments showed that the oil dispersion by the SL was due to the decreases in the surface and interfacial tensions below the CMC and encapsulation of oil droplets in micelles above the CMC (Saborimanesh 2016). Figure 8 illustrates the schematics of oil dispersion in seawater by SLs below and above the critical micelle concentration (CMC of 38 mg/L).

Moreover, biodegradation of weathered biodiesel, diesel, and light crude oil at 22°C with 80 mg/L SL and without SL revealed the important role of four major phyla in the oil degradation. They included *Firmicutes*, the dominant phylum in biodiesel (100%) and diesel (53%), *Actinobacteria*, the second dominant bacteria in the diesel (47%) and *Proteobacteria* (97%) and *Actinobacteria* (3%), the dominant phyla in the light crude oil which contributed in the degradation of 43 ± 0.7%, 45 ± 5.7% and 39 ± 4.6% of weathered biodiesel, diesel, and light crude oil, respectively, without SLs and 44 ± 5%, 47.5 ± 3.9% and 44 ± 1% of weathered biodiesel, diesel, and light crude oil, respectively, with SL during 28 days of biodegradation period (Saborimanesh and Mulligan 2015). Table 7 presents the dominant indigenous hydrocarbon degraders in the biodiesel, diesel, and light crude oil.

Mechanisms of diesel, biodiesel, and light crude oil degradation with and without SLs determined by the cell surface hydrophobicity test revealed that the bacteria showed a hydrophobic effect when exposed to the SLs. However, in the presence of both oil and SLs, the cell surface hydrophobicities were slightly low (Figure 9). This implies that the hydrocarbon removal in the treatments with SLs during the biodegradation period was due in part to increase in the micellar solubilization by the SLs. However, the mechanisms such as direct contact with the oil and uptake of the hydrophilic (water-like) parts of diesel, biodiesel, and light crude oil by the indigenous bacteria positively influenced the oil biodegradation (Saborimanesh and Mulligan 2015).

Similarly, the effect of RLs on the dispersion and biodegradation of crude oil in seawater was determined under laboratory conditions. The results showed that at 20°C and 35 ppt salinity, addition of RLs increased biodegradation rates of fresh BRENT crude oil from 10% to 82% (Dagnew 2004). Moreover, according to Vasefy (2007), treatments of weathered light and heavy crude oils and diesel fuel with RLs

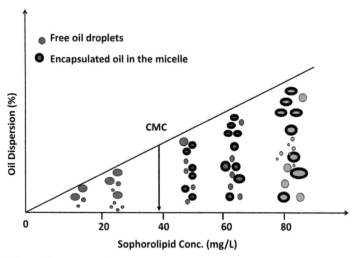

**Figure 8.** Effect of SL concentration, below and above the CMC, on free oil droplet formation and micellar encapsulation (adapted from Saborimanesh 2016).

**Table 7.** Indigenous oil-degrading bacteria present in biodiesel, diesel, and light crude oil (adapted from Saborimanesh and Mulligan 2015).

| Oil Type | Classification | | |
|---|---|---|---|
| | Class | Order | Genus |
| Biodiesel | *Bacilli* | *Bacillales* | *Bacillus* |
| Diesel | *Actinobacteria* | *Actinomycetales* | *Dietzia* |
| | *Bacilli* | *Bacillales* | *Paenibacillus* |
| Light crude oil | *Alphaproteobacteria* | *Sphingomonadales* | *Sphingomonas* |
| | *Actinobacteria* | *Actinomycetales* | *Mycobacterium* |

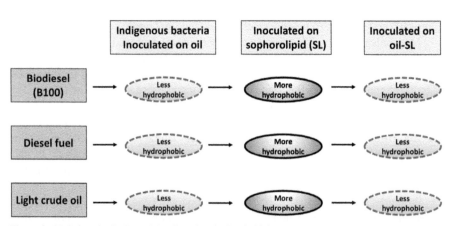

**Figure 9.** Variations in the bacterial cell surface hydrophobicity as grown on the biodiesel, diesel, light crude oil, SLs, and mixtures of oil-SLs during the biodegradation period. Colors represent the state of cell surface hydrophobicity modifications (gray square dots: less hydrophobic; solid black line: more hydrophobic) (adapted from Saborimanesh 2016).

and two commercial biological products (containing bacterial communities and nutrients) at 20°C and salinity of 35 ppt considerably stimulated the biodegradation of tested oils. Results of this study showed that nearly 81% of fresh diesel fuel, 76% of light crude oil, and 64% of heavy crude oil were effectively biodegraded after 28 days and a direct relationship was observed between oil removal and microbial growth (Vasefy 2007).

Application of two BSs, including surfactin, produced by *B. subtilis* ATCC 21332, and RLs, produced by *P. aeruginosa* J4, were studied for biodegradation of diesel-contaminated water (Whang et al. 2009). Results showed that both the indigenous microorganism growth and diesel biodegradation increased in water due to the increase in the solubility of diesel in the presence of BSs.

A study conducted by Thavasi et al. (2006) showed that a glycolipid biosurfactant produced by *Azotobacter chroococcum* (a marine nitrogen-fixing bacterium) had a significant role in degradation (58%) and emulsification of various hydrocarbons including crude oil, diesel, waste motor oil, kerosene, naphthalene, anthracene, and xylene following a 120 h biodegradation period. Two important points were highlighted in this study. First, the biosurfactant was synthesized at the early stationary phase possibly as a secondary metabolite (Thavasi et al. 2006, Rahman et al. 2002b), and secondly, a direct correlation between oil degradation and bacterial cell growth was observed (Thavasi et al. 2006). Additionally, it was proposed that the oil emulsification by the produced BS and modifications in the cell surface hydrophobicity (e.g., CSH of 21.4%) were the main oil degradation mechanisms. In more recent studies, the effect of biosurfactants produced by *Lactobacillus delbrueckii*, *Bacillus megaterium*, *Corynebacterium kutscheri* and *Pseudomonas aeruginosa* and fertilizers on biodegradation of crude oil was examined. The results showed that despite the slightly higher biodegradation of crude oil by the application of biosurfactants and fertilizers, the application of biosurfactant alone significantly increased the biodegradation of crude oil without added fertilizers. This implied that not only the cost of bioremediation process can be minimized if only biosurfactants are used for the crude oil biodegradation but also it could be beneficial if there are concerns regarding the dilution or wash away of water-soluble fertilizers used during bioremediation of aquatic environments (Thavasi et al. 2011a, b).

Saeki et al. (2009) conducted a laboratory research to determine the effectiveness of a JE1058BS BS produced by *Gordonia* sp. strain JE-1058, for (i) dispersion and biodegradation of weathered crude oil spill by indigenous oil-degrading microorganisms at sea and (ii) cleanup of crude oil-contaminated sea sand. Results showed that the JE1058BS dispersed the weathered oil due to reduction of surface and interfacial tensions. Moreover, a maximum of 80% sea sand cleanup was achieved when a 40 ppm of the BS was used. The toxicity results also revealed that in comparison to No. 2 fuel oil (96h-$LC_{50}$ of 5.56 ppm), the JE1058BS had low toxicity on *Menidia beryllina* (a fish living in estuaries and freshwater environments known as "inland silverside") (e.g., 96h-$LC_{50}$ of 91.70 ppm) and did not negatively influence the toxicity of the dispersed oil (e.g., 96h-$LC_{50}$ of 8.68 ppm). Table 8 summarizes the dispersion effectiveness of some chemical and biological surfactants.

**Table 8.** Comparison of chemical and BSs used for dispersion of oil spills (adapted from Moles et al. 2002, Shin et al. 2004, Li et al. 2010).

| Type of Oil | Weathering State | Dispersant | Salinity (ppt) | Temp (°C) | pH | Mixing | Dispersion Effectiveness | Reference |
|---|---|---|---|---|---|---|---|---|
| [1]ANS | Fresh | Corexit 9500 | 22 | 3 | – | 150 rpm | < 10% | Moles et al. (2002) |
| ANS | Fresh | Corexit 9500 | 22 | 10 | – | 150 rpm | < 10% | |
| ANS | Fresh | Corexit 9500 | 22 | 22 | – | 150 rpm | 15.8% | |
| ANS | Fresh | Corexit 9500 | 32 | 3 | – | 150 rpm | < 10% | |
| ANS | Fresh | Corexit 9500 | 32 | 10 | – | 150 rpm | 22.3% | |
| ANS | Fresh | Corexit 9500 | 32 | 22 | – | 150 rpm | 18.4% | |
| ANS | Fresh | Corexit 9527 | 22 | 3 | – | 150 rpm | < 10% | |
| ANS | Fresh | Corexit 9527 | 22 | 10 | – | 150 rpm | < 10% | |
| ANS | Fresh | Corexit 9527 | 22 | 22 | – | 150 rpm | 35.2% | |
| ANS | Fresh | Corexit 9527 | 32 | 3 | – | 150 rpm | < 10% | |
| ANS | Fresh | Corexit 9527 | 32 | 10 | – | 150 rpm | 15.3% | |
| ANS | Fresh | Corexit 9527 | 32 | 22 | – | 150 rpm | 30.5% | |
| ANS | 20% | Corexit 9500 | 22 | 3 | – | 150 rpm | < 10% | |
| ANS | 20% | Corexit 9500 | 22 | 10 | – | 150 rpm | < 10% | |
| ANS | 20% | Corexit 9500 | 22 | 22 | – | 150 rpm | < 10% | |
| ANS | 20% | Corexit 9500 | 32 | 3 | – | 150 rpm | < 10% | |
| ANS | 20% | Corexit 9500 | 32 | 10 | – | 150 rpm | < 10% | |
| ANS | 20% | Corexit 9500 | 32 | 22 | – | 150 rpm | < 10% | |
| ANS | 20% | Corexit 9527 | 22 | 3 | – | 150 rpm | < 10% | |
| ANS | 20% | Corexit 9527 | 22 | 10 | – | 150 rpm | < 10% | |

*Table 8 contd. ....*

| | | | | | | | | |
|---|---|---|---|---|---|---|---|---|
| ANS | 20% | Corexit 9527 | 22 | 22 | – | 150 rpm | <10% | |
| ANS | 20% | Corexit 9527 | 32 | 3 | – | 150 rpm | <10% | |
| ANS | 20% | Corexit 9527 | 32 | 10 | – | 150 rpm | <10% | |
| ANS | 20% | Corexit 9527 | 32 | 22 | – | 150 rpm | <10% | |
| [2]IFO180 | Fresh | Corexit 9500 | [3]SW | 16 | – | Breaking waves (intensive mixing) | 90% | Li et al. (2010) |
| IFO180 | Fresh | SPC 1000 | SW | 16 | – | Breaking waves (intensive mixing) | 50% | |
| IFO180 | Fresh | Corexit 9500 | SW | 10 | – | Breaking waves (intensive mixing) | 3% | |
| IFO180 | Fresh | SPC 1000 | SW | 10 | – | Breaking waves (intensive mixing) | 6% | |
| IFO180 | Fresh | Corexit 9500 | SW | 10–17 | – | Regular wave condition | <15% | |
| IFO180 | Fresh | SPC 1000 | SW | 10–17 | – | Regular wave condition | <15% | |
| Phenanthrene (PN) | Fresh | [5]RLs | [4]DI | 25 | 4 | Orbital shaker | 4 mg PN/240 mg/l RLs | Shin et al. (2004) |
| Phenanthrene | Fresh | RLs | DI | 25 | 4.5 | Orbital shaker | 7 mg PN/240 mg/l RLs | |

*Table 8 contd. ...*

*...Table 8 contd.*

| Type of Oil | Weathering State | Dispersant | Salinity (ppt) | Temp (°C) | pH | Mixing | Dispersion Effectiveness | Reference |
|---|---|---|---|---|---|---|---|---|
| Phenanthrene | Fresh | RLs | DI | 25 | 5 | Orbital shaker | 6 mg PN/240 mg/l RLs | |
| Phenanthrene | Fresh | RLs | DI | 25 | 5.5 | Orbital shaker | 5 mg PN/240 mg/l RLs | |
| Phenanthrene | Fresh | RLs | DI | 25 | 6 | Orbital shaker | 3 mg PN/240 mg/l RLs | |
| Phenanthrene | Fresh | RLs | DI | 25 | 7 | Orbital shaker | 1 mg PN/240 mg/l RLs | |
| Phenanthrene | Fresh | RLs | DI | 25 | 7.5 | Orbital shaker | 1 mg PN/240 mg/l RLs | |

[1] Alaska North Slope crude oil
[2] Heavy fuel oil
[3] SW: seawater
[4] DI: deionized water
[5] RLs: Rhamnolipids

## Role of BSs in Oil Degradation and Remediation of Soil

Soil contamination occurs naturally or through activities such as oil spills or leakages from oil storage tanks and pipelines (Ezeji et al. 2007) and causes damaging effects on the contaminated environments. The damaging effects of oil contamination on soil depend on the quantity and types of spilled oil and environmental or climatic conditions (Ezeji et al. 2007, Wang et al. 2013). The negative impacts of oil on the coverage of soil surface (Ezeji et al. 2007) and "the structure, function, and ecosystem service values of marshes of the Momoge National Nature Reserve in Jilin Province, China" during oil exploration were reported by investigators (Wang et al. 2013). The study highlighted several impacts including (i) higher concentrations of total petroleum hydrocarbons in the oil-contaminated marsh soil (especially near the oil wells) in comparison to the unaffected marsh soil, (ii) lower soil water contents in the oil-contaminated marshes (especially in the fall), (iii) increase in the soil pH up to 8 in the oil-contaminated marsh soil, and (iv) lower phosphorus concentrations in the contaminated soil.

Due to the damaging effects of contaminated oil to organisms and potential leaching to groundwater, cleanup methods (categorized as *in situ* and *ex situ*) have been developed to treat the oil-contaminated soil. These methods include physical (e.g., soil excavation), chemical (e.g., soil washing and soil vapor extraction (SVE)), thermal (e.g., incineration), and biological processes such as natural and accelerated biodegradation (Ezeji et al. 2007). In the following section, the successful application of BSs for the chemical and biological treatments of oil-contaminated soil will be discussed.

Soil washing is an *ex situ* aqueous-based soil separation technique that is used to remove non-volatile organic and inorganic (e.g., heavy metals) pollutants from the soil. The pollutants are either mechanically separated from the soil based on their particle size (e.g., via gravity separation) or dissolved or suspended in chemical or microbial-based reagents (known as wash water) (Urum and Pekdemir 2004). Two main mechanisms including mobilization and solubilization were proposed for enhancement of soil washing with chemical and or biological surfactants (Vigon and Rubin 1989, Cheah et al. 1998, Deshpande et al. 1999, Mulligan et al. 2001, Urum and Pekdemir 2004). Table 9 summarizes the phenomena that result in mobilization and solubilization of oil from soil.

Removal of hydrocarbons from the oil-contaminated soil by BSs has been the focus of research in the last two decades (Urum et al. 2003, Singh and Cameotra 2013). Studies showed that BS concentration, temperature, mixing condition, pH, and contact time are important factors that decides the efficiency of soil washing by BSs since they impact the chemical structures of BSs and their interactions (Singh and Cameotra 2013).

A study conducted by Singh and Cameotra (2013) revealed that surfactin and fengycin BSs synthesized by *B. subtilis* A21 showed significant effectiveness in removal of petroleum hydrocarbons and heavy metals (iron, lead, nickel, cadmium, copper, cobalt, and zinc) from the soil. Approximately, 64.4%, 44.2%, 35.4%, 40.3%, 32.2%, 26.2%, and 32.07% of petroleum hydrocarbons, cadmium, cobalt, lead, nickel, copper, and zinc, respectively, were removed from the soil by these biosurfactants.

**Table 9.** Phenomena associated with mobilization and solubilization of oil from soil (adapted from Urum and Pekdemir 2004).

| Mechanisms | Surfactant Concentration | Phenomenon |
|---|---|---|
| Mobilization | Below the CMC | Reduction of surface and interfacial tension between air-water, oil-water, and soil-water systems |
| | | Reduction of capillary force |
| | | Wettability |
| | | Reduction of contact angle |
| Solubilization | Above the CMC | Encapsulation of pollutants in micellar aggregates |

Another study conducted by Urum and Pekdemir (2004) examined the ability of several biosurfactants including aescin, lecithin, RLs, saponin and tannin in crude oil contaminated soil washing. The BSs foaming, solubilization, sorption to soil, emulsification, and surface and interfacial tension properties were evaluated at different interrelated phases including (i) soil-water, (ii) water-oil, and (iii) oil-soil systems, and were measured and compared with a chemical surfactant, sodium dodecyl sulphate (SDS). Significant amounts of crude oil were removed from the contaminated soil with various concentrations of BSs. For example, 42% and 80% of oil were removed, respectively, from the soil by lecithin, rhamnolipid, and SDS. Mobilization was suggested as the main mechanism for the oil removal as a result of surface and interfacial tensions reduction, while it was found that solubilization and emulsification had no impact on oil removal (Urum and Pekdemir 2004).

The potential application of extracellular bioemulsans and/or cell-bound BSs synthesized by *Gordonia* sp. strain BS29, isolated from a diesel-contaminated soil using aliphatic hydrocarbons as the main source of carbon, for remediation of contaminated soil was studied by Franzetti et al. (2009). Two types of treatments including (i) bioremediation of aliphatic and aromatic hydrocarbons-contaminated soils and (ii) soil washing of contaminated soil by crude oil, PAHs, and heavy metals were examined in this study. Results revealed that BS29 bioemulsans had various effects on the bioremediation and soil washing so that the biodegradation of recalcitrant branched hydrocarbons slightly improved with BS29, while considerable crude oil and PAH removal from the soil were obtained with the BS29 bioemulsans. The study suggested the promising application of BS29 bioemulsans as washing agents for remediation of hydrocarbon-contaminated soils (Franzetti et al. 2009). Moreover, heavy oil removal efficiency of RLs, surfactin BSs, Tween 80, and Triton X-100 surfactants from refinery oil-contaminated soil with low and high contents of total petroleum hydrocarbons showed better capability of RLs and surfactin BSs in removal of TPHs from the polluted soil than chemical surfactants. For example, treatments of low contents of TPH contaminated soil (e.g., ca. 3,000 mg TPH/kg dry soil) with a 0.2 mass% of RLs, surfactin, Tween 80, and Triton X-100, respectively, resulted in the removal of 23%, 14%, 6%, and 4%, respectively, of TPHs from the contaminated soil, while the TPH removal efficiency increased to 63%, 62%, 40%, and 35%, respectively, as the oil-contamination increased in the soil (ca. 9000 mg TPH/kg dry soil) treated with biological and chemical surfactants (Lai et al. 2009).

Accelerated *ex situ* biodegradation of gasoline-contaminated soil by mixed bacterial communities and different additives including RLs produced by *Pseudomonas* sp. DS10-129 was studied during a 90-day remediation period. Results showed that 67% and 87% of hydrocarbons were biodegraded in the first 60 days of treatment of the contaminated soil with additives and RLs, respectively (Rahman et al. 2002a).

The potential application of *B. subtilis* DM-04 and *P. aeruginosa* M and NM strains for biodegradation of petroleum contaminated soil was studied by Das and Mukherjee (2007b). The study highlighted several points including (i) the uptake of crude petroleum oil hydrocarbons as a sole source of carbon and energy by bacterial strains (ii) considerable TPH biodegradation in treatments with *P. aeruginosa* M and NM consortia and *B. subtilis* strain following 120 days of remediation period, (iii) better TPH degradation observed with *P. aeruginosa* strains than by the *B. subtilis* strain, and (iv) significant microbial growth and microbial BS synthesis in the oil contaminated soil. The study confirmed the effectiveness of both *B. subtilis* DM-04 and *P. aeruginosa* M and NM strains for *in situ* oil-contaminated soil bioremediation (Das and Mukherjee 2007b).

BSs have also been successfully used for the treatment of polycyclic aromatic hydrocarbons (PAHs). PAHs are produced during petroleum refining (Park et al. 1990). The toxicity and carcinogenicity of PAHs make them priority environmental pollutants because these compounds are not easily soluble in aqueous systems and can strongly adsorb to solids in soils and sediments (Congiu and Ortega-Calvo 2014), thus demanding effective treatments. The application of RLs (JBR425) in two forms including liquid and foam treatments of PAHs and heavy metal (Pb, Zn and Ni) contaminated freshwater sediments from sector 103 of the Port of Montreal, Canada with various initial concentrations of contaminants was examined in a study conducted by Alavi (2011). Foam or a liquid solution of JBR425 was injected in to dewatered non-dried sediments in flushing column tests. Results revealed that a 99% quality foam with pH 6.8 produced by 0.5% RL solution after 20 pore volumes significantly removed PAHs from the soil through mobilization mechanism. Moreover, a 0.5% RL foam (99% quality) at pH 10.0 showed the highest heavy metal removal activity. For example, 53.3% of Ni, 56.8% of Pb and 55.2% of Zn were removed with this foam. This study revealed that the RLs at pH 6.8 showed better removal efficiencies for PAHs and at pH 10 for heavy metal removal (Alavi 2011).

The effect of RLs on biodegradation of PAHs under desorption-limiting conditions was studied by Congiu and Ortega-Calvo (2014). Results revealed that factors such as soil aging and organic content can negatively influence the solubilization and biodegradation of slow desorption fractions of PAHs (e.g., phenanthrene and pyrene) from the soil. Moreover, it was found that RLs considerably increased solubilization and biodegradation of aged phenanthrene and pyrene. However, it was more effective in biodegradation of unaged PAHs. Intra-aggregate diffusion of the contaminants during aging was proposed as the main reason for the low effectiveness of RL on the biodegradation of aged PAHs and higher biodegradation of soil-sorbed PAHs was found to be due to two mechanisms including (i) micellar solubilization, which improved the aquatic availability of PAHs to microbial cells and (ii) partitioning of RLs into soil organic matter that increased the rate of PAHs desorption from the polluted soil (Congiu and Ortega-Calvo 2014).

In addition to the application of BSs for petroleum hydrocarbon and PAH-contaminated soil, a study conducted by Rufino et al. (2013) showed that the Rufisan BS synthesized by a yeast showed good surface activity and reduced the surface tension of the medium to 25.3 mN/m. Its application for treatment of motor oil-contaminated soil through static and kinetic assays under laboratory conditions showed that nearly 30% to 98% of motor oil was removed, respectively, by both the crude Rufisan BS and the purified/isolated BS at its critical micelle concentration (CMC, 0.03% w/v). As soil type and BS concentration had an insignificant effect on the oil removal rate, oil displacement was suggested as the main mechanism of oil removal. Figure 10 shows the removal of motor oil from clay, silty, and sandy sands with chemical and BS (Rufino et al. 2013).

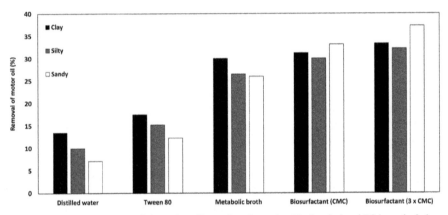

**Figure 10.** Removal of motor oil from clay, silty, and sandy sands with chemical and BS in packed glass columns (adapted from Rufino et al. 2013).

## Conclusions

Oily sludge and wastewater generation are inevitable products of crude oil and petroleum production. Petroleum hydrocarbon pollutants negatively impact human health and the environment and thus selection of treatment and remediation processes that are effective and pose less environmental challenges can minimize the damaging effects of such pollutants. As a single process or technology cannot completely and effectively remove oil pollutants, thus a variety of physicochemical and biological technologies have been developed to date (Hu et al. 2013). In recent decades, the emphasis was on the utilization or co-application of pure bio-reagents such as BEs and BSs for crude oil production at different stages of oil production and processing and for the disposal of oily wastewater and sludge.

It has been demonstrated that petroleum pollutant recovery and removal are possible by the application of BEs and BSs as alternatives to chemical reagents due to their surface active and emulsifying properties, low toxicity, biodegradability, unlimited applicability and relative low production cost. Studies showed that for effective application of BSs, they should be selected based on pollutant characteristics and properties, treatment capacity, costs, regulatory requirements, and time constraints. Moreover, understanding of the mechanisms of interaction between BSs and hydrocarbons or contaminated environment can assist in selection of the appropriate BSs for oil recovery and remediation and overcoming the chemical and environmental limitations.

# References

Al-Bahry, S.N., Y.M. Al-Wahaibi, A.E. Elshafie, A.S. Al-Bemani, S.J. Joshi, H.S. Al-Makhmari and H.S. Al-Sulaimani. 2013. Biosurfactant production by *Bacillus subtilis* B20 using date molasses and its possible application in enhanced oil recovery. J. Int. Biodeterior. Biodegrad. 81: 141–46.

Alavi, A. 2011. Remediation of a Heavy Metal and PAH-Contaminated Sediment by a Rhamnolipid Foam. M.Sc. thesis, Building, Civil and Environmental Engineering, Concordia University, Montreal, Quebec, Canada.

Alexandre, V.M.F., T.M.S. de Castro, L.V. de Araújo, V.M.J. Santiago, D.M.G. Freire and M.C. Cammarota. 2016. Minimizing solid wastes in an activated sludge system treating oil refinery wastewater. Chem. Eng. Process. 103. Elsevier B.V. 53–62.

Amani, H., M.M. Müller, C. Syldatk and R. Hausmann. 2013. Production of microbial rhamnolipid by *Pseudomonas aeruginosa* MM1011 for *ex situ* enhanced oil recovery. Appl. Biochem. Biotechnol. 170(5): 1080–93.

Ball andrew, S., R.J. Stewart and K. Schliephake. 2012. A review of the current options for the treatment and safe disposal of drill cuttings. Waste Manag. Res. 30(5): 457–73.

Banat, I.M., N. Samarah, M. Murad, R. Horne and S. Banerjee. 1990. Biosurfactant production and use in oil tank clean-up. World J. Microbiol. Biotechnol. 7: 80–88.

Banat, I.M. 1995. Biosurfactants production and use in microbial enhanced oil recovery and pollution remediation: A review. Bioresour. Technol. 51: 1–12.

Bodour, A.A., C. Guerrero-Barajas, B.V. Jiorle, M.E. Malcomson, A.K. Paull, A. Somogyi, L.N. Trinh, R.B. Bates and R.M. Maier. 2004. Structure and characterization of flavolipids, a novel class of biosurfactants produced by *Flavobacterium* sp. strain MTN11. Appl. Environ. Microbiol. 70(1): 114–20.

Bognolo, G. 1999. Biosurfactants as emulsifying agents for hydrocarbons. Colloids Surfaces A Physicochem. Eng. Asp. 152: 41–52.

Bouchez-Naïtali, M., H. Rakatozafy, R. Marchal, J.-Y. Leveau and J.-P. Vandecasteele. 1999. Diversity of bacterial strains degrading hexadecane in relation to the mode of substrate uptake. J. Appl. Microbiol. 86: 421–28.

Bourquin, W.A., D.G. Ahearn and S.P. Meyers. 1974. Impact of the use of microorganisms on the aquatic environment. In Symposium-Workshop, Pensacola Beach, Florida. National Environmental Research Center, Corvallis, Oreg. (USA) No. EPA-660/3-75-001; CONF-740498.

Cameotra, S.S. and P. Singh. 2008. Bioremediation of oil sludge using crude biosurfactants. Int. Biodeterior. Biodegrad. 62: 274–80.

Cerón-Camacho, R., R. Martínez-Palou, B. Chávez-Gómez, F. Cuéllar, C. Bernal-Huicochea, J. De-la-Cruz Clavel and J. Aburto. 2013. Synergistic effect of Alkyl-O-Glucoside and -Cellobioside biosurfactants as effective emulsifiers of crude oil in water. A Proposal for the Transport of Heavy Crude Oil by Pipeline. Fuel 110: 310–17.

Chai, L.J., F. Zhang, Y.H. She, I.M. Banat and D. Hou. 2015. Impact of a microbial-enhanced oil recovery field trial on microbial communities in a low-temperature heavy oil reservoir. Nat. Environ. Pollut. Technol. 14(3): 455–62.

Cheah, E.P.S., D.D. Reible, K.T. Valsaraj, W.D. Constant, B.W. Walsh and L.J. Thibodeaux. 1998. Simulation of soil washing with surfactants. J. Hazard. Mater. 59: 107–22.

Chen, Q., M. Bao, X. Fan, S. Liang and P. Sun. 2013. Rhamnolipids enhance marine oil spill bioremediation in laboratory system. Mar. Pollut. Bull. 71(1-2). Elsevier Ltd: 269–75.

Clark, K.A. and D.S. Pasternack. 1932. Hot water separation of bitumen from alberta bituminous sand. Ind. Eng. Chem. 24(12): 1410–16.

Clayton, J.R., J.R. Payne and J.S. Farlow. 1993. Section 3: Factors affecting chemical dispersion of oil and its measurement. pp. 14–53. In Oil Spill Dispersants: Mechanisms of Action and Laboratory Tests. CRC Press.

Congiu, E. and J.-J. Ortega-Calvo. 2014. Role of desorption kinetics in the rhamnolipid-enhanced biodegradation of polycyclic aromatic hydrocarbons. Environ. Sci. Technol. 48(18): 10869–77.

Cooper, D.G. and D.A. Paddock. 1984. Production of a biosurfactant from *Torulopsis bombicola*. Appl. Environ. Microbiol. 47(1): 173–76.

CSIC. 2003. Presence of Heavy Metals in the Sinking Zone of Prestige Tanker and Composition of Metals and Complexers of Coastal Emulsified Oil. Centro Superior de Investigaciones Científicas, Technical Report No. 02.

Dagbert, C., T. Meylheuc and M.-N. Bellon-Fontaine. 2006. Corrosion behaviour of AISI 304 stainless steel in presence of a biosurfactant produced by *Pseudomonas fluorescens*. Electrochim. Acta 51: 5221–27.

Dagnew, M. 2004. Rhamnolipid Assisted Dispersion and Biodegradation of Crude Oil Spilled on Water. M.Sc. thesis, Building, Civil and Environmental Engineering, Concordia University, Montreal, Quebec, Canada.

Das, K. and A.K. Mukherjee. 2007a. Comparison of lipopeptide biosurfactants production by *Bacillus subtilis* strains in submerged and solid state fermentation systems using a cheap carbon source: Some industrial applications of biosurfactants. Process Biochem. 42(8): 1191–99.

Das, K. and A.K. Mukherjee. 2007b. Crude petroleum-oil biodegradation efficiency of *Bacillus subtilis* and *Pseudomonas aeruginosa* strains isolated from a petroleum-oil contaminated soil from North-East India. Bioresour. Technol. 98(7): 1339–45.

Daverey, A. and K. Pakshirajan. 2010. Sophorolipids from *Candida bombicola* using mixed hydrophilic substrates: Production, purification and characterization. Colloids Surf B Biointerfaces. 79(1). Elsevier B.V.: 246–53.

David, J., R. Gupta, C. Mohandass, S. Nair, P.A. LokaBharathi and D. Chandramohan. 1995. Candidates for the Development of Consortia Capable of Petroleum Hydrocarbon Degradation in Marine Environment. Second International Oil Spill Research and Development Forum, IMO, UK.

De Almeida, D.G., R. de, C.F. Soares Da Silva, Juliana M. Luna, Raquel D. Rufino, Valdemir A. Santos, Ibrahim M. Banat and Leonie A. Sarubbo. 2016. Biosurfactants: Promising molecules for petroleum biotechnology advances. Front. Microbiol. 7: 1718.

Deshpande, S., B.J. Shiau, D. Wade, D.A. Sabatini and J.H. Harwell. 1999. Surfactant selection for enhancing *ex situ* soil washing. Water Res. 33(2): 351–60.

Ding, M., W.-H. Jia, Z.-F. Lv and Si-Li Ren. 2014a. Improving bitumen recovery from poor processing oil sands using microbial pretreatment. Energy & Fuels.

Ding, M., Y. Zhang, J. Liu, W. Jia, B. Hu and S. Ren. 2014b. Application of microbial enhanced oil recovery technology in water-based bitumen extraction from weathered oil sands. AIChE J. 60(8): 2985–93.

Duyvesteyn, W.P.C., J.R. Budden and M.A. Picavet. 1999. Extraction of Bitumen from Bitumen Froth and Biotreatment of Bitumen Froth Tailings Generated from Tar Sands. US5968349 A, issued 1999.

Elshafie, A.E., S.J. Joshi, Y.M. Al-Wahaibi, A.S. Al-Bemani, S.N. Al-Bahry, D. Al-Maqbali and I.M. Banat. 2015. Sophorolipids production by *Candida bombicola* ATCC 22214 and its potential application in microbial enhanced oil recovery. Front. Microbiol. 6(1324): 1–11.

Energy.Gov, Office of Fossil Energy. 2016. Enhanced Oil Recovery. 2016. URL http://energy.gov/fe/science-innovation/oil-gas-research/enhanced-oil-recovery.

Ezeji, U.E., S.O. Anyadoh and V.I. Ibekwe. 2007. Clean up of crude oil-contaminated soil. Terrestrial and Aquat. Environ. Toxicol. 1(2): 54–59.

Fingas, M. 2011. Chapter 3: Introduction to oil chemistry and properties. pp. 51–59. *In*: Mervin Fingas (ed.). Oil Spill Science and Technology. Gulf Professional Publishing.

Franzetti, A., P. Caredda, C. Ruggeri, P.L. Colla, E. Tamburini, M. Papacchini and G. Bestetti. 2009. Potential applications of surface active compounds by *Gordonia* sp. strain BS29 in soil remediation technologies. Chemosphere 75(6). Elsevier Ltd: 801–7.

Franzetti, A., E. Tamburini and I.M. Banat. 2010. Applications of biological surface active compounds in remediation technologies. pp. 121–34. *In*: Biosurfactants. Springer New York.

Gargouri, B., F. Karray, N. Mhiri, F. Aloui and S. Sayadi. 2011. Application of a Continuously Stirred Tank Bioreactor (CSTR) for bioremediation of hydrocarbon-rich industrial wastewater effluents. J. Hazard. Mater. 189(1–2). Elsevier B.V.: 427–34.

Ghorbanian, M., G. Moussavi and M. Farzadkia. 2014. Investigating the performance of an up-flow anoxic fixed-bed bioreactor and a sequencing anoxic batch reactor for the biodegradation of hydrocarbons in petroleum-contaminated saline water. Int. Biodeterior. Biodegrad. 90. Elsevier Ltd: 106–14.

Gosselin, P., S.E. Hrudey, M.A. Naeth, A. Plourde, R. Therrien, G.V.D. Kraak and Z. Xu. 2010. Environmental and health impacts of canada's oil sands industry. RSC.

Gross, S.A., H.J. Avens, A.M. Banducci, J. Sahmel, J.M. Panko and B.E. Tvermoes. 2013. Analysis of BTEX groundwater concentrations from surface spills associated with hydraulic fracturing operations. J. Air Waste Manage. Assoc. 63(4): 424–32.

Gutnick, D.L. and E. Rosenberg. 1977. Oil tankers and pollution: A microbiological approach. Annu. Rev. Microbiol. 31: 379–96.

Hasanshahian, M. and G. Emtiazi. 2008. Investigation of alkane biodegradation using the microtiter plate method and correlation between biofilm formation, biosurfactant production and crude oil biodegradation. Int. Biodeterior. Biodegrad. 62(2): 170–78.

Hayes, M.E., K.R. Hrebenar, P.L. Murphy, L.E. Futch Jr., J.F. Deal and P.L. Bolden Jr. 1987. Combustion of viscous hydrocarbons. US Patent 4,684,372, issued 1987.

Heiss-Blanquet, S., Y. Benoit, C. Maréchaux and F. Monot. 2005. Assessing the role of alkane hydroxylase genotypes in environmental samples by competitive PCR. J. Appl. Microbiol. 99(6): 1392–1403.

Hirata, Y., M. Ryu, K. Igarashi, A. Nagatsuka, T. Furuta, S. Kanaya and M. Sugiura. 2009a. Natural synergism of acid and lactone type mixed sophorolipids in interfacial activities and cytotoxicities (a). J. Oleo Sci. 58(11): 565–72.

Hirata, Y., M. Ryu, Y. Oda, K. Igarashi, A. Nagatsuka, T. Furuta and M. Sugiura. 2009b. Novel characteristics of sophorolipids, yeast glycolipid biosurfactants, as biodegradable low-foaming surfactants. J. Biosci. Bioeng. 108(2). The Society for Biotechnology, Japan: 142–46.

Hollebone, B. 2011. Chapter 4: Measurement of oil physical properties. pp. 63–86. *In*: Mervin Fingas (ed.). Oil Spill Science and Technology. Gulf Professional Publishing.

Hu, G., J. Li and G. Zeng. 2013. Recent development in the treatment of oily sludge from petroleum industry: A review. J. Hazard. Mater. 261. Elsevier B.V.: 470–90.

Huang, G., H. Yu, G. Li, C. An. and J. Wei. 2017. Science. Development of an innovative bioremediation technology for oil-sands tailing waste treatment. University of Regina, Canada, PTRC No: 002-00017-UOR.

Inoue, S., Y. Kimura and M. Kinta. 1980. Dehydrating purification process for a fermentation product. US Patent 4,197,166, issued 1980.

IPIECA. 2010. Petroleum Refining Water/wastewater Use and Management. www.ipieca.org.

Javan Roshtkhari, S. 2016. Application of rhamnolipid and microbial activities for improving the sedimentation of oil sand tailings. Ph.D thesis, Building, Civil and Environmental Engineering, Concordia University, Montreal, Quebec, Canada.

Joshi-Navare, K., P. Khanvilkar and A. Prabhune. 2013. Jatropha oil derived sophorolipids: Production and characterization as laundry detergent additive. Biochem. Res. Int. 2013.

Ju, Z., J. Li, L. Chen and R.W. Thring. 2011. Remediation of refinery oily sludge using isolated strain and biosurfactant. Proceedings of 2011 Int. Symp. Water Resour. Environ. Prot. 3(425): 1649–53.

Kaczorek, E., L. Chrzanowski, A. Pijanowska and A. Olszanowski. 2008. Yeast and bacteria cell hydrophobicity and hydrocarbon biodegradation in the presence of natural surfactants: rhamnolipids and saponins. Bioresour. Technol. 99(10): 4285–91.

Kaczorek, E. 2012. Effect of external addition of rhamnolipids biosurfactant on the modification of gram positive and gram negative bacteria cell surfaces during biodegradation of hydrocarbon fuel contamination. Polish J. Environ. Stud. 21(4): 901–9.

Kaczorek, E., T. Jesionowski, A. Giec and A. Olszanowski. 2012. Cell surface properties of *Pseudomonas stutzeri* in the process of diesel oil biodegradation. Biotechnol. Lett. 34(5): 857–62.

Kanicky, J.R., S. Pandey and D.O. Shah. 2002. Chapter 11: Surface chemistry in the petroleum industry. pp. 252–67. *In*: Handbook of Applied Surface and Colloid Chemistry. John Wiley & Sons, Ltd.

Kirkeleit, J., T. Riise, M. Bråtveit and B.E. Moen. 2008. Increased risk of acute myelogenous leukemia and multiple myeloma in a historical cohort of upstream petroleum workers exposed to crude oil. Cancer Causes Control: CCC 19(1): 13–23.

Kirpalani, D.M. and A. Matsuoka. 2008. CFD approach for simulation of bitumen froth settling process—Part I: Hindered settling of aggregates. Fuel 87(3): 380–87.

Kitamoto, D., H. Isoda and T. Nakahara. 2002. Functions and potential applications of glycolipid biosurfactants from energy-saving materials to gene delivery carriers. J. Biosci. Bioeng. 94(3): 187–201.

Kloos, K., J. Charles and M. Schloter. 2006. A new method for the detection of Alkane-monooxygenase homologous genes (alkB) in soils based on PCR-hybridization. J. Microbiol. Methods 66(3): 486–96.

Kohno, T., Y. Sugimoto, K. Sei and K. Mori. 2002. Design of PCR primers and gene probes for general detection of alkane-degrading bacteria. Microbes Environ. 17(3): 114–21.

Kontogiannis, J.E. and C.J. Barnett. 1973. The effect of oil pollution on survival of the tidal pool copepod, tigriopus californicus. Environ. Pollut. (1970) 4(1). Elsevier: 69–79.

Lai, C.C., Y.C. Huang, Y.H. Wei and J.-S. Chang. 2009. Biosurfactant-enhanced removal of total petroleum hydrocarbons from contaminated soil. J. Hazard. Mater. 167(1-3): 609–14.

Lazar, I., A. Voicu, C. Nicolescu, D. Mucenica, S. Dobrota, I.G. Petrisor, M. Stefanescu and L. Sandulescu. 1999. The use of naturally occurring selectively isolated bacteria for inhibiting paraffin deposition. J. Pet. Sci. Eng. 22(1-3): 161–69.

Lazar, I., I.G. Petrisor and T.E. Yen. 2007. Microbial Enhanced Oil Recovery (MEOR). J. Pet. Sci. Technol. 25(11): 1353–66.

Leonard, S.A. and J.A. Stegemann. 2010. Stabilization/solidification of petroleum drill cuttings: Leaching studies. J. Hazard. Mater. 174(1-3): 484–91.

Li, P., D. Zhao, Y. Zhang, L. Sun, H. Zhang, M. Lian and B. Li. 2015. Oil-field wastewater treatment by hybrid Membrane-Aerated Biofilm Reactor (MABR) system. Chem. Eng. J. 264. Elsevier B.V.: 595–602.

Li, Z., Kenneth Lee, T. King, M.C. Boufadel and A.D. Venosa. 2010. Effects of temperature and wave conditions on chemical dispersion efficacy of heavy fuel oil in an experimental flow-through wave tank. Mar. Pollut. Bull. 60(9). Elsevier Ltd: 1550–59.

Long, Y., T. Dabros and H. Hamza. 2002. Stability and settling characteristics of solvent-diluted bitumen emulsions. Fuel 81(15): 1945–52.

Mamer, M. 2007. Oil Sands Tailings Technology: Understanding the Impact to Reclamation. Suncor Energy Inc.

Mannina, G., A. Cosenza, D.D. Trapani, M. Capodici and G. Viviani. 2016. Membrane bioreactors for treatment of saline wastewater contaminated by hydrocarbons (Diesel Fuel): An experimental pilot plant case study. Chem. Eng. J. 291. Elsevier B.V.: 269–78.

Moles, A., L. Holland and J. Short. 2002. Effectiveness in the laboratory of Corexit 9527 and 9500 in dispersing fresh, weathered and emulsion of alaska north slope crude oil under subarctic conditions. Spill Sci. Technol. Bull. 7(5-6): 241–47.

Moussavi, G. and M. Ghorbanian. 2015. The biodegradation of petroleum hydrocarbons in an upflow sludge-blanket/fixed-film hybrid bioreactor under nitrate-reducing conditions: Performance evaluation and microbial identification. Chem. Eng. J. 280. Elsevier B.V.: 121–31.

Moussavi, G., S. Shekoohiyan and K. Naddafi. 2017. The accelerated enzymatic biodegradation and COD removal of petroleum hydrocarbons in the SCR using active bacterial biomass capable of *in-situ* generating peroxidase and biosurfactants. Chem. Eng. J. 308. Elsevier B.V.: 1081–89.

Mulligan, C.N., R.N. Yong and B.F. Gibbs. 2001. Surfactant-enhanced remediation of contaminated soil: A review. Eng. Geol. 60(1-4): 371–80.

Mulligan, C.N. 2005. Environmental applications for biosurfactants. Environ. Pollut. 133(2): 183–98.

Mulligan, C.N. and S. Javan Roshtkhari. 2016. Application of rhamnolipid and microbial activities for improving the sedimentation of oil sand tailings. J. Bioremediation Biodegrad. 7(4): 8.

Mullins, O.C., E.Y. Sheu, A. Hammami and A.G. Marshall. 2007. Asphaltenes, Heavy Oils and Petroleomics. 1sted. Springer-Verlag New York.

Okafor, N. 2011. Chapter 7: Pollution by petroleum in oceans and seas: Role of microorganisms in oil degradation and remediation. In Environmental Microbiology of Aquatic and Waste Systems, 1sted., 307. Springer Netherlands.

Özdemir, G. and U. Malayoglu. 2004. Wetting characteristics of aqueous rhamnolipids solutions. Colloids Surf B Biointerfaces 39(1-2): 1–7.

Park, K.S., R.C. Sims and R.R. Dupont. 1990. Transformation of PAHs in soil systems. J. Environ. Eng. 116(3): 632–40.

Patin, S.A. 1999. Ecotoxicological characteristics of related chemicals and wastes from the offshore oil industry. In Environmental Impact of the Offshore Oil and Gas Industry, 254–98. East Northport, NY: EcoMonitor Publishers.

Pereira, J.F.B., E.J. Gudiña, R. Costa, R. Vitorino, J.A. Teixeira, J.A.P. Coutinho and L.R. Rodrigues. 2013. Optimization and characterization of biosurfactant production by *Bacillus subtilis* isolates towards microbial enhanced oil recovery applications. Fuel 111. Elsevier Ltd: 259–68.

Pérez-Cadahía, B., A. Lafuente, T. Cabaleiro, E. Pásaro, J. Méndez and B. Laffon. 2007. Initial study on the effects of prestige oil on human health. Environ. Int. 33(2): 176–85.

Perfumo, A., I. Rancich and I.M. Banat. 2010. Possibilities and challenges for biosurfactants uses in petroleum industry. pp. 135–45. *In*: Biosurfactants. Springer New York.

Pornsunthorntawee, O., N. Arttaweeporn, S. Paisanjit, P. Somboonthanate, M. Abe, R. Rujiravanit and S. Chavadej. 2008. Isolation and comparison of biosurfactants produced by *Bacillus subtilis* PT2 and *Pseudomonas aeruginosa* SP4 for microbial surfactant-enhanced oil recovery. Biochem. Eng. J. 42(2): 172–79.

Prego, Ricardo and Antonio Cobelo-García. 2004. Cadmium, copper and lead contamination of the seawater column on the prestige shipwreck (NE Atlantic Ocean). Anal. Chim. Acta 524(1-2 SPEC. ISS.): 23–26.

Rahman, K.S.M., I.M. Banat, J. Thahira, Tha Thayumanavan and P. Lakshmanaperumalsamy. 2002a. Bioremediation of gasoline contaminated soil by a bacterial consortium amended with poultry litter, coir pith and rhamnolipid biosurfactant. Bioresour. Technol. 81(1): 25–32.

Rahman, K.S.M., Thahira J. Raman, Stephen McClean, Roger Marchant and Ibrahim M. Banat. 2002b. Rhamnolipid biosurfactant production by strains of *Pseudomonas aeruginosa* using low-cost raw materials. Biotechnol. Prog. 18(6): 1277–81.

Rao, F. and Q. Liu. 2013. Froth treatment in athabasca oil sands bitumen recovery process: A review. Energy & Fuels 27(12): 7199–7207.

Rocha, C.A., D. Gonzalez, M.L. Iturralde, U.L. Lacoa and F.A. Morales. 2000. Production of oily emulsions mediated by a microbial tenso-active agent. US Patent 6,060,287, issued 2000.

Ron, E.Z. and E. Rosenberg. 2002. Biosurfactants and oil bioremediation. Curr. Opin. Biotechnol. 13(3): 249–52.

Rosenberg, E., E. Englander, A. Horowitz and D. Gutnick. 1975. Bacterial growth and dispersion of crude oil in an oil tanker during its ballast voyage. US201302738809. Ecol. Res. Ser EPA US Environ. Prot. Agency.

Rosenberg, E. 1993. Exploiting microbial growth on hydrocarbons—new markets. Trends Biotechnol. 11(10): 419–24.

Rosenberg, Ee, R. Legman, A. Kushmaro, E. Adler, H. Abir and E.Z. Ron. 1996. Oil bioremediation using insoluble nitrogen source. J. Biotechnol. 51(3): 273–78.

Rufino, R.D., J.M. Luna, P.H.C. Marinho, C.B.B. Farias, S.R.M. Ferreira and L.A. Sarubbo. 2013. Removal of petroleum derivative adsorbed to soil by biosurfactant rufisan produced by *Candida lipolytica*. J. Pet. Sci. Eng. 109: 117–22.

Saborimanesh, N. and C.N. Mulligan. 2015. Effect of sophorolipid biosurfactant on oil biodegradation by the natural oil-degrading bacteria on the weathered biodiesel, diesel and light crude oil. J. Bioremediat Biodegrad. 6(6): 8.

Saborimanesh, N. 2016. Dispersion and Bacterial Degradation of Weathered Diesel, Biodiesel and Light Crude Oil in Seawater by Sophorolipid Biosurfactant. Ph.D. Thesis, Building, Civil and Environmental Engineering, Concordia University, Montreal, Quebec, Canada.

Saeki, H., M. Sasaki, K. Komatsu, A. Miura and H. Matsuda. 2009. Oil spill remediation by using the remediation agent JE1058BS that contains a biosurfactant produced by *Gordonia* sp. strain JE-1058. Bioresour. Technol. 100(2). Elsevier Ltd: 572–77.

Sajna, K.V., R.K. Sukumaran, H. Jayamurthy, K.K. Reddy, S. Kanjilal, R.B.N. Prasad and A. Pandey. 2013. Studies on biosurfactants from *Pseudozyma* sp. NII 08165 and their potential application as laundry detergent additives. Biochem. Eng. J. 78(0). Elsevier B.V.: 85–92.

Shennan, J.L. and J.L. Levi. 1987. Biosurfactants and Biotechnology. In Biosurfactants and Biotechnology, edited by Kosaric, 344. CRC Press.

Shin, K.H., K.W. Kim and E.A. Seagren. 2004. Combined effects of pH and biosurfactant addition on solubilization and biodegradation of phenanthrene. Appl. Microbiol. Biotechnol. 65(3): 336–43.

Siddique, T., P.M. Fedorak and J.M. Foght. 2006. Biodegradation of short-chain N-alkanes in oil sands tailings under methanogenic conditions. Environ. Sci. Technol. 40(17): 5459–64.

Silva, R. de C.F.S., D.G. Almeida, R.l.D. Rufino, J.M. Luna, V.A. Santos and L.A. Sarubbo. 2014. Applications of biosurfactants in the petroleum industry and the remediation of oil spills. Int. J. Mol. Sci. 15(7): 12523–42.

Singer, M.E.V. and W.R. Finnerty. 1990. Physiology of biosurfactant synthesis by *Rhodococcus* species H13-A. Can. J. Microbiol. 36(11): 741–45.

Singh, A.K. and S.S. Cameotra. 2013. Efficiency of lipopeptide biosurfactants in removal of petroleum hydrocarbons and heavy metals from contaminated soil. Environ. Sci. Pollut. Res. 20(10): 7367–76.

Smith, M.T. 2010. Advances in understanding benzene health effects and susceptibility. Annu Rev Public Heal. 31: 133–48.

Solomon, G.M. and S. Janssen. 2010. Health effects of the gulf oil spill. J. Am. Med. Assoc. 304(10): 1118.

Soudmand-asli, A., S.S. Ayatollahi, H. Mohabatkar, M. Zareie and S.F. Shariatpanahi. 2007. The *in situ* microbial enhanced oil recovery in fractured porous media. J. Pet. Sci. Eng. 58(1-2): 161–72.

Sponza, D.T. and O. Gok. 2011. Effects of sludge retention time and biosurfactant on the treatment of polyaromatic hydrocarbon (PAH) in a petrochemical industry wastewater. Water Sci. Technol. 64(11): 2282–92.

Suncor Energy Inc. 2015. Suncor, Sustainability, Environment, Water. 2015. URL http://www.suncor. com/.

Tellez, G.T., N. Nirmalakhandan and J.L. Gardea-Torresdey. 2002. Performance evaluation of an activated sludge system for removing petroleum hydrocarbons from oilfield produced water. Adv. Environ. Res. 6(1): 455–70.

Thavasi, R., S. Jayalakshmi, T. Balasubramanian and I.M. Banat. 2006. Biodegradation of crude oil by nitrogen fixing marine bacteria *Azotobacter chroococcum*. Res. J. Microbiol. 1(5): 401–8.

Thavasi, R., S. Jayalakshmi and I.M. Banat. 2011a. Application of biosurfactant produced from peanut oil cake by *Lactobacillus delbrueckii* in biodegradation of crude oil. Bioresour. Technol. 102(3): 3366–72.

Thavasi, R., S. Jayalakshmi and I.M. Banat. 2011b. Effect of biosurfactant and fertilizer on biodegradation of crude oil by marine isolates of *Bacillus megaterium*, *Corynebacterium kutscheri* and *Pseudomonas aeruginosa*. Bioresour. Technol. 102: 772–778.

Tong, Kun, Y. Zhang, G. Liu, Z. Ye and P.K. Chu. 2013. Treatment of heavy oil wastewater by a conventional activated sludge process coupled with an immobilized biological filter. Int. Biodeterior. Biodegrad. 84. Elsevier Ltd: 65–71.

Tyagi, R.D., F.T. Tran and A.K.M.M. Chowdhury. 1993. A pilot study of biodegradation of petroleum refinery wastewater in a polyurethane-attached RBC. Process. Biochem. 2(28): 75–82.

Urum, K., T. Pekdemir and M. Gopur. 2003. Optimum conditions for washing of crude oil-contaminated soil with biosurfactant solutions. Process Saf. Environ. Prot. 81: 203–9.

Urum, K. and T. Pekdemir. 2004. Evaluation of biosurfactants for crude oil contaminated soil washing. Chemosphere 57(9): 1139–50.

US National Renewable Energy Laboratory. 2009. Biodiesel Handling and Use Guide: Fourth Edition (Revised). United States National Renewable Energy Laboratory (NREL), Golden, CO.

Van Bogaert, I.N.A. and W. Soetaert. 2010. Sophorolipids. pp. 178–210. *In*: Soberón-Chávez Gloria (ed.). Biosurfactants from Genes to Applications. Berlin: Springer Science & Business Media.

Vasefy, F. 2007. Capability of Rhamnolipid and Two Biological Products in Bioremediation of Oil in Marine Environment. M.Sc. thesis, Building, Civil and Environmental Engineering, Concordia University, Montreal, Canada.

Vaughn, S.F., R.W. Behle, C.D. Skory, C.P. Kurtzman and N.P.J. Price. 2014. Utilization of sophorolipids as biosurfactants for postemergence herbicides. Crop Protect. 59(0). Elsevier Ltd: 29–34.

Vigon, B.W. and A.J. Rubin. 1989. Practical considerations in the surfactant-aided mobilization of contaminants in aquifers. J. (Water Pollut. Control Fed.) 61(7): 1233–40.

Wang, Y., J. Feng, Q. Lin, X. Lyu, X. Wang and G. Wang. 2013. Effects of crude oil contamination on soil physical and chemical properties in momoge wetland of China. Chinese Geogr. Sci. 23(6): 708–15.

Ward, O.P. 2010. Chapter 5: Microbial biosurfactants and biodegradation. pp. 65–74. *In*: Ramkrishna Sen (ed.). Biosurfactants. Springer New York.

Whang, L.M., P.W.G. Liu, C.C. Ma and S.S. Cheng. 2009. Application of rhamnolipid and surfactin for enhanced diesel biodegradation-effects of pH and ammonium addition. J. Hazard. Mater. 164(2-3): 1045–50.

Wong, J.M. and Y.-T Hung. 2004. Treatment of oilfield and refinery wastes. pp. 144–216. *In*: Lawrence K. Wang, Yung-Tse Hung, Howard H. Lo, Constantine Yapijakis and Kathleen Hung Li (eds.). Handbook of Industrial and Hazardous Wastes Treatment. Marcel Dekker, Inc.

Yakimov, M.M., M.M. Amro, M. Bock, K. Boseker, H.L. Fredrickson, D.G. Kessel and K.N. Timmis. 1997. The potential of *Bacillus licheniformis* strains for *in situ* enhanced oil recovery. J. Pet. Sci. Eng. 18(1-2): 147–60.

Yan, P., M. Lu, Y. Guan, W. Zhang and Z. Zhang. 2011. Remediation of oil-based drill cuttings through a biosurfactant-based washing followed by a biodegradation treatment. Bioresour. Technol. 102(22). Elsevier Ltd: 10252–59.

Yan, P., M. Lu, Q. Yang, H.-L. Zhang, Z.-Z. Zhang and R. Chen. 2012. Oil recovery from refinery oily sludge using a rhamnolipid biosurfactant-producing *Pseudomonas*. Bioresour. Technol. 116. Elsevier Ltd: 24–28.

Yarbrough, H.F. and V.F. Coty. 1983. Microbial enhanced oil recovery from the upper crustaceous nacatoch formation. In Proceedings of the International Conference on Microbial Enhancement of Oil Recovery, edited by Donaldson and Zajic. Bill Linvill, Chief, Technology Transfer Branch, Bartlesville, OK.

Youssef, N., D.R. Simpson, M.J. McInerney and K.E. Duncan. 2013. *In-situ* lipopeptide biosurfactant production by *bacillus* strains correlates with improved oil recovery in two oil wells approaching their economic limit of production. Int. Biodeterior. Biodegrad. 81. Elsevier Ltd: 127–32.

Zajic, J.E. and D.F. Gerson. 1978. Microbial extraction of bitumen from athabasca oil sand. pp. 145–161. *In*: O. P. Strausz (ed.). Oil Sands and Oil Shale Chemistry. Verlag Chemie, New York.

Zhang, F., Y.H. She, H.-M. Li, X.-T. Zhang, F.-C. Shu, Z.-L. Wang, L.-J. Yu and D.-J. Hou. 2012. Impact of an indigenous microbial enhanced oil recovery field trial on microbial community structure in a high pour-point oil reservoir. Appl. Microbiol. Biotechnol. 95(3): 811–21.

Zhang, Y. and R.M. Miller. 1992. Enhanced octadecane dispersion and biodegradation by a *Pseudomonas* rhamnolipid surfactant (Biosurfactant). Appl. Environ. Microbiol. 58(10): 3276–82.

Zhang, Y. and R.M. Miller. 1994. Effect of a *Pseudomonas* rhamnolipid biosurfactant on cell hydrophobicity and biodegradation of octadecane. Appl. Environ. Microbiol. 60(6): 2101–6.

Zheng, C., M. Wang, Y. Wang and Z. Huang. 2012. Optimization of biosurfactant-mediated oil extraction from oil sludge. Bioresour. Technol. 110. Elsevier Ltd: 338–42.

# 10

# Natural and Microbial Biosurfactants' Use in the Food Industry

*Jenyffer Medeiros Campos,*[1] *Ibrahim M. Banat*[2] *and*
*Leonie Asfora Sarubbo*[3,*]

## Introduction

Surfactants are amphipathic compounds containing both a hydrophilic and a hydrophobic moiety capable of partitioning at interfaces such as oil/water, gas/liquid or liquid/solid interfaces. Surfactants' applications are usually classified according to their use or applications. These include usage as detergents and cleaners (54%), as auxiliaries for paper, leather and textile (13%), in cosmetics and pharmaceuticals (10%), in chemical processes (10%), in the food industry (3%), in agriculture (2%) and in miscellaneous others (8%) (Bourdichon et al. 2012). The use of such surfactants is essential to food industry as emulsifiers that can provide high stability when emulsion are produced or required yet at low cost (Nitschke and Costa 2007).

One of the key environmental and societal challenges for our continued life on earth is finding ways to reduce our dependence on the limited supplies of fossil fuels (oil, coal, gas), from which chemical surfactants are produced (produced mainly from crude oil), moving toward the use of sustainable and renewable sources to supply our needs. Dependence on plant and animal sources to supply industrial materials

[1] Av. Prof. Artur de Sá s/n - Cidade Universitária, Recife, PE, Brazil.
[2] School of Biomedical Sciences, Faculty of Life and Health Sciences, University of Ulster, Coleraine, BT52 1SA, N. Ireland, UK.
[3] Centro de Ciências e Tecnologia, Universidade Católica de Pernambuco (UNICAP), Rua do Príncipe, Boa Vista, Recife, PE, Brazil.
    Emails: jenyffermcampos@gmail.com; im.banat@uslter.ac.uk
* Corresponding author: leonie@unicap.br

and fine chemicals, however, are also to some degree non-sustainable since they can be significantly affected by meteorological events, political disturbances and/or limited non-genetically modified resource. Considering the huge genetic diversity within microbial communities, they offer significant alternative novel compounds such as biosurfactants (BSs) capable of replacing their chemical counterparts that are currently used in food and many other industries.

The use of BSs has been advocated in the last few decades and its applications have been expanded as a function of their properties including low toxicity (Campos et al. 2015), biodegradability (Chrzanowski et al. 2012) and specificity (Banat et al. 2014). Several authors described their potential application in combating environmental pollution (Pacwa-Plociniczak et al. 2011, Marchant and Banat 2011a, Sousa et al. 2014), as an ingredient in food processing (Campos et al. 2013), in health and biomedicine (Banat et al. 2014), in agriculture (Sachdev and Cameotra 2013) and as a chemical constituent in the cosmetic industries (Silva et al. 2014, Vijayakumar and Saravanan 2015), mainly related to its dispersant capacity and ability to act as an emulsifier and stabilizer. Despite the high production cost of BSs which limits their use on industrial scale, several literature output has been published over the years, describing the use of cost effective raw materials in BS production, such as animal fat (Santos et al. 2013), waste vegetable frying oils (Batista et al. 2010), vegetable fats (Gusmão et al. 2010), molasses (Lazaridou et al. 2002), soapstock materials (Maneerat 2005), corn steep liquor (Luna et al. 2011, Rufino et al. 2011), dairy industry waste product (Dubey et al. 2005) and cassava flour wastewater (Nitche and Pastore 2004), in addition to optimization of the growth conditions including temperature, pH, oxygen and agitation speed to achieve higher yields (Santos et al. 2016).

Surface active property is well distributed in natural foods mainly composed of combinations of carbohydrates, lipids and proteins. Natural surfactants such as lectins, saponins, lysozyme and casein are often used as ingredients in various types of foods such as milk, eggs and soy to confer properties such as emulsifying, gelling and dispersion. However, regarding the use of BSs in foods, there is a limitation on recent studies on its applicability. This chapter brings together recent data from the literature on properties and applications of BSs in food industry.

## Natural Surfactants and their use in the Food Industry

McClements and Gumus (2016), in their review, compared the properties of emulsifiers isolated from various natural sources that may be utilized within the food industry. They include: Quillaja saponins, phospholipids (lecithin, lysolecithin), proteins (whey protein, α-lactalbumin, β-lactoglobulin, bovine serum albumin, lactoferrin, caseinates, αs-casein, β-casein, egg proteins, ovalbumin, lysozyme, legume proteins (soy, pea, lentil, chickpea, faba bean, etc.), gelatin, and polysaccharides (gum arabic, beet pectin, and citrus pectin).

BSs or natural surfactants are amphipathic molecules that can be used as food additives since these biomolecules play an important role in determining the products microstructure and in affecting its physical and textural stability. BSs act as dispersants, foamers, stabilizers, emulsifiers, controlling the texture, rheology and the crystallization phenomena of food products (Kralova and Sjöblom 2009).

Naturally occurring surfactants are also used in the preparation of a wide range of food products such as dressings, salad creams, mayonnaise, deserts, etc. Many foods can form different kinds of colloids and drops. The aggregation of individual particles forms three dimensional "gel" structures. The level of aggregation depends on the attractive and repulsive van der Waals forces. The resulting structures are formed by the association of surfactants or polymers. Depending on food formulations compositions, these forces can be electrostatic originating from charged interfaces or steric from adsorbed polymers. Ionic surfactants are responsible for the repulsive forces; surfactants with polymeric polar groups often give steric repulsions, which may be strengthened by electrostatic effects if the polar group is also charged. The interactions between proteins and hydrocolloids are very important as they lead to understanding interfacial properties and bulk rheology of the systems (Tadros 2005). Recently, it has become highly desired to explore naturally occurring surfactants (BSs) and polymers (biopolymers) as novel stabilizers for the nanosuspension formulations in functional food applications (Wan et al. 2016).

Structural stability and texture of a product are usually influenced by amphiphilic molecules such as emulsifiers, foamers and dispersing agents. Examples of amphiphilic molecules are glycolipids, lipoproteins, polysaccharides, polar lipids, etc. Emulsifiers reduce interfacial tension between the two immiscible liquids phases leading to their dispersion and stabilizes this dispersion by reducing coalescence and separation between the two immiscible phases known as "rapture" (Fietcher 1992).

## *Saponins*

Quillaia extracts (synonyms: bois de Panama, quillaja extracts, Panama bark extracts, Quillay bark or soapbark extracts) are typically obtained by aqueous extraction of crushed inner bark or wood of pruned stems and branches of *Quillaja saponaria* Molina, native to China, Bolivia, Chile and Peru. Quillaia saponins are structurally different from the saponins derived from other plant species (FAO 2004). Compared to synthetic surfactants, saponins are known for their biodegradability and low toxicity and have therefore attracted increasing interest within the food industry (Yang et al. 2013). Saponins natural surfactants are small amphiphilic molecule that have hydrophilic and hydrophobic moieties distributed within a single molecule (McClements and Gumus 2016); thus, quillaia saponins have a wide range of industrial applications. Saponins are mostly made of glycosides that contain 1, 2, or 3 sugar chains attached to the aglycones, otherwise known as sapogenins, which makes the non-polar parts of the molecule (Oleszek and Hamed 2010).

Saponins have been used as natural surfactants in foods mainly to control the microbial spoilage of food as preservatives. They have been recently used as a natural small molecule surfactant in beverage emulsions to replace synthetic surfactants such as polysorbates because of consumer preferences for natural substance ingredients (Cheok et al. 2014). The presence of hydrophobic and hydrophilic moieties in saponins molecules (the aglycone and sugar residues) accounts for its ability to reduce surface tension at phase boundaries (Mironenko et al. 2010). The interaction of Quillaia saponins and cholesterol in foods results in the formation of both monolayer and micelles, whose critical micelle concentration (CMC) depends on temperature and salt concentration (Mitra and Dunga 2000).

Interest in these compounds has significantly increased recently due to some attractive properties acting as foaming agents in beverages and emulsifiers in foods, as well as their applications in cholesterol-reduction and flavor enhancement (Murakami 1996, San Martin and Briones 1999). *Quillaia saponins* has been used in making cider, cocktail mixes, frozen dairy products, cream soda, baked goods, candies, gelatin and puddings. Some industrial applications include the production of mayonnaise, enhancement of oil-soluble flavors for candies, dissolving propolis and red coloring material, for soy sauce, as antioxidants and as leavening agents (FAO 2004). McClements and Gumus (2016) stated that the formed emulsions are not stable in extreme conditions of acidic and ionic strength, although they are stable under heating. Rosa et al. (2016) used an extract from Brazilian ginseng roots as a BS which is rich in saponins.

## *Lecithins*

Lecithins are hydrophobic molecules with complex structure and internal composition. Phospholipids and sphingolipids (sphingosine and phytosphingosine) from lecithins have traditionally been extracted from products such as oat, soya, eggs, wheat, etc. In general, one can claim that although lecithins form a lamelar liquid crystalline structures in water, it will be difficult to use them as emulsifiers for stabilization of either oil-in-water or water-in-oil (Kralova and Sjöblom 2009) and can form fairly small droplets at low levels using high pressure homogenization (McClements and Gumus 2016).

Naturally occurring lecithin from egg yolk and various proteins from milk are used for the preparation of many food products such as creams, mayonnaise, salad dressings, deserts, etc. Many products with various degrees of purity are extracted from the crude oily lecithin. Products such as plastic lecithin, deoiled lecithin, phosphatidyl inositol, phosphatidyl choline, phosphatidyl choline-enriched lecithin, phosphatidyl ethanolamine, phosphatidyl serine and phosphatidic acid of mixture of fatty acids are ready to use in the market (Kralova and Sjöblom 2009). These amphiphiles have been included in the food products category called nutraceuticals (Yang et al. 2013).

## *Lipopeptides*

Non-phosphorous amphiphilic molecules accompany phospholipids. The surface properties of molecules such as digalactosyl-diglycerides from cereals (oats, wheat, and soya) have been investigated. It is believed that the reduction of the extraction costs will allow the future application of water-in-oil and oil-in-water emulsions formed from digalactosyl-diglycerides and trigalactosyl-diglycerides (Kralova and Sjöblom 2009).

Animal and vegetable oils contains triglycerides, which are their main components, and small quantities of phospholipids. The complex lecithins molecules, although essentially hydrophobic, are used the most in aqueous solutions. The production of cocoa powder is an important example for the use of lecithin in the food industry. Cocoa powder is a popular ingredient in a number of dry and in instant mixes for drinks and desserts that is easy to store and quick to prepare. The lecithin molecule is a mixture of phosphatides, made up of hydrophilic and hydrophobic parts which are surface-active. The hydrophilic part of the lecithin molecule is directed to the outside of the cocoa particle, attracting water molecules in the solution while the hydrophobic part anchors itself to the cocoa butter present in the cocoa powder. In this way, the cocoa particle maintains an outer surface that has hydrophilic properties. Lecithin therefore delivers excellent dispersing, wetting and emulsifying properties suitable as an ingredient in drinks and desserts and in many dry mix bakery products, combining ingredients to produce the necessary, instant high-quality food products (Kourkoutas and Banat 2008).

## *Proteins*

Significant research has been carried out to investigate the microstructures of emulsions involved in ice cream, milk and dairy products which usually contain proteins, such as whey proteins, casein lactoglobulins, bovine albumins, gelatins, lysozymes and ovalbumins, etc. Such proteins provide stabilization, although full coverage of the interface is required and when surfaces are poorly covered, destabilization may occur through bridging flocculation. Regarding protein films' stability, the presence of interfacial shear viscosity may retard the process of film thinning. Therefore, it was important to monitor the viscoelastic properties of adsorbed protein films at the oil/water interface. Several food compositions include a combination of monomeric and macromolecular emulsifiers and evaluating the contribution of each emulsifier at the interface has become important (Kralova and Sjöblom 2009).

Lysozymes, lactoglobulins, ovalbumins and bovine serum albumins are common examples of native emulsifiers used in dairy products like ice cream, milk and related products. Much effort has been put into creating "artificial" glycoproteins through exploitation of the Maillard reaction, and involves covalent bond formation between lysine side chains in the protein and the reducing end group of short and long carbohydrate chains (N-glycosylation). These have been found with improved emulsifying, foaming and heat stability properties over non-conjugated mixtures of the protein and carbohydrate, and are better functional ingredients than the protein alone (Kato 2002).

In their review, Kralova and Sjöblom (2009) investigated the area of proteins as food emulsifiers to better understand the three stages in the stabilization of interfaces by proteins through coagulation, adsorption and denaturation. Activation energy is required to achieve each state and when it is overcome, a lowering of interfacial energy occurs and small molecule surfactants can effectively enhance protein-stabilization and dispersions, decreasing the protein surface concentration equilibrium. Some enzymatic reactions have been carried out on selected proteins in order to improve their surface activity mainly through introducing hydrophobic substituents within their structures.

Xu et al. (2005) investigated fat particle structure and stability when affected by surfactants, protein and fat substitutes within food emulsions. The two main conclusions which formed their investigations were: (1) That water-soluble or oil-soluble surfactant-stabilized food emulsions' are quite unstable under shear stress which results in developing very poor fat particle structure as a result of shear stress; (2) the stability of food emulsions is significantly affected by proteins through the formation of an adsorption layer on the surface of fat particles and micro-layering of protein submicelles around fat particles. They also concluded that the fat particle packing structure and stability of food emulsions can be manipulated by varying the protein concentration and ratio of protein to surfactants (Xu et al. 2005).

He and coworkers (2011, 2013) also reported that soybean protein isolate (SPI), whey protein isolate (WPI) and b-lactoglobulin, could be used as safe stabilizers in nanoemulsion and nanosuspension formulations. In addition a few medical-related applications have also been suggested using microbial surfactants, including several potential uses in formulations of nano-sized drug delivery vectors (Rodrigues et al. 2015).

Albumins, like bovine serum albumin, are also important components in protein–surfactant interactions as they have the ability to bind and transport small molecules (including amphiphiles). Thus, it is not particularly surprising that bovine serum albumin can bind to rhamnolipids molecules with a consequent increase in thermal stability. Rhamnolipids didn't have any effect on the bovine serum albumin structure, showing no evidence for cooperative binding accompanied by unfolding, i.e., behaving like non-ionic surfactants in their protein binding. Their interactions were therefore dominated by entropy, i.e., hydrophobic interactions, rather than electrostatic interactions, despite the presence of the carboxylate group (Otzen 2015).

## *Hydrocolloids*

The stabilization of emulsions has been typically achieved using polysaccharides such as chitosan, alginate, pectin, dextran, carrageenan, xanthan, scleroglucan, carboxymethylcellulose and hydroxypropyl methylcellulose (Kralova and Sjöblom 2009).

Plant-based gums are polysaccharides originating from different parts of plants (e.g., tree exudates, tuber/roots, seeds, plant cell walls, seaweeds). The growing interest in gums is due to their diverse structural properties and metabolic functions in food, cosmetic, textile, pharmaceutical and biomedical products, since they can be used as dietary fiber, gelling agents, texture modifiers, emulsifiers, thickeners, stabilizers, coating agents and packaging films (Albuquerque et al. 2016).

Cellulose, chitin, agar and starch are the most common polysaccharides used in industry. Regarding their broad range of applications, the preference for use of natural polymers is presumed over comparable synthetic materials due to their availability, non-toxicity and low cost. For instance, most of the natural gums are safe enough for oral consumption in the form of food additives or drug carriers. In addition, synthesizing natural polymers as nanomaterials enhances the industrial applicability due to its larger surface, besides the intrinsic properties mentioned above (Albuquerque et al. 2016).

Stable emulsions can be produced by the surfactant-biopolymer complex formed when chitosan is mixed with sorbitan esters. The chitosan-surfactant complexes depend on the length and degree of saturation of the surfactant hydrocarbon chains which influences the chemical structure of the sorbitan ester and the rheological properties of the formed emulsion. Although Chitosan is a natural polysaccharide, it can be synthetically produced by the deacetylation of chitin to produce poly-(1,4-a-D-glucopyranose) molecules, being a Generally Recognized as Safe (GRAS) product, listed for use in food preparations (Kralova and Sjöblom 2009).

The considerably growing interest in gums is due to their diverse structural properties and metabolic functions in food, cosmetic, textile, pharmaceutical and biomedical products (Nishinari et al. 2000), since they can be used as dietary fiber, gelling agents, texture modifiers, thickeners, emulsifiers, coating agents stabilizers, and packaging films (McClements 2005). Studies dealing with gums as texture modifying agents are continuing to increase in the food and culinary industries, especially for gelling, thickening and emulsifying purposes. Some hydrocolloids are used as gelling agents, for example, to increase the satiety sensation in the stomach, brought about by different triggering factors. Also in relation to the food industry, the most common polysaccharides used for the production of edible films are cellulose, chitosan, agar, starch, and no less important, galactomannans, since their mechanical and thermal properties have been widely exploited for biotechnological application (Albuquerque et al. 2016). Rhamnolipids emulsifying capabilities were also recently reported to be comparable to that of *Quillaja saponins* natural surfactant already in use by the food industry (Bai and McClements 2016).

The guar gum galactomannans are neutral polysaccharides isolated from the seeds of *Cyamopsis tetragonolobus* which are considered highly water soluble hydrocolloids, providing highly viscous and stable aqueous solutions that have several functions associated with their physicochemical and biological properties, which in fact offers a wide variety of applications. Compared to other gums, galactomannans exhibit excellent retention of viscosity even at low concentrations. In the food industry, galactomannans as guar and locust bean gum have a widespread use based on their ability to thicken and stabilize many food products, acting as mass-efficient aqueous thickeners, nutritional supplements and components in mixed gels (Albuquerque et al. 2016).

## Microbial BSs and their use in Foods

Yeasts, bacteria and some filamentous fungi are capable of producing BSs. Bacteria of the genera *Pseudomonas* and *Bacillus* are known as great BS producers. *Saccharomyces cerevisiae* and *Candida* are among the most commonly studied yeasts for the production of BSs (Santos et al. 2016). A BS is a surfactant produced extracellularly or as part of the cell membrane by bacteria, yeasts, and fungi from various substrates including sugars, oils, alkanes, among others. BSs have a high tendency to accumulate between fluid phases, thus reducing surface and interfacial tension at the surface and respectively. Microbial surfactants can be classified according to their origin and chemical composition. The properties of these amphipathic molecules make them suitable for many food applications, i.e., they can be used as multipurpose ingredients which exhibit emulsifier, antimicrobial and antiadhesive activities (Campos et al. 2013).

The most suitable property for microbial surfactants in the food industry is the ability to form stable emulsions, which can significantly improve texture, creaminess, and aesthetic appeal of dairy products. Surface active molecules can also delay staling, solubilize flavor oils and improve organoleptic properties in bakery and ice cream formulations and as fat stabilizers during cooking. Rhamnolipids have been suggested as possible ingredient to improve dough characteristics of bakery products; however, this has been hampered by the inability to use a compound derived from a Group II type opportunistic pathogen such as *P. aeruginosa* as a food ingredient. Other BSs obtained from yeasts or *Lactobacilli* that have been long recognized as safe food ingredients and are already used in several food-processing technologies have been suggested instead (Nitschke and Costa 2007). The strong surface and interfacial activity and capacity to act as a wetting and foaming agent, the emulsion forming and stabilization performances, in addition to some antimicrobial and anti-adhesive activities make the use of such lipopeptide BSs in food stuffs formulation highly desirable (Nitschke and Costa 2007, Muthusamy et al. 2008).

BS applications in foods are scarcely reported. Table 1 lists the studies reported in the literature for the use of BS in food products, the first of which were in the 1990s.

Later work investigated emulsions' stability of mayonnaises formulations using different combinations of gums and a BS produced from *Candida utilis* (Figure 1). The black circles in Figure 1 show the separation of phases that occurred in emulsions after four weeks of refrigeration. The treatment identified with number five was the most stable emulsion with 0.7% (w/v) of BS and 0.2% (w/v) Guar gum (Campos et al. 2015).

Surface active products from *Candida* are quite common and many species have been suggested for beneficial use in foods, including *C. maltosa*, *C. intermedia*, *C. versatilis*, *C. etchellsii* and *C. zeylanoides* (Bourdichon et al. 2012). Liposan, a water-soluble emulsifier that is synthesized by *C. lipolytica*, has been used with other

**Table 1.** Food products formulated with BSs.

| BSs | Microorganism | Type of Food | Reference |
|---|---|---|---|
| Not determined | *Candida utilis* | Salad dressing | Sheperd and Rockey (1995) |
| Mannoproteins | *S. cerevisiae* | Mayonnaise formulation | Torabizadeh et al. (1996) |
| Triacylglycerol lipase and native soluble lipase | *Candida antarctica* | Commercial coconut milk | Neta et al. (2012) |
| Various types | *Yarrowia lipolytica* | Fermented dairy and meat products | Groenewald et al. (2014) |
| Complex carbohydrate-lipid-protein | *Candida utilis* | Mayonnaise | Campos et al. (2015) |
| Lipopeptides | *Bacillus subtilis* SP1 | Cookies | Zouari et al. (2016) |

**Figure 1.** Emulsion stability of mayonnaises formulated with different combinations of carboxymethyl cellulose, guar gum and BS (0.7%) during four weeks of refrigeration.

oils in the cosmetic and food industries for producing stable oil/water emulsions (Santos et al. 2016). *Yarrowia lipolytica* is known for its marked proteolytic and lipolytic activities which are primarily detected in foods with high content of fat and/ or protein, particularly in fermented dairy products and meat. Additional benefits of *Y. lipolytica* include a reduction in fruit ripening times, with its associated extended shelf life and economic benefits, particularly for some cheeses (Groenewald et al. 2014).

Zouari et al. (2016) added *B. subtilis* SPB1 BS at 0.1% w/w to sesame peels flour at varying proportions to normal white flour in five cookies' dough formulations. They observed improved dough texture and more pronounced effects of the bioemulsifiers in comparison to glycerol monostearate which is a commercial emulsifier usually used as a positive control.

Exopolysaccharides (EPSs) are polymers usually produced by microorganisms to form a slimy biofilm layer loosely attached to the cell surface or completely excreted into the surrounding environment. Several microbial EPSs are widely used as food additives by the food industries as they are generally recognized as safe components acting as bioemulsifiers, bioflocculants, pharmaceuticals and chemical products. In addition, bacterial EPSs have been shown activity as immunostimulatory, immunomodulatory, anti-inflammatory, anti-viral, anti-tumor and antioxidant agents (Han et al. 2015). Recent studies showed that natural diterpenoid steviol glycosides, such as stevioside, possess significant surface activity and could also be developed into a new type of BS (Wan et al. 2014a, b). Modifications in the microbial cell membranes, such as change in proteins composition or increase of the cell wall hydrophobicity due to the reduction of lipopolysaccharides, caused by BS can promote higher accessibility to hydrocarbons by these cells. This is often enhanced by the dispersion of the hydrocarbon through its encapsulation in micelles, spherical or irregular vesicles and lamellar structures (Aparna et al. 2012).

The properties of polysaccharides such as high molecular weight, hydrophilicity and oil-water interfaces make them interesting stabilizing agents. The formation of gels that act as emulsifiers is promoted by the presence of the polysaccharide gum in the aqueous phase.

When BSs are released in aqueous solutions, their monomers usually organize spherically into micelles in a way that the hydrophobic portion is turned to the center, composing the nucleus and the hydrophilic part is turned to the sphere surface, making an interface with water. Thus, the surfactant reduces the surface tension between water and oil and contributes to micelle formation, increasing hydrocarbon exposure to bacteria and oxygen and favoring, for example, hydrocarbon biodegradation (Soberón-Chávez and Maier 2010). In a micelle, on the hydrophilic end of the BS, which makes an interface with water, a double, compact, electric layer that surrounds the external surface of the micelle sphere is formed and is called the Stern Layer (Tondo et al. 2010).

BSs can act either as emulsifiers or de-emulsifiers. An emulsion is a heterogeneous system consisting of at least one immiscible liquid dispersed in another, in the form of droplets, the diameter of which generally does not exceeds 100 µm, and there are three main types of emulsions that are important in foods (Banat et al. 2010): (1) Oil-in-water (o/w) emulsions are comprised of oil droplets suspended in an aqueous phase and are the basis of water-based products. They can be found in creams like ice creams mixes, mayonnaises cream liqueur and many others. Their properties are affected by the surfactants and the components present in the water phase; (2) Water-in-oil (w/o) emulsions are the opposite-water droplets suspended in a continuous oil phase. The stability of these emulsions depends more on the properties of fat and also on the surfactant used; (3) Water-in-oil-in-water (w/o/w) emulsions that are o/w emulsions whose droplets themselves contain water droplets (i.e., are w/o emulsion) (Benichou et al. 2004, Tadros 2005, Morais et al. 2008).

Emulsions have minimal stability, but the addition of BSs can lead to an emulsion that remains stable for months or even years (Banat et al. 2010). The use of BS in food emulsions are often added to the knowledge of its formation, properties and stabilization of the structure, such as whole milk, cream, mayonnaises and dressings, butter, coffee creamers, etc. (Kralova and Sjöblom 2009).

A good emulsifier shows ability to adsorb at the oil-water interface and to avoid the coalescence of the fine droplets formed, while a good stabilizer shows capacity to maintain the droplets separated from each other during storage.

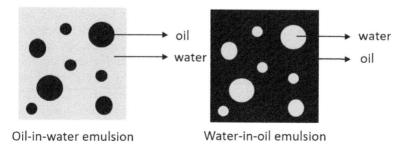

**Figure 2.** Oil-in-water and water-in-oil emulsions.

Emulsions are inherently thermodynamically unstable and yet can be kinetically stabilized by the addition of emulsifiers which can be adsorbed onto the interfaces between the immiscible compounds. The stability of the produced emulsion can be affected by oil type, interfacial tension, the continuous phase viscosity, the presence of steric and electrostatic barriers, phase volume ratio, droplet size, salt concentration, pH and temperature. Emulsions stability can either be a desirable outcome for some applications or an undesirable occurrence for others such as oil separation and recovery which may make it necessary to promote their breakage. Emulsions originating from food products are very complex structures since they are colloidal systems formed by combinations of lipids, proteins and carbohydrates. Food emulsions are difficult to stabilize and may result in different types of arrangements as gels, foams, dispersions, etc.

Although there are some studies reporting on oil separation, it remains essential to explore such methods, as well as the involved microorganisms to ensure becoming more competitive in the use of biotechnological techniques for the production of oils (Moreira et al. 2016).

## Future BSs Potential Demand in the Food Industry

The high demand for products and process ingredient can have a significant impact on the cost of manufacturing end-product, and supply source. This has been evident in recent years through price fluctuations, particularly for the bioemulsifiers/stabilizers gum arabic and locust bean gum, which have been mainly due to either a shortfall in supply (low crop yields) or increase in energy/transport costs. It is generally accepted that climate change will significantly reduce our regional and global crop yield outputs in the future. One of the ways we can address the expected future shortfalls in the supply chain of plant BSs finding more reliable alternative sources for these types of ingredients is becoming an important issue. Microbial BSs offer a reliable and sustainable alternative resource to the extraction of these chemicals from plant or animal sources or production through organo-chemical synthesis. To maximise their potential therefore, focused research is needed in order to not only understand their properties but to produce at a competitive price.

Discovering new microbial polysaccharides and surfactants to use as new ingredient additives is becoming a high priority aim to ensure suitable additives with thickening and stabilizing abilities similar to xanthan gum or a new gelling emulsifier like emulsan. This has been driven by a growing desire to reduce dependency on plant emulsifiers, which are presently mostly produced by genetically modified soybean by many industries hoping to maintain a free from GMO label. Lecithin, which is traditionally used by many food industries for example, is becoming difficult to source from non-genetically modified (GM) plants/seeds. This, therefore, is creating some limitations for food industries reluctant to use GM-based products due to public perceptions or regulatory policies in some countries. This is in addition to the advantages of other favourable properties such as antioxidants, anti-adhesive, antimicrobial and biofilm disruption capacity all of which have a driving interest in finding natural alternative sources for amphiphilic molecules suitable for use in new and advanced formulations in food and other industries (Campos et al. 2013). The provision of natural compounds with the desired characteristics equal to such products traditionally available from animals/plants will add a lot of economic, social and environmental benefits.

One method that has been explored to increase the production yields of BSs relates to designing new techniques for over-producing stains and selective-tailored strains for production of particular congeners through the use of recombinant DNA technology to express them *in vitro* using alternative host systems or genetically modified strains (Xu et al. 2005). Whilst such methods are promising for boosting production yields, nonetheless, traditional methods of screening from natural environments, particularly from relatively unexplored niches, still continues to be a reliable tool for discovering novel BSs for commercial exploitation (Banat et al. 2000, Satpute et al. 2009).

The global surfactant market is estimated to be 650,000 tonnes per annum and it is valued at over US$ one billion, with the Food & Drink (FAD) sector comprising 70% of this market. The surfactant market is divided into 2 main units: lecithin (20%; worth $200 million), and other surfactants (50%; worth $500 million). The bakery industry is the single largest user of food emulsifiers (50%; *www.foodproductdesign. com*). The FAD surfactant market has registered continued growth over the last eight years. On the other hand, the global protein ingredients market is divided between animal proteins (2.3M tonnes in 2012) and plant proteins (1.7M tonnes in 2012) with a value of €20 billion per annum *(http://www.nutraceuticalsworld.com/contents/ view_breaking-news/2013-08-13/)*. Of the animal proteins, 49% of the market share is milk proteins, 40% egg proteins, and ≈ 11% gelatin.

The plant protein market is dominated by soy at 56% and wheat protein at 43%. The value of animal proteins is highly volatile, with milk proteins currently trading around ≈ 7€/kg. The cost of egg proteins has risen dramatically over the last two years with the introduction of new EU laws on the keeping of hens in battery cages (*http://www.legaleggs.com/page/eu-legislation*) with the consequence that many food manufacturers are actively looking to replace eggs in their food formulations. Plant proteins have a large cost advantage over animal proteins (as high as 30–40%) but many have issues with off-flavours and inferior functionality; see (*http://www. naturalproductsinsider.com/articles/2013/07/the-global-protein-ingredients-market. aspx*). Food companies can benefit from a novel source of protein-based emulsifier that can replace animal (and plant) protein as a functional ingredient. The global protein market is worth ≈ €20 billion/year of which animal proteins account for approximately €11 billion. Thus, the potential market for a sustainable source of proteins is huge.

Another factor affecting future demand in Europe for example is legislation pushing the market toward more 'natural' and sustainable alternatives to chemically synthesized emulsifiers, driven to an extent by consumer awareness and a conceptual consideration for the environment. New EU directives have set a ban on the use of various commercial food emulsifiers as additives for the organic foods sector; for example, there is a strong demand by the food industry to replace emulsifiers with E-numbers: E-431 to 436 Sorbates & Stearates, and the E-471 to 495 mono-

and di-glycerides of fatty acids. These emulsifiers are synthetically derived from animal sources and are thus not permitted to be used in organic foods according to EU directive 2092/91 (49th Edition). Thus, there is a demand for new and natural alternatives to replace these emulsifiers.

## Conclusion

With the global trend towards the consumption of natural foods and beverages, major research is being directed towards the development of technology-based natural ingredients processes and systems and integrated solutions for the food and beverage industry. In addition, to promote the efficiency of the food supply chain as well as the quality and safety of food to meet the demands of a rapidly growing world population, it is necessary to develop natural ingredients, including BSs with adequate properties for commercial products use in emulsification, gelling and stabilizing. However, there are few researches in this area and most of them are limited by the production costs and by the difficulties of the choice of microorganisms considered as GRAS (Generally Recognized as Safe). Despite these obstacles, we believe in the development of new technologies with less detrimental effects on the environment based on BSs as alternative additives for the food industry.

## Acknowledgements

This study was funded by the Foundation for the Support of Science and Technology of the State of Pernambuco (FACEPE), the National Council for Scientific and Technological Development (CNPq), and the Coordination for the Improvement of Higher Level Education Personnel (CAPES).

## References

Albuquerque, P.B.S., L.C.B.B. Coelho, José A. Teixeira and Maria G. Carneiro-da-Cunha. 2016. Approaches in biotechnological applications of natural polymers. AIMS Mol. Sci. 3: 386–425. DOI: 10.3934/molsci.2016.3.386.

Bai, L. and D.J. McClements. 2016. Formation and stabilization of nanoemulsions using biosurfactants: Rhamnolipids. J. Colloid Interface Sci. 479: 71–79.

Banat, I.M., R.S. Makkar and S.S. Cameotra. 2000. Potential commercial applications of microbial surfactants. Appl. Microbiol. Biotechnol. 53: 495–508.

Banat, I.M., A. Franzetti, I. Gandolfi, G. Bestetti, M.G. Martinotti, L. Fracchia, T.J. Smyth and R. Marchant. 2010. Microbial biosurfactants production, applications and future potential. Appl. Microbiol. Biotechnol. 87: 427–444.

Banat, I.M., M.A.D. De Rienzo and G.A. Quinn. 2014. Microbial biofilms: Biosurfactants as antibiofilm agents. Appl. Microbiol. Biotechnol. 98: 9915–9929.

Batista, R.M., R.D. Rufino, J.M. Luna, J.E.G. Souza and L.A. Sarubbo. 2010. Effect of medium components on the production of a biosurfactant from *Candida tropicalis* applied to the removal of hydrophobic contaminants in soil. Water Environ. Res. 82: 418–425.

Benichou, A., A. Aserin and N. Garti. 2004. Double emulsions stabilized with hybrids of natural polymers for entrapment and slow release of active matters. Adv. Colloid Interface Sci. 108-109: 29–41.

Bourdichon, F., S. Casaregola, C. Farrokh, J.C. Frisvad, M.L. Gerds, W.P. Hammes, J. Harnett, G. Huys, S. Laulund, A. Ouwehand, I.B. Powell, J.B. Prajapati, Y. Seto, E. Ter Schure, A. Van Boven, V.

Vankerckhoven, A. Zgoda, S. Tuijtelaars and E.B. Hansen. 2012. Microbial production and application of sophorolipids. Int. J. Food Microbiol. 154: 87–97.

Campos, J.M., T.L.M. Stamford, L.A. Sarubbo, J.M. Luna, R.D. Rufino and I.M. Banat. 2013. Microbial biosurfactants as additives for food industries. Biotechnol. Progr. 29: 1097–1108.

Campos, J.M., T.L.M. Stamford, R.D. Rufino, J.M. Luna, T.C.M. Stamford and L.A. Sarubbo. 2015. Formulation of mayonnaise with the addition of a bioemulsifier isolated from *Candida utilis*. Toxicol. Reports 2: 1164–1170.

Cheok, C.Y., H.A.K. Salman and R. Sulaiman. 2014. Extraction and quantification of saponins: a review. Food Res. Int. 59: 16–40.

Dubey, K.V., A.A. Juwarkar and S.K. Singh. 2005. Adsorption-desorption process using woodbased activated carbon for recovery of biosurfactant from fermented distillery wastewater. Biotechnol. Prog. 21: 860–867.

Food and Agriculture Organization (FAO) Quillaia Extracts, Type 1 and Type 2. Chemical and Technical Assessment (CTA). In: http://www.fao.org/fileadmin/templates/agns/pdf/jecfa/cta/61/QUILLAIA. pdf Accessed in 29/08/16.

Groenewald, M., T. Boekhout, C. Neuvéglise, C. Gaillardin, P.W.M. van Dijck and M. Wyss. 2014. *Yarrowia lipolytica*: Safety assessment of an oleaginous yeast with a great industrial potential. Crit. Rev. Microbiol. 40: 187–206. DOI: 10.3109/1040841X.2013.770386.

Gusmão, C.A.B., R.D. Rufino and L.A. Sarubbo. 2010. Laboratory production and characterization of a new biosurfactant from *Candida glabrata* UCP 1002 cultivated in vegetable fat waste applied to the removal of hydrophobic contaminant. World J. Microbiol. Biotechnol. 26: 1683–1692.

He, W., Y. Tan, Z. Tian, L. Chen, F. Hu and W. Wu. 2011. Food protein-stabilized nanoemulsions as potential delivery systems for poorly water-soluble drugs: preparation, *in vitro* characterization, and pharmacokinetics in rats. Int. J. Nanomedicine 6: 521–533.

He, W., Y. Lu, J. Qi, L. Chen, F. Hu and W. Wu. 2013. Food proteins as novel nanosuspension stabilizers for poorly water-soluble drugs. Int. J. Pharm. 441: 269–278.

Kato, A. 2002. Industrial application of maillard-type protein-polysaccharide conjugates. Food Sci. Technol. Res. 8: 193–199.

Kourkoutas, Y. and I.M. Banat. 2008. Biosurfactant production and application. pp. 505–515. *In*: A.P. Pandey (ed.). The Concise Encyclopedia of Bioresource Technology. Haworth Reference Press, Philadelphia, PL, USA.

Kralova, I. and J. Sjöblom. 2009. Surfactants used in food industry: A review. J. Dispers. Sci. Technol. 30: 1363–1383.

Lazaridou, A., T. Roukas, C.G. Biliaderis and H. Vaikousi. 2002. Characterization of pullulan produced from beet molasses by *Aureobasidium pullulans* in a stirred tank reactor under varying agitation. Enzyme Microbiol. Technol. 31: 122–132.

Li, S. and N.P. Shah. 2014. Antioxidant and antibacterial activities of sulphated polysaccharides from *Pleurotus eryngii* and *Streptococcus thermophiles*. Food Chem. 165: 262–70.

Luna, J.M., R.D. Rufino, C.D.C. Albuquerque, L.A. Sarubbo and G.M. Campos-Takaki. 2011. Economic optimized medium for tenso-active agent production by *Candida sphaerica* UCP 0995 and application in the removal of hydrophobic contaminant from sand. Int. J. Mol. Sci. 12: 2463–2476.

Maneerat, S. 2005. Production of biosurfactants using substrates from renewable-resources. Songklanakarin J. Sci. Technol. 27: 675–683.

McClements, D.J. 2005. Food Emulsions: Principles, Practices, and Techniques (2nd ed.). CRC Press. Boca Raton, Florida, FL, USA.

McClements, D.J. and C.E. Gumus. 2016. Natural emulsifiers—Biosurfactants, phospholipids, biopolymers, and colloidal particles: Molecular and physicochemical basis of functional performance. Adv. Colloid Interface Sci. 234: 3–26. Doi: 10.1016/j.cis.2016.03.002.

Mironenko, N., T. Brezhneva, T. Poyarkova and V. Selemenev. 2010. Determination of some surface-active characteristics of solutions of triterpene saponin derivatives of oleanolic acid. Pharm. Chem. J. 44: 157–160.

Mitra, S. and S.R. Dungan. 2000. Micellar properties of quillaja saponin 2. Effect of solubilized cholesterol on solution properties. Colloids Surf. B: Biointerfaces 17: 117–133.

Morais, J.M.O.D.H. Santos, J.R.L. Nunes, C.F. Zanatta and P.A. Rocha-Filho. 2008. W/O/W multiple emulsion obtained by one-step emulsification method and evaluation of the involved variables. J. Dispersion Sci. Technol. 29: 63–69.

Moreira, T.C.P., V.M. Silva, A.K. Gombert and R.L. Cunha. 2016. Stabilization mechanisms of oil-in-water emulsions by *Saccharomyces cerevisiae*. Colloids Surf. B: Biointerfaces 143: 399–405.

Muthusamy, K., S. Gopalakrishnan, T.K. Ravi and P. Sivachidambaram. 2008. Biosurfactants: Properties, commercial production, and application. Curr. Sci. 94: 736–747.

Neta, N.A.S, J.C.S. Santos, S.O. Sancho, S. Rodrigues, L.R.B. Gonçalves, L.R. Rodrigues and J.A. Teixeira. 2012. Enzymatic synthesis of sugar esters and their potential as surface-active stabilizers of coconut milk emulsions. Food Hydrocoll. 27: 324–331.

Nishinari, K., H. Zhang and S. Ikeda. 2000. Hydrocolloid gels of polysaccharides and proteins. Curr. Opin. Colloid Interface Sci. 5: 195–201.

Nitschke, M. and S.G.V.A.O. Costa. 2007. Biosurfactants in food industry. Trends Food Sci. Technol. 18: 252–259.

Oleszek, W. and A. Hamed. 2010. Saponin-based surfactants. pp. 239–248. *In*: M. Kjellin and Johansson (eds.). Surfactants from Renewable Resources. John Wiley & Sons Ltd., United Kingdom.

Otzen, D.E. 2015. Proteins in a brave new surfactant world. Curr. Opin. Colloid Interface Sci. 20: 161–169.

Pacwa-Plociniczak, M., G.A. Plaza, Z. Piotrowska-Seget and S.S. Cameotra. 2011. Environmental applications of biosurfactants: Recent advances. Int. J. Mol. Sci. 13: 633–654.

Rodrigues, L.R. 2015. Microbial surfactants: Fundamentals and applicability in the formulation of nano-sized drug delivery vectors. J. Colloid Interface Sci. 449: 304–316.

Rufino, R.D., J.M. Luna, G.I.B. Rodrigues, G.M. Campos-Takaki, L.A. Sarubbo and S.R.M. Ferreira. 2011. Application of a yeast biosurfactant in the removal of heavy metals and hydrophobic contaminant in a soil used as slurry barrier. Appl. Environ. Soil Sci. Vol. 2011, Article ID 939648, 7 pages, Doi: 10.1155/2011/939648.

Sachdev, D.P. and S.S. Cameotra. 2013. Biosurfactants in agriculture. Appl. Microbiol. Biotechnol. 97: 1005–1016.

Santos, D.K.F., R.D. Rufino, J.M. Luna, V.A. Santos, A.A. Salgueiro and L.A. Sarubbo. 2013. Synthesis and evaluation of biosurfactant produced by *Candida lipolytica* using animal fat and corn steep liquor. J. Pet. Sci. Eng. 105: 43–50.

Satpute, S.K., A.G. Banpurkar, P.K. Dhakephalkar, I.M. Banat and B.A. Chopade. 2009. Methods for investigating biosurfactants and bioemulsifier: a review. Crit. Rev. Biotechnol. 30: 127–144.

Sheperd, R., J. Rockey, I.W. Sutherland and S. Roller. 1995. Novel bioemulsifiers from microorganisms for use in foods. J. Biotechnol. 40: 207–217.

Silva, R.C.F.S., D.G. Almeida, J.M. Luna, R.D. Rufino, V.A. Santos and L.A. Sarubbo. 2014. Applications of biosurfactants in the petroleum industry and the remediation of oil spills. Int. J. Mol. Sci. 15: 12523–12542.

Soberón-Chávez, G. and R.M. Maier. 2010. Biosurfactants: a general overview. pp. 1–11. *In*: G. Soberón-Chávez (ed.). Biosurfactants: From Genes to Applications. Springer, Münster, Germany.

Tadros, T.F. 2005. Applied Surfactants, Principles and Application. Wiley-VCH. Weinheim, Germany.

Tondo, D.W., E.C. Leopoldino, B.S. Souza, G.A. Micke, A.C.O. Costa, H.D. Fiedler. C.A. Bunton and F. Nome. 2010. Synthesis of a new zwitterionic surfactant containing an Imidazolium ring. Evaluating the chameleon-like behavior of zwitterionic micelles. Langmuir 26: 15754–15760.

Torabizadeh, H., S.A. Shojaosadati and H.A. Tehrani. 1996. Preparation and characterisation of bioemulsifier from *Saccharomyces cerevisiae* and its application in food products. Lebensm-Wiss u-Technol. 29: 734–737.

Vijayakumar, S. and V. Saravanan. 2015. Biosurfactants-types, sources and applications. Res. J. Microbiol. 10: 181–192.

Wan, Z.L., J.M. Wang, L.Y. Wang, Y. Yuan and X.Q. Yang. 2014a. Complexation of resveratrol with soy protein and its improvement on oxidative stability of corn oil/water emulsions. Food Chem. 161: 324–331.

Wan, Z.L., L.Y. Wang, J.M. Wang, Y. Yuan and X.Q. Yang. 2014b. Synergistic foaming and surface properties of a weakly interacting mixture of soy glycinin and biosurfactant stevioside. J. Agri. Food Chem. 62: 6834–6843.

Wan, Zhi-Li, L.Y. Wang, X.Q. Yang, J.M. Wang and L.J. Wang. 2016. Controlled formation and stabilization of nanosized colloidal suspensions by combination of soy protein and biosurfactant stevioside as stabilizers. Food Hydrocoll. 52: 317–328.

Xu, W., A. Nikolov and T.W. Darsh. 2005. Shear-induced fat particle structure variation and the stability of food emulsions: II. Effects of surfactants, protein and fat substitute. J. Food Eng. 66: 107–116.

Yang, Y., M.E. Leser, A.A. Sher and D.J. McClements. 2013. Formation and stability of emulsions using a natural small molecule surfactant: Quillaja saponin (QNaturale®). Food Hydrocoll. 30: 589–596.

Zouari, R., S. Besbes, S. Ellouze-Chaabouni and D. Ghribi-Aydi. 2016. Cookies from composite wheat–sesame peels flours: Dough quality and effect of *Bacillus subtilis* SPB1 biosurfactant addition. Food Chem. 194: 758–769.

# 11

# Biosurfactants in Cosmetic, Biomedical and Pharmaceutical Industry

*Letizia Fracchia,*[1] *Chiara Ceresa*[1] and *Ibrahim M. Banat*[2,]*

## Introduction

Biosurfactants (BSs)—biological surface-active agents—are a structurally heterogeneous group of compounds characterized by the presence of a hydrophilic region that can be a carbohydrate, amino acid, cyclic peptide, phosphate, carboxylic acid or alcohol and of a hydrophobic region that can be composed by a saturated, unsaturated, linear or branched long-chain of fatty acids or hydrocarbon acids. These secondary metabolites BSs are produced by different genera of bacteria (*Bacillus, Lactobacillus, Pseudomonas, Burkholderia, Mycobacterium, Rhodococcus, Arthrobacter, Nocardia, Gordonia, Acinetobacter*) and fungi (*Candida, Starmerella, Trichosporon, Saccharomyces, Pseudozyma, Ustilago*) on different substrates such as sugars, oils, alkanes, and wastes, and are generally localized on microbial cell surfaces or excreted extracellularly (Desai and Banat 1997, Franzetti et al. 2009, Banat et al. 2010, Morita et al. 2011).

In nature, BSs play an essential role in many biological processes of the producing microorganisms as quorum-sensing molecules to preserve survival, uptake-promoters of poorly soluble substrates, immune modulators, virulence factors, antimicrobial agents, heavy metals binding and motility promoters (Fracchia et al. 2012).

---

[1] Department of Pharmaceutical Sciences, Università del Piemonte Orientale "A. Avogadro", Largo Donegani 2, 28100, Novara, Italy.
Email: letizia.fracchia@uniupo.it

[2] School of Biomedical Sciences, University of Ulster, Coleraine, BT52 1SA, N. Ireland, UK.

* Corresponding author: im.banat@ulster.ac.uk

BSs are mainly classified according to their chemical structure and microbial origin. They can be divided into two main classes: low molecular weight, including glycolipids, lipopeptides and phospholipids, that efficiently lower surface tension and interfacial tension, and high molecular weight, comprising polymeric and particulate surfactants, that are more efficient as emulsion-stabilizing agents (Smyth et al. 2010a, b). These surface-active compounds are commonly and respectively referred to as BSs and bioemulsifiers (BEs). However, the same molecule can show, in many cases, both properties and these terms are often used interchangeably. The best-studied BSs are the glycolipids such as rhamnolipids (RLs), trehalolipids (TLs), sophorolipids (SLs) and mannosylerythritol lipids (MELs) and the lipopeptides such as surfactin and fengycin. Among BEs, the most known is emulsan, produced mainly by *Acinetobacter*.

The orientation and the behavior of these molecules on surfaces and at interphases allow them the ability to reduce surface and interfacial tensions at liquid/ liquid, liquid/gas or liquid/solid interfaces forming microemulsions and micelles between different phases that affect the properties of the original surface (Chen et al. 2010a, b). In addition, they display several interesting biological properties such as antibacterial, antifungal and anti-adhesive activities. In particular, BSs have the ability to disturb cell membrane integrity and permeability (Ortiz et al. 2009, Zaragoza et al. 2009, Sánchez et al. 2010) and to affect microorganisms' adhesion to biotic and abiotic surfaces.

Interest in BSs has also been increased by their wide range of functional properties (such as emulsification, wetting, foaming, cleansing, phase separation, surface activity, and reduction in viscosity of heavy liquids), and the diverse biosynthetic capabilities of microbes (either physiological or induced) (Singh et al. 2007, Satpute et al. 2010, Makkar et al. 2011). Furthermore, compared to their chemical counterparts, BSs are characterized by a series of commercial advantages as lower toxicity, higher biodegradability, greater environmental compatibility, better foaming properties, stable activity at extremes pH, salinity and temperatures, making them "green alternatives" (Makkar et al. 2011). Other striking advantages are the possibility to produce them from cheap renewable substrates from various industries (Banat et al. 2014a) and to alter their chemical composition by tailoring them via genetic engineering or biological/biochemical techniques, to meet specific functional requirements.

In the past twenty years, large number of researches and patents have been focused on their potential use in numerous industrial and environmental fields, such as bioremediation, enhanced oil recovery, cosmetics, pharmaceutics, healthcare, detergents, agrochemical, food, paint, textile (Table 1) (Lourith and Kanlayavattanakul 2009, Perfumo et al. 2010, Banat et al. 2010, Kanlayavattanakul and Lourith 2010, Müller et al. 2012, Morita et al. 2013) and several commercial products are already available in the market as natural substitutes of surfactants derived from petrochemical or oleochemical sources (Fracchia et al. 2014, Morita et al. 2015, Vecino et al. 2017).

This chapter will describe recent advances in BSs application in cosmetic, pharmaceutical and biomedical fields and explore the existence of marketable products and patents related to above applications.

**Table 1.** List of microorganisms producing BSs and their biological activities (adapted from Mnif and Ghribi 2015a, b, Shekhar et al. 2015).

| Biosurfactants | Producing Microorganisms and Molecules | Biological Activity |
|---|---|---|
| Lipopeptides | *Bacillus subtilis*—surfactin | Antimicrobial, antiviral, antiadhesive, insecticidal, anti-tumor activities, enhancement of hydrocarbon biodegradation |
| | *B. subtilis*—fengycin, iturin | Antifungal activity |
| | *B. pumilus*—pumilacidin | Antifungal activity |
| | *B. lycheniformis*—lychenisin | Surface activity, oil displacement and hemolytic activity |
| | *Serratia marcescens*—serrawettin | Surface-active property |
| | *Pseudomonas fluorescens*—viscosin | Surface-active compound |
| | *Arthrobacter* sp.—arthrofactin | Surface-active compound, antiadhesive, antifungal and swarming activities |
| | *Paenibacillus* sp.—polymyxins | Antimicrobial, immuno-modulating activities |
| | *Pseudomonas corrugata*—cormycin | Phytotoxicity, antibiosis, and hemolytic activity |
| | *Scopulariopsis brevicaulis*—scopularides | Excellent anti-cancer activity and weak anti-bacterial activity |
| | *Glarea lozoyensi*—pneumocandins | Anticandida and antipneumocystis activities |
| | *Aspergillus* sp.—echinocandins | Antifungal activity |
| | *Streptomyces* sp.—daptomycin | Antimicrobial, emulsifying and hemolytic activities |
| Glycolipids | *P. aeruginosa*—rhamnolipids | Biocontrol agent, anti-adhesive activity against several bacterial and yeast strains, solubilization and mobilization of hydrocarbons, dyes solubilization, anti-tumor activity, emulsifying activity and washing of hydrocarbon |
| | *Rhodococcus* sp.—trehalolipids | Inhibition of phospholipase A2, hemolytic activity, membrane-permeabilizing activity |
| | *Wickerhamiella domercqiae, Starmerella bombicola, Trichosporon asahii*—sophorolipids | Surface-active property, anticancer activity, emulsifying activity, reducing and stabilizing agent |
| | *Sympodiomycopsis paphiopedili, Cryptococcus humicola, Pseudozyma* sp.—cellobiose lipids | Antimicrobial activity, fungicidal activity, biocontrol agent, surface active property |
| | *Candida antarctica, Ustilago* sp., *Pseudozyma* sp.—mannosylerythritol lipids | Hydrocarbon solubilization, hemolytic activity, surface-active and self-assembling properties, ceramide-like skin-care property, antimicrobial, immunological and neurological properties |
| Polymeric surfactants | *Arthrobacter calcoaceticus*—emulsan | Emulsifying activity, macrophage activation, adjuvant agent |
| | *Candida tropicalis*—mannan lipid protein | |
| | *C. lipolytica*—liposan | Emulsification of edible oils |

## BSs and BEs as Natural and Biocompatible Compounds for Cosmetic Industry Applications

A great assortment of everyday products such as detergents for skin and hair, dental hygiene products, perfumes and deodorants, skin care formulations and make-up are used worldwide, and surfactants are among the most representative compounds contained in these products. Most of chemical surfactants available in the market are derived from either petrochemical or oleochemical sources. However, it is well known that these chemical-based surfactants can adversely affect the environment and both fauna and flora. In the past years, the demand for natural products has grown considerably in view of greater environmental and health awareness, and the products labelled as 'natural' or 'organic' have attracted increasing number of consumers. For these reasons, many companies are trying to replace some or all of their chemical surfactant components with similar compounds of natural origin principally produced by microorganisms from sustainable feedstock or of plant origin (Marchant and Banat 2012). Beside biodegradability and biocompatibility, the commercial success of these natural compounds are, nevertheless, related to their ability to display equal or higher performance in comparison with the chemical-based products. In this sense, BSs are microbially produced compounds with great potential in the formulation of cosmetic products due to their multifunctional properties and both sustainability and biodegradability (Corley 2007).

Attractive features of BSs for application in the cosmetic industry include emulsification, wetting, foaming, cleansing, phase separation, dispersing and surface activities (Figure 1) (Fracchia et al. 2014, Ceresa et al. 2016).

Very recently, Vecino et al. (2017) have described BSs' properties such as the critical micelle concentration (CMC), the hydrophilic-lipophilic balance (HLB), and the ionic performance as among the most important to determine the type or use of BSs in cosmetic preparations. In particular, while CMC is a general indicator of BS efficiency, HLB is a parameter that defines BSs' emulsifying ability, i.e., whether they are better stabilizer of oil in water emulsions (high HLB) or water in oil emulsions (low HLB). Moreover, the anionic BSs have the greatest wetting, foaming and emulsifying properties as compared with the cationic or non-ionic ones (Vecino et al. 2017).

In general, the commercial success of a new surfactant in dermocosmetic formulations resides in its ability to meet a series of demands and functional performance to the consumer such as agreeable appearance and feeling during usage and long-term preservation of beneficial effects (Varvaresou and Iakovou 2015). In this context, BSs are claimed to be biocompatible, and thus suitable for pharmaceutical, biomedical and cosmetic applications (Fracchia et al. 2015, Satpute et al. 2016b). However, due to their well-known intrinsic activity on cell membranes, their cytotoxicity on cell lines and on animals must be further investigated before applications to humans are envisaged. To date, several researchers have evaluated the toxicity towards animal or human cell lines revealing that BSs are nontoxic or less toxic than synthetic surfactants. Cochis et al. (2012) investigated the cytotoxicity of three lipopeptide BSs on mouse fibroblast cells and human keratinocytes at the concentrations that actively inhibited *C. albicans* biofilm adhesion on resins

**Control**     **AC7BS**

**Figure 1.** Emulsifying activity of lipopeptide AC7BS crude extract tested on kerosene. AC7BS was solubilized in PBS at the concentration of 0.5 mg/mL; control consisted of PBS only.

and silicon materials for prosthetic devices (from 78 to 312.5 µg/mL). Viability of fibroblast and keratinocyte ranged between 99% and 79% and the BSs, at all concentrations considered, were significantly less cytotoxic than chlorhexidine in both cell models.

Dehghan-Noude et al. (2005) revealed that erythrocyte hemolytic activity of the BS produced by *B. subtilis* ATCC 6633 was lower than that of synthetic cationic (benzalkonium chloride, tetradecyltrimethylammonium bromide and hexadecyltrimethylammonium bromide) and anionic (SDS) surfactants. In another work, BSs from mucoid and non-mucoid strains of *P. aeruginosa* induced dose-dependent hemolysis and coagulation of platelet-poor plasma from goat, although at low levels (negligible), but did not exhibit any harmful effect on chicken heart, lungs, liver, and kidney tissues up to a dose of 200 µg/mL (Das and Mukherjee 2005). In another study, the cytotoxicity of two fractions of a new BS produced by *Sphingobacterium detergens* 6.2S has been evaluated in model fibroblast and keratinocyte cell cultures and found to have notably less cytotoxic effects when compared to the reference surfactant SDS, suggesting low skin irritability (Burgos-Díaz et al. 2013).

BS toxicity in *in vivo* systems was also evaluated by several authors. Hwang et al. (2008) studied the genetic and developmental toxicity of surfactin C produced by a *B. subtilis* strain from Korean soybean paste. Besides not observing any genotoxic effect by bacterial reverse mutation and bone marrow micronucleus assays, surfactin C administered orally to inseminated ICR mice at concentrations up to 500 mg/kg per day showed no maternal toxicity, fetotoxicity and teratogenicity. The same authors evaluated the subacute toxicity of surfactin C in rats for four weeks at 500, 1000 or

2000 mg/kg doses. A decrease in body weight gain and an increase in relative liver weight was observed at the highest concentration. Moreover, oral administration of 1000 and 2000 mg/kg surfactin C resulted in serum ALT (alanine transaminase), AST (aspartate aminotransferase), ALP (alkaline phosphatase) levels to increase hepatocytes necrosis. No toxic effects were observed after the administration of 500 mg/kg and this concentration was suggested as the no-observed-adverse-effect level (NOAEL) of surfactin C (Hwang et al. 2009). More recently, Sahnoun et al. (2014) assessed the *in vivo* potential subacute oral toxicity of SPB1 lipopeptide using a male mice model during 28 days of treatment. The 50% lethal dose ($LD_{50}$) was approximately 475 mg/kg, and a daily dosage lower than 47.5 mg/kg of body weight had no significant adverse effect on both hematological parameters and serum biochemical data, suggesting the possible use of SPB1 BS as an additive in food, cosmetic and pharmaceutical applications. Several studies have proven the environmental compatibility and biodegradability of BSs (Fiebig et al. 1997, Kim et al. 2002b, Mohan et al. 2006, Hirata et al. 2009, Pei et al. 2009, Lima et al. 2011) and low toxicity towards marine species (Ivshina et al. 1998, Edwards et al. 2003, Kolwzan et al. 2008).

## BS-containing marketable products and patents in the cosmetic industry

Numerous applications of BSs and BEs in the cosmetic and healthcare industries have been described in the literature and several of them have already reached their commercial status (Lourith and Kanlayavattanakul 2009, Banat et al. 2010, Kanlayavattanakul and Lourith 2010, Morita et al. 2013, Fracchia et al. 2014, Morita et al. 2015). BS/BE-containing marketable products and patents are readily found on the web.

Glycolipids, such as sophorolipids (SLs), rhamnolipids (RLs) and mannosylerythritol lipids (MELs), are the most commonly used BSs in cosmetics. In particular, SLs show good skin compatibility and excellent moisturizing properties, and RLs are natural surfactants and emulsifiers that often replace chemical-based surfactants in cosmetic products (Shoeb 2013).

### SLs

SLs are nonionic surfactants with emulsifying, foaming, wetting and high deterging properties often applied as primary multifunctional cosmetic ingredients (Varvaresou and Iakovou 2015). Thanks to their hygroscopic properties, SL-derivatives containing propylene glycol have been included in moisturizers or softeners (Faivre and Rosilio 2010). Moreover, their antimicrobial properties make them suitable for their incorporation in formulations for the treatment of acne, dandruff and body odours (Lourith and Kanlayavattanakul 2009). More recently, SLs with detergent properties have been formulated in a combination of non-traditional oils with glycerol instead of the more expensive glucose using the strain *Starmerella bombicola* (ATCC 22214) (Bhangale et al. 2014).

The Japanese company Kao Co. Ltd. commercializes SLs as humectants for cosmetic makeup brands such as Sofina. The product has application in lipstick and as moisturizer for skin and hair products (Inoue et al. 1979a, b), eye shadow as well as in compressed powder cosmetics and in aqueous solutions (Kawano et al. 1981).

The French company Soliance (http://www.groupesoliance.com), which develops natural-based active ingredients for the cosmetic industry, has created Sopholiance S, a formulation containing SLs produced by *Starmerella bombicola* from rapeseed oil fermentation, with antibacterial and sebo-regulator activities for applications in deodorants, face cleansers, shower gels, make up removers, and for the treatment of acne-prone skin. The same company has also developed Hydreïs, a barrier repair and anti-aging oligosaccharide derived from the selective hydrolysis of the exopolysaccharide produced by a bacterial strain isolated from a soil subjected to extreme water stress.

The Korean biotech company MG Intobio Co. (http://mgintobio.en.makepolo.com) commercializes Sopholine cosmetics, a brand name for functional soaps, mask sheets and other cleansing products containing SLs active against *Propionibacterium acnes* and *S. aureus* for acne and atopic dermatitis treatment.

In 1999, a patent on a SL compound used for skin treatment and healing, as activator of macrophages, fibrinolytic agent, and for skin desquamation and depigmentation has been described (Maingault 1999). An antimicrobial formulation composed by fruit acids, a surfactant, and a SL, very active against *E. coli, Salmonella,* and *Shigella* has been patented for cleaning fruits, vegetable, skin, and hair (Pierce and Heilman 2001). LG Household & Healthcare Ltd. has patented a composition comprising SLs with excellent antimicrobial activity against *Propionibacterium acne, Staphylococcus, Micrococcus, Corynebacterium,* as well as moisturizing and softening effects on the skin (Han et al. 2002). More recently, a new cosmetic composition containing SLs in conjunction with a lipolytic agent for reducing the subcutaneous fat overload has been issued (Pellicier and Andre 2005).

## RLs

RLs are glycolipid BSs composed of one or two rhamnose moieties bonded to up to three groups of hydroxyl fatty acids with a chain length from eight up to fourteen, produced mainly by the pathogen *Pseudomonas aeruginosa* on glucose, glycerol or triglycerides (Lourith and Kanlayavattanakul 2009). Their excellent surface properties, known antimicrobial activity against bacteria and fungi and non-irritant nature—as indicated by EPA—make them suitable for healthcare and cosmetic applications (Abalos et al. 2001, Haba et al. 2003). Purified RLs are commercialized by companies like Urumqi Unite Bio-Technology Co. Ltd., Jeneil Biosurfactant Co. LLC, and AGAE Technologies LLC. They have been included in various cosmetic formulations such as deodorants, nail care products, toothpastes, insect repellents, antacids, acne pads, antidandruff products and contact lens solutions (Maier and Soberon-Chavez 2000, Muhammad and Mahsa 2014). Moreover, it was demonstrated that the emulsification ability of RLs improved the antimicrobial activity of essential oils (EO) by increasing their availability against *C. albicans* and *S. aureus* (MRSA), thus suggesting RL-based emulsions as a promising approach to the development of EO delivery systems (Haba et al. 2014).

Cosmetic formulation containing RLs for the skin treatments have been patented as anti-wrinkle and antiaging products (Piljac and Piljac 1999), as antimicrobial agents in commercial skin care cosmetics, in personal hygiene and care products (sprays, soaps, shampoo, and creams) and for animal cleaning (Desanto 2008). More recently, Desanto (2015) patented improvements of RLs cosmetic applications, including higher RL purities (complete removal of byproducts and of *Pseudomonas* producer strain after production), different formulations, more effective carriers and different application rates with consequent decrease of costs. Cosmetic formulations containing RLs (from 1% to 5%) such as different kind of shampoos, body wash, shower gel and oils and liquid soap have been patented by Evonik Industries AG (Brandt and Hartung 2014). Other important applications in the cosmetic industry have been patented, in the past years, with RLs for use in making liposomes and emulsions (Ishigami and Suzuki 1997, Ramisse et al. 2000).

## MEL

MELs are glycolipids mainly produced by the yeasts of the genus *Pseudozyma* at favourable production conditions, showing high structural diversity, self-assembling properties, and versatile biochemical functions (Fukuoka et al. 2007). In particular, their protective effects toward skin such as the moisturizing properties, antioxidant activity, and their ability to activate fibroblast and papilla cells and repair damaged hair suggest a potential use as cost-effective cosmetic ingredients (Morita et al. 2013, 2015). Thanks to their amphiphilic nature, MELs showed moisturizing activity on ceramide-3, an essential component of the stratum corneum (Kitamoto et al. 2009). By means of a three-dimensional cultured human skin model, it was demonstrated that MELs had recovery effects, comparable to those of ceramide-3, on sodium dodecylsulfate (SDS)-damaged cells (Yamamoto et al. 2012). Furthermore, MEL-B displayed excellent water retention properties by increasing considerably stratum corneum water content and suppressing perspiration on skin surface.

Several patents for skin care formulations containing MELs have been issued in Japan. MELs are used as the surfactant in anti-wrinkle cosmetics including tocopherol phosphate (Kato and Tsuzuki 2008), as the active ingredient in skin care cosmetics for skin roughness improvement (Masaru et al. 2007, Kitagawa et al. 2008) and as constituents in an antiaging agent (Suzuki et al. 2010).

The Japanese company Daito Kasei Kogyo Co., Ltd. started to provide new cosmetic pigments coated with MELs with excellent moisture retention properties, in place of synthetic chemicals such as silicones, suitable for makeup products such as foundations, eyeshadow and lipsticks or basic skin care (Morita et al. 2013, Kitagawa et al. 2015). The Korean company LABIO Co. Ltd. (http://www.labio.kr/) specialized in the production of fermented biotech ingredients, commercializes MELavo, a product composed of MELs with broad emulsification, moisturizing, and anti-aging benefits recommended for skin, body, and hair care applications.

## Lipopeptide BSs

Lipopeptide BSs, thanks to their exceptional surface properties and diverse biological activities that include antimicrobial and antibiofilm (Sun et al. 2006, Ceresa et al.

2016), anti-wrinkle and moisturizing activities (Kanlayavattanakul and Lourith 2010), are increasingly being included in cosmetic products.

Surfactin, the most effective lipopeptide in terms of interfacial properties, is commercially produced by Shaanxi TOP Pharm Chemical Co. Ltd., Cangzhou Pangoo International Trade Co. Ltd., and by other companies such as the Japanese Kaneka Corporation as sodium surfactin for use in cosmetics and personal care products. Surfactin has been included by the Japanese company SHOWA DENKO in their cosmetic products including a skin preparation, containing tocopherol and ascorbic acid derivatives (Kato et al. 2007) and in different forms including oil in water emulsified compositions with moisture retention and emollient properties for skin care cosmetics (Yoneda et al. 2005). Lipopeptides were also used as emulsifiers for external skin preparations such as transparent cosmetics with a sequestering function and low skin-irritating property (Yoneda et al. 1999). More recently, a U.S. patent on cosmetic composition containing surfactin in a concentration between 25–100 µm has been issued for anti-aging treatment (Lu et al. 2016). Other main applications for lipopeptides have been patented in anti-wrinkle cosmetics (Guglielmo and Montanari 2003), in the treatment and prevention of skin stretch marks (Montanari and Guglielmo 2008) and in cleansing products exhibiting excellent washability, with extremely low skin-irritating property (Yoneda 2006).

### Lipopolysaccharide BS

Emulsan, a lipopolysaccharide BEs produced by *Acinetobacter calcoaceticus* have been incorporated in a moisturizing cream and a lotion for topical application to skin or scalp with ability to remove sebum and to interfere with microbial adhesion on skin or hair, and in shampoos and soaps against acne and eczema (Hayes 1991). More recently, other compositions containing emulsan from *Acinetobacter* have been patented for the formation and stabilization of oil-in-water emulsions (Gutnik and Bach 2002).

## Biomedical and Pharmaceutical Applications of BSs

The applications of BSs in biomedical and pharmaceutical industries have increased in the past two decades (Banat et al. 2000, Rodrigues et al. 2006b, Bhadoriya et al. 2013, Fracchia et al. 2015). Their interesting biological properties, such as antibacterial, antifungal, antiviral activities and anti-adhesive and antibiofilm properties are the most relevant for biomedical and pharmaceutical related applications. Additionally, BSs can be used as gene or drug delivery systems, wound healing, anticancer and immune-modulator compounds.

### BSs as antimicrobial compounds

Antimicrobial agents are routinely used to control and treat bacterial and fungal infections. Nowadays, the growing number of pathogens that are resistant to conventional drugs represents a serious problem in the health care system and highlights the urgent need for new antimicrobial molecules (Peters et al. 2010).

In this context, microbial metabolites represent one of the major sources of bioactive compounds (Alvin et al. 2014, Monciardini et al. 2014). In particular, BSs have attracted the attention of the scientific community due to their potent antibacterial and antifungal properties (Banat et al. 2000, Marchant et al. 2012, Fracchia et al. 2014). The mechanisms of action of antimicrobial activities of BSs are membrane structural changes, disruption of protein conformations, metabolite leakages, cell lysis and alteration of vital membrane functions (Horn et al. 2012, Mandal et al. 2013, Cortés-Sánchez et al. 2013).

Lipopeptides are the most commonly reported class of BSs with antimicrobial activity (Cochrane and Vederas 2014). Well-known antimicrobial lipopeptides include surfactin, iturin, fengycin, mycosubtilins and bacillomycins produced by *Bacillus subtilis* (Vater et al. 2002), daptomycin from *Streptomyces roseosporus* (Baltz et al. 2005), polymyxin B, pumilacidin and lichenysin produced by *Bacillus polymyxa*, *Bacillus pumilus* and *Bacillus licheniformis*, respectively (Landman et al. 2008), and viscosin from *Pseudomonas fluorescens* (Saini et al. 2008).

The antimicrobial activity of a lipopeptide isolated from *Streptomyces amritsarensis* sp. Nov. was evaluated on a broad spectrum of bacteria and fungi. The purified lipopeptide was effective against *Bacillus subtilis*, *Staphylococcus epidermidis*, *Mycobacterium smegmatis* strains and a methicillin resistant *Staphylococcus aureus* (MRSA) with Minimal Inhibitory Concentration values (MICs) within a range between 10 µg/mL and 45 µg/mL. No activity was observed against any of the Gram-negative bacteria and fungi. In addition, the researchers revealed that lipopeptide had no cytotoxic and mutagenic effects against Chinese hamster ovary (CHO) cell line, which is an important prerequisite for drug development (Sharma et al. 2014).

In 2015, two linear lipopeptides, Cavinafungin A and B, and a cyclic depsilipopeptide, Colisporifungin, with antifungal activity were discovered. Cavinafungins displayed a broad-spectrum effect by inhibiting the growth of different *Candida* species within 24 h with low MICs (0.5–4 µg/mL) and *Aspergillus fumigatus* of about 80% in 48 h. In comparison, Colisporifungin did not display any antifungal activity when used alone at 8 µg/mL against all the tested strains but potentiated Caspofungin antifungal effect against *A. fumigatus* and *C. albicans* of about 5.3-fold and 2-fold, respectively (Ortiz-Lopez et al. 2015).

Very recently, Yang et al. (2016) demonstrated the potential applicability of the purified lipopeptide brevibacillin in the pharmaceutical field against Gram-positive bacteria, including *S. aureus* MRSA, vancomycin-resistant *Lactobacillus plantarum* and *Enterococcus faecalis*.

The marine strain *Aneurinibacillus aneurinilyticus* SBP-1 is the producer of a new lipopeptide BS with a promising broad spectrum of antimicrobial activity. Aneurinifactin showed low MICs against *Klebsiella pneumoniae* (4 µg/mL), *E. coli* and *S. aureus* (8 µg/mL), *P. aeruginosa*, *B. subtilis* and *V. cholera* (16 µg/mL) suggesting that this molecule can be considered as a good candidate for various biomedical applications (Balan et al. 2017).

Glycolipids are the other principal class of BSs with antimicrobial properties. RLs from *P. aeruginosa* (Benincasa et al. 2004), SLs from *Candida bombicola* (Kim et al. 2002a, Diaz de Rienzo et al. 2015) and MELs (MEL-A and MEL-B) from *Candida antarctica* (Kitamoto et al. 1993) are the well explored glycolipids. Further, a glycolipid BS from *Halomonas* sp. BS4 showed antibacterial activity against *S. aureus*, *K. pneumoniae*, *Streptococcus pyrogenes* and *Salmonella typhi* and antifungal activity against *Aspergillus niger*, *Fusarium* sp., *Aspergillus flavus* and *Trichophyton rubrum* (Donio et al. 2013).

Joshi-Navare and Prabhune (2013) observed a synergistic effect while using a combination of SLs with selected antibiotics. A strain of *S. aureus* was inhibited by treatment with 15 µg/mL tetracycline alone after 6 h exposure, but completely inhibited within 4 h when combined with 300 µg/mL SLs. Similar results were obtained when Cefaclor antibiotic was administered in combination with SLs on *E. coli*. Furthermore, observation by Scanning Electron Microscopy (SEM) demonstrated the presence of pores and membrane damage on cells treated with the mixtures of the two molecules, leading to enhanced leakage of the cytoplasmic content and accumulation of cell debris.

Haba et al. (2014) investigated the potential use of emulsions containing RLs and essential oils against strains of *S. aureus* MRSA and *C. albicans*. Results demonstrated that the presence of RLs increased the antimicrobial activity of essential oils, promoting their dispersion, availability and thus promoting the inhibitory effect against the two tested strains.

Very recently, SLs were used to develop biocidal sophorolipids-grafted gold monolayers against both Gram-positive strains (*Enteroccocus faecalis*, *Staphylococcus epidermidis*, *Streptococcus pyogenes*) and Gram-negative strains (*Escherichia coli*, *Pseudomonas aeruginosa* and *Salmonella typhymurium*). The exposure to these surfaces caused a significant reduction in viability of all bacterial strains due to membrane damage as shown by fluorescent labelling and SEM-FEG analysis (Valotteau et al. 2017).

In another study, the biocidal properties of BSs produced by *Lactobacillus rhamnosus* (BSLR) and *Lactobacillus jensenii* (BSLJ) were assessed *in vitro* against clinical Multidrug Resistant (MDR) strains of *Acinetobacter baumannii*, *E. coli*, and *S. aureus* (MRSA). Both BSs were effective in killing all three MDR pathogens at 50 mg/mL. In particular, the antibacterial effect of BSLR BS was about 96–97% against *A. baumannii*, 72–85% against *E. coli*, 80–93% against *S. aureus* whereas BSLJ exhibited almost 100% activity against all the tested strains (Sambanthamoorthy et al. 2014).

More recently, a glycolipid from the marine strain *Staphylococcus saprophyticus* SBPS 15 revealed a promising antimicrobial activity against several human pathogenic bacterial and fungal clinical isolates (Mani et al. 2016). A marked reduction was observed both on Gram-negative (*E. coli*, *Klebsiella pneumoniae*, *Pseudomonas aeruginosa*, *Vibrio cholerae*, *Salmonella paratyphi*) and Gram-positive (*S. aureus* and *Bacillus subtilis*) strains, with low MICs (4–64 µg/mL). Regarding antifungal activity, the BS inhibited the growth of *Aspergillus niger* with a MIC of 16 µg/mL and of *C. albicans* and *Cryptococcus neoformans* with MICs of 32 µg/mL.

In addition to their antimicrobial activities, BSs were described as constituents of several preparations suitable for the treatment and prevention of microbial infections (Desanto 2008, Luo 2014, Yin 2014, Stadler 2016).

Desanto (2011) patented RLs-based formulations as an antimicrobial and antifungal agent for the cleaning, disinfecting and deodorizing of living and working environments. In particular, these formulations seem to be specifically useful for the treatment of surfaces for medical procedures, chemical testing, food preparation, daycare centers and hospitals.

Luo et al. (2014) patented a group of new cyclic lipopeptides, locillomycins A, B and C, extracted from a culture of *B. subtilis* Bs916 strain, with strong inhibitory activity against pathogenic fungi, bacteria and viruses. The patent describes the preparation method for antifungal, antibacterial and antiviral medicaments containing the above three bioactive molecules. In the same year, the combined use of ramoplanin and RLs have been patented for pharmaceutical applications. In particular, the invention provides the methods for treating vancomycin-resistant *Enterococcus, Clostridium difficile*, or multidrug-resistant *Clostridium difficile* infections by the administration of a therapeutically effective dose of ramoplanin in combination with one or more RLs. Furthermore, pharmaceutical compositions, a pharmaceutically acceptable carrier, as well as formulations for preventing bacterial infection were described (Yin et al. 2014). Stadler et al. (2016) patented the applicability of structurally different long chain glycolipids to avoid perishing or microbial contamination of materials in pharmaceutical, nutraceutical and cosmetic fields.

However, in spite of the high number of publications focused on antimicrobial activity of BSs, patents related to their usage in pharmaceutical, biomedical and health improvement related industries remains quite limited (Fracchia et al. 2015). Nowadays, only few BSs entered into the pharmaceutical market as antibiotic, such as daptomycin (Robbel and Marahiel 2010), caspofungin (Ngai et al. 2011), micafungin (Emiroglu 2011), and anidulafungin (George and Reboli 2012).

Daptomycin (Cubicin®-Cubist Pharmaceuticals) is a branched cyclic lipopeptide of non-ribosomal origin, produced by *S. roseosporus* as a member of an antibiotic complex (Robbel and Marahiel 2010). In 2003, it has been approved and successfully launched for the treatment of complicated skin structure infections caused by methicillin-susceptible and -resistant *Staphylococcus aureus* strains (MSSA and MRSA), *Streptococcus pyogenes, S. agalactiae, S. dysgalactiae* subsp. *equisimilis*, as well as vancomycin-susceptible *E. faecalis*. In 2006, it has been approved for the treatment of bacteremia and endocarditis caused by *S. aureus* MSSA and MRSA strains (Eisenstein et al. 2010). Daptomycin was also effective against other clinically resistant pathogens, such as coagulase-negative *staphylococci* (CoNS), vancomycin-resistant *Enterococci*, glycopeptide-intermediate-susceptible *S. aureus* (GISA) and penicillin-resistant *Streptococcus pneumoniae* (Tally et al. 1999). It was hypothesized that its marked antibacterial activity may be due to the interaction of daptomycin/calcium complex with the anionic phospholipid phosphatidylglycerol that induces daptomycin oligomerization and leads to its penetration into cell membrane (Muraih et al. 2011).

The echinocandins caspofungin, micafungin, and anidulafungin are low-toxic synthetically modified lipopeptides, originally isolated from the fermentation broths of *Glarea lozoyensis, Coleophoma empetri*, and *Aspergillus nidulans*, respectively (Wagner et al. 2006). Echinocandins inhibit the synthesis of $\beta$-(1,3)-glucan, an essential carbohydrate component for the fungal cell wall, leading to cell wall deterioration and consequent cell death (Yao et al. 2012).

Caspofungin (Cancidas®-Merck & Co., Inc.) was approved since 2001 for the treatment of invasive aspergillosis in adults, particularly in patients refractory or intolerant to standard therapy and for esophageal and invasive candidiasis. Since 2008, caspofungin has been approved for the same indications in pediatric patients (Ngai et al. 2011). Micafungin (Mycamine®-Astellas Pharma US, Inc.) has been used since 2008 in adults, adolescents and children (including neonates) against invasive candidosis, in the prophylaxis of *Candida* infection in patients undergoing allogeneic haematopoietic stem cell transplantation or in patients with suspected granulocytopenia and in the treatment of oesophageal candidosis in adults (Kofla and Ruhnke 2011). Anidulafungin (Eraxis®, Pfizer, Inc.) was also approved in 2007 to treat invasive candidiasis, candidemia and esophageal candidiasis (Sabol and Gumbo 2008).

## BSs as anti-adhesive and anti-biofilm agents

Microbes in their natural environment are predominantly associated with or attached to surfaces where they live in biofilm form (Li and Tian 2012). Microbial colonization of medical devices and the related biofilm formation represent one of the major problems in the health-care systems. The European Centre for Disease Prevention and Control reports that approximately 4,100,000 hospital patients are affected by healthcare-associated infections (HAI) every year, and that the majority of the cases are strictly associated with the presence of bacterial and/or fungal biofilms on implant surfaces (Høiby et al. 2015).

Biofilms are complex biological structures consisting of multicellular communities embedded in a hydrated extracellular polymeric matrix and separated by a network of open water channels (Flemming et al. 2016). Their architecture is the ideal environment for cell-cell interactions, such as the intercellular exchange of genetic material, communication signals and metabolites, necessary for the community sustenance and growth (Giaouris et al. 2015). In humans, biofilms protect cells from antimicrobial agents and host immune response making them highly resistant compared to their planktonic counterpart (Delcaru et al. 2016) and may represent a common starting point for microbial dissemination (Siala et al. 2016).

Biofilm formation is a multi-stage process that starts with microbial adhesion, followed by extracellular matrix production and accumulation, growth, development and maturation. Anti-biofilm strategies are aimed to either interfere with microbial adhesion thus preventing biofilm formation, or to dislodge preformed biofilms, although, in this case, the risk of microbial dispersion in other human body regions exists (Wang et al. 2011). In the last two decades, an increased number of approaches have been developed for biofilm prevention and treatment. Microbial attachment can be counteracted by introducing antimicrobial agents throughout the bulk of the material (Ahmed et al. 2006). However, despite the long-lasting antimicrobial nature of the resulting material, the presence of the antibacterial agent may alter its stability or processability, compromising the potential use in the clinical practice

(Ahmed et al. 2007, Valappil et al. 2007). On the other hand, antimicrobial agents can be used to coat medical surfaces with a little impact on bulk properties of pre-existing biomaterials but these techniques are time consuming and not always successful (Vasilev et al. 2009, de Sainte Claire 2009, Francolini and Donelli 2010). Furthermore, surface coatings releasing biocides can promote the development of microbial biocide resistance and present potential toxicity towards human cells (Hegstad et al. 2010). Surface modifications can be obtained by physical methods such as discharge treatment and ultraviolet (UV) radiation exposure that decrease microbial adhesion and biofilm formation, increasing material hydrophilicity (Pinto et al. 2010). Another possibility consists in the use of chemical methods, such as etching, oxidation, hydrolysis, plasma treatment and the introduction of reactive functional groups onto surfaces in order to modify cell-material interactions or to activate surfaces for subsequent grafting reactions (Holmes and Trabrizian 2015). These approaches present high efficiency and are associated with low toxicity and minimal development of bacterial resistance (Bazaka et al. 2012). However, these types of modification have a temporary effect, due to the reversibility of the chemical-physical treatments, thus leading to hydrophobicity recovery of surfaces (Makamba et al. 2003).

In this context, BSs have recently emerged as a potential new generation of anti-adhesive and anti-biofilm agents with enhanced biocompatibility. Due to their predisposition to partition at the interfaces and altering the chemical and physical condition of the environment in which biofilms develop, BSs interfere with microbial adhesion and limiting biofilm formation (Rodrigues et al. 2006a, Kiran et al. 2010, Biniarz et al. 2015, Diaz de Rienzo et al. 2016a). In particular, BSs can act directy on microorganisms by disrupting cell membranes, leading to increased permeability with metabolite leakage and, eventually, cell lysis. The alteration of the physical bacterial membrane structure and of protein conformation affects important membrane functions such as energy generation and transport. Moreover, BSs reduce cell surface hydrophobicity, thus limiting their adherence to the surfaces (Satpute et al. 2016a).

The study of the anti-adhesive and anti-biofilm activity of BS is commonly carried out in two different ways: in co-incubation or pre-coating conditions. Co-incubation assays are generally applied as screening or preliminary methods for the assessment of biological properties of BSs, as well as to measure their ability to dislodge pre-formed biofilms. On the other hand, pre-coating assays are commonly used to evaluate the real efficacy and the possible applicability of BSs as medical devices coating agents for the prevention of microbial adhesion and biofilm formation (Banat et al. 2014b).

A novel cyclic lipopeptide (Coryxin) from *Corynebacterium xerosis* NS5 has been described for its inhibitory and disruptive activities against bacterial biofilm in co-incubation assays. Coryxin showed the highest antiadhesive activity against *S. aureus* (77.4%) and a lower effect against *Streptococcus mutans* (72%), *E. coli* (65%) and *P. aeruginosa* (52%). In addition, the treatment of 24 h-old biofilms in polystyrene microtiter plates resulted in biofilm dislodgement up to 82.5% for *S. aureus*, 80% for *S. mutans*, 66% for *E. coli* and 30% for *P. aeruginosa* (Dalili et al. 2015). Moryl et al. (2015) evaluated the anti-biofilm activity of lipopeptide BSs produced by the *Bacillus subtilis* I'1a, using 32 uropathogenic strains belonging to 12 different species of Gram-negative and Gram-positive bacteria. In co-incubation conditions, these lipopeptides significantly reduced biofilm formation in 24 uropathogenic strains and dislodged the mature biofilms of 18 strains with a biomass reduction of about 81%. In another study, the efficacy of lipopeptides was investigated against biofilm formation of *C. albicans* strains on medical catheter devices. Co-incubation with AC7 biosurfactant (AC7BS) reduced the number of adherent cells and biofilm formation in a concentration-dependent manner, with a maximum inhibition of 69%. On the other hand, the absorption of AC7BS on silicone surface decreased the hydrophobicity of the material making it less attractive for microbial attachment (Ceresa et al. 2016).

Pradhan et al. (2014) reported a new glycolipid obtained from *Lysinibacillus fusiformis* S9 with remarkable anti-biofilm activity against pathogenic *E. coli* and *S. mutans*. In particular, the BS showed a concentration-dependent anti-biofilm activity and was able to completely prevent biofilm formation at a concentration of 40 µg/mL.

Ceresa et al. (2015) recently described the ability of a BS, produced by *Lactobacillus brevis* CV8LAC, to reduce *C. albicans* adhesion biofilm formation on silicone elastomeric disks. In co-incubation assays, CV8LAC BS (CV8LACBS) significantly reduced biofilm formation by about 90%, whereas in pre-coating conditions fungal adhesion and biofilm formation showed 60% inhibition. In addition, the growth of both planktonic and sessile cells of *C. albicans* was not affected, indicating that CV8LAC BS had no anti-fungal activity but significantly altered cell-surface interactions making the surface less supportive for microbial adhesion.

Pontes et al. (2016) demonstrated that SLs from *Starmerella bombicola* were able to prevent bacterial biofilms formation on silicone catheter tubes. Results revealed that the anti-biofilm activity of SL-coated tubes ($\geq$ 0.38 mg/mL) was more pronounced against *S. aureus* (100%) than against *E. coli* (60%). Surfaces functionalized with commercial surfactant were also tested against the same bacteria but were less effective. Moreover, no significant differences in cell viability were detected when HaCaT cells were grown on SL-coated silicone in comparison to untreated material.

The anti-biofilm properties of RLs from *P. aeruginosa*, in the presence and absence of caprylic acid and ascorbic acid, were recently investigated by Diaz de Rienzo et al. (2016b) and compared with the anionic alkyl sulphate surfactant Sodium dodecyl sulphate (SDS). They concluded that RLs were effective antimicrobial agents in the presence of caprylic acid at pH 5 as they induced cell death and disruption of biofilm better than conventional antimicrobials such as SDS at the concentration used in this study. The effects of different structures of RLs produced by the non-pathogenic *Burkholderia thailandensis* E264 were also investigated on oral pathogens bacterial strains by Elshikh et al. (2016, 2017a). They significantly inhibited the biofilm formation for *Streptococcus oralis*, *Actinomyces naeslundii*, *Neisseria mucosa* in a range of 60–70%, and the biofilm of *Streptococcus sanguinis* by 90%, in both co-incubation and pre-coating conditions. In addition, these RLs were able to dislodge pre-existing 12 h-old biofilms in a range of 50–80% for all the tested strains (Elshikh et al. 2017b).

## Application of BSs in wound healing

Wound healing is a very dynamic process that involves a cascade of cellular events which starts after a tissue injury or wound occurs. The main aim of this process is to reestablish the damaged dermal and epidermal tissues replacing dead cells and tissue scaffolding structure. This process is strictly controlled by complex biochemical and physiological changes involving soluble active mediators such as growth factors, chemokines and their overlapped interactions with extracellular matrices (Goh et al. 2016). This healing process can be hindered by many factors which include poor immune system, insufficient oxygenation or microbial infection often leading to complications and delays in healing (Mekonnen et al. 2013). Many chemical drugs have been traditionally used to enhance this healing process; however, the search for bio-active compounds with such health benefits properties have remained an area of great interest.

BSs have an advantage of low irritancy and compatibility with human skin (Rodrigues et al. 2006b). They are ideal compounds to increase $O_2$ saturation on the surface of healing wounds, providing a natural barrier to infection and aiding the migration of cells in the healing response. To date, there are few studies investigating the potential of these compounds to accelerate wound healing *in vivo* (Piljac et al. 2008, Stipcevic et al. 2006). RLs have been implicated in increasing the rate of healing in human decubitus ulcers and decrease both the migratory response (Saini et al. 2008) and proliferation (Thanomsub et al. 2006) of human carcinoma cell lines (unpublished data from our group), while SLs have been implicated in anti-sepsis activity in a rat model of disseminated intra-peritoneal infection (Hardin et al. 2007). Gupta et al. (2017) demonstrated that the glycolipid biosurfactant from *Bacillus licheniformis* SV1 was able to promote proliferation of 3T3/NIH fibroblast cells preserving cytocompatibility. Futhermore, in a wound rat model, the use of BS ointment resulted in a prompt re-epithelialization and fibroblast cell proliferation in the early phase and, subsequently, in a quicker collagen deposition in later phases of wound healing process. In 2016, an invention was issued comprising methods for the preparation of fabric-based or hydrogel-based wound dressing substrates, incorporating gas vesicles, rhamnolipids and sophorolipids with beneficial properties in wound healing (Ju et al. 2016).

There are few reports on the importance of these molecules in biomedical applications; however, they are often carried out using crude extracts containing bioactive congeners, necessitating purification. It is important, therefore, to examine such capabilities using highly purified products or separated specific congeners. In a recent study, SLs produced by the yeast *Starmerella bombicola* were fractioned and purified to obtain highly pure acidic C18:1 congeners with antimicrobial activity against the nosocomial infective agents *Enterococcus faecalis* and *Pseudomonas aeruginosa*. At SL concentrations of < 0.5 mg/mL, no cytotoxicity on endothelial and keratinocyte-derived cell lines was observed. Moreover, *in vivo* applications of

SL-containing creams on mouse did not affect the time course of healing wounds and led to skin tissue regeneration with no evidence of inflammation (Lydon et al. 2017).

Separating, purifying, identifying and testing these novel biological compounds is therefore important to establish their biomedical applications in comparison to current gold standards for decreasing both microbial load and increasing the rate and extent of epithelialization during the wound healing.

## Antiviral activity of BSs

The antiviral activity of surfactin and its analogues was first described in 1990 (Naruse et al. 1990). The major effects were observed against enveloped viruses (herpes viruses and retroviruses) suggesting that inhibitory effects were the result of physico-chemical interactions between lipopeptides and virus envelopes. In particular, it was discovered that lipopeptides lead to the disintegration of viral capsid and lipid envelope by ion channels formation and loss of viral proteins involved in virus adsorption and/or penetration (Jung et al. 2000, Seydlová and Svobodová 2008).

Hoq et al. (1997) observed that mice treated with trehalose 6,6'-dimycolate (TDM) became more resistant to intranasal infection by influenza virus. Furthermore, the authors demonstrated that the BS induced the proliferation of T-lymphocytes bearing gamma/delta T-cell receptors and hypothesized that accumulation of these T-cells in granulomatous lungs may be associated with the maintenance of acquired resistance to lethal influenza virus infection.

In 2006, the antiviral properties of lipopeptides produced by *B. subtilis* fmbj were tested *in vitro* against Pseudorabies Virus, Porcine Parvovirus, Newcastle Disease Virus and Infectious Bursal Disease Virus (Huang et al. 2006). The mixture of surfactin and fengycin directly inactivated all cell-free virus stocks and effectively inhibited infection and replication of Newcastle Disease Virus and Infectious Bursal Disease Virus but had no effect on Pseudorabies Virus and Porcine Parvovirus.

SLs and RLs alginate complex were classified as antiviral agents against human immunodeficiency virus (Shah et al. 2005) and herpes simplex virus types 1 and 2, respectively (Remichkova et al. 2008). Among SLs from *Candida bombicola* and its structural analogs, SL diacetate ethyl ester derivative was the most potent virucidal agent with a reduction of the viral titer by more than 5.2 log units within 2 min (Shah et al. 2005). The RL PS-17 and in combination with the polysaccharide alginate significantly inhibited the cytopathic effect of herpes virus in the Madin-Darby bovine kidney cell line, by limiting in a dose-dependent manner viral replication at concentrations lower than the critical micelle concentration (Remichkova et al. 2008).

## Potential anticancer activity of BSs

BSs have been recently indicated as potential antitumour agents due to their ability to control a variety of mammalian cell functions such as signal transduction, cell differentiation and cell immune response (Cortés-Sánchez et al. 2013, Gudiña et al. 2013).

Serratamolide AT514, from *Serratia marcescens,* has been reported as a potent apoptosis inducer of several cell lines derived from various human tumours and B-chronic lymphocytic leukemia cells via mitochondria-mediated apoptotic pathway and interference with Akt/NF-kB survival signals (Escobar-Díaz et al. 2005). In another study, viscosin from *Pseudomonas libanensis* M9-3, inhibited the migration of the metastatic prostate cancer cell line PC-3M without visible toxicity effects (Saini et al. 2008).

Antitumoural activities of surfactin were extensively described. Kim et al. (2007) showed that this lipopeptide blocked LoVo cell proliferation by the induction of pro-apoptotic activity and cell cycle arrest by inhibiting extracellular-related protein kinase and phosphoinositide 3-kinase/Akt activation. Cao et al. (2010) further demonstrated that surfactin induces apoptosis in human breast cancer MCF-7 cells via ROS/JNK-mediated mitochondrial/caspase pathway. More recently, lipopeptides produced by a marine *Bacillus circulans* DMS-2 showed an interesting dose-dependent cytotoxic activity against colon cancer cell lines HCT 15 and HT 29 with the highest inhibition (90%) after 24 h of treatment (Sivapathasekaran et al. 2010).

Interesting antitumour activities have also been reported for glycolipids. Zhao et al. (2000) discovered the involvement of glycolipids in the growth arrest and apoptosis of mouse malignant melanoma B16 cells. The treatment with increasing concentrations of MELs induced the entering of B16 cells in the sub-G0/G1 phases and triggered a series of apoptotic events such as condensation of chromatin and DNA fragmentation. In another study, it was observed that the exposure of PC12 cells to mannosylerythritol lipid (MEL) enhanced the activity of acetylcholinesterase and interrupted the cell cycle at the G1 phase, with a resulting outgrowth of neurites and partial cellular differentiation by the stimulation of ERK phosphorylation (Wakamatsu et al. 2001). In 2006, the cytotoxic effects of SLs from *Wickerhamiella domercqiae* on cancer cells of H7402, A549, HL60 and K562 were investigated. The authors observed a dose-dependent inhibition ratio on cell viability at SL concentrations < 62.5 µg/mL due to cell cycle block at G1 phase, activation of caspase-3 and increase of $Ca^{2+}$ concentration in the cytoplasm (Chen et al. 2006).

Fu et al. (2008) treated human pancreatic carcinoma cells with increasing concentrations of natural SLs or selected derivatives and demonstrated that the effect on cell necrosis was dependent on the type of SL. Methyl ester derivative mediated much greater levels of cytotoxicity (63 ± 5%) compared to other derivatives (ethyl ester diacetate, 36 ± 6%, ethyl ester monoacetate, 18 ± 7%) and natural mixture (20 ± 4%) in a concentration-dependent manner. In comparison, lactonic SL diacetate- and acidic SL-mediated toxicity was inversely proportional with the dose. Similarly, Shao et al. (2012) investigated the effect of different SL molecules on human esophageal cancer cell lines. Stronger inhibition was shown for SLs with higher degrees of acetylation in comparison to mono acetylated lactonic SL, for which a double concentration was necessary to obtain the same inhibitory effect. In addition, SL with one double bond in the fatty acid part had the strongest cytotoxic effect, whereas the antitumour activity of acidic SLs was scarce.

Although varying levels of anticancer and antiproliferative activities are reported in literature for SL mixtures, it is quite difficult to both correctly interpret or make conclusions due to the significant variations in these extracts' compositions depending on the producing strains, purity and structural variation. Callaghan et al. (2016) recently reported producing very pure SLs (96% pure C18:1 lactonic) using *Starmerella bombicola*. This product can dose-dependently reduce the viability of colorectal cancer, as well as normal human colonic and lung cell lines *in vitro*. However, oral administration of lactonic SLs (50 mg/kg for 70 days) to Apc$^{min+/-}$ mice resulted in an increase in the number and size of intestinal polyps indicating an exacerbation of disease progression in this model and urging caution concerning the interpretation of *in vitro* studies by examining potential anti-tumour effects of both purified and non-purified preparations.

## BSs as immuno-modulator agents

The immunosuppressive capabilities of surfactin on the immune-stimulatory functions of macrophages are attributable to the block of the NK-κB (nuclear factor kappa-light-chain-enhancer of activated B cells), MAPK (mitogen-activated protein kinases) and Akt (serine/threonine kinase Akt, also known as protein kinase) cell-signaling pathway (Park and Kim 2009). A glycolipid complex from *Rhodococcus ruber* was found to activate the production of IL-1β and TNF-α cytokines without modifying the production of IL-6 (Kuyukina et al. 2007). Furthermore, Hagler et al. (2007) demonstrated that SLs decreased sepsis related mortality *in vivo* at 36 hours in a rat model of septic peritonitis by modulation of nitric oxide, adhesion molecules and cytokine production and decreased IgE production *in vitro* in U266 cells by affecting plasma cell activity. Furthermore, the authors investigated the mechanisms involved in SL-mediated downregulation of IgE *in vitro* in a cellular model of allergic disease and discovered their ability to downregulate important genes involved in IgE pathobiology in a synergistic manner. Wong et al. (2016) patented glycosphingolipids (GSLs) compositions and methods to modulate invariant natural killer T (iNKT) cells in humans, stimulate the production of cytokines/chemokines and activate downstream immune cells bridging the innate and adaptive immunity.

Surfactin showed interesting anti-inflammatory activities due to its inhibitory properties on phospholipase A2, on the release of Interleukin (LK-6) and the overproduction of nitric oxide (Tang et al. 2010). Park et al. (2010) explored the mechanisms by which surfactin induced anti-inflammatory actions in relation to serious gum infection caused by *Porphyromonas gingivalis*. The authors also observed that surfactin significantly reduced the pro-inflammatory cytokines, including interleukin IL-6, IL-12 and IL-1β and tumour necrosis factor-α, through suppression of nuclear factor κB activity in *P. gingivalis*. Zhang et al. (2015) investigated the anti-inflammatory mechanism of surfactin in lipopolysaccharide (LPS)-stimulated macrophages. Surfactin significantly down-regulated LPS-induced IFN-γ and IL-6 production and decreased iNOS mRNA expression in a dose-dependent manner through the inhibition of the NF-κB and TLR4 cell signaling pathways. Surfactin, thus, was able to both inhibit the degradation of IκB-α and the transfer of NF-κB p65 into the nucleus of macrophages induced by LPS in both a time-dependent and dose-dependent manner and to downregulate LPS-induced TLR4 protein expression of macrophages (Zhang et al. 2015). In 2013, Muehlradt and coworkers patented pharmaceutical compositions containing lipopeptides or lipoprotein molecules for treating dysregulated inflammatory diseases.

A relatively new area of application for BSs is their use as immunological adjuvants. Bacterial lipopeptides constitute potent nontoxic and nonpyrogenic immunological adjuvants when mixed with conventional antigens. A marked enhancement of humoral immune response was obtained with the low molecular mass antigens Iturin AL, herbicolin A, and microcystin (MLR) coupled to poly-L-lysine (MLR-PLL) in rabbits and in chickens (Rodrigues et al. 2006b). In 2007, PFIZER PRODUCTS INC patented some compositions and methods for preparing stable adjuvant diluent stock solutions and final adjuvant solutions comprising glycolipids, weak acids, alcohols, nonionic surfactants and buffers. In 2016, the use of lipopeptides or lipoproteins as an adjuvant in therapeutic or prophylactic vaccinations was patented by Guzman and Muhlradt.

## BSs as drug or gene delivery agents

The development of novel drug delivery systems (DDSs) with improved efficacies has made a significant impact on the ability to treat human diseases (Fanun 2012). In the past few years, the attention has been focused on the research of ideal excipients for the design of safer microemulsions. In particular, non-ionic surfactants such as sucrose esters have been widely employed in microemulsion formulation (Chansanroj et al. 2010, Csizmazia et al. 2012).

The emulsification, foaming, detergency, and dispersion properties of BSs appear to be good candidates in the field of drug and gene transfection delivery systems (Kitamoto 2008, Liu et al. 2010, Zhang et al. 2010, Faivre and Rosilio 2010, Gudiña et al. 2013).

In 1988, RL liposomes were patented as microcapsules for drugs, proteins, nucleic acids, dyes, and other compounds, thanks to their suitable affinity for organisms, extended stability and shelf life (Gama et al. 1990).

The applicability of polymeric acylated BSs was patented for their use in topically applied dermatologic products containing pharmaceutical ingredients such as antimicrobial agents, anti-acne agents, and external analgesics (Owen et al. 2007). In 2016, Nam and co-inventors (2016) and Adami and co-inventors (2016) patented the use of a series of glycolipids and lipopeptides for the synthesis of nanoparticles employable in drug or gene delivery. Nguyen et al. (2010) described the use of SLs and RLs mixed with lecithins to prepare biocompatible micro-emulsions whereas Nicoli et al. (2010) demonstrated the efficacy of fengycin and surfactin to promote skin accumulation and transdermal penetration of acyclovir leading to a 2-fold increase in its concentration.

In 2002, Kitamoto et al. demonstrated the increased efficacy of liposomes based on BSs in gene transfection, in comparison to commercially available cationic liposomes. In particular, MELs have been successfully employed for that purpose (Igarashi et al. 2006, Ueno et al. 2007a, Kitamoto et al. 2009, Nakanishi et al. 2009) due to their ability to induce a highly efficient membrane fusion between liposomes and the plasma membrane of the target cells (Ueno et al. 2007b, Inoh et al. 2010).

## Conclusion

BSs represent a group of emerging surface-active molecules, which have numerous cosmetic, biomedical and pharmaceutical application potentials with inherent antimicrobial (bacterial, fungal and viral) properties and the ability to act as antiadhesive on surfaces and as disruptive and dispersant molecules in biofilm structures. Their uses, either on their own or as adjuvants to other antimicrobial or chemotherapeutic agents, may represent a possible way to forward in tackling infections, biofilms formation and microbial proliferation in the future. In addition, recent findigs on BSs' abilities to act as anticancer, immunomodulating, wound healing and drug delivery agents has increased the interest of researcher to investigate BSs for above mentioned activities.

## References

Abalos, A., A. Pinazo, M.R. Infante, M. Casals, F. Garcia and A. Manresa. 2001. Physicochemical and antimicrobial properties of new rhamnolipids produced by *Pseudomonas aeruginosa* AT10 from soy bean oil refinery wastes. Langmuir 17: 1367–1371.

Adami, R.C., M.E. Houston and R.E. Johns. 2016. Lipopeptides for delivery of nucleic acids. United States Patent US2016058870 (A1), March 03.

Ahmed, I., D. Ready, M. Wilson and J.C. Knowles. 2006. Antimicrobial effect of silver-doped phosphate-based glasses. J. Biomed. Mater. Res. A 79: 618–626.

Ahmed, I., E.A. Abou Neel, S.P. Valappil, S.N. Nazhat, D.M. Pickup, D. Carta, D.L. Carroll, R.J. Newport, M.E. Smith and J.C. Knowles. 2007. The structure and properties of silver-doped phosphate-based glasses. J. Mat. Sci. 42: 9827–9835.

Balan, S.S., C.G. Kumar and S. Jayalakshmi. 2017. Aneurinifactin, a new lipopeptide biosurfactant produced by a marine *Aneurinibacillus aneurinilyticus* SBP-11 isolated from Gulf of Mannar: Purification, characterization and its biological evaluation. Microbiol. Res. 194: 1–9.

Baltz, R.H., V. Miao and S.K. Wrigley. 2005. Natural products to drugs, daptomycin and related lipopeptide antibiotics. Nat. Prod. Rep. 22: 717–741.

Banat, I.M., R.S. Makkar and S.S. Cameotra. 2000. Potential commercial applications of microbial surfactants. Appl. Microbiol. Biotechnol. 53: 495–508.

Banat, I.M., A. Franzetti, I. Gandolfi, G. Bestetti, M.G. Martinotti, L. Fracchia, T.J. Smyth and R. Marchant. 2010. Microbial biosurfactants production, applications and future potential. Appl. Microbiol. Biotechnol. 87: 427–444.

Banat, I.M., S.K. Satpute, S.S. Cameotra, R. Patil and N.V. Nyayanit. 2014a. Cost effective technologies and renewable substrates for biosurfactants' production. Front. Microbiol. 5: 697.

Banat, I.M., M.A. Diaz De Rienzo and G.A. Quinn. 2014b. Microbial biofilms: biosurfactants as antibiofilm agents. Appl. Microbiol. Biotechnol. 98: 9915–9929.

Bhangale, A.P., S.D. Wadekar, S.B. Kale and A.P. Pratap. 2014. Sophorolipids synthesized using non-traditional oils with glycerol and studies on their surfactant properties with synthetic surfactant. Tenside Surfactants Detergents 51: 387–396.

Bazaka, K., M.V. Jacob, R.J. Crawford and E.P. Ivanova. 2012. Efficient surface modification of biomaterial to prevent biofilm formation and the attachment of microorganisms. Appl. Microbiol. Biotechnol. 95: 299–311.

Benincasa, M., A. Abalos, I. Oliveira and A. Manresa. 2004. Chemical structure, surface properties and biological activities of the biosurfactant produced by *Pseudomonas aeruginosa* LBI from soapstock. Antonie Van Leeuwenhoek 85: 1–8.

Bhadoriya, S.S., N. Madoriya, K. Shukla, R. Patel and M.S. Parihar. 2013. Biosurfactants: A new pharmaceutical additive for solubility enhancement and pharmaceutical development. Biochem. Pharmacol. 2: 113.

Biniarz, P., G. Baranowska, J. Feder-Kubis and A. Krasowska. 2015. The lipopeptides pseudofactin II and surfactin effectively decrease *Candida albicans* adhesion and hydrophobicity. Antonie Van Leeuwenhoek 108: 343–353.

Brandt, K.D. and C. Hartung. 2014. Cosmetics containing rhamnolipids. EP2786742 A1, October 08.

Burgos-Díaz, C., R. Martín-Venegas, V. Martínez, C.E. Storniolo, J.A. Teruel, F.J. Aranda, A. Ortiz, Á. Manresa, R. Ferrer and A.M. Marqués. 2013. *In vitro* study of the cytotoxicity and antiproliferative effects of surfactants produced by *Sphingobacterium detergens*. Int. J. Pharm. 453: 433–440.

Callaghan, B., H. Lydon, S.L.K.W. Roelants, I.N.A. Van Bogaert, R. Marchant, I.M. Banat and C.A. Mitchell. 2016. Lactonic sophorolipids increase tumor burden in Apc$^{min+/-}$ mice. PLoS One 11: e0156845.

Cameron, D.R., V.A. Boyd, R.A. Leese, W.V. Curran, D.B. Borders, P.W.M. Sgarbi, M. Nodwell, Y. Chen, Q. Jia and D. Dugourd. 2005. Compositions of lipopeptides antibiotic derivatives and methods of use thereof. World Patent WO2005000878 A3, June 16.

Cao, X.H., A.H. Wang, C.L. Wang, D.Z. Mao, M.F. Lu, Y.Q. Cui and R.Z. Jiao. 2010. Surfactin induces apoptosis in human breast cancer MCF-7 cells through a ROS/JNK-mediated mitochondrial/caspase pathway. Chem. Biol. Interact. 183: 357–362.

Cappello, S., A. Crisari, R. Denaro, F. Crescenzi, F. Porcelli and M.M. Yakimov. 2011. Biodegradation of a bioemulsificant exopolysaccharide (EPS2003) by marine bacteria. Water Air Soil Pollut. 214: 645–652.

Ceresa, C., F. Tessarolo, I. Caola, G. Nollo, M. Cavallo, M. Rinaldi and L. Fracchia. 2015. Inhibition of *Candida albicans* adhesion on medical-grade silicone by a *Lactobacillus*-derived biosurfactant. J. Appl. Microbiol. 18: 1116–1125.

Ceresa, C., M. Rinaldi, V. Chiono, I. Carmagnola, G. Allegrone and F. Letizia. 2016. Lipopeptides from *Bacillus subtilis* AC7 inhibit adhesion and biofilm formation of *Candida albicans* on silicone. Antonie van Leeuwenhoek 109: 1375–1388.

Chansanroj, K. and G. Betz. 2010. Sucrose esters with various hydrophilic–lipophilic properties: Novel controlled release agents for oral drug delivery matrix tablets prepared by direct compaction. Acta Biomater. 6: 3101–3109.

Chen, J., X. Song, H. Zhang and Y. Qu. 2006. Production, structure elucidation and anticancer properties of sophorolipid from *Wickerhamiella domercqiae*. Enzyme Microb. Technol. 39: 501–506.

Chen, M.L., J. Penfold, R.K. Thomas, T.J. Smyth, A. Perfumo, R. Marchant, I.M. Banat, P. Stevenson, A. Parry, I. Tucker and I. Grillo. 2010a. Mixing behaviour of the biosurfactant, rhamnolipid, with a conventional anionic surfactant, sodium dodecyl benzene sulfonate. Langmuir 26: 17958–17968.

Chen, M.L., J. Penfold, R.K. Thomas, T.J.P. Smyth, A. Perfumo, R. Marchant, I.M. Banat, P. Stevenson, A. Parry, I. Tucker and I. Grillo. 2010b. Solution self-assembly and adsorption at the air–water interface of the monorhamnose and dirhamnose rhamnolipids and their mixtures. Langmuir 26: 18281–18292.

Chuping, L., C. Zhiyi, L. Xuehui, G. Junyao, W. Xiaoyu and L. Youzhou. 2014. New cyclic lipopeptides locillomycin A, B and C and preparation method thereof. Cinese Patent CN103524600 (A), January 22.

Cochis, A., L. Fracchia, M.G. Martinotti and L. Rimondini. 2012. Biosurfactants prevent *in vitro* Candida *albicans* biofilm formation on resins and silicon materials for prosthetic devices. Oral Surg. Oral Med. Oral Pathol. Oral Radiol. 113: 755–761.

Cochrane, S.A. and J.C. Vederas. 2014. Lipopeptides from *Bacillus* and *Paenibacillus* spp.: A gold mine of antibiotic candidates. Med. Res. Rev. 36: 4–31.

Corley, J.W. 2007. All that is good—Naturals and their place in personal care. pp. 7–12. *In*: A.C. Kozlowski (ed.). Naturals and Organics in Cosmetics: From R & D to the Market Place. Allured Businnes Media, Carol Stream, IL.

Cortés-Sánchez Ade, J., H. Hernández-Sánchez and M.E. Jaramillo-Flores. 2013. Biological activity of glycolipids produced by microorganisms: new trends and possible therapeutic alternatives. Microbiol. Res. 168: 22–32.

Csizmazia, E., G. Eros, O. Berkesi, S. Berkó, P. Szabó-Révész and E. Csányi. 2012. Ibuprofen penetration enhance by sucrose ester examined by ATR–FTIR *in vivo*. Pharm. Dev. Technol. 17: 125–128.

Dalili, D., M. Amini, M.A. Faramarzi, M.R. Fazeli, M.R. Khoshayand and N. Samadi. 2015. Isolation and structural characterization of Coryxin, a novel cyclic lipopeptide from *Corynebacterium xerosis* NS5 having emulsifying and anti-biofilm activity. Colloids Surf. B Biointerfaces 135: 425–432.

Das, K. and A.K. Mukherjee. 2005. Characterization of biochemical properties and biological activities of biosurfactants produced by *Pseudomonas aeruginosa* mucoid and non-mucoid strains isolated from hydrocarbon-contaminated soil samples. Appl. Microbiol. Biotechnol. 69: 192–199.

Dehghan-Noude, G., M. Housaindokht and B.S. Bazzaz. 2005. Isolation, characterization, and investigation of surface and hemolytic activities of a lipopeptide biosurfactant produced by *Bacillus subtilis* ATCC 6633. J. Microbiol. 43: 272–276.

Desai, J.D. and I.M. Banat. 1997. Microbial production of surfactants and their commercial potential. Microbiol. Mol. Biol. Rev. 61: 47–64.

de Sainte Claire, P. 2009. Degradation of PEO in the solid state, a theoretical kinetic model. Macromolecules 42: 3469–3482.

Desanto, K. 2015. High purity rhamnolipid cosmetic application. World Patent WO2015030702 A3, July 09.

Delcaru, C., I. Alexandru, P. Podgoreanu, M. Grosu, E. Stavropoulos, M.C. Chifiriuc and V. Lazar. 2016. Microbial biofilms in urinary tract infections and prostatitis: etiology, pathogenicity, and combating strategies. Pathogens 5: 65.

Deleu, M., R. Brasseur, M. Paquot, H. Legros, S. Dufour, P. Jacques, J. Destain, P. Thonart, A. Brans, B. Joris and J.-M. Frere. 2004. Novel use of lipopeptides preparations. World Patent WO2004002510 A1, January 8.

Desanto, K. 2008. Rhamnolipid-based formulations. World Patent WO 2008013899 A3, October 16.

Diaz De Rienzo, M.A., I.M. Banat, B. Dolman, J. Winterburn and P.J. Martin. 2015. Sophorolipid biosurfactants, antibacterial activities and characteristics. New Biotechnology 32: 720–726.

Diaz de Rienzo, M.A., P. Stevenson, R. Marchant and I.M. Banat. 2016a. Antibacterial properties of biosurfactants against selected Gram-positive and -negative bacteria. FEMS Microbiol. Lett. 363: fnv224.

Diaz de Rienzo, M., P. Stevenson, R. Marchant and I.M. Banat. 2016b. *P. aeruginosa* biofilm disruption using microbial biosurfactants. J. Appl. Microbiol. 120: 868–876.

Dominowski, P.J., R.M. Mannan and M. Sangita. 2016. Novel glycolipid adjuvant compositions. World Patent CA2634888 (C), June 21.

Donio, M.B.S., F.A. Ronica, V.T. Viji, S. Velmurugan, J.S.C.A. Jenifer, M. Michaelbabu, P. Dhar and T. Citarasu. 2013. Halomonas sp. BS4, A biosurfactant producing halophilic bacterium isolated from solar salt works in India and their biomedical importance. SpringerPlus 2: 149–159.

Edwards, K.R., J.E. Lepo and M.A. Lewis. 2003. Toxicity comparison of biosurfactants and synthetic surfactants used in oil spill remediation to two estuarine species. Mar. Pollut. Bull. 46: 1309–1316.

Eisenstein, B.I., F.B. Jr. Oleson and R.H. Baltz. 2010. Daptomycin: from the mountain to the clinic, with essential help from Francis Tally, MD. Clin. Infect. Dis. 50: S10–5.

Elshikh, M., R. Marchant and I.M. Banat. 2016. Biosurfactants: promising bioactive molecules for oral-related health applications. FEMS Microbiol. Lett. 2016 363: fnw213.

Elshikh, M., S. Funstona, A. Chebbib, S. Ahmeda, R. Marchanta and I.M. Banat. 2017a. Rhamnolipids from non-pathogenic *Burkholderia thailandensis* E264: Physicochemical characterization, antimicrobial and antibiofilm efficacy against oral hygiene related pathogens. N. Biotechnol. 36: 26–36.

Elshikh, M., I. Moya-Ramírez, H. Moens, S. Roelants, W. Soetaert, R. Marchant and I.M. Banat. 2017b. Rhamnolipids and Lactonic Sophorolipids: Natural antimicrobial surfactants for oral hygiene. J. Appl. Microbiol. 23: 1111–1123.

Emiroglu, M. 2011. Micafungin use in children. Expert Rev. Anti Infect. Ther. 9: 821–834.

Escobar-Diaz, E., E.M. López-Martín, M. Hernández del Cerro, A. Puig-Kroger, V. Soto-Cerrato, B. Montaner, E. Giralt, J.A. García-Marco, R. Pérez-Tomás and A. Garcia-Pardo. 2005. AT514, a cyclic depsipeptide from *Serratia marcescens*, induces apoptosis of B chronic lymphocytic leukemia cells: interference with the Akt/NF kB survival pathway. Leukemia 19: 572–579.

Faivre, V. and V. Rosilio. 2010. Interest of glycolipids in drug delivery: From physicochemical properties to drug targeting. Expert Opin. Drug Deliv. 7: 1031–1048.

Fanun, M. 2012. Microemulsions as delivery systems. Curr. Opin. Colloid Interface Sci. 17: 306–313.

Fiebig, R., D. Schulze, J.-C. Chung and S.-T. Lee. 1997. Biodegradation of polychlorinated biphenyls (PCBs) in the presence of a bioemulsifier produced on sunflower oil. Biodegradation 8: 67–75.

Flemming, H.-C., J. Wingender, U. Szewzyk, P. Steinberg, S.A. Rice and S. Kjelleberg. 2016. Biofilms: an emergent form of bacterial life. Nat. Rev. Microbiol. 14: 563–575.

Fracchia, L., M. Cavallo, M.G. Martinotti and I.M. Banat. 2012. Biosurfactants and bioemulsifiers: Biomedical and related applications-present status and future potentials. pp. 325–370. *In*: D.N. Ghista (ed.). Biomedical Science, Engineering and Technology. InTech, Rijeka.

Fracchia, L., C. Ceresa, A. Franzetti, M. Cavallo, I. Gandolfi, J. Van Hamme, P. Gkorezis, R. Marchant and I.M. Banat. 2014. Industrial applications of biosurfactants. pp. 245–267. *In*: N. Kosaric and F.V. Sukan (eds.). Biosurfactant—Production and Utilization—Processes, Technologies, and Economics. CRS Press—Taylor & Francis Group, Boca Raton.

Fracchia, L., J.J. Banat, M. Cavallo, C. Ceresa and I.M. Banat. 2015. Potential therapeutic applications of microbial surface-active compounds. AIMS Bioengineering 2: 144–162.

Francolini, I. and G. Donelli. 2010. Prevention and control of biofilm-based medical-device-related infections. FEMS Immunol. Med. Microbiol. 59: 227–238.

Franzetti, A., P. Caredda, C. Ruggeri, P. La Colla, E. Tamburini, M. Papacchini and G. Bestetti. 2009. Potentials applications of surface active compounds by *Gordonia* sp. strain BS29 in soil-remediation technologies. Chemosphere 75: 801–807.

Fu, S.L., S.R. Wallner, W.B. Bowne, M.D. Hagler, M.E. Zenilman, R. Gross and M.H. Bluth. 2008. Sophorolipids and their derivatives are lethal against human pancreatic cancer cells. J. Surg. Res. 148: 77–82.

Fukuoka, T., T. Morita and H. Konishi. 2007. Structural characterization and surface-active properties of a new glycolipid biosurfactant, mono-acylated mannosylerythritol lipid, produced from glucose by *Pseudozyma antarctica*. Appl. Microbiol. Biotechnol. 76: 801–810.

Gama, Y., T. Hongu, Y. Ishigami, H. Nagahora and M. Yamaguchi. 1990. Rhamnolipid liposomes. U.S. Patent US4902512A, February 20.

George, J. and A.C. Reboli. 2012. Anidulafungin, When and how? The clinician's view. Mycoses 55: 36–44.

Giaouris, E., E. Heir, M. Desvaux, M. Hébraud, T. Møretrø, S. Langsrud, A. Doulgeraki, G.J. Nychas, M. Kačániová, K. Czaczyk, H. Ölmez and M. Simões. 2015. Intra- and inter-species interactions within biofilms of important foodborne bacterial pathogens. Front. Microbiol. 6: 841.

Goh, M.C., Y. Hwang and G. Tae. 2016. Epidermal growth factor loaded heparin-based hydrogel sheet for skin wound healing. Carbohydr. Polym. 147: 251–260.

Gudiña, E.J., V. Rangarajan, R. Sen and L.R. Rodrigues. 2013. Potential therapeutic applications of biosurfactants. Trends Pharmacol. Sci. 34: 667–675.

Guglielmo, M. and D. Montanari. 2003. Cosmetic preparation with anti-wrinkle action. World Patent WO 2003000222 A2, January 3.

Gupta, S., N. Raghuwanshi, R. Varshney, I.M. Banat, A.K. Srivastava, P.A. Pruthi and V. Pruthi. 2017. Accelerated *in vivo* wound healing evaluation of microbial glycolipid containing ointment as a transdermal substitute. Biom. Pharmacother. 94: 1186–1196.

Gutnick, D.L. and H.R. Bach. 2002. Compositions containing bioemulsifiers and a method for their preparation. World Patent WO 2002048327 A2, June 20.

Guzman, C.A. and P. Muhlradt. 2016. Use of a lipopeptide or lipoprotein as an adjuvant in therapeutic or prophylactic vaccinations. United States Patent US2016256542 (A1), September 08.

Haba, E., A. Pinazo, O. Jauregui, M.J. Espuny, M.R. Infante and A. Manresa. 2003. Physicochemical characterization and antimicrobial properties of the rhamnolipids products by *Pseudomonas aeruginosa* 47T2 NCIMB 40044. J. Surfactants Deterg. 6: 155–161.

Haba, E., S. Bouhdid, N. Torrego-Solana, A.M. Marqués, M.J. Espuny, M.J. García-Celma and A. Manresa. 2014. Rhamnolipids as emulsifying agents for essential oil formulations: Antimicrobial effect against *Candida albicans* and methicillin-resistant *Staphylococcus aureus*. Int. J. Pharm. 476: 134–141.

Han, S.G., J.C. Kim and Y.S. Kim. 2004. Cosmetics composition comprising sophorolipids. Korean Patent KR20040033376 (A), April 28.

Hagler, M., T.A. Smith-Norowitz, S. Chice, S. Wallner, D. Viterbo, C.M. Mueller, R. Gross, M. Nowakowski, R. Schulze, M.E. Zenilman and M.H. Bluth. 2007. Sophorolipids decrease IgE production in U266 cells by downregulation of BSAP (Pax5), TLR2, STAT3 and IL-6. J. Allergy Clin. Immunol. 119: S263.

Hardin, R., J. Pierre, R. Schulze, C.M. Mueller, S.L. Fu, S.R. Wallner, A. Stanek, V. Shah, R.A. Gross, J. Weedon, M. Nowakowski, M.E. Zenilman and M.H. Bluth. 2007. Sophorolipids improve sepsis survival: Effects of dosing and derivatives. J. Surg. Res. 142: 314–319.

Hayes, M.E. 1991. Personal care products containing bioemulsifiers. U.S. Patent US4999195A, March 12.

Hegstad, K., S. Langsrud, B.T. Lunestad, A.A. Scheie, M. Sunde and S.P. Yazdankhah. 2010. Does the wide use of quaternary ammonium compounds enhance the selection and spread of antimicrobial resistance and thus threaten our health? Microb. Drug Resist. 16: 91–104.

Hirata, Y., M. Ryu, Y. Oda, K. Igarashi, A. Nagatsuka, T. Furuta and M. Sugiura. 2009. Novel characteristics of sophorolipids, yeast glycolipid biosurfactants, as biodegradable low-foaming surfactants. J. Biosci. Bioeng. 108: 142–146.

Høiby, N., T. Bjarnsholt, C. Moser, G.L. Bassi, T. Coenye, G. Donelli, L. Hall-Stoodley, V. Holá, C. Imbert, K. Kirketerp-Møller, D. Lebeaux, A. Oliver, A.J. Ullmann, C. Williams; ESCMID Study Group for Biofilms and Consulting External Expert W. Zimmerli. 2015. ESCMID guideline for the diagnosis and treatment of biofilm infections 2014. Clin. Microbiol. Infect. Suppl. 1: S1–S25.

Holmes, C. and M. Tabrizian. 2015. Surface functionalization of biomaterials. pp. 187–206. *In*: A. Vishwakarma, P. Sharpe, S. Shi and M. Ramalingam (eds.). Stem Cell Biology and Tissue Engineering in Dental Sciences. Academic Press-Elsevier.

Hoq, Md. M., M. Suzutani, T. Toyoda, G. Horijike, I. Yoshida and M. Azuma. 1997. Role of γδ TCR lymphocytes in the augmented resistance of trehalose-6,6'-dimycolate-treated mice to influenza virus infection. J. Gen. Virol. 78: 1597–1603.

Horn, J.N., J.D. Sengillo, D. Lin, T.D. Romo and A. Grossfield. 2012. Characterization of a potent antimicrobial lipopeptide via coarse-grained molecular dynamics. BBA Biomembranes 1818: 212–218.

Huang, X., Z. Lu, H. Zhao, X. Bie, F.X. Lü and S. Yang. 2006. Antiviral activity of antimicrobial lipopeptide from *Bacillus subtilis* fmbj against Pseudorabies Virus, Porcine Parvovirus, Newcastle Disease Virus and Infectious Bursal Disease Virus *in vitro*. Int. J. Pept. Res. Ther. 12: 373–377.

Igarashi, S., Y. Hattori and Y. Maitani. 2006. Biosurfactant MEL-A enhances cellular association and gene transfection by cationic liposome. J. Control. Release 112: 362–368.

Inoh, Y., T. Furuno, N. Hirashima, D. Kitamoto and M. Nakanishi. 2010. The ratio of unsaturated fatty acids in biosurfactants affects the efficiency of gene transfection. Int. J. Pharm. 398: 225–230.

Inoue, S., Y. Kimura and M. Kinta. 1979a. Process for producing a glycolipid methyl ester. German Patent DE2905252A, August 23.

Inoue, S., Y. Kimura and M. Kinta. 1979b. Process for producing a glycolipid ester. German Patent DE2905295A1, August 30.

Ishigami, Y. and S. Suzuki. 1997. Development of biochemicals—Functionalization of biosurfactants and natural dyes. Prog. Org. Coat. 31: 51–61.

Ivshina, I.B., M.S. Kuyukina, J.C. Philp and N. Christofi. 1998. Oil desorption from mineral and organic materials using biosurfactant complexes produced by *Rhodococcus* species. World J. Microbiol. Biotechnol. 14: 711–717.

Joshi-Navare, K. and A. Prabhune. 2013. A Biosurfactant-Sophorolipid Acts in Synergy with Antibiotics to Enhance their Efficiency. BioMed Research International Article ID 512495.

Ju, L.K., S.S. Dashtbozorg and N. Vongpanish. 2016. Wound dressings with enhanced gas permeation and other beneficial properties. U.S. Patent # 9,468,700 B2.

Jung, M., S. Lee and H. Kim. 2000. Recent studies on natural products as anti-HIV agents. Curr. Med. Chem. 7: 649–661.

Kanlayavattanakul, M. and N. Lourith. 2010. Lipopeptides in cosmetics. Int. J. Cosmet. Sci. 32: 1–8.

Kato, E., T. Tsuzuki and E. Ogata. 2007. Agent for skin external use containing tocopherol derivative, ascorbic acid derivative and surface active agent having lipopeptide structure. United States Patent US 20070232687 A1, October 04.

Kato, E. and T. Tsuzuki. 2008. Dermatological anti-wrinkle agent. World Patent WO2008001921A2, January 3.

Kawano, J., T. Utsugi, S. Inoue and S. Hayashi. 1981. Powered compressed cosmetic material. U.S. Patent US4305931 A, December 15.

Kim, H.-S., J.-W. Jeon, S.-B. Kim, H.-M. Oh, T.-J. Kwon and B.-D. Yoon. 2002a. Surface and physico-chemical properties of a glycolipid biosurfactant, mannosylerythritol lipid, from *Candida antarctica*. Biotechnol. Lett. 24: 1637–1641.

Kim, K., D. Yoo. Y. Kim, B. Lee, D. Shin and E.-K. Kim. 2002b. Characteristics of sophorolipid as an antimicrobial agent. J. Microbiol. Biotechnol. 12: 235–241.

Kim, S.Y., J.Y. Kim, S.H. Kim, H.J. Bae, H. Yi, S.H. Yoon, B.S. Koo, M. Kwon, J.Y. Cho, C.E. Lee and S. Hong. 2007. Surfactin from *Bacillus subtilis* displays antiproliferative effect via apoptosis induction, cell cycle arrest and survival signaling suppression. FEBS Lett. 581: 865–871.

Kiran, G.S., B. Sabarathnam and J. Selvin. 2010. Biofilm disruption potential of a glycolipid biosurfactant from marine *Brevibacterium casei*. FEMS Immunol. Med. Microbiol. 59: 432–438.

Kitagawa, M., K. Nishimoto and T. Tanaka. 2015. Cosmetic pigments, their production method, and cosmetics containing the cosmetic pigments. United States Patent US9181436B2, November 11.

Kitamoto, D., H. Yanagishita, T. Shinbo, T. Nakane, C. Kamisawa and T. Nakahara. 1993. Surface active properties and antimicrobial activities of mannosylerythritol lipids as biosurfactants produced by *Candida antarctica*. J. Biotechnol. 29: 91–96.

Kitamoto, D., H. Isoda and T. Nakahara. 2002. Functions and potential applications of glycolipid biosurfactants-from energy-saving materials to gene delivery carriers. J. Biosci. Bioeng. 94: 187–201.

Kitamoto, D., T. Morita, T. Fukuoka, M. Konishi and T. Imura. 2009. Self-assembling properties of glycolipid biosurfactants and their potential applications. Curr. Opin. Colloid Interface Sci. 14: 315–328.

Kofla, G. and M. Ruhnke. 2011. Pharmacology and metabolism of anidulafungin, caspofungin and micafungin in the treatment of invasive candidosis: review of the literature. Eur. J. Med. Res. 16: 159–166.

Kolwzan, B., J. Biazik, A. Czarny, E. Zaczyńska and E. Karpenko. 2008. Assessment of toxicity of biosurfactants produced by *Pseudomonas* sp. ps-17 (Pol). *In*: Ecotoxicology in Environmental Protection Management (Pol). PZITS Oddz. Dolnosl., Wroclaw.

Kuyukina, M.S., I.B. Ivshina, S.V. Gein, T.A. Baeva and V.A. Chereshnev. 2007. *In vitro* immune-modulating activity of biosurfactant glycolipid complex from *Rhodococcus ruber*. Bull. Exp. Biol. Med. 144: 326–330.

Landman, D., C. Georgescu, D.A. Martin and J. Quale. 2008. Polymyxins revisited. Clin. Microbiol. Rev. 21: 449–465.

Li, Y.H. and X. Tian. 2012. Quorum sensing and bacterial social interactions in biofilms. Sensors 12: 2519–42538.

Lima, T.M.S., L.C. Procópio, F.D. Brandão, A.M.X. Carvalho, M.R. Tótola and A.C. Borges. 2011. Biodegradability of bacterial surfactants. Biodegradation 22: 585–592.

Liu, J., A. Zou and B. Mu. 2010. Surfactin effects on the physiochemical property of PC liposome. Colloids Surf. A Physicochem. Eng. Asp. 361: 90–95.

Lourith, N. and M. Kanlayavattanakul. 2009. Natural surfactants used in cosmetics: glycolipids. Int. J. Cosmet. Sci. 31: 255–261.

Lu, J.-K., H.-M. Wang and X.-R. Xu. 2016. Applications of surfactin in cosmetic products. United States Patent US2016030322 A1, February 04.

Lydon, H.L., N. Baccile, B. Callaghan, R. Marchant, C.A. Mitchell and I.M. Banat. 2017. Adjuvant antibiotic activity of acidic sophorolipids with potential for facilitating wound healing. Antimicrob. Agents Chemother. 61: e02547–16.

Maier, R.M. and G. Soberon-Chavez. 2000. *Pseudomonas aeruginosa* rhamnolipids: Biosynthesis and potential applications. Appl. Microbiol. Biotechnol. 54: 625–633.

Maingault, M. 1999. Utilization of sophorolipids as therapeutically active substances or cosmetic product, in particular for the treatment of the skin. United States Patent US5981497A, November 9.

Makamba, H., J.H. Kim, K. Lim, N. Park and J.H. Hahn. 2003. Surface modification of poly(dimethyl siloxane) microchannels. Electrophoresis 24: 3607–3619.

Makkar, R.S., S.S. Cameotra and I.M. Banat. 2011. Advances in utilization of renewable substrates for biosurfactant production. AMB Express 1: 5.

Mandal, S.M., A.E. Barbosa and O.L. Franco. 2013. Lipopeptides in microbial infection control, scope and reality for industry. Biotechnol. Adv. 31: 338–345.

Mani, P., G. Dineshkumar, T. Jayaseelan, K. Deepalakshmi, C. Ganesh Kumar and S. Senthil Balan. 2016. Antimicrobial activities of a promising glycolipid biosurfactant from a novel marine *Staphylococcus saprophyticus* SBPS 15. 3 Biotech. 6: 163.

Marchant, R. and I.M. Banat. 2012. Microbial biosurfactants: Challenges and opportunities for future exploitation. Trends Biotechnol. 30: 558–565.

Masaru, K., S. Michiko and Y. Shuhei. 2007. Skin care cosmetic and skin and agent for preventing skin roughness containing biosurfactants. World Patent WO2007060956 A1, May 31.

Mekonnen, A., T. Sidamo, K. Asres and E. Engidawork. 2013a. *In vivo* wound healing activity and phytochemical screening of the crude extract and various fractions of *Kalanchoe petitiana A. Rich* (Crassulaceae) leaves in mice. J. Ethnopharmacol. 145: 638–646.

Mnif, I. and D. Ghribi. 2015a. Glycolipid biosurfactants: Potential related biomedical and biotechnological applications. Carbohydr. Res. 416: 59–69.

Mnif, I. and D. Ghribi. 2015b. Review lipopeptides biosurfactants: Mean classes and new insights for industrial, biomedical, and environmental applications. Biopolymers 104: 129–147.

Mohan, P.K., G. Nakhla and E.K. Yanful. 2006. Biokinetics of biodegradability of surfactants under aerobic, anoxic and anaerobic conditions. Water Res. 40: 533–540.

Monciardini, P., M. Iorio, S. Maffioli, M. Sosio and S. Donadio. 2014. Discovering new bioactive molecules from microbial sources. Microb. Biotechnol. 7: 209–220.

Montanari, D. and M. Guglielmo. 2008. Cosmetic composition for the treatment and/or prevention of skin stretch marks. World patent W02008080443 A2, July 10.

Morita, T., Y. Ishibashi, N. Hirose, K. Wada, M. Takahashi, T. Fukuoka, T. Imura, H. Sakai, M. Abe and D. Kitamoto. 2011. Production and characterization of a glycolipid biosurfactant, mannosylerythritol lipid B, from sugarcane juice by *Ustilago scitaminea* NBRC 32730. Biosci. Biotechnol. Biochem. 75: 1371–1376.

Morita, T., T. Fukuoka, T. Imura and D. Kitamoto. 2013. Production of mannosylerythritol lipids and their application in cosmetics. Appl. Microbiol. Biotechnol. 97: 4691–4700.

Morita, T., T. Fukuoka, T. Imura and D. Kitamoto. 2015. Mannosylerythritol lipids: production and applications. J. Oleo Sci. 64: 133–141.

Moryl, M., M. Spętana, K. Dziubek, K. Paraszkiewicz, S. Różalska and G.A. Płaza. 2015. Antimicrobial, antiadhesive and antibiofilm potential of lipopeptides synthesised by *Bacillus subtilis*, on uropathogenic bacteria. Acta Biochim. Pol. 62: 725–732.

Muehlradt, P., T. Barkhausen and T. Tschernig. 2013. Pharmaceutical compositions for treating dysregulated inflammatory diseases. United Stated Patent US2013079274 (A1), March 28.

Muhammad, I.-M. and S.-S. Mahsa. 2014. Rhamnolipids: well-characterized glycolipids with potential broad applicability as biosurfactants. Industrial Biotechnology 10: 285–291.

Müller, M.M., J.H. Kügler, M. Henkel, M. Gerlitzki, B. Hörmann, M. Pöhnlein, C. Syldatk and R. Hausmann. 2012. Rhamnolipids-Next generation surfactants? J. Biotechnol. 162: 366–380.

Muraih, J.K., A. Pearson, J. Silverman and M. Palmer. 2011. Oligomerization of daptomycin on membranes. Biochim. Biophys. Acta 1808: 1154–1160.

Nakanishi, M., Y. Inoh, D. Kitamoto and T. Furuno. 2009. Nano vectors with a biosurfactant for gene transfection and drug delivery. J. Drug Deliv. Sci. Technol. 19: 165–169.

Nam, J., J.J. Moon, R. Kuai and A.A. Schwendeman. 2016. Compositions and methods for delivery of biomacromolecule agents. World Patent WO2016154544 (A1), September 29.

Naruse, N., O. Tenmyo, S. Kobaru, H. Kamei, T. Miyaki, M. Konishi and T. Oki. 1990. Pumilacidin, a complex of new antiviral antibiotics: production, isolation, chemical properties, structure and biological activity. J. Antibiot. 43: 267–280.

Ngai, A.L., M.R. Bourque, R.J. Lupinacci, K.M. Strohmaier and N.A. Kartsonis. 2011. Overview of safety experience with caspofungin in clinical trials conducted over the first 15 years, A brief report. Int. J. Antimicrob. Agents 38: 540–544.

Nguyen, T.T.L., A. Edelen, B. Neighbors and D.A. Sabatini. 2010. Biocompatible lecithin-based microemulsions with rhamnolipid and sophorolipid biosurfactants: Formulation and potential applications. J. Colloid Interface Sci. 348: 498–504.

Nicoli, S., M. Eeman, M. Deleu, E. Bresciani, C. Padula and P. Santi. 2010. Effect of lipopeptides and iontophoresis on acyclovir skin delivery. J. Pharm. Pharmacol. 62: 702–708.

Ortíz-López, F.J., M.C. Monteiro, V. González-Menéndez, J.R. Tormo, O. Genilloud, G.F. Bills, F. Vicente, C. Zhang, T. Roemer, S.B. Singh and F. Reyes. 2015. Cyclic colisporifungin and linear cavinafungins, antifungal lipopeptides isolated from *Colispora cavincola*. J. Nat. Prod. 78: 468–475.

Ortiz, A., J.A. Teruel, M.J. Espuny, A. Marqués, A. Manresa and F.J. Aranda. 2009. Interactions of a bacterial biosurfactant trehalose lipid with phosphatidylserine membranes. Chem. Phys. Lipids 158: 46–53.

Owen, D.L.F. and L.C. Paul. 2007. Polymeric biosurfactants. World Patent WO2007143006 A3, February 28.

Park, S.Y. and Y.-H. Kim. 2009. Surfactin inhibits immune-stimulatory function of macrophages through blocking NK-κB, MAPK and Akt pathway. Int. Immunopharmacol. 9: 886–893.

Park, S.Y., Y.H. Kim, E.K. Kim, E.Y. Ryu and S.J. Lee. 2010. Heme oxygenase-1 signals are involved in preferential inhibition of pro-inflammatory cytokine release by surfactin in cells activated with *Porphyromonas gingivalis* lipopolysaccharide. Chem. Biol. Interact. 188: 437–445.

Pei, X., X. Zhan and L. Zhou. 2009. Effect of biosurfactant on the sorption of phenanthrene onto original and $H_2O_2$-treated soils. J. Env. Sci. 21: 1378–1385.

Pellicier, F. and P. Andre. 2005. Cosmetic use of sophorolipids as subcutaneous adipose cushion regulating agents and slimming application. World Patent WO 2004108063 A3, January 27.

Peters, B.M., M.E. Shirtliff and M.A. Jabra-Rizk. 2010. Antimicrobial peptides: Primeval molecules or future drugs? PLoS Pathog. 6: e1001067.

Pierce, D. and T.J. Heilman. 2001. Germicidal composition. United States Patent 6262038, July 17.

Piljac, T. and G. Piljac. 1999. Use of rhamnolipids in wound healing, treating burn shock, atherosclerosis, organ transplants, depression, schizophrenia and cosmetics. Patent WO1999043334 A1, September 02.

Piljac, A., T. Stipcevic, J. Piljac-Zegarac and G. Piljac. 2008. Successful treatment of chronic decubitus ulcer with 0.1% dirhamnolipid ointment. J. Cutan. Med. Surg. 12: 142–146.

Pinto, S., P. Alves, C.M. Matos, A.C. Santos, L.R. Rodrigues, J.A Teixeira and M.H. Gil. 2010. Poly(dimethyl siloxane) surface modification by low pressure plasma to improve its characteristics towards biomedical applications. Colloids Surf. B Biointerfaces 81: 20–26.

Pontes, C., M. Alves, C. Santos, M.H. Ribeiro, L. Gonçalves, A.F. Bettencourt and I.A. Ribeiro. 2016. Can Sophorolipids prevent biofilm formation on silicone catheter tubes? Int. J. Pharm. 513: 697–708.

Pradhan, A.K., N. Pradhan, L.B. Sukla, P.K. Panda and B.K. Mishra. 2014. Inhibition of pathogenic bacterial biofilm by biosurfactant produced by *Lysinibacillus fusiformis* S9. Bioprocess Biosyst. Eng. 37: 139–149.

Prasad, B., Dr. H.P. Kaur and Dr. S. Kaur. 2015. Potential biomedical and pharmaceutical applications of microbial surfactants. World J. Pharm. Pharm. Sci. 4: 1557–1575.

Ramisse, F., C. Delden and S. Gidenne. 2000. Decreased virulence of a strain of *Pseudomonas aeruginosa* O12 overexpressing a chromosomal type 1 β-lactamase could be due to reduced expression of cell-to-cell signalling dependent virulence factors. FEMS Immunol. Med. Microbiol. 28: 241–245.

Remichkova, M., D. Galabova, I. Roeva, E. Karpenko, A. Shulga and A.S. Galabov. 2008. Anti-herpesvirus activities of *Pseudomonas* sp. S-17 rhamnolipid and its complex with alginate. J. Biosci. 63: 75–81.

Robbel, L. and M.A. Marahiel. 2010. Daptomycin, a bacterial lipopeptide synthesized by a nonribosomal machinery. J. Biol. Chem. 285: 27501–27508.

Rodrigues, L., I.M. Banat, J. Teixeira and R. Oliveira. 2006b. Biosurfactants: potential applications in medicine. J. Antimicrob. Chemother. 57: 609–618.

Rodrigues, L.R., I.M. Banat, H.C. van der Mei, J.A. Teixeira and R. Oliveira. 2006a. Interference in adhesion of bacteria and yeasts isolated from explanted voice prostheses to silicone rubber by rhamnolipid biosurfactants. J. Appl. Microbiol. 100: 470–480.

Sabol, K. and T. Gumbo. 2008. Anidulafungin in the treatment of invasive fungal infections. Ther. Clin. Risk Manag. 4: 71–78.

Sahnoun, R., I. Mnif, H. Fetoui, R. Gdoura, K. Chaabouni, F. Makni-Ayadi, C. Kallel, S. Ellouze-Chaabouni and D. Ghribi. 2014. Evaluation of *Bacillus subtilis* SPB1 lipopeptide biosurfactant toxicity towards Mice. Int. J. Pept. Res. Ther. 20: 333–340.

Saini, H.S., B.E. Barragán-Huerta, A. Lebrón-Paler, J.E. Pemberton, R.R. Vázquez, A.M. Burns, M.T. Marron, C.J. Seliga, A.A. Gunatilaka and R.M. Maier. 2008. Efficient purification of the biosurfactant viscosin from *Pseudomonas libanensis* strain M9-3 and its physicochemical and biological properties. J. Nat. Prod. 71: 1011–1015.

Sambanthamoorthy, K., X. Feng, R. Patel, S. Patel and C. Paranavitana. 2014. Antimicrobial and antibiofilm potential of biosurfactants isolated from lactobacilli against multi-drug-resistant pathogens. BMC Microbiol. 14: 197.

Sánchez, M., F.J. Aranda, J.A. Teruel, M.-J. Espuny, A. Marqués, Á. Manresa and A. Ortiza. 2010. Permeabilization of biological and artificial membranes by a bacterial dirhamnolipid produced by *Pseudomonas aeruginosa*. J. Colloid Interface Sci. 341: 240–247.

Satpute, K., A.G. Banpurkar, I.M. Banat, J.N. Sangshetti, R.R. Patil and W.N. Gade. 2016a. Multiple roles of biosurfactants in biofilms. Curr. Pharm. Des. 22: 1429–1448.

Satpute, S.K., A.G. Banpurkar, P.K. Dhakephalkar, I.M. Banat and B.A. Chopade. 2010. Methods for investigating biosurfactants and bioemulsifiers: A review. Crit. Rev. Biotechnol. 30: 127–144.

Satpute, S.K., G.R. Kulkarni, A.G. Banpurkar, I.M. Banat, N.S. Mone, R.H. Patil and S.S. Cameotra. 2016b. Biosurfactant/s from Lactobacilli species: Properties, challenges and potential biomedical applications. J. Basic Microbiol. 56: 1–19.

Sekhon, K.K., S. Khanna and S.S. Cameotra. 2012. Biosurfactant production and potential correlation with esterase activity. J. Pet. Environ. Biotechnol. 3: 133.

Seydlová, G. and J. Svobodová. 2008. Review of surfactin chemical properties and the potential biomedical applications. Cent. Eur. J. Med. 3: 123–133.

Shah, V., G.F. Doncel, T. Seyoum, K.M. Eaton, I. Zalenskaya, R. Hagver, A. Azim and R. Gross. 2005. Sophorolipids, microbial glycolipids with anti-human immunodeficiency virus and sperm-immobilizing activities. Antimicrob. Agents Chemother. 49: 4093–4100.

Shao, L., X. Song, X. Ma, H. Li and Y. Qu. 2012. Bioactivities of sophorolipid with different structures against human esophageal cancer cells. J. Surg. Res. 173: 286–291.

Shekhar, S., A. Sundaramanickam and B. Tangavel. 2015. Biosurfactant producing microbes and their potential applications: a review. Crit. Rev. Env. Sci. Technol. 45: 1522–1554.

Shoeb, E., F. Akhlaq, U. Badar, J. Akhter and S. Imtiaz. 2013. Classification and industrial applications of biosurfactants. Part-I: Natural and Applied Sciences. Academic Research International 4: 243–252.

Siala, W., S. Kucharı´kova, A. Braem, J. Vleugels, P.M. Tulkens, M.-P. Mingeot-Leclercq, P. Van Dijck and F. Van Bambeke. 2016. The antifungal caspofungin increases fluoroquinolone activity against *Staphylococcus aureus* biofilms by inhibiting N-acetylglucosamine transferase. Nat. Commun. 7: 13286.

Singh, A., J.D. Van Hamme and O.P. Ward. 2007. Surfactants in microbiology and biotechnology. Part 2. Application aspects. Biotechnol. Adv. 25: 99–121.

Sivapathasekaran, C., P. Das, S. Mukherjee, J. Saravanakumar, M. Mandal and R. Sen. 2010. Marine bacterium derived lipopeptides: characterization and cytotoxic activity against cancer cell lines. Int. J. Pept. Res. Ther. 16: 215–222.

Smyth, T.J.P., A. Perfumo, S. McClean, R. Marchant and I.M. Banat. 2010a. Isolation and analysis of lipopeptides and high molecular weight biosurfactants. pp. 3689–3704. *In*: K.N. Timmis (ed.). Handbook of Hydrocarbon and Lipid Microbiology. Springer-Verlag, Berlin, Heidelberg.

Smyth, T.J.P., A. Perfumo, S. McClean, R. Marchant and I.M. Banat. 2010b. Isolation and analysis of low molecular weight microbial glycolipids. pp. 3705–3723. *In*: K.N. Timmis (ed.). Handbook of Hydrocarbon and Lipid Microbiology. Springer-Verlag, Berlin, Heidelberg.

Stadler, M., J. Bitzer, B. Köpcke, K. Reinhardt and J. Moldenhauer. 2016. Long chain glycolipids useful to avoid perishing or microbial contamination of materials. Singapore Patent SG10201604666U (A), July 28.

Stipcevic, T., A. Piljac and G. Piljac. 2006. Enhanced healing of full-thickness burn wounds using di-rhamnolipid. Burns 32: 24–34.

Sun, L., Z. Lu, X. Bie, F. Lu and S. Yang. 2006. Isolation and characterization of a co-producer of fengycins and surfactins, endophytic *Bacillus amyloliquefaciens* ES-2, from Scutellaria baicalensis Georgi. World J. Microbiol. Biotechnol. 22: 1259–1266.

Suzuki, M., M. Kitagawa, S. Yamamoto, A. Sogabe, D. Kitamoto, T. Morita, T. Fukuoka and T. Imura. 2010. Activator including biosurfactant as active ingredient, mannosyl erythritol lipid, and production method publication. United States Patent US20100168405A1, July 1.

Tally, F.P., M. Zeckel, M.M. Wasilewski, C. Carini, C.L. Berman, G.L. Drusano and FB Jr. Oleson. 1999. Daptomycin, A novel agent for Gram-positive infections. Expert Opin. Investig. Drugs 8: 1223–1238.

Tang, J.S., F. Zhao, H. Gao, Y. Dai, Z.H. Yao, K. Hong, J. Li, W.C. Ye and X.S. Yao. 2010. Characterization and online detection of surfactin isomers based on HPLC-MS analyses and their inhibitory effects on the overproduction of nitric oxide and the release of TNF-α and IL-6 in LPS-induced macrophages. Mar. Drugs 8: 2605–2618.

Thanomsub, B., W. Pumeechockchai, A. Limtrakul, P. Arunrattiyakorn, W. Petchleelaha, T. Nitoda and H. Kanzaki. 2006. Chemical structures and biological activities of rhamnolipids produced by *Pseudomonas aeruginosa* B189 isolated from milk factory waste. Bioresour. Technol. 97: 2457–2461.

Ueno, Y., N. Hirashima, Y. Inoh, T. Furuno and M. Nakanishi. 2007a. Characterization of biosurfactant containing liposomes and their efficiency for gene transfection. Biol. Pharm. Bull. 30: 169–172.

Ueno, Y., Y. Inoh, T. Furuno, N. Hirashima, D. Kitamoto and M. Nakanishi. 2007b. NBD-conjugated biosurfactant (MEL-A) shows a new pathway for transfection. J. Control Release 123: 247–253.

Valappil, S.P., D.M. Pickup, D.L. Carroll, C.K. Hope, J. Pratten, R.J. Newport, M.E. Smith, M. Wilson and J.C. Knowles. 2007. Effect of silver content on the structure and antibacterial activity of silver-doped phosphate-based glasses. Antimicrob. Agents Chemother. 51: 4453–4461.

Valotteau, C., I.M. Banat, C. Mitchell, H. Lydon, R. Marchant, F. Babonneau, C. Pradier, N. Baccile and V. Humblot. 2017. Antibacterial properties of sophorolipid-modified gold surfaces against Gram positive and Gram negative pathogens. Colloids Surf. B Biointerfaces 157: 325–334.

Varvaresou, A. and K. Iakovou. 2015. Biosurfactants in cosmetics and biopharmaceuticals. Lett. Appl. Microbiol. 61: 214–223.

Vasilev, K., J. Cook and H.J. Griesser. 2009. Antibacterial surfaces for biomedical devices. Expert Rev. Med. Devices 6: 553–567.

Vater, J., B. Kablitz, C. Wilde, P. Franke, N. Mehta and S.S. Cameotra. 2002. Matrix-assisted laser desorption ionization—Time of flight mass spectrometry of lipopeptide biosurfactants in whole cells and culture filtrates of *Bacillus subtilis* C-1 isolated from petroleum sludge. Appl. Environ. Microbiol. 68: 6210–6219.

Vecino, X., J.M. Cruz, A.B. Moldes and L.R. Rodrigues. 2017. Biosurfactants in cosmetic formulations: trends and challenges. Crit. Rev. Biotechnol. 12: 1–16.

Wagner, C., W. Graninger, E. Presterl and C. Joukhadar. 2006. The echinocandins, comparison of their pharmacokinetics, pharmacodynamics and clinical applications. Pharmacology 78: 161–177.

Wakamatsu, Y., X. Zhao, C. Jin, N. Day, M. Shibahara, N. Nomura, T. Nakahara, T. Murata and K.K. Yokoyama. 2001. Mannosylerythritol lipid induce characteristics of neuronal differentiation in PC12 cells through an ERK-related signal cascade. Eur. J. Biochem. 268: 374–383.

Wang, R., B.A. Khan, G.Y.C. Cheung, T.L.H. Bach, M. Jameson-Lee, K.F. Kong, S.Y. Queck and M. Otto. 2011. *Staphylococcus epidermidis* surfactant peptides promote biofilm maturation and dissemination of biofilm-associated infection in mice. J. Clin. Invest. 121: 238–248.

Wong, C.-H., A.L. Yu, T.-N. Wu and K.-H. Lin. 2016. Human iNKT cell activation using glycolipids. World Patent WO2016040369 (A3), September 15.

Xihou, Y. 2014. Formulations combining ramoplanin and rhamnolipids for combating bacterial infection. United States Patent US2014294925 (A1), October 02.

Yamamoto, S., T. Morita, T. Fukuoka, T. Imura, S. Yanagidani, A. Sogabe, D. Kitamoto and M. Kitagawa. 2012. The moisturizing effects of glycolipid biosurfactants, mannosylerythritol lipids, on human skin. J. Oleo Sci. 61: 407–412.

Yang, X., E. Huang, C. Yuan, L. Zhang and A.E. Yousef. 2016. Isolation and structural elucidation of brevibacillin, an antimicrobial lipopeptide from *Brevibacillus laterosporus* combating drug-resistant Gram-positive bacteria. Appl. Env. Microbiol. 82: 2763–2772.

Yoneda, T., E. Masatsuji, T. Tsuzuki, K. Furuya, M. Takama, Y. Miyota and I. Shinobu. 1999. Surfactant for use in external preparations for skin and external preparation for skin containing the same. World patent WO1999062482 A1, December 9.

Yoneda, T., N. Ito and T. Yoneda. 2005. Oil-in-water emulsified composition and external preparation for skin and cosmetics using the composition. World patent WO2005089708 A1, September 29.

Yoneda, T. 2006. Cosmetic composition comprising a and a lipopeptides. United States Patent US0222616 A1, October 5.

Youn-Hwan, H., P. Byung-Kwon, L. Jong-Hwan, K. Myoung-Seok, S. In-Bae, P. Seung-Chun and Y. Hyo-In. 2008. Evaluation of genetic and developmental toxicity of surfactin C from *Bacillus subtilis* BC1212. J. Health Sci. 54: 101–106.

Youn-Hwan, H., K. Myoung-Seok, S. In-Bae, P. Byung-Kwon, L. Jong-Hwan, P. Seung-Chun and Y. Hyo-In. 2009. Subacute (28 day) toxicit of surfactin C, a lipopeptide produced by *Bacillus subtilis*, in rats. J. Health Sci. 55: 351–355.

Zaragoza, A., F.J. Aranda, M.J. Espuny, J.A. Teruel, A. Marqués, A. Manresa and A. Ortiz. 2009. Mechanism of membrane permeabilization by a bacterial trehalose lipid biosurfactant produced by *Rhodococcus* sp. Langmuir 25: 7892–7898.

Zhang, Y., H. Li, J. Sun, J. Gao, W. Liu, B. Li, Y. Guo and J. Chen. 2010. Dc-chol/dope cationic liposomes: A comparative study of the influence factors on plasmid pDNA and Si RNA gene delivery. Int. J. Pharm. 390: 198–207.

Zhang, Y., C. Liu, B. Dong, X. Ma, L. Hou, X. Cao and C. Wang. 2015. Anti-inflammatory activity and mechanism of surfactin in lipopolysaccharide-activated macrophages. Inflammation 38: 756–764.

Zhao, X., C. Geltinger, S. Kishikawa, K. Ohshima, T. Murata, N. Nomura, T. Nakahara and K.K. Yokoyama. 2000. Treatment of mouse melanoma cells with phorbol 12-myristate 13-acetate counteracts mannosylerythritol lipid induced growth arrest and apoptosis. Cytotechnology 33: 123–130.

# 12

# Biosynthesis of Glycolipids and their Genetic Engineering

*Sylwia Jezierska, Silke Claus* and *Inge Van Bogaert\**

## Introduction

This chapter focuses on the biosynthesis of carbohydrate-lipid composed biosurfactants (BSs) or 'glycolipids' (GLs) and the genetic engineering of their main producers. In the first part, microorganisms of bacterial origin are discussed, whereas the second part describes fungal GL producers. As many GLs can be considered as secondary metabolites, we will highlight the general genetic organization and regulation of secondary metabolites below and link them to opportunities in the genetic deciphering and engineering of microbial BSs.

The natural role of GLs produced by microbial cells is not fully understood. However, the possibilities are multifold. Their amphiphilic nature makes them suitable to facilitate the attachment and subsequent transport of water-insoluble substrates into the cell (Kitamoto et al. 2002). For example, the mycolic acid layer or the lipid moiety of trehalose lipids (TLs) inserts in the cell-wall, conferring more hydrophobicity to the bacterial cell surface. In this way, surrounding hydrophobic substrates are more readily taken up (Rakatozafy and Marchal 1999). Hence, the synthesis of TLs is often described as growth-associated cellular adaptation to the presence of hydrocarbons. With this capacity, the bacterial trehalose lipids cannot be considered to be secondary metabolites (Lang and Philp 1998). Where alkane metabolism is a prerequisite for TL production, it is not a general feature linked to BS synthesis (Philp et al. 2002). Rhamnolipids (RLs) and sophorolipids (SLs), for example, can be assembled solely from water-soluble substrates (Wittgens et

Centre for Synthetic Biology, Faculty of Bioscience Engineering, Coupure Links 653, 9000 Gent, Belgium.
\* Corresponding author: Inge.VanBogaert@ugent.be

al. 2011). Because of their assistance in alkane uptake, GLs are of major industrial importance in the field of enhanced oil recovery, bioremediation of hydrocarbon contaminated soils and removal of solids from wastewater (Franzetti et al. 2010).

Furthermore, BSs aid in the environmental adaptation of microorganisms by their essential role in biofilm formation, swarming motility, resistance to toxic compounds, cell signaling and differentiation (Ron and Rosenberg 2001). Often, they are employed in the biological warfare against competitors in the immediate environment by virtue of their antimicrobial activity and pathogenic effects (Van Hamme et al. 2006). These GL characteristics confirm their role as secondary metabolites. Aforementioned industrial application potentials of BSs thus surpasses its traditional use as emulsifying and solubilizing agents for hydrophobic substances, and support their utility as biomedical agents with anti-viral, anti-bacterial, anti-fungal, anti-tumor and anti-inflammatory activities (Rodrigues et al. 2006). Furthermore, GLs are increasingly studied as immunomodulatory agents (Kuyukina et al. 2015).

Interestingly, microorganisms accumulate/produce an extremely rich palette of known and unknown secondary metabolites, the unknown are yet to be explored for their potential application. Secondary metabolites are low-molecular-weight, natural products that do not participate in the central metabolic processes of a living cell. Biosynthesis is correlated with the life cycle of a specific organism and is triggered by certain environmental conditions. The function of these molecules is to help cells cope with stress or compete with surrounding organisms. They are also used as a chemical communication signal. Although secondary metabolites are more abundant in plants, the most famous one, penicillin, is of microbial origin. Besides antibiotics, another group of diverse secondary metabolites produced by a wide variety of microorganisms and the subject of this book chapter are GLs such as rhamnolipids (RLs), trehalose lipids (TLs), sophorolipids (SLs), mannosylerythritol lipids (MELs) and cellobioselipids (CBLs) (Asselineau and Asselineau 1978, Hewald et al. 2006, Teichmann et al. 2007, Reis et al. 2011, Van Bogaert et al. 2013, Roelants et al. 2014).

To be able to exploit the secondary metabolites and their merits even more, numerous studies are dedicated to deciphering their biosynthetic and regulatory mechanisms. While primary metabolism genes are distributed across the genome, genes involved in the synthesis of secondary metabolites are organized in clusters (Demain 1998). Clustering of these secondary metabolite biosynthetic genes is virtually universal and considered a trans-kingdom phenomenon (Osbourn 2010). In this cluster conformation, a minimum amount of steps is needed for the activation and regulation of the biosynthetic machinery. Thereby, the physiological state of the microorganism remains optimally preserved during conditions that challenge this regular physiological state of the cell. It has also been suggested that organization in clusters would be advantageous for horizontal gene transfer among different species and can guarantee the transfer of complete biosynthesis pathways (Walton 2000).

The occurrence of genes in cluster formation can be exploited in the search for new interesting natural products. Software tools are developed to query large sequence databases for gene clusters associated with biomolecules of interest, for example antiSMASH and ClusterFinder (Blin et al. 2013, Cimermancic et al. 2014). Such a sequence-based metagenomics analysis led to the discovery of the genes involved in

BS synthesis by *Serratia marcescens* (Gerc et al. 2014). Nevertheless, the fact that most secondary metabolite genes are structured in gene clusters also poses a great challenge considering that the majority of these clusters is silent under laboratory conditions (Brakhage and Schroeckh 2011). Instead, complex cascades and pathway-specific triggers are required for the activation of the gene and production of specific metabolite (Bibb 2005). As a result, new and potentially interesting biomolecules cannot be produced as long as the triggers for induction have not been identified. An alternative approach is reviewed by O'Connor et al. (2015) on genetic modification of native microbial producers to switch 'on' silent gene clusters.

Secondary metabolite gene clusters are found in both prokaryotes and eukaryotes and may range from only a few to more than 20 genes. By far, most examples are found within filamentous fungi (Osbourn 2010). This discovery implies that fungal secondary metabolite genes share common transcriptional regulatory mechanisms (Shaaban et al. 2010). The co-regulation of the cluster genes is often carried out by one or more transcription factors located within the cluster. Most commonly, these regulators belong to the $Zn_2/Cys_6$ zinc finger proteins such as AflR for aflatoxin/sterigmatocystin biosynthesis in *Aspergillus* spp. (Fernandes et al. 1998). Members of this class of zinc fingers contain two zinc ions bound by six cysteine residues. Also, $Cys_2His_2$ zinc fingers control pathway synthesis, such as Rua1 for biosynthesis of cellobioselipids in *U. maydis* (Teichmann et al. 2010).

Beside the pathway specific regulation, synthesis of secondary metabolites is also controlled by global transcription factors encoded by genes not associated with the biosynthetic gene clusters. This upper level of regulation is a general response to environmental factors such as pH, temperature, and nutrition (Brakhage 2013). This multilevel regulation by both cluster specific regulators briefly discussed above, and broad-domain transcription factors ensures that secondary metabolite pathways can respond to the demands of general cellular metabolism and the presence of specific pathway inducers (Yu and Keller 2005). Interestingly, the order and location of biosynthetic genes within a cluster also appear to be important for their regulation. The first indication of such regulation came from the studies on biosynthetic enzymes of the aflatoxin biosynthetic cluster in *Aspergillus parasiticus*. The expression of gene *ver-1* outside the cluster resulted in 500-fold lower activity (Liang et al. 1997). Moreover, secondary metabolite gene clusters are very often found in subtelomeric localization: regions with a high degree of heterochromatin (transcriptionally silent regions) and prone to variations. Indeed, the key role of secondary metabolites is to readily anticipate to environmental changes and the flexibility of the subtelomeric region makes it the ideal platform for secondary metabolite production (Osbourn 2010). It facilitates non-allelic recombination, DNA inversion, partial deletions, translocations and other rearrangements contributing to adaptive evolution (Farman 2007).

Associated with the subtelomeric location of secondary metabolite clusters is another overall silencing mechanism: Telomere Position Effect (TPE). The extend of TPE can vary depending on the organism, but generally spreads over 20 kb (Palmer and Keller 2010). The positional silencing of the gene expression mentioned here is associated with an ATP-dependent chromatin remodeling, linked to the decoration of specific histone residues by acetyl and methyl groups on lysines or arginines as well as phosphorylation of serines or threonins and ubiquitination of lysines (Strauss and Reyes-Dominguez 2011). Under primary metabolic conditions, repressive heterochromatin domains protect the cluster, preventing transcription of the genes within. Activation of the cluster, on the other hand, requires remodeling the chromatin landscape (Gacek and Strauss 2012). Chromatin-based regulation of secondary metabolite clusters was first proven for *Aspergillus nidulans* by using histone de-acetylase (HDAC) deletion mutants. It was demonstrated that either genetic or chemical inactivation of HDACs can circumvent this repressive phenomenon in several fungi and lead to upregulation of secondary metabolite genes and corresponding enhanced production (Shwab et al. 2007). Only recently, it was discovered that chromosome remodeling is not restricted to the eukaryotic domain (Osbourn 2010). Histone-like proteins modulate transcription of gene sets in Actinomycetes, e.g., *Streptomyces coelicolor,* conferring metabolite production levels that depend on the genomic location of the synthetic genes.

### Rhamnolipids (RLs)

#### Structure and microbial source of RLs

The term 'rhamnolipid' (RL) is designated to those glycolipids which are composed of one or two rhamnose units in combination with two, or more rarely, one or three long chain β-hydroxyalkanoic acids ($C_8$-$C_{16}$) (Abdel-Mawgoud et al. 2010). They were initially discovered in 1947 by Bergström et al. (1947) as exoproducts of the opportunistic pathogen *Pseudomonas aeruginosa.* In general, the mono-RL, L-rhamnosyl-3-hydroxydecanoyl-3-hydroxydecanoate (Rha-$C_{10}$-$C_{10}$) and the di-RL, L-rhamnosyl-L-rhamnosyl-3-hydroxydecanoyl-3-hydroxydecanoate (Rha-Rha-$C_{10}$-$C_{10}$) are the main compounds to be found in liquid cultures (Figure 1; Lang and Wullbrandt 1999). Twenty eight congeners have been described in the mixture synthesized by *P. aeruginosa* (Benincasa et al. 2004).

Observing all RLs producing species—mainly *Pseudomonas* spp. and various *Burkholderia* spp., but also members of the *Actinobacteria* and *Firmicutes*—as many as 60 congeners have been discovered (Abdel-Mawgoud et al. 2010). Considerable variation is observed in the hydrophobic tail, on account of differences in number

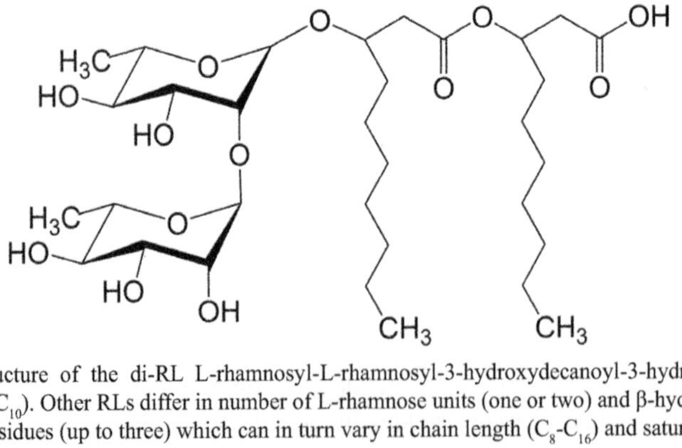

**Figure 1.** Structure of the di-RL L-rhamnosyl-L-rhamnosyl-3-hydroxydecanoyl-3-hydroxydecanoate (Rha-Rha-$C_{10}$-$C_{10}$). Other RLs differ in number of L-rhamnose units (one or two) and β-hydroxy alkanoic acid residues (up to three) which can in turn vary in chain length ($C_8$-$C_{16}$) and saturation.

of fatty acid residues, chain length, saturation of double bonds, methyl groups at the carboxylic end, etc. The hydrophilic head is fairly conserved and can consist of one or two rhamnose units.

## RLs biosynthesis

Both building blocks needed for RL synthesis, rhamnose and β-hydroxyalkanoic acid, are derived from central metabolic pathways. The precursor deoxythymidine di-phospho (dTDP)-L-rhamnose is synthesized from glucose-6-phosphate in five enzymatic steps (Figure 2). The first conversion is executed by the phosphoglucomutase encoded by *AlgC* (Olvera et al. 1999); this enzyme interconverts glucose-6-phosphate to glucose-1-phosphate (Coyne et al. 1994). The remaining four enzymes are encoded by the *rmlBDAC* operon (Rahim et al. 2000). The thymidylyltransferase RmlA catalyzes the reversible bimolecular group transfer of deoxythymidine triphosphate (dTTP) and glucose-1-phosphate (Blankenfeldt et al. 2000). Subsequently, the resulting dTDP-D-glucose is dehydrated by RmlB (dTDP-D-glucose 4,6-dehydratase) to form dTDP-4-oxo-6-deoxy-D-glucose, after which RmlC (dTDP-4-keto-6-deoxy-d-glucose-3,5-epimerase) catalyzes its epimerization into dTDP-4-oxo-6-deoxy-L-mannose (Allard et al. 2000, Giraud et al. 1999). The last step is governed by RmlD (dTDP-4-keto-l-rhamnose reductase) that converts dTDP-4-oxo-6-deoxy-L-mannose to dTDP-L-rhamnose (Giraud et al. 1999). Notably, these rhamnose synthetic enzymes are highly conserved

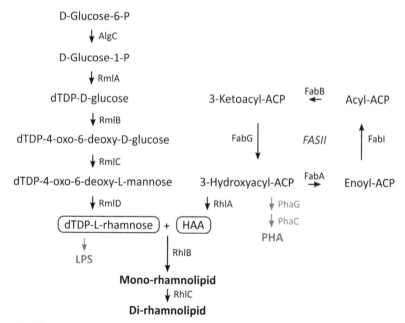

**Figure 2.** Rhamnolipid biosynthetic pathway, modified from Soberón-Chávez et al. 2005, Zhu and Rock 2008. HAA: 3-(3-hydroxyalkanoyloxy)alkanoic acids, ACP: acyl carrier protein. Besides their use for RL biosynthesis, dTDP-L-rhamnose is also a constituent of lipopolysaccharide (LPS) and 3-hydroxyacyl-ACP is also used for synthesis of polyhydroxyalkanoates (PHA).

amongst microorganisms (Graninger et al. 1999, Ma et al. 1997). Besides being the carbohydrate moiety of RLs, L-rhamnose is also a frequently observed constituent of the O-antigen of lipopolysaccharide in Gram-negative bacteria, including *P. aeruginosa*. Furthermore, rhamnose is involved in alginate production and embodies the glycosylation of flagellin in the latter micro-organism (Lindhout et al. 2009, Olvera et al. 1999).

On the other hand, the lipid precursors originate from *de novo* fatty acid synthesis (Figure 2; Zhu and Rock 2008). This implies that in RLs, lipid substrates are not directly incorporated, a merit that is gratefully used for other glycolipids where the lipid backbone can be modified solely by feeding with different lipid substrates (Ashby et al. 2008).[1] Appropriate intermediates from the bacterial type II fatty acid synthetase (FASII) cycle are intercepted by the periplasmic membrane associated protein RhlA. The latter catalyzes the synthesis of 3-(3-hydroxyalkanoyloxy) alkanoic acids (HAAs), starting from two ACP activated 3-hydroxy fatty acids (Déziel et al. 2003, Zhu and Rock 2008). Being the gateway enzyme into RL biosynthesis, RhlA directly competes with PhaG and FabA, key enzymes in the metabolic pathways of polyhydroxyalkanoate (PHA) and fatty acid synthesis, respectively. RhlA exhibits a highly selective activity toward acyl carrier protein (ACP) activated intermediates with a particular length and therefore serves as the molecular ruler determining the length of the RL fatty acid backbone. Accordingly, $C_{10}$ fatty acids are observed the most in the RL mixture produced by *P. aeruginosa* (Soberón-Chávez et al. 2005). In contrast, members of the *Burkholderia* genus prefer to incorporate $C_{14}$ fatty acids in their RLs.

When looking at the core of the RL biosynthetic route, three major sequential steps are involved: (i) RhlA couples two ACP activated 3-hydroxy fatty acids to form HAA, (ii) the first rhamnosyltransferase RhlB transfers dTDP-L-rhamnose to the HAA moiety giving rise to mono-RLs (iii) and optionally, a second rhamnosyltransferase RhlC attaches dTDP-L-rhamnose to the newly synthesized mono-RLs to produce di-RLs (Rahim et al. 2001). As expected for secondary metabolites, RLs are produced upward of the early stationary growth phase. Production is accomplished only when the cell density reaches a minimal threshold or when substrates become limiting. This delayed onset is maintained by elaborative control mechanisms which are discussed in the next paragraph.

The RL biosynthetic pathway mentioned in Figure 2 cannot explain the existence of RL congeners with only one 3-hydroxy fatty acid, though these have been observed frequently (Déziel et al. 1999). Two possible routes have been suggested for the biosynthesis of rhamno-mono-lipids: RhlB may transfer dTDP-L-rhamnose to a single ACP-activated 3-hydroxy fatty acid in an anabolic manner. Contrary, a

---

[1] The addition of particular lipidic substrates can however influence the RL congener distribution (Zhang et al. 2014). For example, the feeding of naphthalene resulted in 80% mono-rhamnolipids whereas usually, the two fatty acid-containing analogue is by far most abundant (Déziel et al. 1999). However, it is not known how the naphthalene substrate affects the biosynthesis. Growth on naphthalene may compromise generating large amounts of di-fatty acids or it may push biosynthesis to at least one other rhamnolipid synthesis pathway.

hypothetical enzyme may hydrolyze the common rhamno-di-lipids in a catabolic manner (Soberón-Chávez et al. 2005). Only recently, Wittgens and colleagues (2016) delivered evidence in favor for the second hypothesis by proving the absolute dependence of RhlB on supply of HAA precursors by RhlA. Further research will shed light on the enzymes responsible for degradation and may permit diversification of the available RL spectrum. The same research group also provided more insight into the *pa1131* gene. Interestingly, it shares an operon with the gene *rhlC* encoding the second rhamnosyltransferase, but its function is still unclear. *In silico* research revealed that its amino acid sequence shares significant homology to TetA, conferring tetracycline resistance. It also contains 11 hydrophobic stretches, presumably spanning a lipid bilayer. From this, it can be suggested that the *pa1131* gene product embodies a putative transporter involved in RL export (Rahim et al. 2001). However, comparable results between *P. putida* strains heterologously expressing *pa1131-rhlC* and a synthetic *rhlABC* operon implied neither a quantitative nor a qualitative role for pa1131 in RL biosynthesis (Wittgens et al. 2016).

## Regulation of RL biosynthesis

Natural RL production is controlled by quorum sensing (QS). This system provides bacteria the means for cell-to-cell interaction, correlated to population density. Small hormone-like molecules (autoinducers) are released and trigger altered gene expression once a minimal threshold stimulatory concentration is reached (Waters and Bassler 2005). Most of the current understanding of the genetic regulation controlling RL production is deduced from studies in the best characterized producer *P. aeruginosa*. An overview of the regulatory machinery in this microorganism is given below and depicted in Figure 3.

The genes *rhlA* and *rhlB*, of which the gene products RhlA and RhlB are responsible for the first and second step in the RL biosynthetic pathway respectively, are colocalized in the *rhlAB* operon (Ochsner et al. 1994). *rhlC*, encoding the second rhamnosyltransferase, forms an operon with *pa1131* and is located elsewhere in the

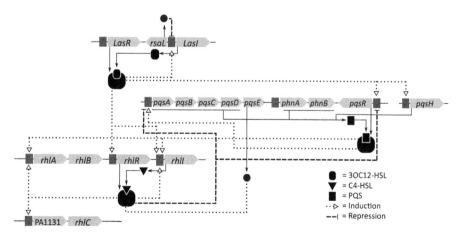

**Figure 3.** Schematic representation of the gene clusters involved in RL production and its regulation in *P. aeruginosa*, after Gallagher et al. 2002, Reis et al. 2011a, Abdel-Mawgoud et al. 2011.

genome (Rahim et al. 2001). The study of Perfumo et al. (2013) showed that *rhlA* and rhlB gene expression is upregulated during the log and early stationary phase of growth while *rhlC* is only expressed in a low level. At the onset of the stationary phase, the tables are turned and *rhlC* is upregulated.

The regulatory mechanism behind this expression pattern is quite elaborate; both operons carry '*lux* box'-like consensus sequences located upstream of their open reading frame, enabling transcriptional control by QS regulators (Pearson et al. 1997). Expression of both operons is subject to the same regulatory machinery (Rahim et al. 2001). At least three different QS systems are involved, the RhlI/RhlR and LasI/LasR systems, both dependent on acyl homoserine lactone (AHL) signal molecules and the *Pseudomonas* quinolone signal (PQS) system driven by 4-hydroxy-2-alkylquinolines (HAQs), the other class of signal molecules produced by *P. aeruginosa.*

Direct transcriptional control of *rhlAB* and *rhlC* is executed by *rhlI*, the *N*-butanoylhomoserine lactone (C4-HSL) autoinducer synthase gene, and *rhlR*, encoding a transcriptional regulator. Coupled to C4-HSL, RhlR acts as a transcriptional activator, while it acts as a repressor in the absence of C4-HSL (Medina et al. 2003). RhlR-C4-HSL also controls transcription of *rhlI*, forming a positive feedback loop. One level higher in the controlling hierarchy is the LasI/LasR QS system. After binding with 3-oxo-dodecanoyl-homoserine lactone (3OC12-HSL) produced by the autoinducer synthase LasI, LasR positively upregulates *rhlI/rhlR*. It also activates *lasI*, hereby securing its own activation via a second positive feedback loop. *rsaL* is located in between *lasR* and *lasI*. Its gene product functions as a general repressor for several hundred QS-controlled genes in order to preserve homeostasis, including *lasI* (Rampioni et al. 2009). The third QS system is intertwined with both the RhlI/RhlR and LasI/LasR systems described above (Dekimpe and Déziel 2009). The HAQ 2-heptyl-3-hydroxy-4-quinolone or PQS influences transcription of *RhlI* in a LasI/LasR independent manner and therefore indirectly activates the *rhlAB-PA1131/rhlC* operons and subsequent RL synthesis. The operons responsible for synthesis of PQS, *pqsABCD* and *phnAB,* are activated by the transcriptional activator PqsR, which in turn is activated by the LasI/LasR QS system. *pqsH*, the gene needed for PQS maturation, is also positively controlled by LasR-C4-HSL (Gallagher et al. 2002). In the midst of the PQS synthetic genes lies *pqsE*. Although its function is unknown up to now, it was proven that its absence did not affect PQS biosynthesis, but it did hamper RL production (Lee and Zhang 2014). It was hypothesized that PqsE enhances RL production by altering RhlR functionality, independent of PqsR and PQS (Farrow et al. 2008). Contradictorily, *pqsR* and *pqsABCD* are susceptible to transcriptional repressor activity executed by RhlR. Hence, the role of the third quorum sensing system is actually determined by the ratio 3OC12-HSL/C4-HSL (McGrath et al. 2004). Several other transcriptional regulators modulate the three QS systems: QscR, VqsR, PtxR, AlgR, DksA, RsmA and GidA; all acting on some level of the regulatory network influencing RL production. For their mode of action, the reader is referred to the review of Reis et al. (2011).

Apart from the intrinsic quorum sensing mechanisms, RL production is also influenced by environmental factors. Déziel et al. (2003) came to a similar conclusion based on a different study conducted two years earlier by Bollinger et al. (2001);

QS-controlled genes such as *rhlAB* do not strictly depend on high cell density in contrast to what is generally believed (Kim et al. 2005). Especially, poor nutrient conditions with an abundance of carbon source can supersede the tightly regulated QS communication and promote the production of RLs (Guerra-Santos et al. 1986). This is part of the clever coping mechanisms of the bacteria; RLs are essential for swarming motility in *P. aeruginosa* (Kohler et al. 2000). Hence, relocation to a new niche with more favorable conditions is encouraged, instead of settling and forming a biofilm.

For example, in a phosphate-deprived environment, expression of RhlR is no longer dependent on LasR, but it is upregulated by multiple transcriptional activators such as Vfr, RpoN (the sigma factor $\sigma^{54}$) and RhlR itself. Under these circumstances, phosphorylated PhoB acts as the transcriptional activator of an array of phosphate-dependent genes (Jensen et al. 2006). This includes *rhlR, pqsABCD* and *pqsR*, again indirectly stimulating RLs production (Bains et al. 2012). Iron availability is also inversely correlated with RLs production because it induces both *lasI/lasR* and *rhlI/rhlR* expression (Déziel et al. 2003, Jensen et al. 2006, Kim et al. 2005).

Another key environmental factor is nitrogen. RLs production improves significantly when *P. aeruginosa* is grown in nitrogen-limited medium (Pearson et al. 1997). This can be explained by the upregulation of *rhlR* by RpoN when nitrogen is scarce (Medina et al. 2003). Several studies show that nitrate as the sole nitrogen source results in an optimal RL production (Soberón-Chávez et al. 2005). The reason for nitrate preference is unknown up to now. Nevertheless, Van Alst and colleagues (2007) studied the link between nitrate metabolism and swarming motility versus biofilm formation in *P. aeruginosa*. Strains mutated in the nitrate response regulator NarX/NarL displayed altered motility. As RLs have a profound role in swarming capacity, they proposed that NarX in the absence of its cognate response regulator NarL may activate a response regulator that (in)directly activates *rhlAB* or alternatively, that NarL may repress *rhlAB* expression. Furthermore, RpoS (sigma factor $\sigma^{38}$) which is transcribed in the late exponential phase or in case of nutrient depletion, indirectly upregulates the *rhlAB* operon (Medina et al. 2003). This again exemplifies the correlation between environmental factors and RL production.

Only very few reports are available about genetic regulation of RLs production in species other than *P. aeruginosa*. In 2009, Dubeau et al. (2009) reported the corresponding genes in *Burkholderia thailandensis* and *Burkholderia pseudomallei*. Two identical gene clusters containing homologues of *rhlA, rhlB* and *rhlC* were described. Strikingly, sequence identities between *B. thailandensis* and *P. aeruginosa rhl* genes are less than 50%. Since *rhlA, rhlB* and *rhlC* are grouped in the same operon, it is hypothesized that *rhlC* is simultaneously expressed with *rhlA* and *rhlC*, resulting in a more efficient addition of the second rhamnose unit (Funston et al. 2016). This explains the higher ratio of di-RLs to mono-RLs observed in *B. thailandensis* (13:1) compared to *P. aeruginosa* (4:1). Additionally, their expression pattern differs from the one that was described for *P. aeruginosa* by Perfumo et al. (2013) where *rhlC* only becomes expressed in the stationary phase. Seeing that gene products of *rhlA, rhlB* and *rhlC* are found in the same timeframe, this could allow for longer RL production in *B. thailandensis* E264 (Funston et al. 2016). Whether both copies of the gene cluster are regulated by the same promoter system is not known yet. Also present in these clusters are genes similar to efflux pumps and transporters, whose products are most likely involved in the transport of RLs outside the cell (Dubeau et al. 2009).

## Genetic engineering

RLs are the most intensively studied among glycolipid BSs. Besides contributing to the physiological functions of their producers, RLs also have biotechnological uses because of their excellent surface activity (Lang and Wullbrandt 1999). RLs can be used as emulsifiers in the food, cosmetic and pharmaceutical industry. Moreover, their antifungal, antimicrobial and antiviral properties make them even more attractive in these fields. Their biocompatibility and microbial degradability provide many opportunities in soil remediation, e.g., oil recovery and heavy metal removal (Mao et al. 2015). Particularly, RLs have proven their potential as replacement for synthetic, petroleum-derived detergents (Maier and Soberón-Chávez 2000). Yet, commercial viability requires competitive production costs compared to the current production processes. While synthetic surfactants are available at $1–3/kg, the actual cost of biotechnologically produced RLs is at least $5–$20/kg when produced in fermenters of 20–100 m³ at 100 g/L (Lang and Wullbrandt 1999). This is consistent with a more recent study, claiming that the cost for BSs production amounts to three to ten times more than the cost to produce a chemical surfactant (Reis et al. 2013). At present, only few companies venture such an enterprise (Randhawa and Rahman 2014). The highest RL production rates are achieved with *P. aeruginosa* DSM 7107 supplied with plant oils as sole carbon source and amount to 112 g/L with a volumetric productivity of 0.424 g/L/h (Lang and Wullbrandt 1999). In general, product yields do not exceed 20 g/L which makes progress towards commercialization difficult. Moreover, concerns were voiced about the accuracy of quantification methods in studies claiming higher product yields. For an extensive review, the reader is referred to Irore et al. (2017).

Efforts have been made to optimize fermentation based on cheap raw materials, e.g., oil mill waste water, molasses or industrial by-products like glycerol in order to reduce costs (Gudiña et al. 2016, Li et al. 2011, Wei et al. 2005, Costa et al. 2009). Genetic engineering of *P. aeruginosa* was also used to enhance BS production. Zhao et al. (2015) increased the *rhlAB* copy number governed by a strong endogenous promoter resulting in a 1.8-fold improvement of production. The wild type strain in this study reached a maximum of 11.65 g/L, but 20.98 g/L RLs were obtained in the same culture conditions after engineering the strain. Furthermore, strains provided with the *Vitreoscilla* hemoglobin (VHb) gene *vgb* exhibited an enhanced production compared to the wild type strain (Kahraman and Erenler 2012). In absolute numbers, this meant an improvement from 6.918 g/L to 7.593 g/L under the same optimized conditions. The rationale behind this general strategy is as follows: harnessing bacteria with VHb boosts the oxygen uptake rate, thereby positively influencing cell density, protein and antibiotic production and bioremediation (Dogan et al. 2006).

However, increasing yields to industrial levels is challenging, because the key biosynthetic genes are subjected to a complex network of regulation. Also, considering the opportunistic pathogenic properties of the best natural producer *P. aeruginosa,* large-scale industrial production with this organism is not desirable. Therefore, the isolation of new RLs producing micro-organisms is a first methodology in the attempt to acquire large-scale industrially-safe production of RLs (Abdel-Mawgoud et al. 2010). Some non-pathogenic *Pseudomonas* spp., such as *P. putida* and *P. alcaligenes,* were investigated but do not approach the productivity of *P. aeruginosa* (Martínez-Toledo and Rodríguez-Vázquez 2011, Oliveira et al. 2009). While process optimization is still at its infancy, experiments with *P. chlororaphis,* another non-pathogenic Pseudomonad, using glucose as sole carbon source result in product titers of approximately 1 g/L (Gunther et al. 2005). Though this result appears in sharp contrast to the product titers mentioned above, it is comparable to the product titers obtained with *P. aeruginosa* when the latter is only provided with glucose as carbon source. Further investigation showed that only mono-RLs were produced by *P. chlororaphis,* indicating that the absence of gene *rhlC. B. thailandensis* was also considered; it reached approximately 1.5 g/L and 5 g/L when, respectively, canola oil or glycerol were supplemented (Marchant et al. 2015, Dubeau et al. 2009). Non-*Pseudomonas* RL producers other than *Burkholderia* spp. include Gram negative bacteria such as *Pseudoxanthomonas* sp. PNL-04, *Acinetobacter calcoaceticus, Enterobacter hormaechei, Pantoea stewartii* and *E. asburiae* (Toribio et al. 2010). RL production is, however, negligible compared to *P. aeruginosa.* The same goes for the Gram positive bacteria *Renibacterium salmonarum* and *Bacillus subtilis* (Christova et al. 2004a, b).

Besides the use of natural non-pathogenic producers, recombinant expression of the entire RL pathway offers another promising alternative methodology. In this way, safe industrial strains can be made without the complex environmental regulation of RL biosynthesis of *P. aeruginosa.* Ochsner and colleagues (1995) introduced the *rhlAB* operon present on an expression vector in *Escherichia coli.* Following their example, Wang et al. (2007) integrated the operon into the genome of *E. coli.* RL production in this host was however minimal, probably because of the limited availability of the required precursor dTDP-L-rhamnose in *E. coli.* Cabrera-Valladares et al. (2006) were able to prove this by heterologously expressing the *rmlBDAC* operon and consequently improving the achieved titers. Using this approach, they were able to produce mono-RLs with *E. coli* in a concentration of 120 mg/L.

Other research groups executed the same strategy, this time using the Generally Recognized As Safe (GRAS) organism *P. putida* as the host organism. Ochsner and colleagues (1995) did the pioneering work by overexpressing the *rhlAB* operon. A recombinant strain of *P. putida* was able to produce 0.6 g/L RLs. Utilizing soybean oil as a substrate, Cha et al. (2008) improved this result tenfold in the same acceptor organism, resulting in 7.3 g/L. Notably, this strain features the highest reported volumetric productivity (101.39 mg/L/h) so far in heterologous hosts. *Burkholderia kururiensis* received the same *rhlAB* operon from *P. aeruginosa.* The recombinant strain reached product formation of 7.4 g/L (Tavares et al. 2013). In an attempt to reduce downstream processing complexity, *P. putida* strains were engineered by Wittgens et al. (2011) which are solely dependent on glucose as carbon source and reached a maximal concentration of 1.5 g/L. This simpler concept will be of importance in future prospects because downstream processing accounts for 70–80% of the entire production costs in case of BSs production (Mukherjee et al. 2006, Randhawa and Rahman 2014). The same research group implemented a driven-by-demand metabolic engineering strategy, which led to 2.8 g/L; this is the highest titer achieved so far using a recombinant RL producer with glucose as sole carbon source. In brief, the organism's carbon metabolism was redistributed with synthetic promoters to increase both the rhamnose and *de novo* fatty acid synthesis by 300% and 50%, respectively (Tiso et al. 2016).

### Trehalose lipids (TLs): Trehalose mycolates and succinoyl trehalose lipids

#### Structure and microbial source of TLs

A second group of bacterial glycolipids is represented by trehalose lipids, or trehalolipids. They were first identified and purified in 1956 as a virulence factor of *Mycobacterium tuberculosis,* the so-called 'cord factor' (Noll et al. 1956). This name originated from of the molecule's contribution to the formation of long and slender *M. tuberculosis* cells. The main producers of TLs are found in the actinobacterial genera *Arthrobacter, Nocardia, Corynebacterium, Gordonia, Mycobacterium* and, most prominently, *Rhodococcus* (Desai and Banat 1997). The sugar moiety of these glycolipids consists of the disaccharide trehalose (two glucose units linked by an α-1,1-glycosidic linkage) and is esterified to α-branched β-hydroxy fatty acids (mycolic acids) or up to eight long acyl chains (Figure 4; Rapp et al. 1979). Among these, the trehalose esters formed by *R. erythropolis* and the 'cord factors' or trehalose dimycolates formed by *M. tuberculosis* are the most extensively studied (Shao 2011). In the membranes of *M. tuberculosis,* the trehalose group is occasionally tailored with a sulfate functionality, rendering TL derivatives called 'sulfoglycolipids' (Domenech et al. 2004).

TLs with mycolic acids fatty acid fraction are referred to as trehalose mycolates and they are 'nonionic'. More than other glycolipids, the hydrophobic fraction of trehalose mycolates is variable. The number of incorporated mycolic acids varies from one to three, resulting in trehalose mono-, di- and trimycolates where the

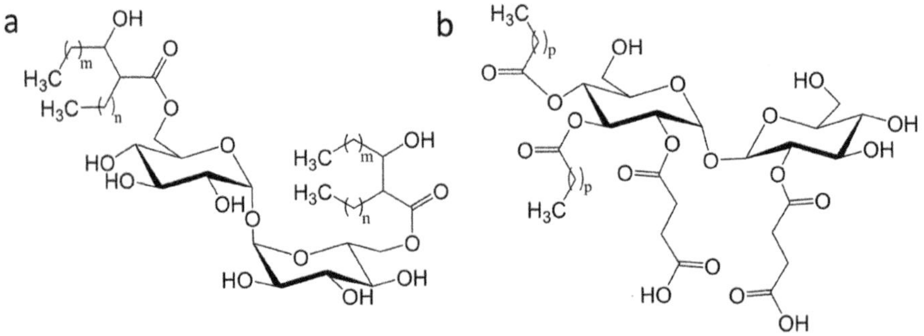

**Figure 4.** (a) Trehalose dimycolate from *Rhodococcus erythropolis* DSM 43215. m + n = 27 to 30. (b) Succinoyl trehalose lipid from *Rhodococcus* sp. SD-74. p = 14, 12, 10 (Tokumoto et al. 2009, Rapp et al. 1979).

trehalose dimycolates is by far the most common (Shao 2011). The length of the mycolic acids varies from 22 to 92 carbon atoms depending on the producing species (Kügler et al. 2015). While mycobacteria usually incorporate branched long-chain hydroxy aliphatic acids, bacteria from genus *Rhodococcus* prefer short-chain fatty acids (Asselineau and Asselineau 1978). Occasionally, the naming of the mycolic acids is done based on their microbial origin; eumycolic (60–92 carbon atoms), corynomycolic (22–36 carbon atoms) or nocardomycolic acid (44–60 carbon atoms) from *Mycobacterium* spp., *Corynebacterium* spp. and *Nocardia* spp., respectively. Apart from differing fatty acid length, their degree of unsaturation and branching vary among microbial species (Shao 2011). TLs from different microbes may also harbor several functional groups such as methoxy-, keto- or epoxy ester groups as well as cyclopropane rings (Barry et al. 1998). The myriad of lipid moieties is so big that 500 distinct molecular species of TLs were reported (Christie 2013).

As stated above, not all TLs contain mycolic acids as their hydrophobic fraction. Another group of TLs is represented by trehalose lipid esters. They have one to four $C_8$-$C_{20}$ saturated acyl chains linked to the trehalose unit (Kügler et al. 2015). Interestingly, the fatty acids are preferentially attached to the secondary hydroxy groups, whereas mycolic acids are preferentially attached to primary hydroxy groups in trehalose mycolates (Kitamoto et al. 2002). Recently, studies revealed that other nonpathogenic Actinobacteria *Tsukamurella spumae* and *Tsukamurella pseudospumae* are also amenable trehalose lipid ester producers (Kügler et al. 2014). TLs produced by the two actinobacteria differ from the others as they only carry two short acyl chains ($C_4/C_6$ or $C_{16}/C_{18}$). *R. erythropolis* also has an outstanding capacity to produce trehalose lipid esters that behave anionic depending on the pH of the solution: succinoyl trehalose lipids (STLs) (Figure 4b; Kim et al. 1990, Tokumoto et al. 2009). In STLs, one or two succinic acids are attached to the trehalose head group giving the molecule an anionic character. In general, nonionic TL titers around 3 g/L are obtained in *R. erythropolis* (Shao 2011), but the highest reported titer in this organism so far for anionic trehalose tetraesters was reached under nitrogen-limiting and resting cell conditions: 32 g/L after 160 h (Kim et al. 1990). *Rhodococcus* sp. SD-74 is able to produce the succinoyl derivatives extracellularly with titers of 40 g/L after 10 days of culturing with *n*-hexadecane under high alkaline and osmotic conditions (Uchida et al. 1989a).

## TLs biosynthesis

*Trehalose mycolates*: Based on the identification of intermediates in the trehalose dimycolate synthesis of *R. erythropolis*, Kretschmer and Wagner (1983) suggested a provisional biosynthetic pathway, depicted in Figure 5. UDP-glucose and glucose-6-phosphate are joined at C1 and C1' by trehalose-6-phosphate synthetase (TPS), forming the carbohydrate head group trehalose-6-phosphate. The mycolic acid residues on the other hand are most likely generated by Claisen-type condensation of two fatty acids. These are derived from β-oxidation degradation intermediates or —especially the long chain fatty acids constituting the major part of the molecule- derived from various other alkane metabolic pathways before condensation (Larkin et al. 2010, Lang and Philp 1998). A carboxylated acyl coenzyme A and an AMP-

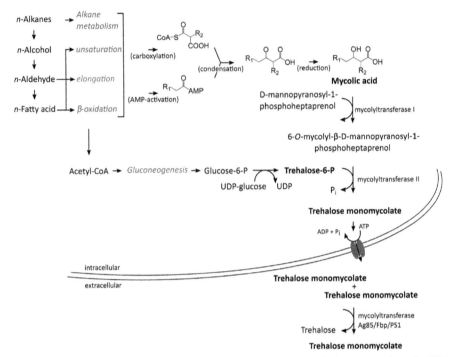

**Figure 5.** Trehalose mono- and dimycolate biosynthesis pathway from *Rhodococcus erythropolis* DSM 43215 as proposed from Kretschmer and Wagner (1982), Lang and Philp (1998) and supplemented with study results in *M. tuberculosis* from Takayama et al. (2005).

activated meroacyl chain are joined to yield a 3-oxo intermediate, which could then be reduced to form mycolic acid (Kuyukina and Ivshina 2010, Takayama et al. 2005). Subsequently, the mycolic acids are linked by a stepwise esterification to the trehalose moiety, resulting in the final TL. Although not much is known about the trehalose acylases that carry out the final step in rhodococci, a detailed description of the process in *M. tuberculosis* was given in 2005 by Takayama et al. (2005): the cytoplasmic mycolyltransferase I forms 6-*O*-mycolyl-β-D-mannopyranosyl-1-phosphoheptaprenol by transferring D-mannopyranosyl-1-phosphoheptaprenol to a mycolic acid residue. The proposed membrane-associated mycolyltransferase II then translocates the mycolic acid group to trehalose-6-phosphate, forming trehalose monomycolate-phosphate. Dephosphorylation by a membrane-associated phosphatase subsequently results in formation of trehalose monomycolate that is immediately transported outside the cell by an ABC transporter cassette. Once there, the extracellular mycolyltransferase Ag85/Fbp/PS1 will combine two trehalose monomycolate molecules resulting in the final product trehalose dimycolate or it will attach a fraction of the trehalose monomycolates to the peptidoglycan-arabinogalactan complex of the cell wall resulting in arabinogalactan-mycolates (Nguyen et al. 2005, Takayama et al. 2005).

It is essential to know by which biosynthetic pathways TLs are produced in order to optimize their production, or even expand their structural variety with biotechnological solutions. Unraveling the underlying genetic link is therefore imperative. However, as opposed to many other glycolipids described in the remainder of this chapter, not much is known about the genetics behind trehalose mycolate biosynthesis (Shao 2011). Fueled by the industrial potential of TLs, this knowledge-gap will most likely be closed in the imminent future as more molecular tools and whole genome sequences for *Rhodococcus* spp. become available (Larkin et al. 2010).

*Succinoyl trehalose lipids*: Biosynthesis of this anionic type of TLs is only observed when cultivated on *n*-alkanes under growth-limiting conditions (Lang and Philp 1998). Using a transposon mutagenesis system, Inaba and colleagues (2013) identified three genes as essential for STL precursor delivery in *Rhodococcus* species: *alkB, fda* and *tlsA*. Based on these results, they proposed a hypothetical STL biosynthetic pathway (Figure 6). Knowledge about the genes involved in the core pathway, however, remains hitherto absent. Whether the overall biosynthetic pathway is similar in other STL producing species is unknown but can be assumed (Inaba et al. 2013).

## Regulation of TLs biosynthesis

*Trehalose mycolates*: As pointed out in the introduction, certain glycolipids play a pivotal role in cellular adaptation to the presence of *n*-alkanes (Kitamoto et al. 2002). Experiments in *R. ruber* demonstrated that TL production starts at the $C_{11}$ substrate length and reaching its maximal level at $C_{16}$ (Philp et al. 2002). The transcript levels of two out of four predicted alkane-1-monooxygenases genes (*alkB*) in *R. erythropolis* were elevated after addition of hexadecane and diesel oil (Laczi et al. 2015).

This enzyme is responsible for the first step in *n*-alkane dissimilation. In analogy to the proposed STL biosynthetic pathway (Figure 6), it is also most likely

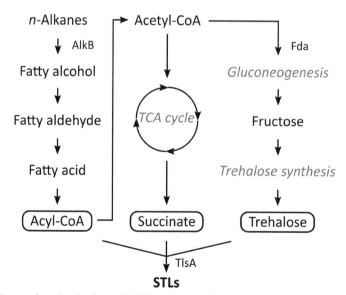

**Figure 6.** Proposed succinoyl trehalose lipid biosynthesis pathway according to Inaba et al. (2013). TCA = tricarboxylic acid cycle, CoA = Coenzyme A.

the first enzyme leading to trehalose mycolates. Interestingly, a putative TetR-type transcriptional regulator gene is located in the proximity of the *alkB* operons. Its genomic location together with its simultaneous upregulation leads to the assumption that this TetR-type protein is in transcriptional control of *alkB* (Laczi et al. 2015). Indirect evidence is given by Cappelletti et al. (2011). Searching for homologous genes in the genomic region of the *alkB* cluster of *Rhodococcus* sp. BCP1 led to the discovery of two putative transcriptional regulators at both sides. Immediately downstream of the cluster, an incomplete open reading frame showed 80% homology to a TetR-like transcriptional regulator found in *Rhodococcus* sp. RHA1. Further research is needed to have conclusive proof of the role for a TetR-like transcriptional regulator in the alkane-induced response.

However, some reports in *R. erythropolis* indicate the production of trehalose mycolates based on soluble substrates such as glycerol, glucose and ethanol (Ciapina et al. 2006, Pirog et al. 2004). This indicates that the presence of hydrocarbons does not govern all biosynthesis regulation. Furthermore, it indicates that the regulation varies among TL producers, even at the genus level (Pirog et al. 2013). The type of carbon source also affects the capability of the producer to secrete the TLs in the extracellular environment (Franzetti et al. 2010). While they are usually associated to the cell wall, the addition of glycerol as sole carbon source led to the secretion of the BSs (Ciapina et al. 2006). Furthermore, the hydrophilic head is susceptible to structural differences caused by the applied water-soluble carbon source: the addition of fructose led to the formation of fructose-1,6-dimycolates (Itoh and Suzuki 1974).

Just as with the other glycolipids described throughout this book chapter, there is an intricate dependence between environmental factors such as nitrogen source, salt composition or use of yeast extract and TL outcome (Kügler et al. 2014). Franzetti et al. (2009) were the first to apply a fractional factorial design to study the cultural factors affecting TL production by the genus *Gordonia*. Proper statistical tools identified the concentration of *n*-hexadecane, phosphate and sodium chloride as the parameters significantly determining the TL biosynthesis. With this knowledge, they succeeded to improve the BS concentration 5-fold. Supported by the findings of Madry et al. (1979) in *Streptomyces*, another actinomycete, they speculated that the accumulation of glucose-6-phosphate under high phosphorous conditions led to an increased activity of TPS. Pirog et al. (2013) added the organic acids fumarate and citrate to *R. erythropolis IMV Ac-5017* culture medium as 'biosynthesis precursor' rather than as secondary source of carbon. In these low amounts (0.1–0.2%), the organic acids succeeded in increasing the activity of isocitrate lyase (which provides precursors of gluconeogenesis), phosphoenolpyruvate synthetase (key to gluconeogenesis) and trehalose phosphate synthase (trehalose mycolate synthetic enzyme). Ultimately, this led to an increased surfactant concentration of 1.5–1.7 fold. Unfortunately, absolute numbers in g/L are not given.

### Succinoyl trehalose lipids (STLs)

*Rhodococcus* sp. strain SD-74 reached maximum productivity of STLs when grown on *n*-hexadecane. The transcriptional level of the gateway enzyme AlkB is upregulated by this substrate, opening up the metabolic flux. The transcriptional level of the enzyme completing the pathway, TlsA, is also upregulated by the presence of *n*-hexadecane (Inaba et al. 2013). However, more detailed genetic and biochemical studies need to be performed to uncover the regulatory circuitry. As opposed to what is described for RL synthesis, the carbon chain length and number of unsaturated bonds of the fatty acids in STLs can be steered by the added hydrocarbon substrate (Uchida et al. 1989b). This direct incorporation is most likely possible via terminal oxidation of the carbon source (Tokumoto et al. 2009). This 'plug-in'-compatibility is of industrial importance because it is often difficult to produce specifically defined BS varieties (Inaba et al. 2013). Product mixtures rather than pure products add to the cost of downstream processing, which already substantially prevents market penetration for microbial BSs (Makkar et al. 2011).

## Genetic engineering

The TL category of glycolipids differs from the other glycolipids; instead of being secreted in the extracellular environment, they are mostly withheld in the cell wall (Vergne and Daffé 1998). The industrial importance of this BS is therefore somewhat inferior. The productivity is not as high as other BSs because they are often restricted to the two-dimensional space of the cell wall and their recovery is more difficult (Kitamoto et al. 2002). Generally obtained yields are below 3 g/L, albeit the utilization of experimental design techniques; surface response methodology has led to increased yields up to 10 g/L (Mutalik et al. 2008, Shao 2011, Franzetti et al. 2009). In order to make production processes for TLs cost-competitive, three general strategies are evaluated: the use of cheap and waste substrates, optimization of fermentative and recovery conditions and the development of overproducing strains (Franzetti et al. 2010). Notwithstanding its high potential, the third option remains fairly unexplored for this type of glycolipid because not much is known about the involved genetics of the respective producers. Only one study reports about the use of recombinant TL producers: Dogan et al. (2006) inserted the VHb gene *vgb* in *Gordonia amarae*. The engineered strain showed a 2.1-fold (60 mg/L) and 4 fold (90 mg/L) increase in extracellular TL production when limited and normal aeration were applied, respectively. A 1.4-fold (22 mg/L) and 2.1-fold (20 mg/L) improvement was observed in TL that remained in the cell.

*Rhodococcus* sp. SD-74 was engineered in an attempt to increase production levels of STLs (Inaba et al. 2013). By cloning the strong alkane-inducible promoter of *AlkB* in front of *tlsA* on an overexpression plasmid, Inaba et al. (2013) succeeded to create a mutant with an STL production of nearly 200% compared to the wild type. Unfortunately, STL overproduction was presented in this study in relative instead absolute quantities. However, the authors state that the increasing production volume represent 40 g/L when converted into grams at theoretical yields. Overall, metabolic engineering approaches towards TL production improvement are very limited due to lack of knowledge about the accountable genetics. However, further research is desired not only to increase TL production, but also to optimize their export and growth on water-soluble substrates (White et al. 2013). The latter will simplify downstream processing which in turn aids commercial viability.

## Sophorolipids (SL)

### Structure and microbial source of SLs

SLs belong to the most promising glycolipids from a commercial point of view. They consist of sophorose as the hydrophilic part and a terminal or subterminal hydroxylated fatty acid of 16 or 18 carbon atoms as the hydrophobic tail. Sophorose is a glucose disaccharide with an unusual β-1,2-glycosidic bond that, incorporated in SLs, can be further acetylated at the 6'and/or 6" positions. SLs can occur in the open, acidic form or the closed, lactonized form with an internal esterification between the carboxylic end and the 4" of the sophorose head (Figure 7). Exceptionally, lactonization occurs via the 6'- or 6"-position (Tulloch et al. 1962, 1967, 1968). SLs are produced by several non-pathogenic yeasts. The first organism reported to be able to synthesize hydroxylated fatty acids linked to a sophorose unit was the anamorphic ascomycetous yeast *Candida apicola*, initially identified as *Torulopsis magnolia* (Gorin et al. 1961). Later on, Spencer et al. (1970) demonstrated SL production by *Starmerella bombicola* (former *Torulopsis bombicola*, anamorph *Candida bombicola*). By applying a phylogenetic approach on species closely related to *S. bombicola*, other sophorolipid producers such as *Candida batistae* (Konishi et al. 2008a), *C. riodocensis*, *C. stellata*, *Candida* sp. NRR Y-27208 (Kurtzman et al. 2010) and *C. floricola* (Imura et al. 2010) were identified. Although all previously mentioned SL producers belong to the *Starmerella* clade (Kurtzman et al. 2010), the ability of SL biosynthesis was also suggested for less related fungi such as: *Cyberlindnera samutprakarnensis* (Poomtien et al. 2013), *Pichia anomala* (Thaniyavarn et al. 2008) and is extensively described for the basidiomycetous yeast *Pseudohyphozyma bogoriensis,* formerly known as *Rhodotorula bogoriensis* (Tulloch et al. 1968). While there are several reports on the ascomycetous yeast *Wickerhamiella domercqiae* var. sophorolipid to be a novel excellent SL producer (Chen et al. 2006, Liu et al. 2016), it was demonstrated by a molecular analysis of rRNA sequences that this species is in fact the well-known *S. bombicola*. The reclassification of *W. domercqiae* var. sophorolipid as *S. bombicola* CGMCC 1576 was only possible when its whole genome sequences became publicly available (Li et al. 2016).

The best studied and most widely used sophorolipid-producing yeast is *S. bombicola*. This novel name for *C. bombicola* was introduced by Rosa and

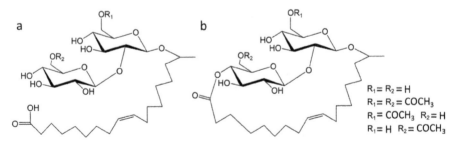

**Figure 7.** Structure of typical SLs produced by *S. bombicola* (a) acidic SL (b) lactonic SL.

Lachance referring to the teleomorph (sexual form) of *C. bombicola* (Rosa and Lachance 1998). The genus name *Starmerella* was presented in honor to William T. Starmer, in recognition of his contribution to the ecology and evolution of yeasts associated with plants and insects.

The structure of SLs synthesized by yeast from the *Starmerella* clade is nearly identical, and differences are limited to acidic and lactonic form ratios (Table 1) or the fatty acid hydroxylation preferences. For instance, *S. bombicola* and *C. apicola* mainly produce ω-1 hydroxyl fatty acids to be incorporated whereas *C. batistae* and *Candida NRRL Y-27208* favor the terminally hydroxylated one. On the contrary, SLs produced by *R. bogoriensis* display two major structural variations: the predominant form contains a $C_{22}$ hydroxyl fatty acid which is internally hydroxylated at the 13th carbon atom. Furthermore, lack of terminal or subterminal hydroxylation disfavors internal esterification, resulting in solely acidic branched SLs; thus, no lactonic forms are observed (Tulloch et al. 1968).

## SL biosynthesis

*R. bogoriensis* was the first organism for which the SL biosynthesis was investigated. It was demonstrated that SLs are formed by the action of glucosyl- and acetyltransferases. The synthesis involves a transfer of glucose residues from a UDP-glucose to a hydroxylated fatty acid. In the first glycosylation, a glucolipid is formed to which a second glucose is added. The resulting SL can subsequently be acetylated and is predominantly secreted as a di-acetylated SL. Although the early biochemical studies with *R. bogoriensis* cell lysates indicated a presence of

**Table 1.** Sophorolipid mixture profile of different yeasts from the *Starmerella* clade. Adapted from (de Oliveira et al. 2014).

| Microorganism | Acidic SLs | | | Lactonic SLs | | | Titer g/L | References |
|---|---|---|---|---|---|---|---|---|
| | Non-Ac | Mono-Ac | Di-Ac | Non-Ac | Mono-Ac | Di-Ac | | |
| *S. bombicola* ATCC 22214 | + | N/D | ++ | N/D | N/D | +++ | 300 | (Davila et al. 1997) |
| *C. stellata* | + | ++ | +++ | N/D | N/D | N/D | 11.9 | (Kurtzman et al. 2010) |
| *C. riodocensis* | + | ++ | +++ | N/D | +++ | N/D | 8.3 | (Kurtzman et al. 2010) |
| *C. apicola* NRRL-Y-2481 | + | + | + | + | ++ | +++ | 52.7 | (Kurtzman et al. 2010) |
| *C. batistae* CBS 8550 | N/D | N/D | +++ | N/D | + | + | 6 | (Konishi et al. 2008a) |
| *C. kuoi* (NRR Y-27208) | + | ++ | +++ | N/D | ++ | N/D | 20.1 | (Kurtzman et al. 2010, Price et al. 2012) |
| *C. floricola* | N/D | N/D | ++ | N/D | N/D | ++ | 22.9 | (Imura et al. 2010, Konishi et al. 2016) |

+ minor product; ++ lower quantities; +++ major product; N/D not detected two different glucosyltransferases, the potential protein candidates could not be isolated from each other (Esders and Light 1972). Therefore, the existence of a single multifunctional enzyme catalyzing both glucosylation reaction in *R. bogoriensis* was not ruled out (Breithaupts and Light 1982). Even though the first insights into the biosynthesis of yeast glycolipids was already revealed in the early seventies, it was only in 2007 that light was shed on the genetics of SL synthesis and a hypothetical pathway was proposed (Van Bogaert et al. 2007). Finally, in 2013, Van Bogaert et al. (2013) presented the complete SL biosynthetic pathway of *S. bombicola* and their respective genes. As would be expected for secondary metabolites, the genes encoding essential enzymes involved in the biosynthesis of SLs are organized in one cluster located near the telomere (Figure 8). Only one out of the six genes, the lactone esterase responsible for the lactonization of SLs, is found elsewhere in the genome (Ciesielska et al. 2014, 2016).

Biosynthesis of SLs occurs in the stationary phase under nitrogen limitation, and it is dissociated from cellular growth (Davila et al. 1992). Ideally, both a hydrophobic and hydrophilic carbon source is present in the SL production medium (Cooper and Paddock 1984). If there is no hydrophobic substrate such as fatty acids supplemented in the medium, the fatty acid backbone of the SL can be synthesized *de novo* from acetyl-CoA (derived from the glycolysis pathway for instance; Van Bogaert et al. 2008), yet direct incorporation is more efficient (Linton 1991). The first step in SLs biosynthesis (Figure 9) is terminal ($\omega$) or sub-terminal ($\omega$-1) hydroxylation of the fatty acid by means of an NADPH-dependent, cytochrome P450 monooxygenase (Cyp52M1; Van Bogaert et al. 2009b). The hydroxylation-activated fatty acid is then glycosylated in a stepwise manner by the action of glucosyltransferase I (UgtA1; Saerens et al. 2011a) and glucosyltransferase II (UgtB1; Saerens et al. 2011b) requiring UDP-activated glucose as the glucosyl donor. The first glucose molecule is linked to the hydroxyl fatty acid by its C1' position while the second one is added in position C2' to the formed glucolipid. *In vitro* donor specificity study showed strong specificity of both glucosyltransferases towards UDP-glucose, but also a minor activity was observed towards UDP-activated galactose and UDP-glucuronic acid. Furthermore, 17-hydroxylated octadecanoic acid is the preferred acceptor for UgtA1 which is in line with the fact that SLs in a wild-type mixture harbor a $C_{18}$ hydroxy fatty acid tail. The product of the second glycosylation, a non-acetylated SL, can be further decorated with one or two acetyl groups at the sophorose moiety. Acetylation is mediated by a specific acetyltransferase (At) and can occur either at the C6' or C6'' position (Saerens et al. 2011b). A crucial role in the SL biosynthesis pathway is attributed to the SL plasma membrane transporter belonging to the

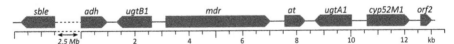

**Figure 8.** The SL biosynthetic gene cluster of *S. bombicola* containing a *cyp52M1* monooxygenase gene, two glucosyltransferase genes *ugtA1* and *ugtB1*, an acetyltransferase gene *at* and an ABC transporter gene *mdr* Gene encoding lactone esterase *sble* is located at the distance of 2.5 Mb from the cluster (Van Bogaert et al. 2013, Ciesielska et al. 2016).

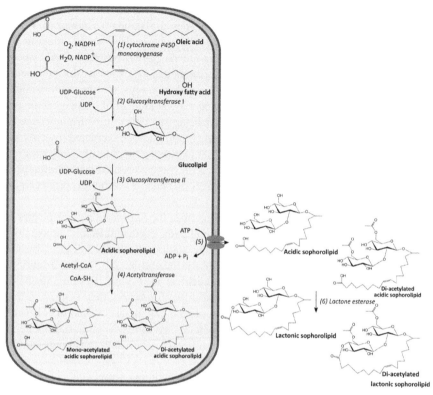

**Figure 9.** Biosynthesis pathway for the production of sophorolipids in *S. bombicola*. (1) Cytochrome P450 monooxygenase Cyp52M1, (2) UDP-glucosyltransferase UgtA1, (3) UDP-glucosyltransferase UgtB1, (4) Acetyltransferase At, (5) Sophorolipid transporter Mdr, (6) Lactone esterase Sble.

ABC transporter protein superfamily. Efficient transport is one of the main reasons why high and commercially relevant SL titers are achieved. This active exporter is responsible for secretion of the acidic sophorolipids to the extracellular space where they can undergo another final modification: lactonization. The lactonization process occurs through an internal esterification between the carboxyl group of the fatty acid and a sophorose hydroxyl at the C4" mediated by cell wall-bound lactone esterase. It is suggested that acetylated congeners are preferred over non-acetylated ones (Ciesielska et al. 2014, 2016).

## Regulation of SL biosynthesis

The central regulation drivers in SL biosynthesis in both *S. bombicola* and *R. bogoriensis* are carbon and nitrogen source, their ratio, as well as the cellular stage of the culture (Cutler and Light 1979, Casas and García-Ochoa 1999). For *C. apicola,* not only the C/N ratio but also the absolute quantity of N was proposed to play a role in SL biosynthesis regulation (Hommel et al. 1994). Since nitrogen plays a crucial role in the regulation of SL biosynthesis and as yeast extract (YE) is the most commonly used nitrogen source, its optimal concentration for a synthesis of SLs has been tested: whereas 5 g/L is considered a standard amount of YE in SL production medium, Zhou and Kosaric (1993) obtained higher SL titers using 2–3 g/L of YE. However, Casas and García-Ochoa (1999) reported 1 g/L YE to be optimal. These apparently contradictory results can be explained by the use of different overall media compositions by the individual groups, so other factors might be involved. Nevertheless, in general, applying a lower amount of nitrogen disfavors the biomass formation and cells quickly reach the stationary growth phase due to nitrogen depletion. Consequently, the excess carbon source is directed towards SL formation. Moreover, it was demonstrated that increasing ammonium concentration under constant YE concentration (1 g/L) affects the biomass formation and influences the ratio of the two hydroxyl fatty acid isomers ($\omega$ and $\omega$-1) of the SL produced by *C. apicola* (Hommel et al. 1994).

Though the glycolipid synthesis in strains from the *Starmerella* clade occurs in the stationary phase, glucosyltransferases are already expressed in mid exponential phase and their activity increases toward the stationary phase, followed by the upregulation of their corresponding genes (Ciesielska et al. 2013). These findings are consistent with the conclusions of Albrecht et al. (1996). Besides the basal constitutive expression of enzymes involved in SL formation, they suggested that nitrogen (discussed above) and phosphate depletion can indirectly lead to enhanced SL production. Under nitrogen starvation conditions, synthesis of cellular components such as protein, DNA or RNA terminates while the carbon source is continuously directed into lipid formation.

Lipid biosynthesis is a cytosolic process that requires acetyl-CoA as the core precursor. In non-oleaginous yeasts, acetyl-CoA originates from mitochondria whereas in oleaginous yeast citrate is an acetyl donor for fatty acid biosynthesis (Evans and Ratledge 1985). The nitrogen depletion results in a decrease of the cellular AMP concentration, important to keep specific activities of NAD+- and NADP+-dependent isocitrate dehydrogenase. Inactivation of this enzyme interrupts the oxidative role of the tricarboxylic acid cycle which leads to accumulation of isocitrate and subsequently citrate in the mitochondria. Both acids are then transported into the cytosol where citrate is cleaved by adenosine triphosphate (ATP): citrate lyase generating acetyl-CoA and oxaloacetate, the precursor for fatty acid synthesis (Albrecht et al. 1996). The accumulation of citrate activates the acetyl-CoA carboxylase, an enzyme that catalyzes irreversible carboxylation of cytosolic acetyl-CoA to malonyl-CoA, the building block for new fatty acids. The formed fatty acid can then be directed to SL biosynthesis.

Interestingly, upon addition of sodium citrate total BS yield was higher, and an increased amount of lactonic sophorolipids was observed. Citrate also has a positive effect on biomass formation since the medium supplemented with 2.5 to 10 g/L at the beginning of the cultivation resulted in a slightly increased biomass. Taken together, the addition of citrate causes slightly increased biomass formation, significant alteration of the produced SL mixture and better overall yield. The observed impact of additional citrate can be linked to its abovementioned inhibitory effect on NAD+- and NADP+-dependent isocitrate dehydrogenase (Evans and Ratledge 1985) and as a consequence, yeast metabolism is pooled towards fatty acid biosynthesis. Stüwer et al. (1987) proposed that the controlling role of citrate on SL lactonization in the wild-type *S. bombicola* is attributed to a pH triggered regulation effect. This theory has been countered when the lactone esterase (Sble) was discovered (Roelants et al. 2015). A strain overexpressing *sble* under control of the constitutive *pki* promoter mainly produces lactonic SLs in the absence of citrate (at low pH values). Yet, an (indirect) regulatory effect of the extracellular pH at the native promotor of the *sble* gene could not be excluded. It has been further suggested that these pH effects trigger *cis*- or *trans*-regulatory elements acting on the SL biosynthetic genes (Roelants et al. 2015). Although SL producing enzymes are organized in a gene cluster which indicates co-regulation, no transcription factor, neither within the cluster nor on the genome, has been identified so far (Van Bogaert et al. 2013, Ciesielska et al. 2013). This is in contrast to the other yeast glycolipid gene clusters discussed in the later sections.

Transcriptomic and proteomic comparison of producing (early stationary) and non-producing (exponential growing) *S. bombicola* cells indicated at strong upregulation of the SL cluster genes in the early stationary phase is linked with clearly higher protein abundance. Changes between 2 and 6 log 2 units were observed (Ciesielska et al. 2013). Furthermore, the same study puts forward a possible enhancer for Cyp52M1 action, a homolog to damage resistant protein 1 (Dap1), which is highly abundant in resting cells. Dap1 is known to play a role in the regulation of cytochrome P450 enzymes participating in the metabolism of lipids and sterols.

## Genetic engineering

Rather than pure compounds, SLs are synthesized as a mixture of slightly different molecules, and these structural variations influence their physicochemical properties and consequently, their potential application. In other words, in order to broaden up the application range of microbial glycolipids, their structural variability needs to be controlled or expanded. Genetic engineering can offer the solution to tailor the entire enzymatic cascade to the needs of a particular application. It also opens up opportunities for synthesis of new-to-nature SLs. Indeed, great work has been done to genetically modify *S. bombicola*. As mentioned before, the predominant SL fatty acid chain length is represented by 16 or 18 carbon atoms and therefore poses limits to the use of SLs in certain cleaning purposes. A shorter fatty acid tail will cause a shift in the hydrophilic/hydrophobic balance of the molecule causing better water solubility and even improved surface lowering capacities in some cases (Brakemeier et al. 1998a). The strict length distribution of the fatty acid is guarded by the substrate specificity of the Cyp52M1 that is clearly towards fatty acids of 22.55–25.05 Å corresponding to 16–18 carbon atoms (Tulloch et al. 1962). One strategy in order to circumvent this strict length preference of Cyp52M1 is to provide substrates structurally resembling stearic acid and in this way deceive the enzyme. Here, the substrates with an internal ester bond are of particular interest since this bond can be cleaved by alkaline hydrolysis and give rise to new-to-nature sophorolipids consisting of a shorter fatty acid chain. Indeed, cultivating *S. bombicola* on alkyl ester dodecyl glutarate, a molecule structurally very similar to stearic acid, led to the production of diacetylated dodecyl glutarate SLs and after post-fermentation alkaline hydrolysis, $C_{12}$ SL with the hydroxyl fatty alcohol as a hydrophobic moiety were recovered (Van Bogaert et al. 2011b). The second strategy is to use already hydroxylated substrates and thus bypass the P450 controlling action completely. Following this approach, ω-hydroxylated medium-chain fatty acid ($C_{12}$) and 1,12-dodecanediol were tested. While these two substrates were readily incorporated in SL biosynthesis, the corresponding fatty acid, dodecanoic acid, was not. However, because of the symmetric character 1,12-dodecanediol, sophorose units could be introduced at both sites, resulting in a heterogeneous mixture of glycolipids (Van Bogaert et al. 2011a).

Medium-chain ($C_{12}$-$C_{14}$) SLs can be obtained by cultivation of the yeast *S. bombicola* on terminally hydroxylated medium chain substrates such as alcohols, diols and hydroxyl fatty acids. However, the production yields are still low since these special substrates are readily metabolized via β-oxidation and used for *de novo* synthesis of hydroxyl $C_{16}$/$C_{18}$ fatty acids. One solution to avoid substrate degradation is to use a medium with high glucose and yeast extract content, and ketons and/ or aldehydes as a secondary carbon source as the latter ones can be enzymatically reduced by yeast to the corresponding alcohols and attached directly to the sophorose molecule (Brakemeier et al. 1998b). Another solution is to knock out the *mfe2* gene, encoding the multifunctional enzyme type 2, and hereby create a *S. bombicola* strain with a disrupted β-oxidation pathway. Blocking this pathway results in a 3-fold improved production of SLs with shorter chain lengths compared to the wild type since the substrates are now directed to SL biosynthesis instead of being degraded (Van Bogaert et al. 2009a). Yet, the dependence on special substrates instead of vegetable oil is not in favor for cost-effective fermentation processes.

Next to the fatty acid chain length, the acetylation pattern is another structural feature influencing physio-chemical properties of SLs. In comparison to acetylated SLs, non-acetylated variants display decreased cytotoxicity, crucial for biomedical applications, and increased water solubility, essential for detergent industry. Therefore the production of solely one type of SLs is important not only from a downstream processing point of view but also from a potential application angle. So far, the best enrichment for non-acetylated SLs without applying any genetic manipulation was obtained by single-step cultivation of *S. bombicola* on whey concentrate and rapeseed oil. The secreted SL mixture was enriched with 45% of the non-acetylated lactonic form (Otto et al. 1999). One would think that simple genetic engineering could offer a solution here. However, the *S. bombicola* deletion mutant lacking acetyltransferase (*at*), necessary for acetylation of the sophorose moiety, not only produces non-acetylated SLs (Saerens et al. 2011b) but surprisingly also bolaamphiphilic SLs (Van Bogaert et al. 2016). The latter molecules consist of a fatty acid chain with a carbohydrate head on both sides (Figure 10). These bolaamphiphilic SLs seem to be present as well in the natural mixture, but in extremely low amounts. The concentration below the detection is most likely the reason why they were never detected until more sensitive techniques became available (Price et al. 2012).

Knocking out the second glucosyltransferase UgtB1 resulted in the absence of SL production. Instead, glucolipids, molecules with only one glucose moiety attached to the fatty acid tail, were recovered from the production broth. These new types of glycolipids are interesting intermediates for several kinds of biocatalytic or chemical conversion reactions. Therefore, the Δ*UgtB1 S. bombicola* is an attractive one-step production platform of glucolipids by conventional fermentation on cheap substrates (Saerens et al. 2011c).

In order to produce solely acidic SLs, Ashby and colleagues applied culture conditions. They used alkyl esters of soybean oil as a fermentation substrate. This approach indeed led to a higher concentration of the open-chain acidic SLs (80%) but at the same time contributed to a drastic decline in the S. *bombicola* productivity (Ashby et al. 2006). However, by a single knockout of the lactone esterase gene, a strain producing pure acidic SLs in concentrations comparable to the wild-type strain was obtained (Ciesielska et al. 2014, 2016). On the other hand, the production of lactonic variants is of industrial desire. Bearing that in mind, Roelants et al. (2016) overexpressed the lactone esterase. The resulting fermentation broth contained over 99% of lactonic SLs in comparison to 57% after wild-type strain fermentation. Thanks to genetic manipulations, the shift from mixed SLs produced by a wild-type strain towards pure acidic or pure lactonic forms is now just a matter of choosing the appropriate production strain, instead of applying complex purification methods.

Single knockout of *le* or *at* led to the production of acidic SLs with a mixed acetylation pattern and non-acetylated lactonic/acidic forms, respectively. Therefore, the next logical step in the path of taking control over SL complexity would be to combine these two knockouts in one strain in order to obtain non-acetylated acidic SLs. Indeed, Van Bogaert et al. (2016) created such a strain. However, besides the expected non-acetylated acidic variants, the production broth contained

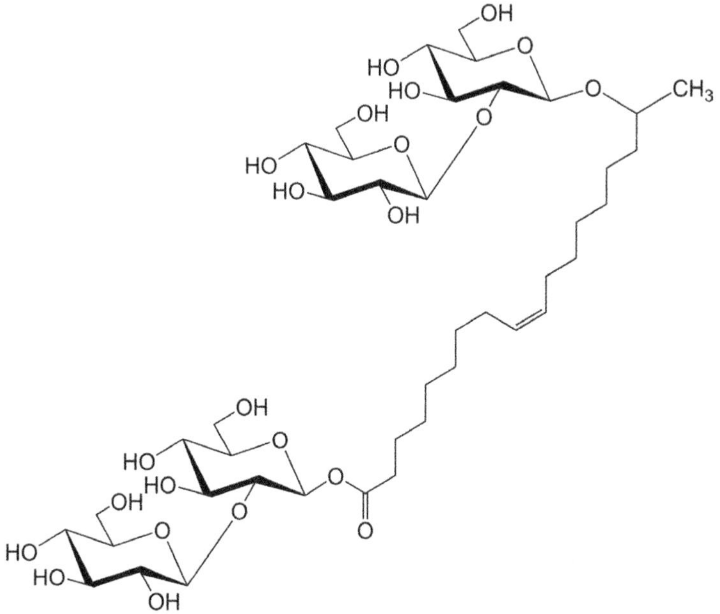

**Figure 10.** Bolaamphiphilic sophorolipid molecule with two sophorose molecules attached to the fatty acid chain.

bolaamphiphilic glycolipids, identical as that for the Δ*at S. bombicola*. Interestingly, further evaluation of the enzymes possibly involved in the biosynthesis of these intriguing structures revealed that the same glucosyltransferases as for the SL synthesis add a third (UgtA1) and fourth (UgtB1) glucose moiety. However, the overall titer for the single *at* knockout is much lower (10.2 g/L) compared to the combined knockouts (27.7 g/L) and the abundance of bolaamphiphilic congeners is also higher in the case of Δ*at*Δ*le* (Van Bogaert et al. 2016).

## *Mannosylerythritol lipids (MELs)*

### *Structural diversity and microbial source of MELs*

Next to the SLs, MELs are other, very promising representatives of microbial glycolipids. MELs consist of the mannosylerythritol disaccharide (4-O-β-D-mannopyranosyl-*meso*-erythritol) which is acylated with a short-chain ($C_2$-$C_8$) and a medium-chain ($C_{10}$-$C_{18}$) fatty acid (Kitamoto et al. 1990). Furthermore, based on the degree of acetylation at C4 and C6 positions of the hydrophilic head, MELs are classified as MEL-A, MEL-B, MEL-C and MEL-D (Figure 11; Kurz et al. 2003b). MEL-A represents the diacetylated variant, while MEL-B and MEL-C are monoacetylated glycolipids at C4 and C6, respectively. The non-acetylated structure is known as MEL-D. Although very similar, these different MEL variants display specific phase behavior and different self-assembled structures in aqueous solution. For instance, while MEL-A shows very low water solubility and/or hydrophilicity, MEL-B and MEL-C display higher hydrophilicity and critical micelle concentration (CMC). Therefore, MEL-B/MEL-C are more suited for use as a washing detergent or as water-in-oil type emulsifiers (Imura et al. 2006).

MELs were first discovered as oily compounds in the culture suspension of a dimorphic basidiomycete *Ustilago maydis*. This fungus has the unique ability to produce two structurally different extracellular glycolipids. MELs and cellobiose lipids (CBLs) which are discussed in the next section (Haskins 1950). Besides the initially identified producer, many other MEL producing species were discovered during the years: *Schizonella melanogramma* (schizonellin; Deml et al. 1980b), *Geotrichum candidum* (Kurz et al. 2003) and *Kurtzmanomyces* sp. (Kakugawa et al. 2002). However, the best MEL producers are found in the *Pseudozyma* clade (Kitamoto et al. 1990); *P. antarctica, P. aphididis, P. rugulosa* and *P. parantarctica*

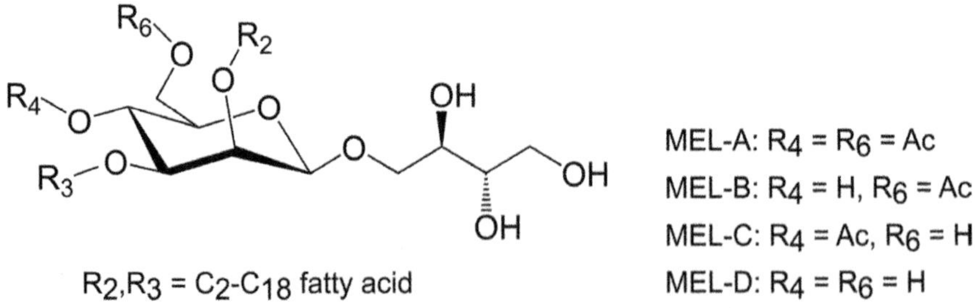

MEL-A: $R_4$ = $R_6$ = Ac
MEL-B: $R_4$ = H, $R_6$ = Ac
MEL-C: $R_4$ = Ac, $R_6$ = H
MEL-D: $R_4$ = $R_6$ = H

$R_2$,$R_3$ = $C_2$-$C_{18}$ fatty acid

**Figure 11.** Structure of different MEL variants (1) MEL-A: fully acetylated (2) MEL-B: monoacetylated at $R_6$ (3) MEL-C: monoacetylated at $R_4$ (4) MEL-D: non-acetylated.

can reach production titers of over 100 g/L depending on the culture conditions (Morita et al. 2008a, c, 2009a, b). So far, the highest yield of 165 g/L was obtained by fed-batch cultivation of *P. aphidis* DSM 14930 on glucose and soybean oil (Rau et al. 2005). A detailed list of MEL-producers, their production titers, and type of produced MELs is given in Table 2.

The high MEL-producers mentioned above secrete mainly MEL-A along with smaller amounts of MEL-B and MEL-C. Extensive studies on MELs indicated that the production pattern is closely related to the taxonomical characteristics of the producer. Indeed, while *P. antarctica*, *P. aphididis*, *P. rugulosa* and *P. parantarctica* are positioned close together on the molecular phylogenetic tree of the *Pseudozyma* genus, *P. tsukubaensis*, a MEL-B producer, and *P. hubeiensis*, a MEL-C producer, are positioned independently of each other and from MEL-A producers (Figure 12). Interestingly, *P. crassa*, producer of MEL-A diastereoisomers (erythritol configuration is opposite to that of classic MELs), is clustered closely together with the MEL-A producers, however on a separate taxonomic branch (Morita et al. 2015).

## Biosynthesis of MELs

As is typical for genes involved in secondary metabolite synthesis, the genes encoding MEL biosynthetic enzymes are organized in a cluster (Figure 13). The *U. maydis* MEL cluster was the first fungal glycolipid gene cluster described and specifies five genes encoding the biosynthetic proteins (Hewald et al. 2006). The first identified enzyme was a glycosyltransferase, Emt1, catalyzing mannosylation derived from GDP-mannose of erythritol at the C4 (Hewald et al. 2005). This reaction is also the first in the MELs biosynthesis pathway (Figure 14) and appears to be stereospecific, since only mannosyl-D-erythritol is generated (Hewald et al. 2005). The resulting disaccharide is subsequently acylated with short- and medium-chain fatty acids at positions C2 and C3, respectively, by acyltransferases Mac1 and Mac2. However, the order of activity remains unclear. In a way, the acylation pattern might imply that Mac1 and Mac2 differ in their regioselectivity as well as in the preference for the length of the acyl-CoA cofactor (Hewald et al. 2006). A study on another MEL producer, *P. antarctica*, revealed that medium length fatty acids are derived from longer ones by partial peroxisomal β-oxidation (chain-shortening pathway; Kitamoto et al. 1998). Transcriptome analysis of the gene sets responsible for fatty acid metabolism in this yeast showed higher expression as compared with the homologous genes in *U. maydis*. The difference might be related to the natural habitat of each fungus. While *P. antarctica* was isolated from leave surfaces where it thrives on the cuticle constituted of a fatty acid polyester, *U. maydis* is a plant pathogen that causes corn smut, so it is more adapted to metabolizing carbohydrates (Morita et al. 2014).

The last step of MEL biosynthesis is acetylation catalyzed by acetyl-CoA-dependent acetyltransferase Mat1. The acetyl groups are added to both C4 and C6 of the mannosyl moiety. Mat1 overexpressed in *E. coli* can perform this reaction *in vitro*, indicating that no further enzymes are involved in MEL acetylation. Since the wild

**Table 2.** List of best MELs producing microorganism, their production titer and cultivation conditions.

| Microorganisms | Type of MEL | Titer (g/L) | Conditions | References |
|---|---|---|---|---|
| *U. maydis* ATCC 1482 | MEL-A + CBLs | 30 | On sunflower oil | (Spoeckner et al. 1999) |
| *U. maydis* NBRC 5346 | MEL-A | 2.62 | On olive oil | (Morita et al. 2009d) |
| *U. scitaminea* NBRC 32730 | MEL-B | 12.8 25.1 | On sucrose On sugar cane juice and urea | (Morita et al. 2009c) (Morita et al. 2011) |
| *U. cynodontis* NBRC 7530 | MEL-C | 1.4 | On soybean oil | (Morita et al. 2008a) |
| *P. antarctica* T-34 | MEL-A (70%) MEL-B MEL-C | 140 | On n-alkanes, batch culture | (Kitamoto et al. 2001) |
| *P. antarctica* JMC 10317 | MEL-A | 12.6 | On glycerol in the presence of sucrose | (Faria et al. 2015) |
| *P. antarctica* PYCC 5048 | MEL-A | 1.3 3.2 | On xylan On D-xylose | (Faria et al. 2015) |
| *P. aphidis* DSM 70725 | MEL-A (major) MEL-B MEL-C | 165 | On soybean oil, fed-batch bioreactor | (Rau et al. 2005) |
| *P. hubeiensis* KM-59 | MEL-C (65%) MEL-A MEL-B | 76.3 | On soybean oil, fed-batch bioreactor | (Konishi et al. 2008b) |
| *P. shanxiensis* | MEL-C | 2.72 | On soybean oil | (Fukuoka et al. 2007b) |
| *P. tsukbaensis* JCM 10324T | MEL-B (100%) with new diastereoizomer of erythritol | 25 | On soybean oil, batch culture | (Fukuoka et al. 2008b) |
| *P. tsukbaensis* 1E5 | MEL-B with new diastereoizomer of erythritol | 73.1 | On olive oil | (Morita et al. 2010b) |
| *P. fusiformata* | MEL-A | < 5 | On soybean oil | (Morita et al. 2007) |
| *P. graminicola* CBS 10092 | MEL-C (85%) | 9.6 | On soybean oil | (Morita et al. 2008d) |
| *P. parantarctica* JCM 11752 | MEL-A | 30 | On soybean oil | (Morita et al. 2007) |
| | Tri-acetylated MEL-A Di-acetylated MEL-A | 106.7 | On soybean oil at 34°C | (Morita et al. 2008c) |
| | MEL-A with mannosylmannitol (34.7%) | 52.4 | On soybean oil and in a presence of 4% mannitol | (Morita et al. 2009b) |

*Table 2 contd. ...*

*...Table 2 contd.*

| Microorganisms | Type of MEL | Titer (g/L) | Conditions | References |
|---|---|---|---|---|
| *P. rugulosa* NBRC 10877 | MEL-A (68%) MEL-B (12%) MEL-C (20%) | 142 | On soybean oil and in a presence of 8% erythritol | (Morita et al. 2006a) |
| *P. siamensis* CBS 9960 | MEL-C (> 84%) | 19 | In the presence of safflower oil | (Morita et al. 2008b) |
| *P. crassa* | Mix of MEL-A, MEL-B and MEL-C diastereoisomers | 4.6 | On glucose and oleic acid | (Fukuoka et al. 2008a) |
| *S. elanogramma* | Schizonellin A (similar to MEL-A) Schizonellin B (similar to MEL-B) | ND* | On glucose | (Deml et al. 1980) |
| *Kurtzmanomyces* sp. | MEL I-11 (similar to MEL-B) | ND* | On soybean oil | (Kakugawa et al. 2002) |

*No data

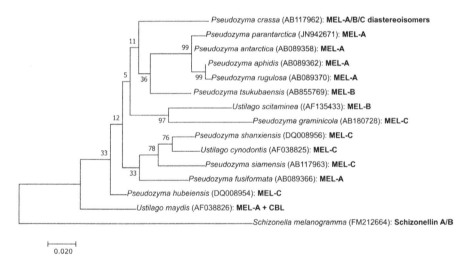

0.020

**Figure 12.** The maximum-likelihood phylogenetic tree constructed using ITS1, 5.8S rRNA gene and ITS2 sequences of the genus *Pseudozyma* and *Ustilago* from (Morita et al. 2009a). The DDBJ/GenBank/ EMBL accession numbers are indicated in parenthesis. The sequences were aligned using MUSCLE, after which the phylogenetic tree was generated with MEGA7 using the bootstrap maximum likelihood method with 1000 replicates. The tree is drawn to scale, with branch lengths measured in the number of substitutions per site. The tree is routed against outgroup *Schizonella melanogramma*.

type *U. maydis* secretes mostly monoacetylated MELs, it has been suggested that the second acetylation reaction occurs significantly slower than the first one. Finally, the fully assembled MELs are recognized by the Mmf1 transporter and secreted. Mmf1 is a plasma membrane transporter belonging to the major facilitator superfamily. It was demonstrated that this exporter is essential for secretion since a Δmmf1 strain did not produce extracellular MELs (Hewald et al. 2005). Moreover, the lack of MEL

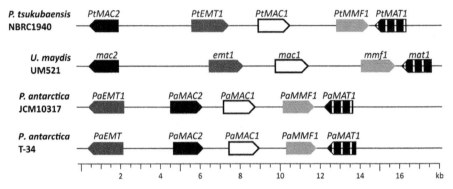

**Figure 13.** The MEL biosynthetic cluster of *P. tsukubaensis* NBRC1940, *U. maydis* UM521, *P. antarctica* JCM10317 and *P. antarctica* T-34. Erythritol/mannose transferase: *emt1*; acyltransferases: *mac1* and *mac2*; acetyltransferase: *mat1*; putative transporter: *mmf1* (Konishi et al. 2016).

production for Δ*mac1* or Δ*mac2* mutants has been linked to probable specificity of the Mmf1 transporter toward glycolipids that are acylated at both positions (Hewald et al. 2006).

Gene clusters homologous to the one in *U. maydis* were also found within the genome sequences of *P. antarctica* T-34 (Morita et al. 2013), *P. antarctica* JCM10317 (Saika et al. 2014), *P. tsukubaensis* NBRC1940 (Saika et al. 2016) *P. hubeiensis* SY62 (Konishi et al. 2013) and *P. aphidis* (Lorenz et al. 2014) indicating a conserved BS metabolism (Figure 13). Despite the genetic similarities among the MEL producers, several structural differences are encountered. Some examples are given below. While the MELs produced by *P. antarctica* and *U. maydis* are similar, MELs secreted by *P. tsukubaensis* differ in disaccharide conformation and the degree of acetylation of the mannose moiety. For instance, *P. tsukubaensis* predominantly produces MEL-B which contains 4-O-β-D-mannopyranosyl-(2R,3S)-erythritol (R-form) as the sugar moiety, while the conventional type of MELs contains 4-O-β-D-mannopyranosyl-(2S,3R)-erythritol (S-form) as the sugar moiety (Saika et al. 2016). Additionally, MELs produced by *P. antarctica* on vegetable oil ($C_{18}$) primarily contain $C_{10}$ and $C_{12}$ fatty acids, while the MELs produced by *P. siamensis*, *P. shanxiensis* and *U. maydis* on the same substrate mainly harbor $C_{16}$ and $C_4$ fatty acids (Spoeckner et al. 1999, Morita et al. 2008a).

## Regulation of MELs biosynthesis

As in the case of SLs, expression of all genes involved in MEL biosynthesis is highly induced under nitrogen limitation conditions. Although no obvious conserved regulatory motifs were found in the promoter regions of the cluster genes, several GATA sequences were identified in all of them. This suggests that a GATA factor homolog of AreA, the general nitrogen regulator from *A. nidulans*, is involved in regulation of MELs biosynthesis (Hewald et al. 2006). Moreover, expression of the glycosyltransferase Emt1 is enhanced by nitrogen limitation: its RNA could already be detected after four hours of incubation in nitrogen starvation medium and attains

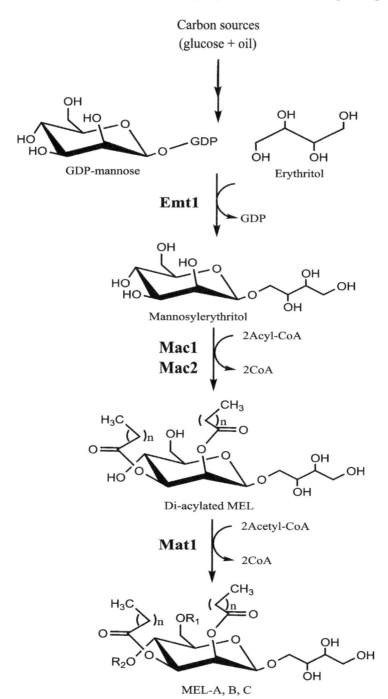

**Figure 14.** MELs biosynthetic pathway. Emt1: erythritol/mannose transferase, Mac1 and Mac2: acyltransferases, Mat1: acetyltransferase (Saika et al. 2016).

maximum activity at 24 h (Hewald et al. 2005). Moreover, RNA-seq of *P. aphidis* showed clear up-regulation of MEL gene cluster under MEL-producing condition (excess of soybean oil in the presence of glucose). Among the cluster genes, *pagEMT1* displayed the highest induction fold of close to 40 times. Interestingly, one gene located within the cluster, *pagMAT1* was not induced under given conditions indicating different regulation mechanism (Günther et al. 2015).

Besides nitrogen, the carbon source can affect MEL biosynthesis, but this time in a species dependent way. On the one hand, a study on *P. antarctica* showed that the expression of none of the five MEL biosynthetic genes is influenced by any type of carbon source (Morita et al. 2014). On the other hand, as mentioned above, transcriptomic analysis of the closely related *P. aphidis* showed that soybean oil has a strong inducing effect on the expression of the MEL cluster genes. Hence, the presence of a hydrophobic carbon source is essential for MEL biosynthesis in this species. In contrast, when *P. antarctica* is grown on glucose as a sole carbon source, significant surface activity is detected (Morita et al. 2006b). These results might, therefore, suggest that the regulatory mechanism of the glycolipid synthesis evolved differently in *P. aphidis* and *P. antarctica*. In conclusion, different MEL variants are produced depending on the organism (see Table 2) which can be explained by different expression levels and/or activity of the enzymes involved in MELs biosynthesis, especially acetyltransferases.

### *Biosynthesis of novel mannosyl glycolipids*

Well-performing MEL producers (over 100 g/L) such as *P. antarctica, P. aphidis,* and *P. rugulosa* mainly synthesize MEL-A. These conventional MELs consist of 4-O-β-D-mannopyranosyl-*meso*-erythritol as a hydrophilic part and two fatty acyl esters as a hydrophobic part (Fukuoka et al. 2007c). Despite the excellent surface active properties along with high production titers, MEL-A glycolipids possess a significant downside in aqueous formulations due to their low hydrophilicity and water solubility, limiting the application potential of these promising BSs. Therefore, the production of novel MELs is of considerable interest and would enable to obtain higher water solubility and hydrophilicity. Indeed, small differences in MEL structure cause a dramatic change in the physicochemical properties of the molecules. In order to produce MELs with different lengths of hydrophobic chains or with different carbohydrate head, alternative substrates as a primary carbon source can be used with altered culture conditions. For instance, tri-acylated MELs are obtained when cultivating *P. antarctica* and *P. rugulosa* on a medium with a high amounts of soybean oil, 80 g/L or more. Interestingly, the yeast did not produce tri-acylated MELs at 40 g/L of soybean oil (Fukuoka et al. 2007a). While the fatty acids in classic MEL-A are composed of only medium-chain ($C_8$-$C_{12}$) acids, the third acyl group in the tri-acylated MELs consists of a long-chain fatty acid ($C_{16}$-$C_{18}$), suggesting an input different from β-oxidation (Kitamoto et al. 1998). The authors suggested that the tri-acylated MELs are derived from conventional di-acylated MELs by esterification of the latter one with a free fatty acid present in the culture medium. In fact, *P. antarctica* as well as *P. rugulosa* secrete lipases and/or esterases when grown on soybean oil which might take part in this event (Morita et al. 2006b). Additional acyl group resulted in higher hydrophobicity of the tri-acylated MELs in comparison to the conventional di-acylated MEL-A, which gives it an advantage in water/ oil emulsification and/or reverse micelle systems (Fukuoka et al. 2007a). While cultivation of *P. antarctica* on high amount of vegetable oil led to the production of MELs with an additional acyl group, cultivation on high glucose concentration and without oil resulted in the presence of mono-acylated MELs. This compound only has one fatty ester and no acetyl groups on the mannose moiety. The removal of functional groups from MELs drastically increased the water solubility as well as hydrophilicity without diminishing the surface-active performance (Fukuoka et al. 2007c). Structures of the tri-, di- and mono-acylated MELs are depicted in Figure 15.

Glycosyltransferase Emt1 is responsible for the formation of sugar moiety of MELs. Its substrate specificity and affinity is important to determine the structure of the hydrophilipic part of the glycolipid and therefore it has a critical effect on its physiochemical properties. It was demonstrated that sugar alcohols sharing a similar configuration as erythritol are directly accepted by Emt1 to form the corresponding sugar moiety. Morita et al. (2009b) showed that cultivation of *P. parantarctica* in the medium containing an excess amount of D-mannitol gave rise to di-acetylated mannosyl-D-mannitol lipids MDML-A (Figure 16). This novel glycolipid possesses mannitol (C6) instead of erythritol (C4). Although the classic MEL-B which consists of monoacetylated mannose shows higher hydrophilicity than MEL-A (diacetylated mannose), the new diacetylated mannosyl-D-mannitol displays even higher hydrophilicity compared to MEL-A and MEL-B (Morita et al. 2009b). Furthermore, the same *P. parantarctica* strain cultivated on olive oil and D-ribitol and D-arabitol led to production of mannosyl-D-ribitol lipids (MDRL-A) and mannosyl-D-arabitol lipids (MDAL-A), respectively (Figure 16; Morita et al. 2012). While the MDML-A shows similar CMC and surface tension to classic MEL-A, the MDRL-A and MDAL-A containing a C5 sugar alcohol showed higher surface activities compared to conventional MELs. It was suggested that the number and pattern of the hydroxyl group at the alcohol moiety plays an important role in the adsorption at air-water interface. In order to further investigate the influence of carbohydrate configuration on MEL properties, Morita et al. (2015) cultivated *P. tsukubaensis* in medium containing an excess amount of optical isomers of sugar alcohols such as L-arabitol, for instance. The resulting culture contained a novel glycolipid, namely, mono-acetylated mannosyl-L-arabitol lipid (MLAL-B) which showed unique self-assembling properties and higher CMC than the diastereomer type MEL-B (Morita et al. 2015).

Besides the source and concentration of a carbon source, temperature appeared to have an influence on production titers and type of MELs. The optimal temperature for MELs biosynthesis in high yield producers including *P. rugulosa*, *P. antarctica* and *P. aphididis* is 30°C. However, *P. parantarctica* secreted greater amount of classic di-acylated MEL-A as well as novel tri-acylated MEL-A at 34°C than at 30°C and overall production titer reached over 100 g/L; this is in contrast to conventional MELs producers where the yield decreased with an increase of the cultivation temperature. Interestingly, higher production in *P. parantarctica* correlates with

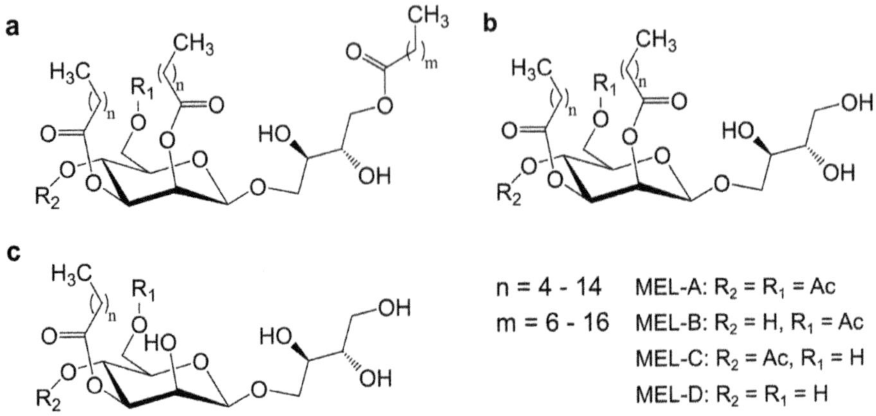

**Figure 15.** The structure of (a) tri-, (b) di- and (c) monoacylated MELs produced by *P. antarctica*.

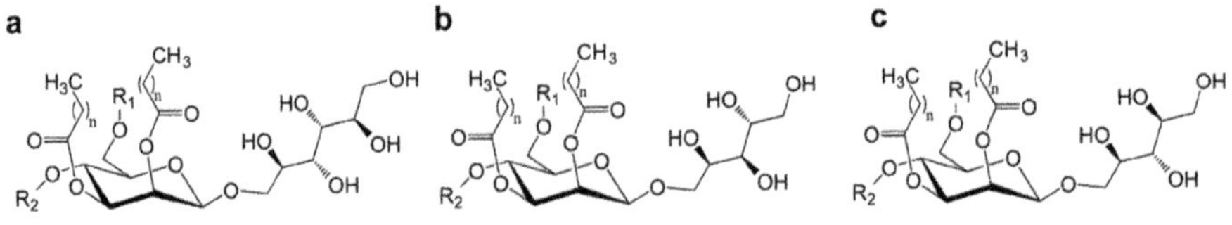

$n = 4 - 14$  MEL-A: $R_2 = R_1 = Ac$
$m = 6 - 16$  MEL-B: $R_2 = H$, $R_1 = Ac$
MEL-C: $R_2 = Ac$, $R_1 = H$
MEL-D: $R_2 = R_1 = H$

MDML-A    MDAL-A    MDRL-A

**Figure 16.** The structure of (a) Mannosyl-mannitol lipid (MDML-A), (b) Mannosyl-arabitol lipid (MDAL-A) and (c) Mannosyl-ribitol lipid (MDRL-A) produced by *P. parantarctica* (Morita et al. 2009b, 2012).

higher lipase activity, the enzyme catalyzing the hydrolysis of soybean oil (Morita et al. 2008c).

### Genetic engineering

Similar to SLs, developments of new producers and structural variety could broaden up the application range. Besides searching for new natural MELs producers and testing various substrates, genetic engineering approaches are applied in order to improve and optimize MEL production. The first example of genetic engineering of MEL production was a deletion of the gene encoding the acetyltransferase, *mat1,* in *U. maydis* which resulted in the exclusive synthesis of de-acetylated MELs (MEL-D) without a negative effect on the overall production capacity (Hewald et al. 2006). Alternatively, this non-acetylated MEL can be derived by enzymatic synthesis from MEL-B (Fukuoka et al. 2011). Knockout of the *U. maydis Emt1* gene encoding, the glycosyltransferase, led to a total abolishment of MEL synthesis and thus creation of a strain producing exclusively CBLs. However, overexpression of this gene with the strong arabinose-inducible *crg* promoter did not improve MEL yield (Hewald et al. 2005). Expressed sequence tags (EST) analysis of *P. antarctica* under MEL production conditions revealed a putative gene encoding a mitochondrial ADP/ATP carrier protein, PaAAC1. Overexpression of *paAAC1* not only confirmed its role in MEL biosynthesis but more importantly led to improved production of these glycolipids (Morita et al. 2010a).

Although the genetic studies, including gene manipulation, of *Pseudozyma* sp. have been limited in comparison to *U. maydis*, sequencing of the *P. antarctica* T-34 genome (Morita et al. 2013) and discovery of MEL gene clusters in other top *Pseudozyma* producers will certainly provide novel opportunities for development of industrial strains.

## Cellobiose lipids (CBLs)

### Structure and microbial source of CBLs

Ustilagic acid (UA) produced by *Ustilago maydis* was the first CBL described (Haskins 1950). UA consists of a cellobiose moiety O-glycosidically linked to the terminal hydroxyl group of tri (2,15,16-) or di (15,16-) hydroxyl palmitic acid. Additionally, the sugar moiety is esterified with an acetyl group and a short chain β-hydroxy fatty acid ($C_6$ or $C_8$) (Figure 17a). Four major UA variants can be identified when *U. maydis* is cultivated on vegetable oil. The differences are in the length of the fatty acid chain and the presence or absence of a α-hydroxy group on the fatty acid.

While the phytopathogenic basidiomycetous fungus *U. maydis* was the first described producer of UA, there are other fungal species capable of secreting CBLs such as *Pseudozyma graminicola* (Golubev et al. 2008), *P. fusiformata* (Kulakovskaya et al. 2005), *P. flocculosa* (Cheng et al. 2003, Mimee et al. 2005), *Trichosporon porosum* (Kulakovskaya et al. 2010), *Sympodiomycopsis paphiopedili* (Kulakovskaya et al. 2004) and *Cryptotoccus humicola* (Puchkov et al. 2002). Different producing strains also bring certain variations in the hydroxylation pattern of the lipid chain or in the carbohydrate decoration. Flocculosin, a rare CBL produced by *P. flocculosa,* is one interesting example of such diversity (Figure 17).

### Biosynthesis of CBLs

*Ustilagic acid*: Similar to other BSs and secondary metabolites, synthesis of UA occurs under nitrogen starvation and when cells enter stationary growth phase. Although UA was already described in the fifties (Haskins 1950), the discovery of a cytochrome P450 monooxygenase (Cyp1) in 2005 provided the first insight into genetics behind UA biosynthesis (Hewald et al. 2005). Two years later, Teichmann et al. (2007) presented complete CBL gene cluster. Analysis of genes located directly

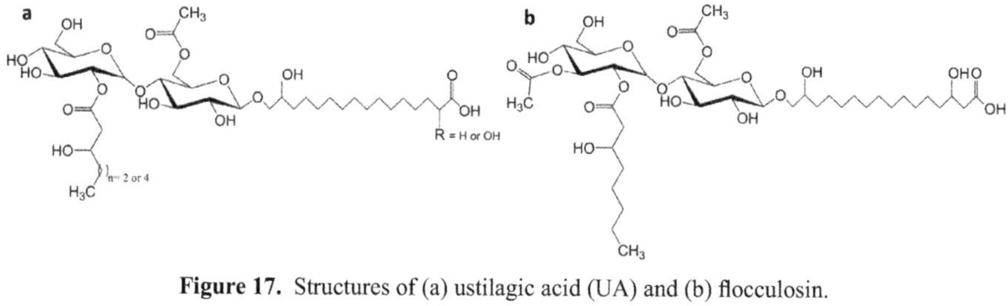

**Figure 17.** Structures of (a) ustilagic acid (UA) and (b) flocculosin.

adjacent to the telomere of chromosome 23 indicated, besides *cyp1*, nine ORFs to be significantly upregulated by nitrogen depletion, which indeed take part in *U. maydis* CBL biosynthesis.

Genes responsible for biosynthesis of *U. maydis* CBL are organized in a 45 kb gene cluster (Figure 18a), located in the subterminal region of chromosome 23, that contains 12 co-regulated ORFs: two cytochrome P540 monooxygenases *cyp1* and *cyp2*, acyltransferase *uat1*, acetyltransferase *uat2*, fatty acid synthase *fas2*, glycosyltransferase *ugt1*, two hydroxylases *uhd1* and *ahd1*, ABC-transporter *atr1*. The regulation of the cluster is mediated by zinc finger transcriptional factor, Rua1, also located within the cluster.

Taking all together, the biosynthesis of CBLs begins with terminal (ω) hydroxylation of palmitic acid catalyzed by Cyp1 (Figure 19a). A second hydroxyl group is introduced by another P450 monooxygenase, Cyp2, at the subterminal (ω-1) position. In the next step, the cellobiose moiety is coupled with the resulting hydroxyl fatty acid by the sequential addition of two glucose molecules by the action of one glucosyltransferase Ugt1. The resulting CBL is subsequently decorated by acetylation at the 6' position of the proximal glucose residue and by acylation of a short-chain β-hydroxy fatty acid at the distal glucose residue. The reactions are carried out by the acetyltransferase Uat2 and the acyltransferase Uat1, respectively. The largest ORF in the cluster, *fas2*, encodes for the fatty acid synthase involved in formation of the backbone of the short-chain β-hydroxy fatty acid that subsequently is β-hydroxylated by Uhd1 hydroxylase. The final step in the biosynthesis of UA is hydroxylation at the α-position of the C16 dihydroxy fatty acid (Ahd1). Once the synthesis has been completed, the UA is recognized by an ABC transporter Atr1 and exported. Atr1 seems to be quite unspecific since it is able to secrete many of the UA derivatives that are produced by the *U. maydis* mutants (Teichmann et al. 2007, 2011a, b).

*Flocculosin*: Flocculosin, as briefly mentioned above, is a glycolipid structurally very similar to UA produced by *U. maydis*, and consists of 3,15,16-trihydroxy palmitic acid O-glycosidically linked to cellobiose (Figure 17b). The sugar head

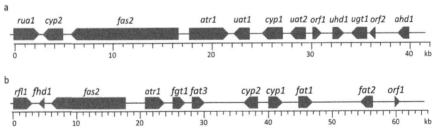

**Figure 18.** Organization of (a) UA biosynthesis gene clusters from *U. maydis* and (b) flocculosin biosynthesis gene cluster from *P. flocculosa*. Both gene clusters consist of two cytochrome P450 monooxygenases *cyp1* and *cyp2*, acyltransferase *uat1/fat1*, acetyltransferase *uat2/fat2*, fatty acid synthase *fas2*, glycosyltransferase *ugt1/fgt1*, short-chain fatty acid hydroxylase *uhd1/fhd1* and ABC-transporter *atr1*. Regulation of both clusters is driven by zinc finger transcriptional factor *rua1/rfl1*. Whereas the flocculosin gene cluster contains additional acetyltransferase gene *fat3*, the additional hydroxylase *ahd1* can be found in the CBL gene cluster (Teichmann et al. 2011a).

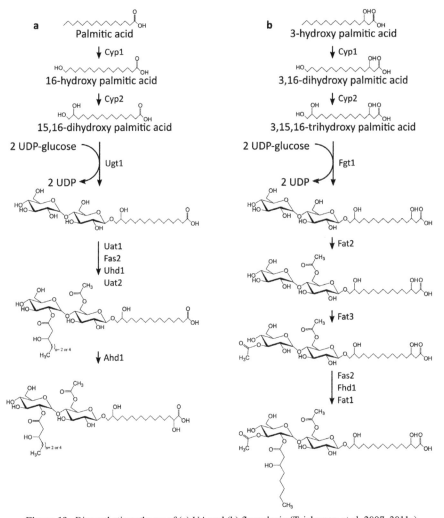

**Figure 19.** Biosynthetic pathway of (a) UA and (b) flocculosin (Teichmann et al. 2007, 2011a).

can be further decorated by two acetyl groups and a short-chain hydroxy fatty acid. Flocculosin is produced by the basidiomycetous fungus *P. flocculosa* and because of its particular activity against powdery mildews, it has been used as a biological fungicide (Cheng et al. 2003). The close phylogenetic relation of *U. maydis* and *P. flocculosa* together with the high structural similarity between flocculosin and UA might suggest a common biochemical pathway. Indeed, searching for homologs of the UA gene cluster led to the discovery of *P. floculosa* cytochrome P450 monooxygenase and acetyltransferase corresponding to the *U. maydis cyp1* and *uat1*, respectively (Marchand et al. 2009). Following the same hypothesis, Teichmann et al. (2011a) identified other genes involved in flocculosin biosynthesis and proved their cluster organization (Figure 18b). The flocculosin gene cluster contains 11 ORFs that share similarity to the corresponding *U. maydis* homologs. However, in contrast to the

*U. maydis* cluster, the flocculosin biosynthesis cluster contains an additional acetyl-transferase gene *(fat3)*, and a gene homologous to the α-hydroxylase Ahd1 could not be identified. These two differences in gene organization are reflected in structure variations between flocculosin and UA (Figure 19b).

## Regulation of CBL biosynthesis

As observed in other BSs production, nitrogen limitation is considered as a general regulator of UA and flocculosin production. However, doubling the ammonium sulfate concentration in the medium, which was expected to delay or inhibit flocculosin production, did not have the expected effect. Instead, N-consumption stopped after 48 h; despite that, it was still present in the medium and synthesis was initiated anyway, probably due to a limitation of another compound (Hammami et al. 2008). Next to nitrogen, the carbon source is another factor that controls the synthesis of glycolipids. For instance, in the MEL and CBL producer *U. maydis*, glucose favors CBL production, whereas fatty acids in the production medium will cause a shift towards MEL synthesis (Spoeckner et al. 1999). Glucose is also a preferred carbon source for both flocculosin production and biomass formation (Hammami et al. 2008).

Production of flocculosin is also affected by cell density. Hammami et al. (2008) set the optimal start-up inoculum size at 0.4 g/L. They suggested that *P. flocculosa* first needs to reach a specific physiological state before the start of producing flocculosin. Cell density higher than the recommended concentration of 0.4 g/L not only stopped the synthesis, but the flocculosin concentration started to decrease rapidly. The latter coincided with an exhaustion of carbon source which can indicate glycolipid consumption. Yet, no enzymes capable of degrading flocculosin have been reported (Hammami et al. 2008).

Genes organized in clusters are very often co-regulated. In *U. maydis,* the activity of all genes involved in the biosynthesis of UA is controlled by Rua1 (regulator of UA production), a zinc finger transcription factor, also located within the cluster. In order to prove the role of Rua1 in biosynthesis, a knockout strain was created. While the knockout resulted in a totally abolished production of UA, MEL production was not affected. Hence, Rua1 appears to act as a cluster-specific regulator. On the other hand, overexpression of this regulator stimulated the biosurfactant production even without nitrogen limitation. Teichmann et al. (2010) suggested that transcription regulation by Rua1 occurs by its direct binding to the conserved sequence elements in the promotor regions of the cluster genes and does not require other transcription factors or additional *U. maydis* proteins. This DNA element serves as an upstream activating sequence (UAS) mediating Rua1-dependent expression: Rua1 binds directly to the UAS of all genes of the cluster, except for *rua1* itself. A similar zinc finger transcription factor Rfl1 regulates the flocculosin gene cluster (Teichmann et al. 2011a).

## Genetic engineering

The elicitation of the *U. maydis* genome sequence was crucial for the discovery of the genetic basis of UA biosynthesis (Kämper et al. 2006). It also enables the genetic manipulation that may result in strains with improved biocontrol potential and/or higher production titers by, for example, enhancing the expression of CBL biosynthetic genes or creation of knockout mutants (Bölker et al. 2008). Deletion of the *U. maydis uat1* resulted in a mix of several UA derivatives differing from the wild type. All secreted glycolipids were lacking the acyl group. Besides non-acylated UA, non-acylated glucolipids (compounds with only one glucose molecule) were also recovered. Additionally, some of the glucolipids were also de-acetylated (Teichmann et al. 2011a). Furthermore, knockout of *uat2* led to the production of UA lacking the acetyl group on the proximal glucose molecule whereas *U. maydis* mutants overexpressing the *P. flocculosa fat3* secrete four derivatives additional to the wild type UA containing extra acetyl groups (Teichmann et al. 2011a). Biosynthesis of UA and flocculosin includes several hydroxylation steps. The long chain fatty acids of both glycolipids can be hydroxylated at three positions: the subterminal, α- and β-position. Therefore, deletion of *ahd1*, responsible for the partial α-hydroxylation and cytochrome P450 monooxygenase, *cyp2* gave rise to novel UAs lacking the α-hydroxyl or subterminal hydroxyl group, respectively. Moreover, synthesis of UA with a non-hydroxylated short chain fatty acid is possible with a Δ*uhd1 U. maydis* mutant strain (Teichmann et al. 2011b). Finally, the codon optimized *ugt1* gene from *U. maydis* was used to replace the second glucosyltransferase (*ugtB1*) in the SL producer *S. bombicola* for heterologous production of CBLs. Although the resulting mutant strain synthesizes mainly glucolipids (70%), new-to-nature CBLs with a SLs-like decoration were observed as well (Roelants et al. 2013).

## Future Perspective

The entanglement of white biotechnology in various industrial sectors is expected to further increase, particularly with the current consumer's drive towards a more bio-based and green economy. Microbial BSs have not reached broad industrial appliance because they are limited by high production costs and relatively low yields. Particularly, the product recovery requires attention since it takes up 70–80% of the entire production costs (Mukherjee et al. 2006). On the other hand, the remaining 20–30% of the total production costs are represented by the raw materials (Mulligan et al. 2014). The use of cheap waste substrates such as oil mill waste water or molasses greatly reduces substrate expenses and can make the biosurfactant production economically viable. On the downside, the formulation of such media is far more complex, unidentified compounds can negatively interfere and downstream processing becomes harder and consequently more expensive. The more deterministic strategy towards the future would be to focus on the use of renewable low-cost raw materials of otherwise negligible value (Makkar et al. 2011). Logically, the focus will lie on the optimization of the corresponding cultivation conditions to obtain higher yields. During recent years, statistical methods and experimental design techniques including factorial design and surface response methodology gained momentum (Rocha et al. 2014, Franzetti et al. 2009, Oliveira et al. 2009). Medium optimization can be even more effective when the statistics are supplemented with mathematical modeling. For example, artificial neural network coupled with genetic algorithm was successfully used for enhancement in BS yield (Pal et al. 2009, Sivapathasekaran et al. 2010).

The most promising evolution in the field of BSs, however, will occur on the molecular level: future advances in genomics, transcriptomics, proteomics and metabolomics will enable the further elucidation of the complete biosynthetic pathways and their regulation. On the one hand, the analysis of genetic background of glycolipid biosynthesis will help to understand the natural role of these molecules in microorganism. On the other hand, insight in the genetics behind BS production paves the way towards the creation of recombinant and hyperproducing mutant strains with the capability to consume the desired value-added (agro-industrial waste) resource streams (Das et al. 2008). Not only will these mutant strains accommodate overexpression of the desired BS, genetic insight is also necessary to broaden up the structural variability of glycolipids by means of the genetic engineering of the producing host. This will allow tinkering with the coding sequence and tailoring the surfactant to specific applications. Development of strains capable of producing tailored glycolipids will not only enhance productivity but also contribute to decreasing production costs associated to downstream processing. Larger structural variability and better productivity are indispensable for further industrialization of microbial glycolipids.

# References

Abdel-Mawgoud, A.M., F. Lépine and E. Déziel. 2010. Rhamnolipids: Diversity of structures, microbial origins and roles. Appl. Microbiol. Biotechnol. 86: 1323–1336.

Abdel-Mawgoud, A.M., R. Hausmann, F. Lépine, M.M. Müller and E. Déziel. 2011. Rhamnolipids: detection, analysis, biosynthesis, genetic regulation, and bioengineering of production. In Biosurfactants, Springer, pp. 13–55.

Albrecht, A., U. Rau and F. Wagner. 1996. Initial steps of sophoroselipid biosynthesis by *Candida bombicola* ATCC 22214 grown on glucose. Appl. Microbiol. Biotechnol. 46: 67–73.

Allard, S.T.M., M.-F. Giraud, C. Whitfield, P. Messner and J.H. Naismith. 2000. The purification, crystallization and structural elucidation of dTDP-D-glucose 4,6-dehydratase (RmlB), the second enzyme of the dTDP-L-rhamnose synthesis pathway from *Salmonella enterica* serovar Typhimurium. Acta Crystallogr. Sect. D 56: 222–225.

Ashby, R.D., D.K.Y. Solaiman and T.A. Foglia. 2006. The use of fatty acid esters to enhance free acid sophorolipid synthesis. Biotechnol. Lett. 28: 253–260.

Ashby, R.D., D.K.Y. Solaiman and T.A. Foglia. 2008. Property control of sophorolipids: influence of fatty acid substrate and blending. Biotechnol. Lett. 30: 1093–1100.

Asselineau, C. and J. Asselineau. 1978. Trehalose-containing glycolipids. Prog. Chem. Fats other Lipids 16: 59–99.

Bains, M., L. Fernández and R.E.W. Hancock. 2012. Phosphate starvation promotes swarming motility and cytotoxicity of *Pseudomonas aeruginosa*. Appl. Environ. Microbiol. 78: 6762–6768.

Barry, C.E., R.E. Lee, K. Mdluli, A.E. Sampson, B.G. Schroeder, R.A. Slayden and Y. Yuan. 1998. Mycolic acids: Structure, biosynthesis and physiological functions. Prog. Lipid Res. 37: 143–179.

Benincasa, M., A. Abalos, I. Oliveira and A. Manresa. 2004. Chemical structure, surface properties and biological activities of the biosurfactant produced by *Pseudomonas aeruginosa* LBI from soapstock. Antonie Van Leeuwenhoek. 85: 1–8.

Bergström, S., H. Theorell and H. Davide. 1947. On a metabolic product of *Ps. pyocyania*. Pyolipic acid active against *M. tuberculosis*. Ark. för kemi, Mineral. och Geol. 23: 1–15.

Bibb, M.J. 2005. Regulation of secondary metabolism in streptomycetes. Curr. Opin. Microbiol. 8: 208–215.

Blankenfeldt, W., M.-F. Giraud, G. Leonard, R. Rahim, C. Creuzenet, J.S. Lam and J.H. Naismith. 2000. The purification, crystallization and preliminary structural characterization of glucose-1-phosphate thymidylyltransferase (RmlA), the first enzyme of the dTDP-L-rhamnose synthesis pathway from *Pseudomonas aeruginosa*. Acta Crystallogr. Sect. D. 56: 1501–1504.

Blin, K., M.H. Medema, D. Kazempour, M.A. Fischbach, R. Breitling, E. Takano and T. Weber. 2013. antiSMASH 2.0—a versatile platform for genome mining of secondary metabolite producers. Nucleic Acids Res. 41: 204–212.

Bölker, M., C.W. Basse and J. Schirawski. 2008. *Ustilago maydis* secondary metabolism-from genomics to biochemistry. Fungal Genet. Biol. 45: 88–93.

Bollinger, N., D.J. Hassett, B.H. Iglewski, J.W. Costerton and T.R. McDermott. 2001. Gene expression in *Pseudomonas aeruginosa*: Evidence of iron override effects on quorum sensing and biofilm-specific gene regulation. J. Bact. 183: 1990–1996.

Brakemeier, A., D. Wullbrandt and S. Lang. 1998a. *Candida bombicola*: Production of novel alkyl glycosides based on glucose/2-dodecanol. Appl. Microbiol. Biotechnol. 50: 161–166.

Brakemeier, A., D. Wullbrandt and S. Lang. 1998b. Microbial alkyl-sophorosides based on 1-dodecanol or 2-, 3- or 4-dodecanones. Biotechnol. Lett. 20: 215–218.

Brakhage, A.A. and V. Schroeckh. 2011. Fungal secondary metabolites—strategies to activate silent gene clusters. Fungal Genet. Biol. 48: 15–22.

Brakhage, A.A. 2013. Regulation of fungal secondary metabolism. Nat. Rev. Microbiol. 11: 21–32.

Breithaupts, T.B. and R.J. Light. 1982. Affinity chromatography and further characterization of the glucosyltransferases involved in hydroxydocosanoic acid sophoroside production in *Candida bogoriensis*. J. Biol. Chem. 257: 9622–9628.

Cabrera-Valladares, N., A.P. Richardson, C. Olvera, L.G. Treviño, E. Déziel, F. Lépine and G. Soberón-Chávez. 2006. Monorhamnolipids and 3-(3-hydroxyalkanoyloxy)alkanoic acids (HAAs) production using *Escherichia coli* as a heterologous host. Appl. Microbiol. Biotechnol. 73: 187–194.

Cappelletti, M., S. Fedi, D. Frascari, H. Ohtake, R.J. Turner and D. Zannoni. 2011. Analyses of both the alkB gene transcriptional start site and alkB promoter-inducing properties of *Rhodococcus* sp. strain BCP1 grown on n-Alkanes. Appl. Environ. Microbiol. 77: 1619–1627.

Casas, J.A. and F. García-Ochoa. 1999. Sophorolipid production by *Candida bombicola*: Medium composition and culture methods. J. Biosci. Bioeng. 88: 488–494.

Cha, M., N. Lee, M. Kim, M. Kim and S. Lee. 2008. Heterologous production of *Pseudomonas aeruginosa* EMS1 biosurfactant in *Pseudomonas putida*. Bioresour. Technol. 99: 2192–9.

Chen, J., X. Song, H. Zhang, Y.B. Qu and J.Y. Miao. 2006. Sophorolipid produced from the new yeast strain *Wickerhamiella domercqiae* induces apoptosis in H7402 human liver cancer cells. Appl. Microbiol. Biotechnol. 72: 52–59.

Cheng, Y., D.J. McNally, C. Labbé, N. Voyer, F. Belzile and R.R. Bélanger. 2003. Insertional mutagenesis of a fungal biocontrol agent led to discovery of a rare cellobiose lipid with antifungal activity. Appl. Environ. Microbiol. 69: 2595–2602.

Christie, W.W. 2013. Rhamnolipids, Sophorolipids, and Other Glycolipid Biosurfactants. http://aocs.files. cms-plus.com/LipidsLibrary/images/Importedfiles/lipidlibrary/Lipids/rhamno/file.pdf.

Christova, N., B. Tuleva, Z. Lalchev, A. Jordanova and B. Jordanov. 2004a. Rhamnolipid biosurfactants produced by *Renibacterium salmoninarum* 27BN during growth on n-Hexadecane. Zeitschrift für Naturforsch. -Sect. C J. Biosci. 59: 70–74.

Christova, N., B. Tuleva and B. Nikolova-Damyanova. 2004b. Enhanced hydrocarbon biodegradation by a newly isolated *Bacillus subtilis* strain. Zeitschrift für Naturforsch. -Sect. C J. Biosci. 59: 205–208.

Ciapina, E.M.P., W.C. Melo, L.M.M. Santa Anna, A.S. Santos, D.M.G. Freire and N. Pereira. 2006. Biosurfactant production by *Rhodococcus erythropolis* grown on glycerol as sole carbon source. pp. 880–886. *In*: James D. McMillan, William S. Adney, Jonathan R. Mielenz, K. Thomas Klasson and N.J. Totowa (eds.). Twenty-Seventh Symposium on Biotechnology for Fuels and Chemicals. Humana Press.

Ciesielska K., B. Li, S. Groeneboer, I. Van Bogaert, Y.C. Lin, W. Soetaert, Y. Van De Peer and B. Devreese. 2013. SILAC-based proteome analysis of *Starmerella bombicola* sophorolipid production. J. Proteome Res. 12: 4376–4392.

Ciesielska K., I.N. Van Bogaert, S. Chevineau, B. Li, S. Groeneboer, W. Soetaert, Y. Van de Peer and B. Devreese. 2014. Exoproteome analysis of *Starmerella bombicola* results in the discovery of an esterase required for lactonization of sophorolipids. J. Proteomics. 98: 159–174.

Ciesielska, K., S.L.K.W. Roelants, I.N.A. Van Bogaert, S. De Waele, I. Vandenberghe, S. Groeneboer, W. Soetaert and B. Devreese. 2016. Characterization of a novel enzyme—*Starmerella bombicola* lactone esterase (SBLE)—responsible for sophorolipid lactonization. Appl. Microbiol. Biotechnol. 100: 9529–9541.

Cimermancic, P., M.H. Medema, J. Claesen, K. Kurita, L.C. Wieland Brown, K. Mavrommatis, A. Pati, P.A. Godfrey, M. Koehrsen, J. Clardy, B.W. Birren, E. Takano, A. Sali, R.G. Linington and M.A. Fischbach. 2014. Insights into secondary metabolism from a global analysis of prokaryotic biosynthetic gene clusters. Cell. 158: 412–421.

Cooper, D.G. and D.A. Paddock. 1984. Production of a biosurfactant from *Torulopsis bombicola*. Appl. Environ. Microbiol. 47: 173–176.

Costa, S.G.V.A.O., F. Lépine, S. Milot, E. Déziel, M. Nitschke and J. Contiero. 2009. Cassava wastewater as a substrate for the simultaneous production of rhamnolipids and polyhydroxyalkanoates by *Pseudomonas aeruginosa*. J. Ind. Microbiol. Biotechnol. 36: 1063–1072.

Coyne, M.J., K.S. Russell, C.L. Coyle and J.B. Goldberg. 1994. The *Pseudomonas aeruginosa* algC gene encodes phosphoglucomutase, required for the synthesis of a complete lipopolysaccharide core. J. Bacteriol. 176: 3500–3507.

Cutler, A.J. and R.J. Light. 1979. Regulation of hydroxydocosanoic acid sophoroside production in *Candida bogoriensis* by the levels of glucose and yeast extract in the growth medium. J. Biol. Chem. 254: 1944–1950.

Das, P., S. Mukherjee and R. Sen. 2008. Genetic regulations of the biosynthesis of microbial surfactants: an overview. Biotechnol. Genet. Eng. Rev. 25: 165–185.

Davila, A.M., R. Marchal and J.P. Vandecasteele. 1997. Sophorose lipid fermentation with differentiated substrate supply for growth and production phases. Appl. Microbiol. Biotechnol. 47: 496–501.

de Oliveira, M.R., D. Camilios-Neto, C. Baldo, A. Magri and M.A.P.C. Celligoi. 2014. Biosynthesis and production of sophorolipids. Int. J. Sci. Technol. Res. 3: 133–146.

Dekimpe, V. and E. Déziel. 2009. Revisiting the quorum-sensing hierarchy in *Pseudomonas aeruginosa*: The transcriptional regulator RhlR regulates LasR-specific factors. Microbiology 155: 712–723.

Demain, A.L. 1998. Induction of microbial secondary metabolism. Int. Microbiol. 1: 259–264.

Deml, G., T. Anke, F. Oberwinkler, B. Max Giannetti and W. Steglich. 1980. Schizonellin A and B, new glycolipids from *Schizonella melanogramma*. Phytochemistry 19: 83–87.

Desai, J.D. and I.M. Banat. 1997. Microbial production of surfactants and their commercial potential. Microbiol. Mol. Biol. Rev. 61: 47–64.

Déziel, E., F. Lépine, D. Dennie, D. Boismenu, O.A. Mamer and R. Villemur. 1999. Liquid chromatography/ mass spectrometry analysis of mixtures of rhamnolipids produced by *Pseudomonas aeruginosa* strain 57RP grown on mannitol or naphthalene. Biochim. Biophys. Acta - Mol. Cell Biol. Lipids 1440: 244–252.

Déziel, E., F. Lépine, S. Milot and R. Villemur. 2003. rhlA is required for the production of a novel biosurfactant promoting swarming motility in *Pseudomonas aeruginosa*: 3-(3-hydroxyalkanoyloxy) alkanoic acids (HAAs), the precursors of rhamnolipids. Microbiology 149: 2005–2013.

Dogan, I., Æ.K.R. Pagilla, Æ.D.A. Webster and B.C. Stark. 2006. Expression of Vitreoscilla hemoglobin in *Gordonia amarae* enhances biosurfactant production. J. Ind. Microbiol. Biotechnol. 33: 693–700.

Domenech, P., M.B. Reed, C.S. Dowd, C. Manca, G. Kaplan and C.E. Barry. 2004. The role of MmpL8 in sulfatide biogenesis and virulence of *mycobacterium tuberculosis*. J. Biol. Chem. 279: 21257–21265.

Dubeau, D., E. Déziel, D.E. Woods and F. Lépine. 2009. *Burkholderia thailandensis* harbors two identical rhl gene clusters responsible for the biosynthesis of rhamnolipids. BMC Microbiol. 9: 263.

Esders, T.W. and R.J. Light. 1972. Glucosyl- and acetyltransferases involved in the biosynthesis of glycolipids from *Candida bogoriensis*. J. Biol. Chem. 247: 1375–1386.

Evans, C.T. and C. Ratledge. 1985. The physiological significance of citric acid in the control of metabolism in lipid-accumulating yeasts. Biotechnol. Genet. Eng. Rev. 3: 349–376.

Faria, N.T., S. Marques, C. Fonseca and F.C. Ferreira. 2015. Direct xylan conversion into glycolipid biosurfactants, mannosylerythritol lipids, by *Pseudozyma antarctica* PYCC 5048T. Enzyme Microb. Technol. 71: 58–65.

Farman, M.L. 2007. Telomeres in the rice blast fungus *Magnaporthe oryzae*: The world of the end as we know it. FEMS Microbiol. Lett. 273: 125–132.

Farrow, J.M., Z.M. Sund, M.L. Ellison, D.S. Wade, J.P. Coleman and E.C. Pesci. 2008. PqsE functions independently of PqsR-*Pseudomonas* quinolone signal and enhances the rhl quorum-sensing system. J. Bacteriol. 190: 7043–7051.

Fernandes, M., N.P. Keller and T.H. Adams. 1998. Sequence-specific binding by *Aspergillus nidulans* AflR, a C6 zinc cluster protein regulating mycotoxin biosynthesis. Mol. Microbiol. 28: 1355–1365.

Franzetti, A., P. Caredda, P. La Colla, M. Pintus, E. Tamburini, M. Papacchini and G. Bestetti. 2009. Cultural factors affecting biosurfactant production by *Gordonia* sp. BS29. Int. Biodeterior. Biodegradation. 63: 943–947.

Franzetti, A., I. Gandolfi, G. Bestetti and T.J.P. Smyth. 2010. Production and applications of trehalose lipid biosurfactants. Eur. J. Lipid Sci. Technol. 112: 617–627.

Fukuoka, T., T. Morita, M. Konishi, T. Imura and D. Kitamoto. 2007a. Characterization of new glycolipid biosurfactants, tri-acylated mannosylerythritol lipids, produced by *Pseudozyma* yeasts. Biotechnol. Lett. 29: 1111–1118.

Fukuoka, T., T. Morita, M. Konishi, T. Imura and D. Kitamoto. 2007b. Characterization of new types of mannosylerythritol lipids as biosurfactants produced from soybean oil by a basidiomycetous yeast, *Pseudozyma shanxiensis*. J. Oleo Sci. 56: 435–442.

Fukuoka, T., T. Morita, M. Konishi, T. Imura, H. Sakai and D. Kitamoto. 2007c. Structural characterization and surface-active properties of a new glycolipid biosurfactant, mono-acylated mannosylerythritol lipid, produced from glucose by *Pseudozyma antarctica*. Appl. Microbiol. Biotechnol. 76: 801–810.

Fukuoka, T., M. Kawamura, T. Morita, T. Imura, H. Sakai, M. Abe and D. Kitamoto. 2008a. A basidiomycetous yeast, *Pseudozyma crassa*, produces novel diastereomers of conventional mannosylerythritol lipids as glycolipid biosurfactants. Carbohydr. Res. 343: 2947–2955.

Fukuoka, T., T. Morita, M. Konishi, T. Imura and D. Kitamoto. 2008b. A basidiomycetous yeast, *Pseudozyma tsukubaensis*, efficiently produces a novel glycolipid biosurfactant. The identification of a new diastereomer of mannosylerythritol lipid-B. Carbohydr. Res. 343: 555–560.

Fukuoka, T., T. Yanagihara, T. Imura, T. Morita, H. Sakai, M. Abe and D. Kitamoto. 2011. Enzymatic synthesis of a novel glycolipid biosurfactant, mannosylerythritol lipid-D and its aqueous phase behavior. Carbohydr. Res. 346: 266–271.

Funston, S.J., K. Tsaousi, M. Rudden, T.J. Smyth, P.S. Stevenson, R. Marchant and I.M. Banat. 2016. Characterising rhamnolipid production in *Burkholderia thailandensis* E264, a non-pathogenic producer. Appl. Microbiol. Biotechnol. 100: 7945–7956.

Gacek, A. and J. Strauss. 2012. The chromatin code of fungal secondary metabolite gene clusters. Appl. Microbiol. Biotechnol. 95: 1389–1404.

Gallagher, L.A., S.L. McKnight, M.S. Kuznetsova, E.C. Pesci and C. Manoil. 2002. Functions required for extracellular quinolone signaling by *Pseudomonas aeruginosa*. J. Bacteriol. 184: 6472–6480.

Gerc, A.J., N.R. Stanley-Wall and S.J. Coulthurst. 2014. Role of the phosphopantetheinyltransferase enzyme, PswP, in the biosynthesis of antimicrobial secondary metabolites by *Serratia marcescens* Db10. Microbiology 160: 1609–17.

Giraud, M.-F., H.J. McMiken, G.A. Leonard, P. Messner, C. Whitfield and J.H. Naismith. 1999. Overexpression, purification, crystallization and preliminary structural study of dTDP-6-deoxy-L-lyxo-4-hexulose reductase (RmlD), the fourth enzyme of the dTDP-l-rhamnose synthesis pathway, from *Salmonella enterica* serovar Typhimu. Acta Crystallogr. Sect. D. 55: 2043–2046.

Golubev, V.I., T.V. Kulakovskaia, A.S. Shashkov, E.V. Kulakovskaia and N.V. Golubev. 2008. Antifungal cellobiose lipid secreted by the epiphytic yeast *Pseudozyma graminicola*. Microbiology 77: 201–206.

Gorin, P.A.J., J.F.T. Spencer and A.P. Tulloch. 1961. Hydroxy fatty acid glycosides of sophorose from *Torulopsis magnoliae*. Can. J. Chem. 39.

Graninger, M., B. Nidetzky, D.E. Heinrichs, C. Whitfield and P. Messner. 1999. Characterization of dTDP-4-dehydrorhamnose 3,5-epimerase and dTDP-4-dehydrorhamnose reductase, required for dTDP-L-rhamnose biosynthesis in *Salmonella enterica* Serovar Typhimurium LT2. Biochemistry 274: 25069–25077.

Gudiña, E.J., A.I. Rodrigues, V. de Freitas, Z. Azevedo, J.A. Teixeira and L.R. Rodrigues. 2016. Valorization of agro-industrial wastes towards the production of rhamnolipids. Bioresour. Technol. 212: 144–150.

Guerra-Santos, L.H., O. Käppeli and A. Fiechter. 1986. Dependence of *Pseudomonas aeruginosa* continous culture biosurfactant production on nutritional and environmental factors. Appl. Microbiol. Biotechnol. 24: 443–448.

Günther, M., C. Grumaz, S. Lorenz, P. Stevens, E. Lindemann, T. Hirth, K. Sohn, S. Zibek and S. Rupp. 2015. The transcriptomic profile of *Pseudozyma aphidis* during production of mannosylerythritol lipids. Appl. Microbiol. Biotechnol. 99: 1375–1388.

Gunther, N.W., A. Nunez, W. Fett and D.K.Y. Solaiman. 2005. Production of rhamnolipids by *Pseudomonas chlororaphis*, a nonpathogenic bacterium. Appl. Environ. Microbiol. 71: 2288–2293.

Hammami, W., C. Labbé, F. Chain, B. Mimee and R.R. Bélanger. 2008. Nutritional regulation and kinetics of flocculosin synthesis by *Pseudozyma flocculosa*. Appl. Microbiol. Biotechnol. 80: 307–315.

Haskins, R.H. 1950. Biochemistry of the Ustilaginales I. Preliminary cultural studies of *Ustilago zeae*. Can. J. Res. 28c: 213–223.

Hewald, S., K. Josephs, M. Bölker and M. Bo. 2005. Genetic analysis of biosurfactant production in Ustilago maydis genetic analysis of biosurfactant production in *Ustilago maydis*. Appl. Environ. Microbiol. 71: 3033–3040.

Hewald, S., U. Linne, M. Scherer, M.A. Marahiel, J. Kämper and M. Bölker. 2006. Identification of a gene cluster for biosynthesis of mannosylerythritol lipids in the basidiomycetous fungus *Ustilago maydis*. Appl. Environ. Microbiol. 72: 5469–5477.

Hommel, R.K., L. Weber, A. Weiss, U. Himmelreich, O. Rilke and H.-P. Kleber. 1994. Production of sophorose lipid by *Candida (Torulopsis) apicola* grown on glucose. J. Biotechnol. 33: 147–155.

Imura, T., N. Ohta, K. Inoue, N. Yagi, H. Negishi, H. Yanagishita and D. Kitamoto. 2006. Naturally engineered glycolipid biosurfactants leading to distinctive self-assembled structures. Chem. - A Eur. J. 12: 2434–2440.

Imura, T., Y. Masuda, H. Minamikawa, T. Fukuoka, M. Konishi, T. Morita, H. Sakai, M. Abe and D. Kitamoto. 2010. Enzymatic conversion of diacetylated sophoroselipid into acetylated glucoselipid: surface-active properties of novel bolaform biosurfactants. J. Oleo Sci. 59: 495–501.

Inaba, T., Y. Tokumoto, Y. Miyazaki, N. Inoue, H. Maseda, T. Nakajima-kambe, H. Uchiyama and N. Nomura. 2013. Analysis of genes for succinoyl trehalose lipid production and increasing production in *Rhodococcus* sp . Strain SD-74. Appl. Environ. Microbiol. 79: 7082–7090.

Irorere, V.U., L. Tripathi, R. Marchant, S. McClean and I.M. Banat. 2017. Microbial rhamnolipid production: a critical re-evaluation of published data and suggested future publication criteria. Appl. Microbiol. Biotechnol. 101: 3941–3951.

Itoh, S. and T. Suzuki. 1974. Fructose-Lipids of *Arthrobacter, Corynebacteria, Nocardia* and *Mycobacteria* grown on fructose. Agric. Biol. Chem. 38: 1443–1449.

Jensen, V., D. Löns, C. Zaoui, F. Bredenbruch, A. Meissner, G. Dieterich, R. Münch and S. Häussler. 2006. RhlR expression in *Pseudomonas aeruginosa* is modulated by the *Pseudomonas* quinolone signal via PhoB-dependent and -independent pathways. J. Bacteriol. 188: 8601–8606.

Kahraman, H. and S.O. Erenler. 2012. Rhamnolipid production by *Pseudomonas aeruginosa* engineered with the *Vitreoscilla* hemoglobin gene. Prikl Biokhim Mikrobiol. 48: 212–217.

Kakugawa, K., M. Tamai, K. Imamura, K. Miyamoto, S. Miyoshi, Y. Morinaga, O. Suzuki and T. Miyakawa. 2002. Isolation of yeast *Kurtzmanomyces* sp. I-11, novel producer of mannosylerythritol lipid. Biosci. Biotechnol. Biochem. 66: 188–91.

Kämper, J., R. Kahmann, M. Bölker, L.-J. Ma, T. Brefort, B.J. Saville et al. 2006. Insights from the genome of the biotrophic fungal plant pathogen Ustilago maydis. Nature. 444: 97–101.

Kim, E.J., W. Wang, W.D. Deckwer and A.P. Zeng. 2005. Expression of the quorum-sensing regulatory protein LasR is strongly affected by iron and oxygen concentrations in cultures of *Pseudomonas aeruginosa* irrespective of cell density. Microbiology 151: 1127–1138.

Kim, J.S., M. Powalla, S. Lang, F. Wagner, H. Lünsdorf and V. Wray. 1990. Microbial glycolipid production under nitrogen limitation and resting cell conditions. J. Biotechnol. 13: 257–266.

Kitamoto, D., S. Akiba, C. Hiok and T. Tabuchi. 1990. Extracellular accumulation of mannosylerythritol lipids by a strain of *Candida antarctica*. Agric. Biol. Chem. 54: 31–36.

Kitamoto, D., H. Yanagishita, K. Haraya and H.K. Kitamoto. 1998. Contribution of a chain-shortening pathway to the biosynthesis of the fatty acids of mannosylerythritol lipid (biosurfactant) in the yeast *Candida antarctica*: Effect of beta-oxidation inhibitors on biosurfactant synthesis. Biotechnol. Lett. 20: 813–818.

Kitamoto, D., T. Ikegami, G.T. Suzuki, A. Sasaki, Y.I. Takeyama, Y. Idemoto, N. Koura and H. Yanagishita. 2001. Microbial conversion of n-alkanes into glycolipid biosurfactants, mannosylerythritol lipids, by *Pseudozyma (Candida) antarctica*. Biotechnol. Lett. 23: 1709–1714.

Kitamoto, D., H. Isoda and T. Nakahara. 2002. Functions and potential applications of glycolipid biosurfactants—from energy-saving materials to gene delivery carriers. J. Biosci. Bioeng. 94: 187–201.

Kohler, T., L.K. Curty, F. Barja, C. Van Delden and J.C. Pechere. 2000. Swarming of *Pseudomonas aeruginosa* is dependent on cell-to-cell signaling and requires flagella and pili. J. Bacteriol. 182: 5990–5996.

Konishi, M., T. Fukuoka, T. Morita, T. Imura and D. Kitamoto. 2008a. Production of new types of sophorolipids by *Candida batistae*. J. Oleo Sci. 57: 359–369.

Konishi, M., T. Morita, T. Fukuoka, T. Imura, K. Kakugawa and D. Kitamoto. 2008b. Efficient production of mannosylerythritol lipids with high hydrophilicity by *Pseudozyma hubeiensis* KM-59. Appl. Microbiol. Biotechnol. 78: 37–46.

Konishi, M., Y. Hatada and J. Horiuchi. 2013. Draft genome sequence of the basidiomycetous yeast-like fungus *Pseudozyma hubeiensis* SY62, which produces an abundant amount of the biosurfactant mannosylerythritol lipids. Genome Announc. 1: 13–14.

Konishi, M., M. Fujita, Y. Ishibane, Y. Shimizu, Y. Tsukiyama and M. Ishida. 2016. Isolation of yeast candidates for efficient sophorolipids production: their production potentials associate to their lineage. Biosci. Biotechnol. Biochem. 80: 2058–64.

Kretschmer, A., H. Bock and F. Wagner. 1982. Chemical and physical characterization of interfacial-active lipids from *Rhodococcus erythropolis* grown on n-Alkanes. Appl. Environ. Microbiol. 44: 864–870.

Kretschmer, A. and F. Wagner. 1983. Characterization of biosynthetic intermediates of trehalose dicorynomycolates from *Rhodococcus erythropolis* grown on n-alkanes. Biochim. Biophys. Actamica Biophys. Acta. 753: 306–313.

Kügler, J.H., C. Muhle-Goll, B. Kühl, A. Kraft, R. Heinzler, F. Kirschhöfer, M. Henkel, V. Wray, B. Luy, G. Brenner-Weiss, S. Lang, C. Syldatk and R. Hausmann. 2014. Trehalose lipid biosurfactants produced by the actinomycetes *Tsukamurella spumae* and *T. pseudospumae*. Appl. Microbiol. Biotechnol. 98: 8905–8915.

Kügler, J.H., M. Roes-hill, C. Le, Syldatk and R. Hausmann. 2015. Surfactants tailored by the class Actinobacteria. Front. Microbiol. 6.

Kulakovskaya, T.V., A.S. Shashkov, E.V. Kulakovskaya and W.I. Golubev. 2004. Characterization of an antifungal glycolipid secreted by the yeast *Sympodiomycopsis paphiopedili*. FEMS Yeast Res. 5: 247–252.

Kulakovskaya, T.V., A.S. Shashkov, E.V. Kulakovskaya and W.I. Golubev. 2005. Ustilagic acid secretion by *Pseudozyma fusiformata* strains. FEMS Yeast Res. 5: 919–923.

Kulakovskaya, T.V., W.I. Golubev, M.A. Tomashevskaya, E.V. Kulakovskaya, A.S. Shashkov, A.A. Grachev, A.S. Chizhov and N.E. Nifantiev. 2010. Production of antifungal cellobiose lipids by *Trichosporon porosum*. Mycopathologia. 169: 117–123.

Kurtzman, C.P., N.P.J. Price, K.J. Ray and T.M. Kuo. 2010. Production of sophorolipid biosurfactants by multiple species of the *Starmerella (Candida) bombicola* yeast clade. FEMS Microbiol. Lett. 311: 140–146.

Kurz, M., C. Eder, D. Isert, Z. Li, E.F. Paulus, M. Schiell, L. Toti, L. Vertesy, J. Wink and G. Seibert. 2003. Ustilipids, acylated beta-D-mannopyranosyl D-erythritols from *Ustilago maydis* and *Geotrichum candidum*. J Antibiot. 56: 91–101.

Kuyukina, M.S. and I.B. Ivshina. 2010. *Rhodococcus* biosurfactants: biosynthesis, properties, and potential applications. In Biology of Rhodococcus, Springer, pp. 291–313.

Kuyukina, M.S., I.B. Ivshina, T.A. Baeva, O.A. Kochina, S.V. Gein and V.A. Chereshnev. 2015. Trehalolipid biosurfactants from nonpathogenic *Rhodococcus* actinobacteria with diverse immunomodulatory activities. N. Biotechnol. 32: 559–568.

Laczi, K., Á. Kis, B. Horváth, G. Maróti, B. Hegedüs, K. Perei and G. Rákhely. 2015. Metabolic responses of *Rhodococcus erythropolis* PR4 grown on diesel oil and various hydrocarbons. Appl. Microbiol. Biotechnol. 99: 9745–9759.

Lang, S. and J.C. Philp. 1998. Surface-active lipids in rhodococci. Antonie van Leeuwenhoek, Int. J. Gen. Mol. Microbiol. 74: 59–70.

Lang, S. and D. Wullbrandt. 1999. Rhamnose lipids—biosynthesis, microbial production and application potential. Appl. Microbiol. Biotechnol. 51: 22–32.

Larkin, M.J., L.A. Kulakov and C.C.R. Allen. 2010. Genomes and plasmids in *Rhodococcus*. pp. 73–90. *In*: Héctor M. Alvarez (ed.). Biology of Rhodococcus. Berlin, Heidelberg: Springer Berlin Heidelberg.

Lee, J. and L. Zhang. 2014. The hierarchy quorum sensing network in *Pseudomonas aeruginosa*. Protein Cell 6: 26–41.

Li, A.H., M.Y. Xu, W. Sun and G.P. Sun. 2011. Rhamnolipid production by *Pseudomonas aeruginosa* GIM 32 using different substrates including molasses distillery wastewater. Appl. Biochem. Biotechnol. 163: 600–611.

Li, J., H. Li, W. Li, C. Xia and X. Song. 2016. Identification and characterization of a flavin-containing monooxygenase MoA and its function in a specific sophorolipid molecule metabolism in *Starmerella bombicola*. Appl. Microbiol. Biotechnol. 100: 1307–1318.

Liang, S., T. Wu, R. Lee, F.U.N.S.U.N. Chu and J.E. Linz. 1997. Analysis of mechanisms regulating expression of the ver-1 gene, involved in aflatoxin. Biosynthesis 63: 1058–1065.

Lindhout, T., P.C.Y. Lau, D. Brewer and J.S. Lam. 2009. Truncation in the core oligosaccharide of lipopolysaccharide affects flagella-mediated motility in *Pseudomonas aeruginosa* PAO1 via modulation of cell surface attachment. Microbiology 155: 3449–3460.

Linton, J.D. 1991. Metabolite production and growth efficiency. Antonie Van Leeuwenhoek 60: 293–311.

Liu, X., X. Ma, R. Yao, C. Pan and H. He. 2016. Sophorolipids production from rice straw via SO3 micro-thermal explosion by Wickerhamiella domercqiae var. sophorolipid CGMCC 1576. AMB Express 6: 60.

Lorenz, S., M. Guenther, C. Grumaz, S. Rupp, S. Zibek and K. Sohn. 2014. Genome sequence of the basidiomycetous fungus Pseudozyma aphidis DSM70725, an efficient producer of biosurfactant mannosylerythritol lipids. Genome Announc. 2: e00053–14.

Ma, Y., J.A. Mills, J.T. Belisle, V. Vissa, M. Howell, K. Bowlin, M.S. Scherman and M. McNeil. 1997. Determination of the pathway for rhamnose biosynthesis in mycobacteria: Cloning, sequencing and expression of the *Mycobacterium tuberculosis* gene encoding alpha-D-glucose-1-phosphate thymidylyltransferase. Microbiology 143: 937–945.

Madry, N., R. Sprinkmeyer and H. Pape. 1979. Regulation of tylosin synthesis in *Streptomyces*: Effects of glucose analogs and inorganic phosphate. Eur. J. Appl. Microbiol. Biotechnol. 7: 365–370.

Maier, R.M. and G. Soberón-Chávez. 2000. *Pseudomonas aeruginosa* rhamnolipids: biosynthesis and potential applications. Appl. Microbiol. Biotechnol. 54: 625–633.

Makkar, R.S., S.S. Cameotra and I.M. Banat. 2011. Advances in utilization of renewable substrates for biosurfactant production. AMB Express 1: 5.

Mao, X., R. Jiang, W. Xiao and J. Yu. 2015. Use of surfactants for the remediation of contaminated soils: A review. J. Hazard. Mater. 285: 419–435.

Marchand, G., W. Rémus-Borel, F. Chain, W. Hammami, F. Belzile and R.R. Bélanger. 2009. Identification of genes potentially involved in the biocontrol activity of *Pseudozyma flocculosa*. Phytopathology 99: 1142–1149.

Marchant, R., S. Funston, C.I. Uzoigwe (Chibuzo), P.K.S.M. Rahman (Pattanathu) and I.M. Banat. 2015. Production of biosurfactants from non-pathogenic bacteria. *In*: Kosaric and F.V. Sukan (eds.). Biosurfactants. Surfactant Sci. Ser. 159: 73–82.

Martínez-Toledo, A. and R. Rodríguez-Vázquez. 2011. Response surface methodology (Box-Behnken) to improve a liquid media formulation to produce biosurfactant and phenanthrene removal by *Pseudomonas putida*. Ann. Microbiol. 61: 605–613.

McGrath, S., D.S. Wade and E.C. Pesci. 2004. Dueling quorum sensing systems in Pseudomonas aeruginosa control the production of the *Pseudomonas quinolone* signal (PQS). FEMS Microbiol. Lett. 230: 27–34.

Medina, G., K. Juárez, R. Díaz and G. Soberón-Chávez. 2003. Transcriptional regulation of *Pseudomonas aeruginosa* rhlR, encoding a quorum-sensing regulatory protein. Microbiology 149: 3073–3081.

Mimee, B., C. Labbé, R. Pelletier and R.R. Bélanger. 2005. Antifungal activity of flocculosin, a novel glycolipid isolated from Pseudozyma flocculosa. Antimicrob. Agents Chemother. 49: 1597–1599.

Morita, T., M. Konishi, T. Fukuoka, T. Imura and D. Kitamoto. 2006a. Discovery of *Pseudozyma rugulosa* NBRC 10877 as a novel producer of the glycolipid biosurfactants, mannosylerythritol lipids, based on rDNA sequence. Appl. Microbiol. Biotechnol. 73: 305–313.

Morita, T., M. Konishi, T. Fukuoka, T. Imura and D. Kitamoto. 2006b. Physiological differences in the formation of the glycolipid biosurfactants, mannosylerythritol lipids, between *Pseudozyma antarctica* and *Pseudozyma aphidis*. Appl. Microbiol. Biotechnol. 74: 307–15.

Morita, T., M. Konishi, T. Fukuoka, T. Imura, H.K. Kitamoto and D. Kitamoto. 2007. Characterization of the genus *Pseudozyma* by the formation of glycolipid biosurfactants, mannosylerythritol lipids. FEMS Yeast Res. 7: 286–292.

Morita, T., M. Konishi, T. Fukuoka, T. Imura and D. Kitamoto. 2008a. Identification of *Ustilago cynodontis* as a new producer of glycolipid biosurfactants, mannosylerythritol lipids, based on ribosomal. J. Oleo Sci. 57: 549–556.

Morita, T., M. Konishi, T. Fukuoka, T. Imura and D. Kitamoto. 2008b. Production of glycolipid biosurfactants, mannosylerythritol lipids, by *Pseudozyma siamensis* CBS 9960 and their interfacial properties. J. Biosci. Bioeng. 105: 493–502.

Morita, T., M. Konishi, T. Fukuoka, T. Imura, H. Sakai and D. Kitamoto. 2008c. Efficient production of di- and tri-acylated mannosylerythritol lipids as glycolipid biosurfactants by *Pseudozyma parantarctica* JCM 11752(T). J. Oleo Sci. 57: 557–565.

Morita, T., M. Konishi, T. Fukuoka, T. Imura, S. Yamamoto, M. Kitagawa, A. Sogabe and D. Kitamoto. 2008d. Identification of *Pseudozyma graminicola* CBS 10092 as a producer of glycolipid biosurfactants, mannosylerythritol lipids. J. Oleo Sci. 57: 123–131.

Morita, T., T. Fukuoka, T. Imura and D. Kitamoto. 2009a. Production of glycolipid biosurfactants by basidiomycetous yeasts. Biotechnol. Appl. Biochem. 53: 39–49.

Morita, T., T. Fukuoka, M. Konishi, T. Imura, S. Yamamoto, M. Kitagawa, A. Sogabe and D. Kitamoto. 2009b. Production of a novel glycolipid biosurfactant, mannosylmannitol lipid, by *Pseudozyma parantarctica* and its interfacial properties. Appl. Microbiol. Biotechnol. 83: 1017–1025.

Morita, T., Y. Ishibashi, T. Fukuoka, T. Imura, H. Sakai, M. Abe and D. Kitamoto. 2009c. Production of glycolipid biosurfactants, mannosylerythritol lipids, by a smut fungus, *Ustilago scitaminea* NBRC 32730. Biosci Biotechnol Biochem. 73: 788–792.

Morita, T., Y. Ishibashi, T. Fukuoka, T. Imura, H. Sakai, M. Abe and D. Kitamoto. 2009d. Production of glycolipid biosurfactants, mannosylerythritol lipids, using sucrose by fungal and yeast strains, and their interfacial properties. Biosci. Biotechnol. Biochem. 73: 2352–2355.

Morita, T., E. Ito, T. Fukuoka, T. Imura and D. Kitamoto. 2010a. The role of PaAAC1 encoding a mitochondrial ADP/ATP carrier in the biosynthesis of extracellular glycolipids, mannosylerythritol lipids, in the basitiomycetous yeast Pseudozyma antarctica. Yeast 27: 379–388.

Morita, T., M. Takashima, T. Fukuoka, M. Konishi, T. Imura and D. Kitamoto. 2010b. Isolation of basidiomycetous yeast *Pseudozyma tsukubaensis* and production of glycolipid biosurfactant, a diastereomer type of mannosylerythritol lipid-B. Appl. Microbiol. Biotechnol. 88: 679–688.

Morita, T., Y. Ishibashi, N. Hirose, K. Wada, M. Takahashi, T. Fukuoka, T. Imura, H. Sakai, M. Abe and D. Kitamoto. 2011. Production and characterization of a glycolipid biosurfactant, mannosylerythritol lipid B, from sugarcane juice by Ustilago scitaminea NBRC 32730. Biosci. Biotechnol. Biochem. 75: 1371–1376.

Morita, T., T. Fukuoka, T. Imura and D. Kitamoto. 2012. Formation of the two novel glycolipid biosurfactants, mannosylribitol lipid and mannosylarabitol lipid, by *Pseudozyma parantarctica* JCM 11752T. Appl. Microbiol. Biotechnol. 96: 931–938.

Morita, T., H. Koike, Y. Koyama, H. Hagiwara, E. Ito, T. Fukuoka, T. Imura, M. Machida and D. Kitamoto. 2013. Genome Sequence of the Basidiomycetous Yeast *Pseudozyma antarctica* T-34, a Producer of the Glycolipid Biosurfactants Mannosylerythritol Lipids. Genome Announc. 1: e0006413-e00064-13.

Morita, T., H. Koike, H. Hagiwara, E. Ito, M. Machida, S. Sato, H. Habe and D. Kitamoto. 2014. Genome and transcriptome analysis of the basidiomycetous yeast *Pseudozyma antarctica* producing extracellular glycolipids, mannosylerythritol lipids. PLoS One. 9: 1–12.

Morita, T., T. Fukuoka, T. Imura and D. Kitamoto. 2015. Mannosylerythritol lipids: Production and applications. J. Oleo Sci. 64: 133–141.

Mukherjee, S., P. Das and R. Sen. 2006. Towards commercial production of microbial surfactants. Trends Biotechnol. 24: 509–515.

Mulligan, C.N., S.K. Sharma and M. Ackmez. 2014. Research Trends and Applications.

Mutalik, S., B. Vaidya, R. Joshi, K. Desai and S. Nene. 2008. Use of response surface optimization for the production of biosurfactant from *Rhodococcus* spp. MTCC 2574. Bioresour. Technol. 99: 7875–7880.

Nguyen, L., S. Chinnapapagari and C.J. Thompson. 2005. FbpA-dependent biosynthesis of trehalose dimycolate is required for the intrinsic multidrug resistance, cell wall structure, and colonial morphology of *Mycobacterium smegmatis* 187: 6603–6611.

Noll, H., H. Bloch, J. Asselineau and E. Lederer. 1956. The chemical structure of the cord factor of *Mycobacterium tuberculosis*. Biochim. Biophys. Acta. 20: 299–309.

O'Connor, S.E. 2015. Engineering of secondary metabolism. Annu. Rev. Genet. 1–24.

Ochsner, U.A., A. Fiechter and J. Reiser. 1994. Isolation, characterization, and expression in *Escherichia coli* of the *Pseudomonas aeruginosa* rhlAB genes encoding a rhamnosyltransferase involved in rhamnolipid biosurfactant synthesis. J. Biol. Chem. 269: 19787–19795.

Ochsner, U.R.S.A., J. Reiser and A. Fiechter. 1995. Production of *Pseudomonas aeruginosa* rhamnolipid biosurfactants in heterologous hosts. Appl. Environ. Microbiol. 61: 3503–3506.

Oliveira, F.J.S., L. Vazquez, N.P. de Campos and F.P. de França. 2009. Production of rhamnolipids by a *Pseudomonas alcaligenes* strain. Process 44: 383–389.

Olvera, C., J.B. Goldberg, R. Sánchez and G. Soberón-Chávez. 1999. The *Pseudomonas aeruginosa* algC gene product participates in rhamnolipid biosynthesis. FEMS Microbiol. Lett. 179: 85–90.

Osbourn, A. 2010. Secondary metabolic gene clusters: evolutionary toolkits for chemical innovation. Trends Genet. 26: 449–457.

Otto, R.T., H.J. Daniel, G. Pekin, K. Müller-Decker, G. Fürstenberger, M. Reuss and C. Syldatk. 1999. Production of sophorolipids from whey: II. Product composition, surface active properties, cytotoxicity and stability against hydrolases by enzymatic treatment. Appl. Microbiol. Biotechnol. 52: 495–501.

Pal, M.P., B.K. Vaidya, K.M. Desai, R.M. Joshi, S.N. Nene and B.D. Kulkarni. 2009. Media optimization for biosurfactant production by *Rhodococcus erythropolis* MTCC 2794: Artificial intelligence versus a statistical approach. J. Ind. Microbiol. Biotechnol. 36: 747–756.

Palmer, J.M. and N.P. Keller. 2010. Secondary metabolism in fungi: Does chromosomal location matter? Curr. Opin. Microbiol. 13: 431–436.

Pearson, J.P., E.C. Pesci and B.H. Iglewski. 1997. Roles of *Pseudomonas aeruginosa* las and rhl quorum-sensing systems in control of elastase and rhamnolipid biosynthesis genes. J. Bacteriol. 179: 5756–5767.

Perfumo, A., M. Rudden, T.J.P. Smyth, R. Marchant, P.S. Stevenson, N.J. Parry and I.M. Banat. 2013. Rhamnolipids are conserved biosurfactants molecules: Implications for their biotechnological potential. Appl. Microbiol. Biotechnol. 97: 7297–7306.

Philp, J., M. Kuyukina, I. Ivshina, S. Dunbar, N. Christofi, S. Lang and V. Wray. 2002. Alkanotrophic *Rhodococcus ruber* as a biosurfactant producer. Appl. Microbiol. Biotechnol. 59: 318–324.

Pirog, T.P., T.A. Shevchuk, I.N. Voloshina and E.V. Karpenko. 2004. Production of surfactants by *Rhodococcus erythropolis* strain EK-1, grown on hydrophilic and hydrophobic substrates. Appl. Biochem. Microbiol. 40: 470–475.

Pirog, T., A. Sofilkanych, T. Shevchuk and M. Shulyakova. 2013. Biosurfactants of *Rhodococcus erythropolis* IMV Ac-5017: Synthesis intensification and practical application. Appl. Biochem. Biotechnol. 170: 880–894.

Poomtien, J., J. Thaniyavarn, P. Pinphanichakarn, S. Jindamorakot and M. Morikawa. 2013. Production and characterization of a biosurfactant from *Cyberlindnera samutprakarnensis* JP52 T. Biosci. Biotechnol. Biochem. 77: 2362–2370.

Price, N.P.J., K.J. Ray, K.E. Vermillion, C.A. Dunlap and C.P. Kurtzman. 2012. Structural characterization of novel sophorolipid biosurfactants from a newly identified species of *Candida* yeast. Carbohydr. Res. 348: 33–41.

Puchkov, E.O., U. Zähringer, B. Lindner, T.V. Kulakovskaya, U. Seydel and A. Wiese. 2002. The mycocidal, membrane-active complex of Cryptococcus humicola is a new type of cellobiose lipid with detergent features. Biochim. Biophys. Acta Biomembr. 1558: 161–170.

Rahim, R., L.L. Burrows, M.A. Monteiro, M.B. Perry and J.S. Lam. 2000. Involvement of the rml locus in core oligosaccharide and O polysaccharide assembly in *Pseudomonas aeruginosa*. Microbiology 146: 2803–2814.

Rahim, R., U.A. Ochsner, C. Olvera, M. Graninger, P. Messner, J.S. Lam and G. Soberón-Chávez. 2001. Cloning and functional characterization of the *Pseudomonas aeruginosa* rhlC gene that encodes rhamnosyltransferase 2, an enzyme responsible for di-rhamnolipid biosynthesis. Mol. Microbiol. 40: 708–718.

Rakatozafy, H. and R. Marchal. 1999. Diversity of bacterial strains degrading hexadecane in relation to the mode of substrate uptake. 421–428.

Rampioni, G., M. Schuster, E.P. Greenberg, E. Zennaro and L. Leoni. 2009. Contribution of the RsaL global regulator to *Pseudomonas aeruginosa* virulence and biofilm formation. FEMS Microbiol. Lett. 301: 210–217.

Randhawa, K.K.S. and P.K.S.M. Rahman. 2014. Rhamnolipid biosurfactants-past, present, and future scenario of global market. Front. Microbiol. 5: 1–7.

Rapp, P., H. Bock, V. Wray and F. Wagner. 1979. Formation, isolation and characterization of trehalose dimycolates from *Rhodococcus erythropolis* grown on n-Alkanes. J. Gen. Microbiol. 115: 491–503.

Rau, U., L.A. Nguyen, H. Roeper, H. Koch and S. Lang. 2005. Fed-batch bioreactor production of mannosylerythritol lipids secreted by *Pseudozyma aphidis*. Appl. Microbiol. Biotechnol. 68: 607–613.

Reis, R.S., A.G. Pereira, B.C. Neves and D.M.G. Freire. 2011. Gene regulation of rhamnolipid production in *Pseudomonas aeruginosa*—A review. Bioresour. Technol. 102: 6377–6384.

Reis, R.S., G.J. Pacheco, A.G. Pereira and D.M.G. Freire. 2013. Biosurfactants: Production and Applications.

Rocha, M.V.P., J.S. Mendes, M.E. Aparecida Giro, V.M.M. Melo and L.R.B. Gonçalves. 2014. Biosurfactant production by *Pseudomonas aeruginosa* MSIC02 in cashew apple juice using a 2^4 full factorial experimental design. Chem. Ind. Chem. Eng. Q. 20: 49–58.

Rodrigues, L., I.M. Banat, J. Teixeira and R. Oliveira. 2006. Biosurfactants: Potential applications in medicine. J. Antimicrob. Chemother. 57: 609–618.

Roelants, S.L.K.W., K.M.J. Saerens, T. Derycke, B. Li, Y.C. Lin, Y. Van de Peer, S.L. De Maeseneire, I.N. a Van Bogaert and W. Soetaert. 2013. *Candida bombicola* as a platform organism for the production of tailor-made biomolecules. Biotechnol. Bioeng. 110: 2494–2503.

Roelants, S.L.K.W., S.L. De Maeseneire, K. Ciesielska, I.N.A. Van Bogaert and W. Soetaert. 2014. Biosurfactant gene clusters in eukaryotes: regulation and biotechnological potential. Appl. Microbiol. Biotechnol. 98: 3449–61.

Roelants, S.L.K.W., K. Ciesielska, S.L. De Maeseneire, H. Moens, B. Everaert, S. Verweire, Q. Denon, B. Vanlerberghe, I.N.A. Van Bogaert, P. Van der Meeren, B. Devreese and W. Soetaert. 2015. Towards the industrialization of new biosurfactants: Biotechnological opportunities for the lactone esterase gene from *Starmerella bombicola*. Biotechnol. Bioeng. 113: 550–559.

Ron, E.Z. and E. Rosenberg. 2001. Natural roles of biosurfactants. 3.

Rosa, C.A. and M.A. Lachance. 1998. The yeast genus Starmerella gen. nov. and Starmerella bombicola sp. nov., the teleomorph of Candida bombicola (Spencer, Gorin & Tullock) Meyer & Yarrow. Int. J. Syst. Bacteriol. 48 Pt 4: 1413–7.

Saerens, K.M.J., S.L. Roelants, I.N. Van Bogaert and W. Soetaert. 2011a. Identification of the UDP-glucosyltransferase gene UGTA1, responsible for the first glucosylation step in the sophorolipid biosynthetic pathway of Candida bombicola ATCC 22214. FEMS Yeast Res. 11: 123–132.

Saerens, K.M.J., L. Saey and W. Soetaert. 2011b. One-step production of unacetylated sophorolipids by an acetyltransferase negative *Candida bombicola*. Biotechnol. Bioeng. 108: 2923–2931.

Saerens, K.M.J., J. Zhang, L. Saey, I.N.A. Van Bogaert and W. Soetaert. 2011c. Cloning and functional characterization of the UDP-glucosyltransferase UgtB1 involved in sophorolipid production by *Candida bombicola* and creation of a glucolipid-producing yeast strain. Yeast 28: 279–292.

Saika, A., H. Koike, T. Hori and T. Fukuoka. 2014. Draft genome sequence of the yeast *Pseudozyma antarctica* type strain JCM10317, a producer of the glycolipid biosurfactants, mannosylerythritol lipids. Genome Announc. 2: 4–5.

Saika, A., H. Koike, T. Fukuoka, S. Yamamoto, T. Kishimoto and T. Morita. 2016. A gene cluster for biosynthesis of mannosylerythritol lipids consisted of 4-O-beta-D-Mannopyranosyl-(2R,3S)-Erythritol as the sugar moiety in a basidiomycetous yeast *Pseudozyma tsukubaensis*. PLoS One 11: 1–16.

Shaaban, M., J.M. Palmer, W.A. EL-Naggar, M.A. EL-Sokkary, E.S.E. Habib and N.P. Keller. 2010. Involvement of transposon-like elements in penicillin gene cluster regulation. Fungal Genet. Biol. 47: 423–432.

Shao, Z. 2011. Trehalolipids. pp. 121–143. *In*: Gloria Soberón-Chávez (ed.). Biosurfactants: From Genes to Applications. Berlin, Heidelberg: Springer Berlin Heidelberg.

Shwab, E.K., J.W. Bok, M. Tribus, J. Galehr, S. Graessle and N.P. Keller. 2007. Histone deacetylase activity regulates chemical diversity in *Aspergillus*. Eukaryot. Cell 6: 1656–64.

Sivapathasekaran, C., S. Mukherjee, A. Ray, A. Gupta and R. Sen. 2010. Artificial neural network modeling and genetic algorithm based medium optimization for the improved production of marine biosurfactant. Bioresour. Technol. 101: 2884–2887.

Soberón-Chávez, G., F. Lépine and E. Déziel. 2005. Production of rhamnolipids by *Pseudomonas aeruginosa*. Appl. Microbiol. Biotechnol. 68: 718–725.

Spencer, J.F.T., P.A.J. Gorin and A.P. Tulloch. 1970. Torulopsis bombicola sp.n. Antonie Van Leeuwenhoek 36: 129–133.

Spoeckner, S., V. Wray, M. Nimtz and S. Lang. 1999. Glycolipids of the smut fungus *Ustilago maydis* from cultivation on renewable resources. Appl. Microbiol. Biotechnol. 51: 33–39.

Strauss, J. and Y. Reyes-Dominguez. 2011. Regulation of secondary metabolism by chromatin structure and epigenetic codes. Fungal Genet. Biol. 48: 62–69.

Stüwer, O., R. Hommel, D. Haferburg and H.P. Kleber. 1987. Production of crystalline surface-active glycolipids by a strain of *Torulopsis apicola*. J. Biotechnol. 6: 259–269.

Takayama, K., C. Wang and G.S. Besra. 2005. Pathway to synthesis and processing of mycolic acids in *Mycobacterium tuberculosis*. Clin. Microbiol. Rev. 18: 81–101.

Tavares, L.F.D., P.M. Silva, M. Junqueira, D.C.O. Mariano, F.C.S. Nogueira, G.B. Domont, D.M.G. Freire and B.C. Neves. 2013. Characterization of rhamnolipids produced by wild-type and engineered *Burkholderia kururiensis*. Appl. Microbiol. Biotechnol. 97: 1909–1921.

Teichmann, B., U. Linne, S. Hewald, M.A. Marahiel and M. Bölker. 2007. A biosynthetic gene cluster for a secreted cellobiose lipid with antifungal activity from *Ustilago maydis*. Mol. Microbiol. 66: 525–533.

Teichmann, B., L. Liu, K.O. Schink and M. Bölker. 2010. Activation of the ustilagic acid biosynthesis gene cluster in *Ustilago maydis* by the C2H2 zinc finger transcription factor. Appl. Environ. Microbiol. 76: 2633–2640.

Teichmann, B., C. Labbé, F. Lefebvre, M. Bölker, U. Linne and R.R. Bélanger. 2011a. Identification of a biosynthesis gene cluster for flocculosin a cellobiose lipid produced by the biocontrol agent *Pseudozyma flocculosa*. Mol. Microbiol. 79: 1483–1495.

Teichmann, B., F. Lefebvre, C. Labbé, M. Bölker, U. Linne and R.R. Bélanger. 2011b. Beta-hydroxylation of glycolipids from *Ustilago maydis* and *Pseudozyma flocculosa* by an NADPH-dependent β-hydroxylase. Appl. Environ. Microbiol. 77: 7823–7829.

Thaniyavarn, J., T. Chianguthai, P. Sangvanich, N. Roongsawang, K. Washio, M. Morikawa and S. Thaniyavarn. 2008. Production of Sophorolipid Biosurfactant by *Pichia anomala*. Biosci. Biotechnol. Biochem. 72: 2061–2068.

Tiso, T., P. Sabelhaus, B. Behrens, A. Wittgens, F. Rosenau, H. Hayen and L.M. Blank. 2016. Creating metabolic demand as an engineering strategy in *Pseudomonas putida*: Rhamnolipid synthesis as an example. Metab. Eng. Commun. 3: 234–244.

Tokumoto, Y., N. Nomura, H. Uchiyama, T. Imura, T. Morita, T. Fukuoka and D. Kitamoto. 2009. Structural characterization and surface-active properties of a succinoyl trehalose lipid produced by *Rhodococcus* sp. SD-74. J. Oleo Sci. 58: 97–102.

Toribio, J., A.E. Escalante and G. Soberón-Chávez. 2010. Rhamnolipids: Production in bacteria other than *Pseudomonas aeruginosa*. Eur. J. Lipid Sci. Technol. 112: 1082–1087.

Tulloch, A.P., J.F.T. Spencer and P.A.J. Gorin. 1962. The fermentation of long-chain compounds by *Torulopsis magnoliae*: I. Structures of the hydroxy fatty acids obtained by the fermentation of fatty acids and hydrocarbons. Can. J. Chem. 40: 1326–1338.

Tulloch, A.P., A. Hill and J.F.T. Spencer. 1967. A new type of macrocyclic lactone from *Torulopsis apicola*. Chem. Commun. 584.

Tulloch, Spencer and Deinema. 1968. A new hydroxy fatty acid sophoroside from *Candida bogoriensis*. Can. J. Chem. 46: 345–348.

Uchida, Y., S. Misawa, T. Nakahara and T. Tabuchi. 1989a. Factors affecting the production of succinoyl trehalose lipids by *Rhodococcus erythropolis* SD-74 grown on n-Alkanes. Agric. Biol. Chem. 53: 765–769.

Uchida, Y., R. Tsuchiya, M. Chino, J. Hirano and T. Tabuchi. 1989b. Extracellular accumulation of mono- and di-succinoyl trehalose lipids by a strain of *Rhodococcus erythropolis* grown on n-alkanes. Agric. Biol. Chem. 53: 757–763.

Van Alst, N.E., K.F. Picardo, B.H. Iglewski and C.G. Haidaris. 2007. Nitrate sensing and metabolism modulate motility, biofilm formation, and virulence in *Pseudomonas aeruginosa*. Infect. Immun. 75: 3780–3790.

Van Bogaert, I., S. Fleurackers, S. Van Kerrebroeck, D. Develter and W. Soetaert. 2011a. Production of new-to-nature sophorolipids by cultivating the yeast *Candida bombicola* on unconventional hydrophobic substrates. Biotechnol. Bioeng. 108: 734–741.

Van Bogaert, I.N.A., K. Saerens, C. De Muynck, D. Develter, W. Soetaert and E.J. Vandamme. 2007. Microbial production and application of sophorolipids. Appl. Microbiol. Biotechnol. 76: 23–34.

Van Bogaert, I.N.A., D. Develter, W. Soetaert and E.J. Vandamme. 2008. Cerulenin inhibits *de novo* sophorolipid synthesis of *Candida bombicola*. Biotechnol. Lett. 30: 1829–1832.

Van Bogaert, I.N.A., J. Sabirova, D. Develter, W. Soetaert and E.J. Vandamme. 2009a. Knocking out the MFE-2 gene of *Candida bombicola* leads to improved medium-chain sophorolipid production. FEMS Yeast Res. 9: 610–617.

Van Bogaert, I.N.A., M. Demey, D. Develter, W. Soetaert and E.J. Vandamme. 2009b. Importance of the cytochrome P450 monooxygenase CYP52 family for the sophorolipid-producing yeast *Candida bombicola*. FEMS Yeast Res. 9: 87–94.

Van Bogaert, I.N.A., J. Zhang and W. Soetaert. 2011b. Microbial synthesis of sophorolipids. Process Biochem. 46: 821–833.

Van Bogaert, I.N.A, K. Holvoet, S.L.K.W. Roelants, B. Li, Y.C. Lin, Y. Van de Peer and W. Soetaert. 2013. The biosynthetic gene cluster for sophorolipids: A biotechnological interesting biosurfactant produced by *Starmerella bombicola*. Mol. Microbiol. 88: 501–509.

Van Bogaert, I.N.A., D. Buyst, J.C. Martins, S.L.K.W. Roelants and W.K. Soetaert. 2016. Synthesis of bolaform biosurfactants by an engineered *Starmerella bombicola* yeast. Biotechnol. Bioeng. 113: 2644–2651.

Van Hamme, J.D., A. Singh and O.P. Ward. 2006. Physiological aspects. Part 1 in a series of papers devoted to surfactants in microbiology and biotechnology. Biotechnol. Adv. 24: 604–20.

Vergne, I. and M. Daffé. 1998. Interaction of mycobacterial glycolipids with host cells. Front. Biosci. 865–876.

Walton, J.D. 2000. Horizontal gene transfer and the evolution of secondary metabolite gene clusters in fungi: an hypothesis. Fungal Genet. Biol. 30: 167–71.

Wang, Q., X. Fang, B. Bai, X. Liang, P.J. Shuler, W.A. Goddard and Y. Tang. 2007. Engineering bacteria for production of rhamnolipid as an agent for enhanced oil recovery. Biotechnol. Bioeng. 98: 842–853.

Waters, C.M. and B.L. Bassler. 2005. Quorum sensing: Communication in bacteria. Annu. Rev. Cell Dev. Biol. 21: 319–346.

Wei, Y., C. Chou and J. Chang. 2005. Rhamnolipid production by indigenous *Pseudomonas aeruginosa* J4 originating from petrochemical wastewater. Biochem. Eng. J. 27: 146–154.

White, D.A., L.C. Hird and S.T. Ali. 2013. Production and characterization of a trehalolipid biosurfactant produced by the novel marine bacterium *Rhodococcus* sp., strain PML026. J. Appl. Microbiol. 115: 744–755.

Wittgens, A., T. Tiso, T.T. Arndt, P. Wenk, J. Hemmerich, C. Müller, R. Wichmann, B. Küpper, M. Zwick, S. Wilhelm, R. Hausmann, C. Syldatk, F. Rosenau and L.M. Blank. 2011. Growth independent rhamnolipid production from glucose using the non-pathogenic *Pseudomonas putida* KT2440. Microb. Cell Fact. 10: 80.

Wittgens, A., F. Kovacic, M.M. Müller, M. Gerlitzki, B. Santiago-Schübel, D. Hofmann, T. Tiso, L.M. Blank, M. Henkel, R. Hausmann, C. Syldatk, S. Wilhelm and F. Rosenau. 2016. Novel insights into biosynthesis and uptake of rhamnolipids and their precursors. Appl. Microbiol. Biotechnol. 101: 2865–2878

Yu, J.-H. and N. Keller. 2005. Regulation of secondary metabolism in filamentous fungi. Annu. Rev. Phytopathol. 43: 437–458.

Zhang, L., J.E. Pemberton and R.M. Maier. 2014. Effect of fatty acid substrate chain length on *Pseudomonas aeruginosa* ATCC 9027 monorhamnolipid yield and congener distribution. Process Biochem. 49: 989–995.

Zhao, F., R. Shi, J. Zhao, G. Li, X. Bai, S. Han and Y. Zhang. 2015. Heterologous production of *Pseudomonas aeruginosa* rhamnolipid under anaerobic conditions for microbial enhanced oil recovery. J. Appl. Microbiol. 118: 379–389.

Zhou, Q.H. and N. Kosaric. 1993. Effect of lactose and olive oil on intra- and extracellular lipids of *Torulopsis bombicola*. Biotechnol. Lett. 15: 477–482.

Zhu, K. and C.O. Rock. 2008. RhlA converts beta-hydroxyacyl-acyl carrier protein intermediates in fatty acid synthesis to the beta-hydroxydecanoyl-beta-hydroxydecanoate component of rhamnolipids in *Pseudomonas aeruginosa*. J. Bacteriol. 190: 3147–3154.

# 13

# Microbial Biosurfactants:
# From Lab to Market

*Sophie L.K.W. Roelants,[1,2,*] Lisa Van Renterghem,[1]*
*Karolien Maes,[2] Bernd Everaert,[2] Emile Redant,[2] Brecht*
*Vanlerberghe,[2] Sofie L. Demaeseneire[1] and Wim Soetaert[1,2]*

## Introduction

The global surfactant market, worth about 36 billion dollars (Markets and Markets 2015), is characterized by hundreds of different structural variants, which are found in a plethora of applications, from construction and food to precision cleaning industries. About half of this huge production volume is used in household and laundry detergents, while the other half is employed in various industries, e.g. (oilfield) chemicals, mining, paints and coatings, textile and paper, agrochemicals, industrial emulsions, construction, food processing, pharmaceuticals, cosmetics, etc. (Grand view research 2016). An emerging class of surfactants is the so called biosurfactants. Biosurfactants or 100% biobased surfactants can be produced chemically or biologically, starting from natural and renewable building blocks, such as sugars and plant oils as substrates, offering a renewable and 100% biobased alternative to the traditional (petrochemically produced) surfactants. Biobased surfactants constitute about 3% of the global surfactant market (Tropsch 2017), a volume which is mainly dominated by chemically produced biosurfactants like methyl ester sulfonates (MESs), alkyl polyglucosides (APGs) and sugar esters. The

---
[1] Ghent University, Faculty of Bioscience engineering, Department of Biotechnology, Centre for Industrial Biotechnology and Biocatalysis (InBio.be), Coupure Links 653, Ghent, Belgium, B-9000.
[2] Bio Base Europe Pilot Plant, R&D Department, Rodenhuizekaai 1, Ghent, Belgium, B-9042.
* Corresponding author: Sophie.Roelants@Ugent.be; sophie.roelants@bbeu.org

biologically produced biosurfactants can be obtained through extraction from plants (e.g., cardanol from cashew nut shells), biocatalysis (e.g., enzymatic sugar esters) and fermentation (e.g., rhamnolipids, sophorolipids, surfactin and xylolipids), better known as microbial biosurfactants. In the latter a biological production process (fermentation), natural building blocks, such as sugars and plant oils or even waste/ side streams, are employed. The ecological advantage associated with such processes, together with the rising awareness towards sustainability, clearly underpins the market potential of biosurfactants also translated in the patenting activity (Shete et al. 2006). In this book, the last class of biosurfactants, i.e., microbial biosurfactants has been described in detail. Although a lot of research has been devoted to this class of biochemicals, a very limited amount is currently available in the market, estimated by the authors to account for only a few thousand metric tons/year and thus below 0.1% of the global surfactant market. This limited market penetration is caused by three main reasons: first of all, microbial biosurfactants are generally produced as complex mixtures, e.g., rhamnolipids can be found as mono- and dirhamnolipids and the chain length of the hydrophobic monomers can vary (Abdel-Mawgoud et al. 2010); *sophorolipids can be found in acidic and lactonic forms and chain lengths, site of hydroxylation, saturations degree, etc. (Davila et al. 1993, Price et al. 2012), *lipopeptides vary in the constituting amino acids and fatty acids (Mnif and Ghribi 2015, Bóka et al. 2016), while *MELs vary in their acetylation and acylation degree (Onghena et al. 2011). This situation is schematically summarized in Figure 1; clearly for all the biosurfactants, mixtures are produced, and when one speaks about, e.g., "sophorolipids", one is speaking about a mixture of between 20 and 100 compounds. The fact that mixtures are produced as such is not the biggest issue though. If a mixture does the job, the industry and end users will be satisfied,

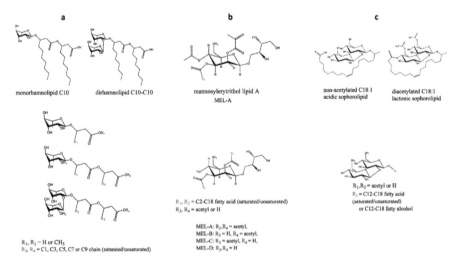

**Figure 1.** Schematic representation of the structural variety of biosurfactants (a) Rhamnolipids, (b) Mannosylerythritol lipids and (c) Sophorolipids. Top: the classic representation of the biosurfactants. Below: a schematic representation of the diversity of molecular structures captured within the biosurfactant types.

irrespective of it being a mixture. However, the fact that the ratio of the respective compounds is prone to variation (culture/growth conditions, substrate variation, medium components, etc.) (Cavalero and Cooper 2003, Soberon-Chavez et al. 2005, Ashby et al. 2006, Fukuoka et al. 2008, Morita et al. 2009, Nitschke et al. 2010), in combination with the fact that the different congeners are often associated with very different properties (e.g., lactonic versus acidic sophorolipids (Roelants et al. 2016), MELs with a variation of the ac(et)ylation degree (Fukuoka et al. 2007, 2008), RLs with one or two rhamnose moieties (Zhang et al. 2013), etc.) can result in highly confusing and undesired situations. One batch of a specific product can perfectly show completely different functionality as compared to another batch of supposedly the same product. The occurrence of such issues is completely unacceptable from a market perspective, where products must comply with the specifications defined by the manufacturer.

A second reason for the small market share is the fact that the molecular variants of microbial biosurfactants, produced at acceptable efficiencies (and thus acceptable production costs) by the respective microorganisms, is currently too low as compared with synthetic alternatives. In formulation business, mixing and combining ingredients to get to a certain functionality requires the availability of choice. Indeed, a range/a portfolio of hundreds of synthetic products is available on the market, while only a handfull of microbial biosurfactants is available.

Last, but not least, price is major issue. This third and last issue is a constraint for many new biotech products and processes, according to several biotech startups (personal communications). This is mostly due to the fact that such new products cannot profit from the economy of scale yet, while the associated production processes (strains, fermentation and purification) have not yet been thoroughly optimized. Moreover, petrochemical production plants are mostly fully depreciated.

To summarized the above in three words: uniformity, diversity and efficiency are key issues of major importance to increase the market share of microbial biosurfactants.

In this book chapter, the approach followed by researchers at InBio.be and BBEPP to resolve these issues to bring (new-to-nature) microbial glycolipid biosurfactants to the market is elaborated. In this so-called 'integrated bioprocess design' (IBD) approach, strain engineering, process development and scale up and dedicated application research are closely interconnected. Iteration between the different 'unit operations' enables early identification of bottlenecks and the definition of solutions along the way. This approach tackles the three bottlenecks mentioned above: uniformity, diversity and efficiency.

## Increasing Molecular Variety and Uniformity through Strain Engineering

The working horse at the core of this 'IBD' strategy is the 'exotic' yeast *Starmerella bombicola* (Rosa and Lachance 1998), 'exotic' because it is not a well-described lab strain for which molecular tools are readily available, like *Saccharomyces cerevisiae*. Molecular tools are required for genetic engineering to be possible.

This first requirement/bottleneck was defined by the research group of Professor Soetaert fifteen years ago as the way forward to increase molecular variety and strain efficiency (decreasing production costs), while reducing product complexity (increase uniformity). His research group, InBio.be, thus set out on the quest to develop such molecular tools for *S. bombicola*, which has resulted in the slow, though eventually exponential expansion of the possibilities. A 'hands on' overview of this endeavor has been recently compiled (Lodens et al. 2017). The choice for this particular yeast strain to be transformed into a platform organism for (new-to-nature biosurfactants) is the fact that *S. bombicola* naturally produces high amounts of the biosurfactant sophorolipids (SLs) (> 200 g/L; 4 g/L.h) (Gao et al. 2013). The biosynthetic pathway of SLs has been elucidated by our lab (see Figure 2) and the contributing enzymes described and characterized (Saerens et al. 2011a, 2011b, 2015, Van Bogaert et al. 2013, Ciesielska et al. 2016).

SLs are a well-known example of glycolipid microbial biosurfactants and are composed of the rare disaccharide sophorose, attached to a (hydroxylated) fatty acid chain, and occur in an 'open' or acidic conformation or 'closed' or lactonic conformation (see Figure 1). *S. bombicola* synthesizes these molecules in high amounts from renewable resources and even waste streams (Daniel et al. 1999, Fleurackers 2006, Felse et al. 2007, Montoneri et al. 2009, Nitschke et al. 2010, Daverey and Sumalatha 2011), which results in substantial industrial interest. Commercialization of SLs has thus been pursued by several companies, like Evonik, Soliance and Wheatoleo, amongst others and application by Henkel, Ecover, Saraya, and Wheatoleo (Roelants et al. 2014). SLs are thus one of the microbial biosurfactant success stories, being one of the few types that have made it to the market. However, some issues are associated with the natural occurring 'sophorolipids'. As mentioned above, they occur as a mixture of lactonic and acidic congeners (see Figure 1), which have very different properties. The ratio of lactonic over acidic varies between the two extremes and is influenced by media and culture conditions (Roelants et al. 2016). Moreover, variation in the acetylation degree of the hydrophilic head group (Saerens et al. 2011b) in the saturation/length of the hydrophobic tail and in its site of hydroxylation (terminal or subterminal, and thus linkage to the sophorose head group) occurs.

Although the biosynthetic enzymes are thus quite promiscuous, they still have a certain preference, e.g., oleic acid is the preferred substrate for the CYP52M1 monooxygenase (Figure 2b, step 1), but substrates of between 16 and 20 carbon atoms can be hydroxylated with this enzyme (Huang et al. 2014) and although both terminal ($\omega$) and subterminal ($\omega$-1) hydroxylation occurs, the enzyme has a preference for subterminal hydroxylation. The glucosyltransferases are more promiscuous as they can accept/glycosylate hydroxylated substrates of between 4 and 20 carbon atoms (Saerens et al. 2015) and the first enzyme (UGTA1) (Figure 2b, step 2) can glycosylate both carboxylic and alcohol functions (Van Bogaert et al. 2016). The abovementioned (biased) variation gives rise to about 20 'major', and no less than over 100 'minor' homologs in the SL mixture. Several strategies have been described to at least control the lactonic/acidic ratio, as this variation is responsible for the largest part of the functionality shift and in the opinion of the authors, explanatory

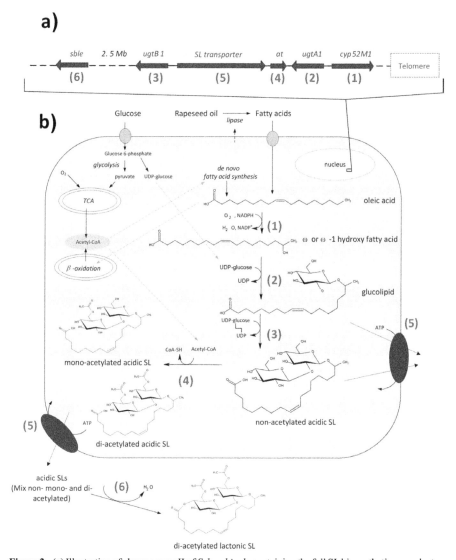

**Figure 2.** (a) Illustration of chromosome II of *S. bombicola* containing the full SL biosynthetic gene cluster (± 11 kb) and the gene responsible for lactonization (*sble*) at the other side of the same chromosome (b). The full SL biosynthetic pathway as elucidated at our lab, consisting of (1) hydroxylation of a fatty acid by a CYP52M1 monooxygenase (2) glucosylation of the FA-OH by the first glucosyltransferase UGTA1 (3) second glucosylation step of the formed glucolipid by a second glucosyltransferase UGTB1 giving rise to an acidic SL, which can be (4) acetylated by the action of an acetyltransferase AT. The different SLs are transported into the extracellular space by a multidrug transporter protein (5). Lactonization (6) occurs extracellularly as the responsible enzyme is secreted.

of the varying statements in the literature concerning SLs physicochemical and biological characteristics. To give an example, lactonic SLs have clear antimicrobial and -viral properties and low foaming potential, while acidic SLs show very low or no antimicrobial activity and foam considerably well. Fermentation and purification based strategies can quite efficiently generate two types of quite uniform SL types, i.e., 1. > 95% diacetylated lacton SLs and 2. 100% non-acetylated acidic SLs, respectively (Lang et al. 2000, Develter and Renkin 2012, Roelants et al. 2016) (see Figures 1c and 3), as two structural variants. The last are deduced from the first by applying alkaline hydrolysis (Rau et al. 1999), as such hydrolyzing all ester functions (acetyl- and lacton functionalities). Several companies also generate/produce a lactonic SL product which is partly hydrolyzed, i.e., generation of a mixture of acidic and lactonic SLs in a controlled way. This is done for two reasons: increasing the water solubility of lactonic SLs and the stimulation of synergistic effects between the two forms.

Although the industry has thus eventually managed to valorize the (complicated) potential of wild type SLs, genetic engineering offers a more elegant and absolute solution to the abovementioned issues. Evonik for example applied this strategy to generate a rhamnolipid producing strain. This allowed them to avoid the use of the opportunistic pathogen *Pseudomonas aeruginosa* for RL production. Moreover, this strategy allows to cicrumvent the complex regulation of RL biosynthesis in wild type producers as such giving rise to more robust processes. Indeed, Genetic engineering indeed also allowed our and some other research groups to generate a range of new *Starmerella bombicola* strains producing (new-to-nature) glycolipids (Saerens et al. 2011b, 2011c, Roelants et al. 2016, Takahashi et al. 2016, Van Bogaert et al. 2016, Van Renterghem et al. 2017). An overview of the most important molecular structures which can be produced at productivities similar as the wild type strain is shown in Figure 3.

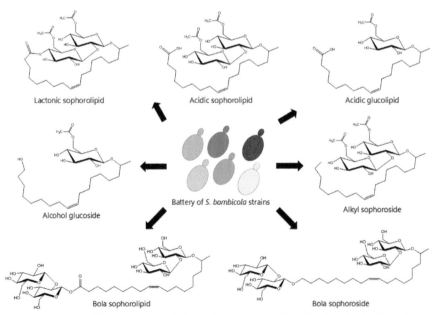

**Figure 3.** Overview of the glycolipid portfolio based on engineered *S. bombicola* strains for the production of new to nature glycolipids at acceptable productivities developed at the University of Ghent (InBio.be). Acetylgroups (in blue) can be present (varying degrees) or absent.

In this figure, it is thus obvious that we have unchained *S. bombicola* as a platform organism for new types of glycolipid biosurfactants. The availability of a battery of very similar glycolipid biosurfactants does not only tackle the 'diversity' criterion, it also provides proof of concept for other microbial biosurfactants. Moreover, this is the first time in our knowledge that there is a realistic possibility to determine in depth structure-function relationships, i.e., investigate the influence of the polar head group (one versus two sugars), the influence of acetylation (no, low, medium or high acetylation degree), the influence of the functionalization of the lipophilic part, etc. on the molecular functionality. The latter will be extremely important for these new types of surfactants to gain importance and interest from the market. Besides these achievements on the first abovementioned key point 'diversity', the second bottleneck, 'uniformity', was also tackled by our research group. For example, by deleting the gene/enzyme responsible for lactonization of acidic SLs (the *Starmerella bombicola* lactone esterase, *sble* (Ciesielska et al. 2016)) a strain was generated which exclusively produces acidic SLs (Roelants et al. 2016). An overexpression strain of the same gene enabled us to control the degree of lactonization in the other direction, i.e., the generation of a strain exclusively producing lactonic sophorolipids, independent of the culture conditions (Roelants et al. 2016). Most of the portfolio depicted in Figure 3 was obtained by tinkering with the SL biosynthetic pathway or, more specifically, in the SL biosynthetic gene cluster (see Figure 2a), e.g., deleting the enzyme attaching the second glucose molecule (UGTB1; Figure 2, step 3) to generate a sophorolipid resulted in the generation of a strain producing glucolipids (Saerens et al. 2011b).

The introduction of a heterologous gene/enzyme (*ugt1* from another biosurfactant producing yeast: *U. maydis*) in this strain then enabled us to produce cellobioselipids with *S. bombicola*, although the latter with limited productivities (Roelants et al. 2013). This was initially also the case for the glucolipid producing strain developed by Saerens et al. (2011b), with productivities of about 0,01 g/L.h. This strain was derived from a spontaneous auxotrophic mutant, G9 (Δ*ura3*) (Van Bogaert et al. 2008), as no rationally engineered *S. bombicola* strain was available at that time. We recently found out that this strain apparently (not unexpectedly) also had mutations in other genomic regions and was as such affected in its entire metabolism (unpublished results). This also affected glycolipid production and was thus the main reason of the very low productivities of this glucolipid producing strain. The use of a rationally designed Δ*ura3* strain resolved this issue and gave rise to a highly productive glucolipid producing strain (unpublished results). The same was attempted for the cellobioselipid producing strain (so introduction of the *ugt1* gene in the new Δ*ugtB1* strain), but in this case did not increase the productivity of cellobioselipids (Geys 2017). Besides this attempt to increase cellobioselipids productivity, the introduction of the *ugt1* gene of *U. maydis* in a *S. bombicola ugtA1* knock out strain (non-SL producing strain (Saerens et al. 2011a)) was also attempted (Geys 2017) to rule out possible competition between the UGT1 and UGTA1 enzymes in the Δ*ugtB1* background. The last experiment showed that the UGT1 enzyme catalyzes both the glucosylation of the hydroxylated fatty acids as of the glucolipid, as also in this case cellobioselipids were produced. However, the efficiency of cellobiose production was not increased as compared to *ugt1* expression in the Δ*ugtB1* background. This indicates that further engineering strategies are required for this enzyme to work with optimal efficiency in *S. bombicola*. The main reason for its suboptimal efficiency in *S. bombicola* can be manifold: one reason could be the fact that the UGT1 enzyme is quite strict in its substrate choice. In *U. maydis*, the main hydrophobic tail incorporated in cellobioselipids is terminally hydroxylated hexadecenoic acid, while the main hydroxylated product produced by *S. bombicola* is ω and ω-1 hydroxy oleic acid, as explained above. Another reason could be related to transcription and/or translational issues, as this was found to be the case for failure of a non-codon optimized *yegfp* gene in *S. bombicola* (Roelants 2013). However, in this case a codon optimized variant of the *ugt1* gene was used. The gene was placed under the control of the *ugtA1* transcriptional denominators (promotor and terminator sequences) inside the *S. bombicola* SL biosynthetic gene cluster, i.e., replacing the *ugtA1* coding sequence (see Figure 2a) by the *ugt1* coding sequence. RNA and/or protein stability could be of importance, while transport related issues could also be a reason for these observations. Indeed, a specific SL transporter is present in the SL biosynthetic gene cluster (see Figure 2b, step 5), which is responsible for SL transport (Van Bogaert et al. 2013). However, the fact that the new compounds depicted in Figure 3 are all produced/secreted at reasonable levels does not suggest this to be a major reason for the very low cellobioselipid production. However, transport issues could attribute to lower productivities as compared to the wild type in the other glycolipid producing strains as we currently reach average productivities of between 0.5 and 1.5 g/L.h with the new *S. bombicola* strains depicted in Figure 3. Last, but not the least, regulatory effects are expected to play a big role. In contrast to other glycolipid production pathways (Teichmann et al. 2010, Roelants et al. 2014), the molecular base of SL/glycolipid biosynthesis in *S. bombicola* has not yet been unraveled. The fact that its genes are highly upregulated during SL biosynthesis suggests very tight regulation (Ciesielska et al. 2013). It is very likely that the modifications in the new strains (inside the SL biosynthetic gene cluster) have a large though unknown effect on the regulatory network affecting its expression. This hypothesis is strengthened by the fact that for most of the new strains downregulation of the biosynthetic enzymes was observed as compared to the wild type (unpublished results). Although the abovementioned accomplishments can be considered as big breakthroughs in the microbial biosurfactant world, it is clear that to further broaden and optimize the production of new to nature glycolipids, in depth metabolic engineering strategies are required. The latter on the other hand requires the further expansion of the molecular toolkit and an in-depth knowledge of the molecular regulation of SL biosynthesis and its related pathways. The last goals are currently thoroughly being investigated at Inbio.be.

To come back to the abovementioned integrated bioprocess design (IBD) strategy, one of the new strains/molecules depicted in Figure 3, i.e., a new type of SLs, so-called bolaform SLs, will be used as an example to guide the reader through the several steps of this IBD approach. These bolaform SLs were quite recently discovered in the wild type SL mixture in minute amounts (Price et al. 2012). The analytical strength in the early years of SL characterization was probably too low to detect these compounds, resulting in their late discovery compared to 'sophorolipids' in general (Spencer et al. 1970, Asmer et al. 1988). Moreover, their limited stability (i.e., the stability of the ester function linking the second sophorose moiety to the hydrophobic linker), on which we will further elaborate below, probably also contributed to this. The last hypothesis is substantiated by reports of the detection of free sophorose in *S. bombicola* cultures (AL-Jasim et al. 2016). In contrast to the structure of classic SLs (i.e., acidic and lactonic SLs) (see Figures 1c and 3), bolaform SLs consist of two sophorose moieties located on each side of the lipophilic chain. Shortly after their discovery by Price et al. (2012), our lab succeeded in generating a strain that almost exclusively produces these bolaform SLs (Van Bogaert et al. 2016). The last was a rather unexpected result as the developed strain was generated with the aim of generating a strain exclusively (uniformly) producing non-acetylated acidic SLs.

When looking at Figure 2, this is logically done by deleting the lactone esterase (*sble* gene, Figure 2b, step 6) and the acetyltransferase (*at* gene, Figure 2b, step 4) genes. However, this new strain (Δ*at* Δ*sble*) now surprisingly produces bola sophorolipids. Due to their unique structure, bolaform amphiphiles are promising for a range of applications. Synthetic bolaforms are, for example, applied for nanomaterial synthesis of the anti-HIV drug Zidovudine® (Jin et al. 2010). Bolaform SLs could represent an interesting biological alternative and/ or addition. Besides such rather high-end applications, the use of the bolaform SLs in detergent applications is another possibility as this is the market where > 90% of industrially produced 'classic' SLs find application. After pursuing the IBD approach to develop and optimize production methods for these new and intriguing molecules (which will be further explained below), it indeed became clear that the molecules have a rather limited stability. This apparently was caused by the fact that the ester function is prone to (spontaneous) hydrolysis. This last fact offers the possibility to efficiently produce the rare sugar sophorose from bola SLs (Soetaert et al. 2013). A process to do so was developed at BBEPP, yielding a highly pure sophorose product (> 99% purity) and non-acetylated acidic SLs from one fermentation product. Although interesting, the above shows that bola SLs might not be the best target for application development as their instability complicates their production (hydrolysis during fermentation, and more significantly, during purification) and would give rise to instable functionality in watery applications. Considering the IBD approach, we thus thought about ways to circumvent this issue (see Figure 4).

A solution was identified by coupling back to the strain/fermentation level, aiming to generate compounds with two glycosidic linkages instead of one glycosidic linkage and one ester linkage, as also depicted in Figure 3, i.e., bola sophorosides or disophorosides. The generation of such compound would also further increase the possible structural variability. One solution was identified in the unit operation 'fermentation' of IBD, i.e., by feeding the new *S. bombicola* strain with fatty alcohols like oleyl alcohol instead of fatty acids like oleic acid/vegetable oil, as the latter are the origin of the ester function in bola SLs. However, the occurrence of alcohol oxidase enzymes in *S. bombicola* results in conversion of fatty alcohols to fatty acids (Hommel and Ratledge 1990), thus again giving rise to the production of bola SLs. Therefore, the strain engineering 'unit operation' of IBD was considered. More precisely, deletion of two putative fatty alcohol oxidase genes was performed in the

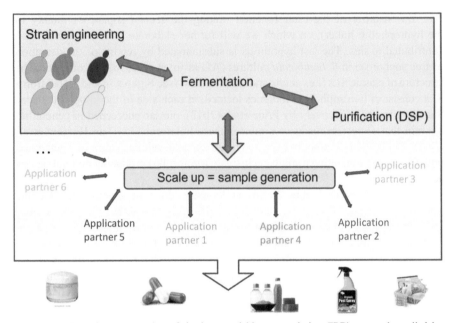

**Figure 4.** Schematic representation of the integrated bioprocess design (IBD) approach applied by InBio.be and BBEPP. Strain engineering, process development (fermentation and purification) and dedicated application research are closely interconnected and drive the movement of the new to nature glycolipids along the innovation chain.

bola sophorolipid producing strain. One of the newly generated strains now indeed efficiently produces bola sophorosides (Van Renterghem et al. 2017).

As mentioned above, after the development of the bola SLs producing strain, the IBD approach was further followed to enable the final valorization of these molecules. First, the fermentation to increase efficiency/productivity was investigated, which will be described below. Secondly and in parallel with this, a purification method was developed and thirdly the application potential of these new molecules was investigated. Emphasis is put on the parameter 'productivity' as the most important parameter influencing efficiency is used below as this was found to be the most important parameter influencing production cost and is described in the third part of this chapter.

## Increasing Efficiency and Uniformity through Process Development

When the bola SL producing strain was transferred to BBEPP to perform process development and -scale up, the medium used for other *S. bombicola* strains (Delbeke et al. 2016, Baccile et al. 2017) was used. The biggest difference between this medium and the lab medium used at InBio.be (Lang et al. 2000) is the use of corn steep liquor as a cheaper nitrogen source compared to yeast extract. We later found out, when performing the techno-economical analysis (TEA) (see below), that the use of such cheaper substrate does not have a significant influence on production costs (unless at very large scale), especially not if productivity (i.e., efficiency) is negatively influenced by its use. The latter was indeed shown to be the case, but this was not known at this point. For this reason, the first fermentations performed at BBEPP thus made use of corn steep liquor as a source of nitrogen instead of yeast extract, which is also reported to be used for industrial scale SL production (Develter and Fleurackers 2012). The initial ratio of bola SLs and acidic SLs using this medium in repeated fermentations only accounted to 40%/60% and low productivities (about 0.08 g/L.h for both glycolipids together, see below). Taking all the above into account, this was a highly undesired result as low efficiency was combined with low uniformity.

Initially, this observed difference with the lab scale fermentations performed at InBio.be was not expected to be derived from the medium, as this medium generated perfectly fine results with other engineered *S. bombicola* strains, exclusively producing lactonic or acidic SLs. The bola SLs producing strain ($\Delta sble \Delta at$) (Van Bogaert et al. 2016) only has one additional gene deletion as compared to the acidic SL producing strain ($\Delta sble$) (Roelants et al. 2016). As the CSL medium is highly microbially 'contaminated' and the fermentations indeed suffered from contamination with other microorganisms, this was initially suspected to be the reason for the bad results. A spore inducing method was thus applied, where the medium was heated to 80°C for 1 hour to induce sporulation, followed by the actual autoclaving action (1.5 h; 121°C). Although this indeed resolved the contamination issue, the ratios and efficiencies were not positively influenced. As adaptation of the feeding (glucose and colza oil) regime and stirring/aeration also did not resolve the issue, finally, the medium described by Lang et al. (2000), which is also used at lab scale, was evaluated. Indeed, using this medium increased the ratio of ns bola SLs/acidic SLs to above 90% and efficiency to an average of about 0.19 g/L.h. As the other medium components also differ slightly between both media, a growth experiment was started where the nitrogen sources of both media were switched, as shown in Table 1.

**Table 1.** Composition of the 'CSL' versus the 'YE' medium and the adapted media to evaluate the effect of yeast extract versus CSL.

|  | CSL | YE | CSL Adapted | YE Adapted |
|---|---|---|---|---|
| Glucose.H$_2$O | 110 | 132 | 110 | 132 |
| Corn steep liquor (CSL) | 5 |  |  | 5 |
| Yeast extract (YE) |  | 4 | 4 |  |
| (NH$_4$)$_2$SO$_4$ | 4 |  | 4 |  |
| NH$_4$Cl |  | 1.5 |  | 1.5 |
| MgSO$_4$.7H$_2$O | 0.5 | 0.7 | 0.5 | 0.7 |
| KH$_2$PO$_4$ | 1 | 1 | 1 | 1 |
| K$_2$HPO$_4$ |  | 0.16 |  | 0.16 |
| NaCl |  | 0.5 |  | 0.5 |
| CaCl$_2$.2H$_2$O |  | 0.27 |  | 0.27 |
| 3Na-citraat.2H$_2$O |  | 5 |  | 5 |

On the left is the media composition described in the literature and used before, on the right the adapted media. The same results were obtained as for the 'original' media, i.e., low uniformity and low efficiency for the medium containing CSL as the nitrogen source. This clearly shows that the difference in the nitrogen source, i.e., CSL versus YE, is at the base of the observations. Again, such effects were never observed for other *S. bombicola* strains, so these observations were quite peculiar. The composition of both media was subsequently thoroughly compared, and growth trails were started to, for example, evaluate the potential effect of lactic acid, which was a big difference in both media. No clear differences were observed for any of these experiments, and the fundamental reason for the varying results is still unknown. However, the very fact that such discrepancies were observed shows that adapting/optimizing the medium composition holds a lot of potential to further increase efficiency and uniformity and, more importantly, unravel the 'background' machinery of glycolipid biosynthesis. This is currently being investigated at our lab in a semi-high throughput set up and by applying DOE. Moreover, -omic analyses are planned to evaluate the differences between these situations at several -omic levels, which should enable us to unravel the molecular basis of the regulation of SL biosynthesis and would be the key to tightly control and steer production of the glycolipid portfolio shown in Figure 3.

Another parameter, which has been shown to increasingly influence the productivity of wild type SLs, is the choice of the hydrophobic carbon source (Kosaric and Vardar-Sukan 2014). It was thus expected to also be the case for this new strain. Historically, colza oil was always used at lab scale as this is the cheapest and easiest to obtain hydrophobic substrate (available in any supermarket), also for the new types of SLs. However, at this point, applying the IBD approach, a more substantiated choice should be made, i.e., the substrate resulting in the lowest production cost should be selected. A more expensive substrate might give rise to higher productivities and would as such result in a lower production cost per kg of bola SLs. We thus set up an experiment evaluating colza oil (Vandemoortele), high oleic sunflower oil (Oleon), oleic acid (Oleon), FAMES (Oleon) and FAEEs (Oleon) in a shake flask set-up, as described before (Roelants et al. 2013). As can be seen in Figure 5, the fatty acid esters (methyl and -ethyl) and free fatty acids (oleic acid) showed the best results, whereas a higher oleic acid content in a plant oil also results in a better process, i.e., considering colza oil versus high oleic sunflower oil. The prices of the used substrates are shown in Table 2; they were implemented in the techno-economic model described below and it was shown that at these price differences, and it was found that the positive influence of the productivities for more expensive substrates (see Figure 5 and Table 2) resulted in a decrease of the production cost of bola SLs.

Although the ethyl- and methylesters scored best, oleic acid was chosen to continue because the possible remaining ethyl- and methylesters and/or release of methanol in the fermentations/end product would be undesirable for cosmetic and/or food applications. Next, the optimal level of glucose was evaluated. For the wild type SL production, usually levels of 120 g/L are used. The bola SLs contain twice as much glucose in comparison to classic SLs (mono-sophorose versus di-sophorose), so it was not unlikely that higher levels would be more suitable for this new strain/

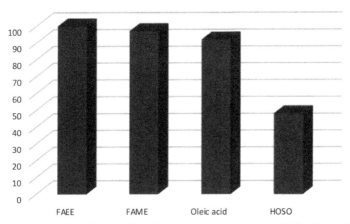

**Figure 5.** Effect of different hydrophobic substrates on the end titers of bola SLs in shake flask experiments. Data are shown relative to the best one.

**Table 2.** Commercial prices per metric ton of the evaluated substrates.

|  | Commercial Price €/ton |
| --- | --- |
| Colza Oil | 750 |
| High Oleic Sunflower Oil 75% (HOSO) | 1600 |
| Fatty acid ethyl esters (FAEE) | 1200 |
| Fatty acid methyl esters (FAME) | 1100 |
| Oleic acid | 2200 |

product. Glucose concentrations between 60 and 180 g/L were thus evaluated, similar to what was done for the hydrophobic carbon sources. Indeed, for the wild type *S. bombicola* strain, the level of 120 g/L was the best for SL production, whereas for the bola SL production strain, a level of 150 g/L glucose resulted in the highest production levels. Related to this, the initial level of hydrophobic substrate added to the fermentations was also evaluated. It was shown that initial addition of the hydrophobic substrate has a positive effect on the biosynthesis and 5 g/L was found to be the optimal level, which was thus further applied.

While optimizing/evaluating the processing conditions, bioreactor experiments were conducted along the way to evaluate the processes on the bioreactor level (7,5 L), implementing the above described adaptations. These experiments were all performed in identical set-ups, as described before (Delbeke et al. 2016), to allow comparison between them. The main results of these experiments are summarized in Table 3 and it is clear that all the changes resulted in a dramatic positive evolution of the bola SL productivity (i.e., a 14 fold increase). As mentioned above, a purification method was also developed at lab scale, similar to that described for acidic SLs (Roelants et al. 2016, Baccile et al. 2017), i.e., microfiltration or centrifugation to remove cells followed by a two-step ultrafiltration (50 and 5 kDa) resulting in a product of high purity (> 90%).

**Table 3.** Summary of the achievements for bola SLs productivity in bioreactor experiments along the innovation chain for the process development part of the IBD approach.

| | Productivity ns Bola SLs (g/L.h) |
|---|---|
| CSL + Colza oil | 0.05 |
| YE + Colza oil | 0.19 |
| YE + High oleic sunflower oil (HOSO) | 0.22 |
| YE + Oleic acid | 0.44 |
| YE + Oleic acid + cell retention | 0.63 |

The last result shown in Table 3 is derived from a continuous fermentation by applying cell retention in the stationary phase of the fermentation, as schematically depicted in Figure 6a. Seeing the high solubility of bola SLs (> 500 g/L) and its successful purification using a two-step ultrafiltration process (Van Renterghem et al. 2017), the idea arose to couple fermentation and purification in a full continuous set up (Figure 6a and b).

Such full continuous set-up would decrease the down-time of the equipment and the coupling of fermentation with purification would be a highly innovative achievement for biosurfactant production. Although full continuous systems can represent some clear advantages, such systems also demand highly robust biosynthesis, e.g., constant productivity has to be maintained throughout the process. This was first investigated by evaluating the continuous fermentation set-up separately (Figure 6), i.e., the broth was continuously filtered over a TFF PES 0.2 µm filter (Pall Kleenpak). The cell retention was initiated after 100 hours, when the bola SL titer amounted to 60 g/L. Influent was added to the fermenter at the same flow rate as the removal of the filtrate. The influent consisted of the production medium, as depicted in Table 1, without phosphate salts nor nitrogen containing compounds (yeast extract and $NH_4Cl$). These compounds are described to favor growth instead of SL production as nitrogen and/or phosphate limitation are described to be crucial for the production of SLs (Davila et al. 1992, Albrecht et al. 1996). Oleic acid was fed in shots of 15 g/L. Product, glucose, salts, solubilized oleic acid and other water-soluble compounds are present in the filtrate. The retentate, consisting of a slightly more concentrated fermentation broth, was sent back to the fermenter. The productivity as shown in Table 3 (0.63 g/L.h) could be maintained for 10 days, whereas, typically, only around 0.37 g/L.h was obtained during fed batch fermentations. However, after 10 days of continuous fermentation, the productivity started to drop. It was not entirely clear what was the reason for this drop as most of the constituents of the medium were fed to the yeasts in the influent. However, phosphate and nitrogen are crucial elements of DNA, RNA and proteins. The influent was thus enriched with the phosphate salts in the concentrations, as mentioned in Table 1. This did not result in a productivity rise. As nitrogen is also a crucial component of DNA, RNA and proteins, a shot of 4 g/L yeast extract (YE) was added 326 h after inoculation to supply the yeast with nutrients, which may be depleted. As YE is a complex medium component containing a lot of nutrients, vitamins and possible cofactors, the addition

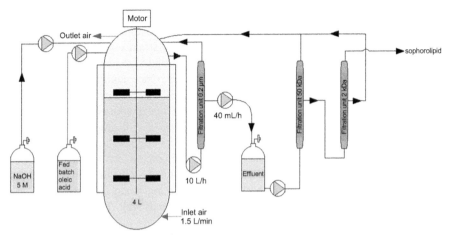

**Figure 6.** Schematic representation of a full continuous set up for bola SLs production, i.e., fermentation and purification (two step filtration). The continuous fermentation with cell retention was successfully performed, but the steady state for constant productivity could only be maintained for 10 days.

could have a direct or indirect influence on production. It was observed that the cell dry weight (CDW) rose from 35 g/L to 48 g/L after the YE addition, so as expected, a growth phase indeed existed. Glucose usage rose, but productivity dropped. This drop of productivity could be attributed to the presence of one component limiting the production and stimulating growth, e.g., nitrogen and/or another nutrient present in YE. Alternatively, nitrogen addition results in growth, while addition of another compound present in YE inhibits production. This hypothesis is currently being further investigated. A continuous set up, as shown in Figure 6, can possibly aid in maintaining productivities as possible limiting factors are constantly reintroduced in the fermentation. The experiment was repeated, but in this case the influent contained the phosphate salts and in this case the productivity only amounted to 0.51 g/L.h, which was again maintained during 10 days after the initiation of the cell retention and again a productivity drop was observed after 10 days of continuous fermentation. We are currently investigating the cause of the productivity drop in detail. Resolving this issue might enable the development of a full continuous system, as shown in Figure 6.

Although it is clear that the productivity of bola SLs should be further increased to allow the production costs to drop to acceptable levels (see below), the initial productivity using the bola SLs base strain was increased from 0.046 g/L.h to 0.63 g/L.h, which has an enormous effect on the production costs. Following the IBD approach, in parallel with these fermentation process actions, the strain engineering unit operation was also further considered. More precisely, more advanced metabolic engineering strategies are applied to increase the inherent productivity of the bola SL producing strain. For this purpose, as mentioned above, a comparative proteomic study was performed analogously, as described by Baccile et al. (2017), to identify possible bottlenecks in this strain. The results clearly show that the proteins responsible for bola SL production (i.e., Figure 2, steps 1, 2 and 3) are all less abundantly expressed in the bola SL producing strain, explaining lower inherent productivities as compared to the *S. bombicola* wild type strain (unpublished results). Overexpression strains using bidirectional promotors were thus generated, but this will be described in detail elsewhere. These strain engineering strategies were performed at the University (InBio.be) in parallel with the above described process development strategies performed at BBEPP. Once superior strains should be available, they can easily be plugged in the innovation chain using the optimized process conditions. This approach thus replaces a linear approach, where strain optimization would have to be finalized for process optimization to be started. Although some precaution should be taken, as strain engineering could also give rise to a change in performance in the optimized fermentation set-up/medium, in our hands, this approach has already been very successful. To give an example, all the above-mentioned adaptations for the bola SL producing strain are currently being implemented for the bola sophoroside producing strain as well. This thus represents a clear and enormous time saving measure.

## Scale up and Techno-economic and Environmental Profiling

To further investigate the developed production processes on the techno-economic, but also on the environmental level, the developed processes were further scaled up to the 100 L and 15 m³ scale. Such scale up not only enabled us to evaluate the feasibility of the processes, but also resulted in the generation of kg scale amounts of the product (bola SLs) for market exploration (see below). Three 100 L fermentations with corresponding DSP were performed to evaluate the reproducibility of the fermentations (batch-to-batch variation) by evaluating the product ratios, 'specs' of the purified products and their functionality in application experiments. The ratios between bola and acidic SLs were highly stable between the three 100 L fermentations, specs were highly similar and no differences in functionality were observed between the products derived from different batches. The process was further scaled up to 15 m³ scale to evaluate the scalability of the process. Although this was challenging, still an average productivity of 0.38 g/L.h was obtained, which was considered as a positive result.

The technical and economic (TE) feasibility of a dedicated bola sophorolipids (SL) producing plant was subsequently assessed. The production cost per kg was generated in a fully scalable Excel model, built in-house at BBEPP. The envisioned plant operates assuming repeated fed-batch fermentations (100 m³ fermenters in parallel) coupled with a batch-wise downstream process. The downstream process results in an overall purification yield of 96%. A complete mass balance was made for each process step with the empirical data obtained during the demonstration processes at 15 m³ scale. The energy balance was modeled using thermodynamic calculations with conservative assumptions. The impact of the fermentation productivity (0.05 to 4 g/L.h) and the yearly production scale (0 and 10 kTon per year) on the total production cost (CAPEX and OPEX) is shown in Figure 7.

Ashby et al. (2013) published a calculated projected production cost of wild type SLs of about 2 euro/kg SLs. However, this group assumed very high production volumes (90000 ton/year). The absolute minimum level of production cost will

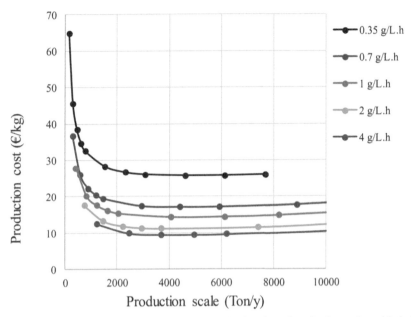

**Figure 7.** Prediction of production costs of bola SLs in function of production scale and bola SL productivity as generated by an in-house developed model. The 0.05 g/L.h option was not included to allow clear reading of the figure. The plateau for this value was found at around 100 euro/kg.

always be dictated by the major substrate cost, i.e., glucose and oil/fatty acids, together accounting to between 1.5 and 3 euro/kg, depending on the used substrates. A clear trade-off between cost (substrate, medium, type of DSP) and efficiency would have to be considered in detail at each specific production scale. Some parameters will be negligible at smaller scale (i.e., CSL versus YE), but might become important at very high scales as such parameters do not scale, like labor for example.

The proposed process and generated data at 15 m³ was also used to calculate the environmental impact of the production and use of bola SLs, similar to what was performed for acidic SLs (Baccile et al. 2017). Similarly, as described for acidic SLs by these authors, the environmental impact of bola SLs was unexpectedly high. Biosurfactants are always considered and described as sustainable and green alternatives to classic surfactants. This is a clear illustration of the fact that biobased/ green products/processes are not automatically the more sustainable solution. However, the largest part (> 87%) of the environmental impact was derived from the use of first generation substrates (glucose and oleic acid) as input for the production processes. This impact is derived from the negative influence on the environment of the agricultural processes associated with the production of these substrates. This is a result also seen for other products derived from renewable resources but is not generally or widely known/accepted.

The use of second generation (2G) substrates or even better, substrates requiring very little fertilization/watering (as the 2G substrates are still indirectly associated with these practices through the generation of the 1G substrates), would supposedly result in a positive effect on the environmental impact if the process efficiencies can remain largely unaffected. The latter was thus evaluated at BBEPP using 2G sugars derived from the bio-refinery company CIMV in combination with a microbial oil and in a second experiment using safflower oil and waste frying oil. The microbial oil was obtained from Neol Bio and was derived from fermentation on lignocellulosic sugars, while the safflower oil was supplied by GO-resources. The latter is able to grow on very marginal lands and requires no or very little watering and/or fertilization. The results will be described in detail elsewhere, but summarizing: this entire (hydrophobic and hydrophilic carbon sources) 2G fermentation for SL production was successful and gave rise to good efficiencies and product of excellent quality. Again, further optimization of the processes would be required to valorize these results. However, seeing the outcomes described above, these results were considered as very positive and innovative results as there have not been a lot of reports about/of full 2G processes for biosurfactant production.

## Application Potential and Exploratory Marketing

Depending on the application, the market demands cheap and characterized products, offering a versatile application range and compatibility with other formulation ingredients. A multitude of companies, e.g., Wheatoleo, Henkel, Cargill, DSM, Synthezyme, Evonik Degussa, Croda, Soliance, Ecover, EOC, Biomedica, Ecolife, Kao Soap Company, Shandong Jinmei Biotechnology, Saraya, etc. are active in research on/about production and/or application of SLs. These companies are mostly active in the detergent field, but also personal care products containing SLs like deodorants and/or shampoos with antimicrobial activity against bacteria causing bad armpit smell and dandruff, respectively, are on the market. As these last biological activities are attributed to the lactonic SLs and not, for example, acidic SLs, it is clear that for new types of compounds, the exercise has to be done all over again, i.e., the application potential has to be evaluated from scratch.

For the bola SLs, we initially only evaluated their use in ecological detergents as a 'usual suspect' application. Although quite good results were obtained for such applications (Van Renterghem et al. 2017), the current production costs are too high for these types of applications to be a realistic option. When looking at Figure 7, the production cost (not commercial selling price) of bola SLs could reach 30 euro/kg active matter at the current productivity ($\sim$ 0.7 g/L.h), once about 430 tons of this product would be produced/commercialized. We expect that this price should go down for B2C companies to 'massively' apply these compounds in their products. However, as the new molecules have an interesting and innovative structure, they might have properties/application potential for more high-end applications (e.g., pharma, cosmetics, nanotechnology). However, as the possible application potential of the bola SLs, and of the entire portfolio shown in Figure 3, is very broad, we are currently applying an exploratory strategy considering a very broad variety of applications/sectors, as shown in Figure 8.

The identification of high end applications benefiting from bola SLs would enable their market introduction at higher prices, after which the economy of scale will kick in, as shown in Figure 7. We are executing this exploratory marketing

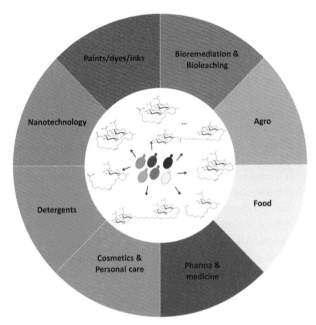

**Figure 8.** Representation of the potential application markets for the glycolipid portfolio shown in Figure 3. All of the mentioned markets/applications fields are currently under evaluation in partnerships with academia and/or industry.

approach in collaboration with other research institutes/labs, but also with industrial partners. We thus combine the determination of basic parameters and properties with specific application data and even toxicology (eco-, cyto- and geno-) of the molecules. Several papers will soon be published about/on our findings for the glycolipid portfolio depicted in Figure 3. This third part of the IBD approach is also clearly integrated with the other unit operations. For the bola SLs, for example, it became clear during scale up and also during application experiments for wash up liquids, that the molecules are not that stable. This subsequently gave rise to the development of the strain generating bola sophorosides, which is expected to give rise to the same functionality in the application experiments, but in combination with a higher stability.

After a suitable application is found, the regulatory aspects should still be taken into account, which will be different depending on the targeted application (e.g., food versus detergents). Tackling this hurdle will be one of the last ones to take before one of the products can make it to the market. This regulatory tract is both very laborious and expensive.

## Final Thoughts

The abovementioned accomplishments clearly show that applying an IBD approach can enable the valorization of new to nature microbial glycolipid biosurfactants. The unique combination of genetic/metabolic engineering strategies, process development, scale-up and application development is key to valorize the abovementioned technology platform.

Although the abovementioned accomplishments can thus be considered as very valuable proof of concepts in the microbial biosurfactant world, it is clear that there is still room for a lot of improvement. First of all, more in-depth metabolic engineering strategies are required to steer the engineering efforts in a rationalized fashion. The latter on the other hand requires the further expansion of the molecular toolkit, in-depth knowledge of the molecular regulation of glycolipid biosynthesis and its related pathways. The last goals are currently thoroughly being investigated at Inbio.be.

The medium composition is another uncultivated source of improvement. Although we have already performed some improvements as described in this chapter, we have not thoroughly dissected the medium optimization option yet. This is another point where we are currently extensively investing efforts by applying DOE and HTP set-ups. All the above-mentioned efforts were performed aiming to reduce the production costs. However, anno 2017 a balance with the environmental profile of the molecules should also be considered. In this research, it was shown that the environmental impact can be decreased by the application of 2G (like) substrates. However, these technologies should first be further optimized to reach similar efficiencies and cost as 1G substrates. A last and very important factor is the further and expanded exploratory marketing research. First, the determination of basic parameters of surfactants/glycolipids defining their application potential in a multitude of sectors/applications will aid the substantiated choice for more in depth, specific application research.

To conclude, we confirm that applying an integrated bioprocess design strategy (IBPD), i.e., considering the entire innovation chain, from genetic engineering through fermentation and downstream processing to final application testing, is the key to develop new strains and processes for the industrial production and commercialization of new biosurfactants and, with expansion, other types of (non-drop in) biochemicals.

## Acknowledgements

This research was funded by the European Horizon 2020 Bio-Based Industries (BBI) Consortium Project Carbosurf (669003), the national IWT project (innovation mandate 140917), the European FP7 Project IB2Market (111043) and the European FP7 Project Biosurfing (289219).

## References

Abdel-Mawgoud, A.M., F. Lépine and E. Déziel. 2010. Rhamnolipids: diversity of structures, microbial origins and roles. Appl. Microbiol. Biotechnol. 86: 1323–1336.

AL-Jasim, A., M. Davis, D. Cossar, T. Miller, P. Humphreys and A.P. Laws. 2016. Isolation of sophorose during sophorolipid production and studies of its stability in aqueous alkali: epimerisation of sophorose to 2-O-β-d-glucopyranosyl-d-mannose. Carbohydr. Res. 421: 46–54.

Albrecht, A., U. Rau and F. Wagner. 1996. Initial steps of sophoroselipid biosynthesis by *Candida bombicola* ATCC 22214 grown on glucose. Appl. Microbiol. Biotechnol. 46: 67–73.

Ashby, R.D., D.K.Y. Solaiman and T.A. Foglia. 2006. The use of fatty acid esters to enhance free acid sophorolipid synthesis. Biotechnol. Lett. 28: 253–260.

Ashby, R.D., A.J. McAloon, D.K.Y. Solaiman, W.C. Yee and M. Reed. 2013. A process model for approximating the production costs of the fermentative synthesis of sophorolipids. J. Surfactants Deterg. 16: 683–691.

Asmer, H.-J., S. Lang, F. Wagner and V. Wray. 1988. Microbial production, structure elucidation and bioconversion of sophorose lipids. J. Am. Oil Chem. Soc. 65: 1460–1466.

Baccile, N., F. Babonneau, I.M. Banat, K. Ciesielska, A.-S. Cuvier, B. Devreese, B. Everaert, H. Lydon, R. Marchant, C.A. Mitchell, S. Roelants, L. Six, E. Theeuwes, G. Tsatsos, G.E. Tsotsou, B. Vanlerberghe, Inge N.A. Van Bogaert and W. Soetaert. 2017. Development of a cradle-to-grave approach for acetylated acidic sophorolipid biosurfactants. ACS Sustain. Chem. Eng. 5: 1186–1198.

Bóka, B., L. Manczinger, A. Kecskeméti, M. Chandrasekaran, S. Kadaikunnan, N.S. Alharbi, C. and A. Szekeres. 2016. Ion trap mass spectrometry of surfactins produced by *Bacillus subtilis* SZMC 6179J reveals novel fragmentation features of cyclic lipopeptides. Rapid Commun. Mass Spectrom. 30: 1581–1590.

Cavalero, D.A. and D.G. Cooper. 2003. The effect of medium composition on the structure and physical state of sophorolipids produced by *Candida bombicola* ATCC 22214. J. Biotechnol. 103: 31–41.

Ciesielska, K., B. Li, S. Groeneboer, I. Van Bogaert, Y.C. Lin, W. Soetaert, Y. Van de Peer and B. Devreese. 2013. SILAC-based proteome analysis of *Starmerella bombicola* sophorolipid production. J. Proteome Res. 12: 4376–4392.

Ciesielska, K., S.L.K.W. Roelants, I.N.A. Van Bogaert, S. De Waele, I. Vandenberghe, S. Groeneboer, W. Soetaert and B. Devreese. 2016. Characterization of a novel enzyme—*Starmerella bombicola* lactone esterase (SBLE)—responsible for sophorolipid lactonization. Appl. Microbiol. Biotechnol. 100: 9529–9541.

Daniel, H.J., R.T. Otto, M. Binder, M. Reuss and C. Syldatk. 1999. Production of sophorolipids from whey: development of a two-stage process with Cryptococcus curvatus ATCC 20509 and *Candida bombicola* ATCC 22214 using deproteinized whey concentrates as substrates. Appl. Microbiol. Biotechnol. 51: 40–45.

Daverey Pakshirajan, K. and S.A. Sumalatha. 2011. Sophorolipids production by *Candida bombicola* using dairy industry wastewater. Clean Technol. Environ. Policy 13: 481–488.

Davila, A.M., R. Marchal and J.P. Vandecasteele. 1992. Kinetics and balance of a fermentation free from product inhibition—sophorose lipid production by *Candida bombicola*. Appl. Microbiol. Biotechnol. 38: 6–11.

Davila, A.M., R. Marchal, N. Monin and J.P. Vandecasteele. 1993. Identification and determination of individual sophorolipids in fermentation products by gradient elution high-performance liquid chromatography with evaporative light-scattering detection. J. Chromatogr. A 648: 139–149.

Delbeke, E.I.P., S.L.K.W. Roelants, N. Matthijs, B. Everaert, W. Soetaert, T. Coenye, Kevin M. Van Geem and Christian V. Stevens. 2016. Sophorolipid amine oxide production by a combination of fermentation scale-up and chemical modification. Ind. Eng. Chem. Res. 55: 7273–7281.

Develter, D. and S. Fleurackers. 2011. Improved sophorolactone production.

Develter, D. and M. Renkin. 2012. Sophorolactone compositions and uses thereof. (E. C. CT, ed.).

Felse, P.A., V. Shah, J. Chan, K.J. Rao and R.A. Gross. 2007. Sophorolipid biosynthesis by *Candida bombicola* from industrial fatty acid residues. Enzyme Microb. Technol. 40: 316–323.

Fleurackers, S.J.J. 2006. On the use of waste frying oil in the synthesis of sophorolipids. Eur. J. Lipid Sci. Technol. 108: 5–12.

Fukuoka, T., T. Morita, M. Konishi, T. Imura and D. Kitamoto. 2007. Characterization of new glycolipid biosurfactants, tri-acylated mannosylerythritol lipids, produced by *Pseudozyma* yeasts. Biotechnol. Lett. 29: 1111–1118.

Fukuoka, T., M. Kawamura, T. Morita, T. Imura, H. Sakai, M. Abe and D. Kitamoto. 2008. A basidiomycetous yeast, Pseudozyma crassa, produces novel diastereomers of conventional mannosylerythritol lipids as glycolipid biosurfactants. Carbohydr. Res. 343: 2947–2955.

Gao, R., M. Falkeborg, X. Xu and Z. Guo. 2013. Production of sophorolipids with enhanced volumetric productivity by means of high cell density fermentation. Appl. Microbiol. Biotechnol. 97: 1103–1111.

Geys, R. 2017. Engineering the metabolism of Starmerella bombicola for the production of tailor-made glycolipids. Ghent University.

Grand view research. 2016. Surfactant Market to Grow at a CAGR of 4.6% from 2015 to 2022.

Hommel, R. and C. Ratledge. 1990. Evidence for 2 fatty alcohol oxidases in the biosurfactant-producing yeast Candida-(Torulopsis)-bombicola. Fems Microbiol. Lett. 70: 183–186.

Huang, F.-C., A. Peter and W. Schwab. 2014. Expression and characterization of CYP52 genes involved in the biosynthesis of sophorolipid and alkane metabolism from *Starmerella bombicola*. Appl. Environ. Microbiol. 80: 766–776.

Jin, Y., N. Qi, L. Tong and D. Chen. 2010. Self-assembled drug delivery systems. Part 5: Self-assemblies of a bolaamphiphilic prodrug containing dual zidovudine. Int. J. Pharm. 386: 268–274.

Kosaric, N. and F. Vardar-Sukan. Biosurfactants: Production and Utilization—Processes, Technologies, and Economics 2014. CRC press, Taylor and Francis group.

Lang, S., A. Brakemeier, R. Heckmann, S. Spöckner and U. Rau. 2000. Production of native and modified sophorose lipids. Chim. Oggi. 18: 76–79.

Lodens, S., M. De Graeve, S.L.K.W. Roelants, S.L. De Maeseneire and W. Soetaert. 2017. Transformation of an exotic yeast species into a platform organism: a case study for engineering glycolipid production in the yeast *Starmerella bombicola*. P. *In*: J. Braman (ed.). Methods in Molecular Biology Series. Synth. Biol. Springer Publishing Co.

Ma, K.-Y., M.-Y. Sun, W. Dong, C.-Q. He, F.-L. Chen and Y.-L. Ma. 2016. Effects of nutrition optimization strategy on rhamnolipid production in a Pseudomonas aeruginosa strain DN1 for bioremediation of crude oil. Biocatal. Agric. Biotechnol. 6: 144–151.

Markets and Markets. 2015. Surfactants Market by Type (Anionic, Non-Ionic, Cationic, and Amphoteric), Substrate (Synthetic, and Bio-based), Application (Detergents, Personal Care, Textile, Elastomers & Plastics, Crop Protection, Food & Beverage) - Global Forecast to 2021.

Mnif, I. and D. Ghribi. 2015. Review lipopeptides biosurfactants: Mean classes and new insights for industrial, biomedical, and environmental applications. Biopolymers 104: 129–147.

Montoneri, E., V. Boffa, P. Savarino, F. Tambone, F. Adani, L. MichelettiC. Gianotti and R. Chiono. 2009. Use of biosurfactants from urban wastes compost in textile dyeing and soil remediation. Waste Manag. 29: 383–389.

Morita, T., T. Fukuoka, M. Konishi, T. Imura, S. Yamamoto, M. Kitagawa, A. Sogabe and D. Kitamoto. 2009. Production of a novel glycolipid biosurfactant, mannosylmannitol lipid, by Pseudozyma parantarctica and its interfacial properties. Appl. Microbiol. Biotechnol. 83: 1017–1025.

Nitschke, M., S.G.V.A.O. Costa and J. Contiero. 2010. Structure and applications of a rhamnolipid surfactant produced in soybean oil waste. Appl. Biochem. Biotechnol. 160: 2066–2074.

Onghena, M., T. Geens, E. Goossens, M. Wijnants, Y. Pico, H. Neels, A. Covaci and F. Lemiere. 2011. Analytical characterization of mannosylerythritol lipid biosurfactants produced by biosynthesis based on feedstock sources from the agrofood industry. Anal. Bioanal. Chem. 400: 1263–1275.

Price, N.P.J., K.J. Ray, K.E. Vermillion, C.A. Dunlap and C.P. Kurtzman. 2012. Structural characterization of novel sophorolipid biosurfactants from a newly identified species of Candida yeast. Carbohydr. Res. 348: 33–41.

Rau, U., R. Heckmann, V. Wray and S. Lang. 1999. Enzymatic conversion of a sophorolipid into a glucose lipid. Biotechnol. Lett. 21: 973–977.

Roelants, S.L.K.W. 2013. *Starmerella bombicola* as a platform organism for the production of biobased compounds. Ghent University.

Roelants, S.L.K.W., K. Saerens, B. Derycke, B. Li, Y.-C. Lin, Y. Van de Peer, S.L. De Maeseneire, I.N.A Van Bogaert and W. Soetaert. 2013. *Candida bombicola* as a platform organism for the production of tailor-made biomolecules. Biotechnol. Bioeng. 110: 494–503.

Roelants, S.L.K.W., S.L. De Maeseneire, K. Ciesielska, I.N.A. Van Bogaert and W. Soetaert. 2014. Biosurfactant gene clusters in eukaryotes: regulation and biotechnological potential. Appl. Microbiol. Biotechnol. 98: 3449–3461.

Roelants, S.L.K.W., K. Ciesielska, S.L. De Maeseneire, H. Moens, B. Everaert, S. Verweire, Q. Denon, B. Vanlerberghe, Inge N.A. Van Bogaert, P. Van der Meeren, B. Devreese and W. Soetaert. 2016. Towards the industrialization of new biosurfactants: Biotechnological opportunities for the lactone esterase gene from *Starmerella bombicola*. Biotechnol. Bioeng. 113: 550–559.

Rosa, C.A. and M.-A. Lachance. 1998. The yeast genus Starmerella gen. nov. and *Starmerella bombicola* sp. nov., the teleomorph of *Candida bombicola* (Spencer, Gorin & Tullock) Meyer & Yarrow. Int. J. Syst. Bacteriol. 48: 1413–1417.

Saerens, K.M.J., S.L.K.W. Roelants, I.N.A. Van Bogaert and W. Soetaert. 2011a. Identification of the UDP-glucosyltransferase gene UGTA1, responsible for the first glucosylation step in the sophorolipid biosynthetic pathway of *Candida bombicola* ATCC 22214. FEMS Yeast Res. 11: 123–132.

Saerens, K.M.J., J. Zhang, L. Saey, I.N.A. Van Bogaert and W. Soetaert. 2011b. Cloning and functional characterization of the UDP-glucosyltransferase UgtB1 involved in sophorolipid production by *Candida bombicola* and creation of a glucolipid-producing yeast strain. Yeast 28: 279–292.

Saerens, K.M.J., L. Saey and W. Soetaert. 2011c. One-step production of unacetylated sophorolipids by an acetyltransferase negative *Candida bombicola*. Biotechnol. Bioeng. 108: 2923–31.

Saerens, K.M.J., I.N.A. Van Bogaert and W. Soetaert. 2015. Characterization of sophorolipid biosynthetic enzymes from *Starmerella bombicola* (G. Daum, ed.). FEMS Yeast Res. 15: fov075.

Shete, A.M., G. Wadhawa, I.M. Banat and B.A. Chopade. 2006. Mapping of patents on bioemulsifier and biosurfactant: A review. J. Sci. Ind. Res. 65: 91–115.

Soberon-Chavez, G., F. Lepine and E. Deziel. 2005. Production of rhamnolipids by Pseudomonas aeruginosa. Appl. Microbiol. Biotechnol. 68: 718–725.

Soetaert, W., I.N.A. Van Bogaert and S.L.K.W. Roelants. 2013. Methods to produce bolaamphiphilic glycolipids.

Spencer, J.F.T., P.A.J. Gorin and A.P. Tulloch. 1970. Torulopsis bombicola sp. n. Antonie Van Leeuwenhoek Int. J. Gen. Mol. Microbiol. 36: 129–133.

Takahashi, F., K. Igarashi and H. Hagihara. 2016. Identification of the fatty alcohol oxidase FAO1 from *Starmerella bombicola* and improved novel glycolipids production in an FAO1 knockout mutant. Appl. Microbiol. Biotechnol. 100: 9519–9528.

Teichmann, B., L.D. Liu, K.O. Schink and M. Bolker. 2010. Activation of the ustilagic acid biosynthesis gene cluster in Ustilago maydis by the C2H2 zinc finger transcription factor Rua1. Appl. Environ. Microbiol. 76: 2633–2640.

Tropsch, J. 2017. A journey to standardization of bio-based surfactants in Europe. Inf. Mag. 28: 18–19.

Van Bogaert, I.N.A., S.L. De Maeseneire, D. Develter, W. Soetaert and E.J. Vandamme. 2008. Development of a transformation and selection system for the glycolipid-producing yeast *Candida bombicola*. Yeast 25: 273–278.

Van Bogaert, I.N.A., K. Holvoet, S.L.K.W. Roelants, B. Li, Y.-C. Lin, Y. Van de Peer and W. Soetaert. 2013. The biosynthetic gene cluster for sophorolipids: a biotechnological interesting biosurfactant produced by *Starmerella bombicola*. Mol. Microbiol. 88: 501–509.

Van Bogaert, I.N.A., D. Buyst, J.C. Martins, S.L.K.W. Roelants and W.K. Soetaert. 2016. Synthesis of bolaform biosurfactants by an engineered *Starmerella bombicola* yeast. Biotechnol. Bioeng. 113: 2644–2651.

Van Renterghem, L., S.L.K.W. Roelants, N. Baccile, K. Uytersprot, M.-C. Taelman, B. Everaert, S. Mincke, S. Ledegen, S. Debrouwer, K. Scholtens, C. Stevens and W. Soetaert. 2017. From lab to market: An integrated bioprocess design approach for new-to-nature biosurfactants produced by *Starmerella bombicola*. Biotechnol. Bioeng. (under review).

Zhang, X., Q. Guo, Y. Hu and H. Lin. 2013. Effects of monorhamnolipid and dirhamnolipid on sorption and desorption of triclosan in sediment-water system. Chemosphere 90: 581–587.

# 14

# The Future of Microbial Biosurfactants and their Applications

*Roger Marchant*

*'Prediction is very difficult especially when it is about the future'*—Nils Bohr

## Introduction

When starting to think about the future prospects for research and exploitation in a particular field, the first essential is to establish the existing context. The previous chapters of this volume will have gone much of the way to establish this context; however, I will spend a little time reiterating one or two points that may be particularly relevant to considerations of the future directions for the whole area.

The total annual worldwide consumption of surfactants of all types has been estimated at between 10–20 million tonnes; the actual figure is unimportant for the current discussion except that it is a very large amount. The other key points are that this consumption is made up of many different types of surfactants, many of which come from non-renewable resources, and are employed in a very diverse range of products from bulk consumer products such as laundry detergents and cleaners and foods to highly specialist items such as pharmaceuticals and cosmetics. What conclusions can we draw from this situation? First, the market is potentially enormous for microbial biosurfactants and the profits from market exploitation could provide the revenue for extensive research and development. Second, there is a great diversity of possible applications for microbial biosurfactants which may fit very well with the diversity of biosurfactant molecules produced by microorganisms. In general terms therefore, the whole situation looks remarkably bright for future

School of Biomedical Sciences, Ulster University, Coleraine, County Londonderry, BT52 1SA, Northern Ireland.
Email: r.marchant@ulster.ac.uk

research and development of microbial biosurfactants in a wide range of commercial products. What we may expect is that specific biosurfactants will be evaluated for their performance in particular product formulations and that, if they are equivalent or better in performance than the surfactants currently used, they will be used as probably partial replacements rather than total replacements, at least in the first instance. As we shall see later, the other major consideration relating to their use will be the cost, especially in relatively cheap bulk consumer products; however, there may be some niche markets where higher costs for microbial biosurfactants can be tolerated. As Nils Bohr identified, making predictions about the future is not only very difficult but is fraught with dangers. As a consequence, I do not propose to make specific predictions about the future for the development and application of microbial biosurfactants (BSs); instead I will take a few specific areas relating to the production and potential use of these molecules and try to identify aspects that merit attention.

## Production and Costs of Microbial BSs

Many of the research publications relating to microbial BSs have reported work carried out at laboratory scale and have focussed on the identity of the products, detailed purification and characterisation, but have paid little attention to the problems of larger scale production and downstream processing. In the final analysis, whether or not a particular BS is used in a product formulation depends initially on the efficacy of the material but ultimately on the cost. A number of recent publications have tried to identify some of the cost factors for economic production of BSs and have concluded that the substrate for the microbial fermentation is a critical cost factor that could be overcome through the use of waste materials and by-products. In many instances, these have been very specific and not generally available wastes, which would have very limited availability and value for large scale industrial production. In actual fact, substrate costs for many of the microbial BS fermentations are quite low and do not contribute significantly to the cost of the final product; this is particularly true when plant oils such as rape seed oil form the major carbon source for the production. The use of waste materials or by-products that by their nature contain complex mixtures of different carbon substrates, together with other contaminants and often show large composition fluctuations from batch to batch, make reproducible control of fermentations difficult to achieve and complicate the subsequent downstream processing. While the use of cheap waste materials as a substrate appeals to academic researchers, it is frowned upon by industry as an unnecessary complication in the quest for a reliable and consistent production process.

The only way in which a true picture can be obtained of how the various inputs contributing to a process contribute to the final cost is through conducting a full LCA (life cycle analysis). Using these methods, the individual contributions of substrate costs, fermentation costs and downstream processing costs can be identified and quantified. Surprisingly, in a recent large industrial project where fermentation production was carried out at 30 $m^3$ scale and 1 tonne of product was processed through supercritical $CO_2$ extraction, the major cost was the energy required to run the fermentation stage for seven days. This result would indicate that the main target for research and development of microbial BS production should be the highest possible product yield in the shortest fermentation time. For the future, more attention needs to be paid to downstream processing and methods for separating individual congeners of the mixture of BSs produced by microorganisms.

## Bioactivity of Microbial BSs

Many of the potential applications for microbial BSs will exploit the specific characteristics of detergency, foaming or emulsification and these will be the functions that will be sought and exploited by industry. In these applications, it may be possible to tolerate the use of the mixed congeners directly produced by the organism without resorting to complex and expensive separation methods even though individual congeners may have different properties. What is becoming increasing clear, however, is that many microbial BSs have bioactivities which may have important applications either as adjuncts to their physicochemical properties or can be exploited as properties in their own right. The types of properties which have been identified are bacteriostactic, bacteriocidal (Díaz de Rienzo et al. 2016), biofilm disruption (Elshikh et al. 2017) and anti-cancer applications. The complication for this area of application, however, is that the different congeners produced by a single organism may have very different bioactivity and indeed may have opposing activities. The microbial biosurfactants that have been most widely studied for their bioactivities are the glycolipids such as the bacterial rhamnolipids (RLs) and the yeast sophorolipids (SLs). Experimentation in this field requires that only highly purified individual BS molecules are used since small contaminations of, for example, bacterial toxins can produce a completely spurious result. The ability of some microbial BSs to act effectively against microorganisms through a bacteriocidal or bacteriostatic or biofilm disruption action in addition to their surface active properties offers appealing additional capability in various cleaning products, particularly those used for surface cleaning. Furthermore, these molecules may have supplementary benefits, for example, through replacing existing surfactants in products like mouth washes and toothpastes where their anti-microbial properties could help to control bacterial biofilms on teeth and the subsequent development of plaque. What is clear, however, is that the antimicrobial properties of microbial glycolipid BSs is very specific not just to a broad category of molecule, for example, RLs but to molecules with alkyl chains of one specific length. This is exemplified by the RLs from *Pseudomonas aeruginosa* which have two $C_{10}$ alkyl chains and have bacteriocidal activity, while the RLs from *Burkholderia thailandensis* with two $C_{14}$ alkyl chains do not show the same bacteriocidal activity (Elshikh et al. 2017). It is clear that the surfactant activity of these molecules is also closely linked to their antibacterial activity. It may also be possible to exploit both the antimicrobial and surfactant characteristics of BSs as a means of creating an adjuvant effect with antibiotics allowing the use of lower concentrations of antibiotic as an aid to avoiding antibiotic resistance in bacteria, or by allowing otherwise toxic antibiotics to be used at sub-toxic doses.

The observed anticancer activity of some BS molecules has opened the opportunity for their application in therapeutic or prophylactic applications (Callaghan et al. 2016). This area of research does require considerably more investigation before we can identify useful real applications. It is once again beset with the problem of congener mixtures produced by the microorganisms, for example the SLs produced by the yeast *Starmerella bombicola* are normally produced in a mixture of two molecular configurations, the acidic form and the lactonic form, and these have different activities against cancer cell lines *in vitro* and indeed behave differently in *in vivo*. Further difficulties stem from what appear to be rather narrow therapeutic windows to achieve differential killing of cancer and normal cells. Clinical use of microbial BSs for cancer treatment would also be constrained by the mode of administration, with topical use for skin cancers readily feasible and perhaps for gut cancers also since the purified glycolipids do not appear to have any major toxicity when ingested. Investigation of the bioactivity of microbial BSs does seem to be a fruitful area for future research in the light of the specificity of activity that may be linked to molecular structure.

## Genetic Manipulation of Producer Organisms

Genetic manipulation of BS producing microorganisms is in its infancy, although there are studies being published examining the regulation of their production. There are perhaps three main approaches and aims for genetic manipulation of BS producers. The first approach has been to clone and express the genes for BS production in a different host, for example, *E. coli* with a view to developing a familiar production system. This has worked to an extent; however, simply cloning the production genes does not seem to yield a host strain producing large quantities of product probably because the metabolic fluxes of the precursors in the host cells are not adequate to support the process. The second approach would be to try and engineer a strain of the original producer organism in which the regulatory mechanisms were by-passed or competing pathways are eliminated (Funston et al. 2017) to allow massive overproduction. In certain bacteria, for example, *Pseudomonas aeruginosa*, RL production is strictly regulated by the quorum sensing system in the cells and although many strains of *P. aeruginosa* have been isolated from the environment, investigated overproducing strains do not seem to have appeared. Whether extensive genetic manipulation of these bacterial systems has taken place under the cover of industrial secrecy is not clear. The third approach, and one which has been adopted for the SLs of *Starmerella bombicola*, is to engineer the cells to produce either a restricted range of the normal congeners or to produce 'new to nature' BSs (Roelants et al. 2016). This is undoubtedly a strategy which will be pursued further in the future. Perhaps one slight caveat here is that while a new molecule can be designed and produced, it is not easy to predict exactly what characteristics the new molecule will have. Genetically manipulated/modified organisms (GMO) for the production of microbial BSs are certainly an acceptable route for many industrial applications; however, there will remain some applications such as the food industry where companies will resist the use of GMOs for the production of ingredients for their products as they strive for 'green labelling.'

# New Producer Organisms?

A quick scan of the published literature will yield many published papers with claims for new BS molecules from new microorganisms. Unfortunately, many of these papers lack the rigour to make them useful (Irorere et al. 2017, Claus and Van Bogaert 2017). There are three key features that any new claim should contain. First, the microorganism being used should be completely and securely identified and available for other investigators to examine (if not covered by commercial protection). Second, the BS being produced should be completely purified and characterised using a range of different analytical methods such as NMR, FTIR, HPLC, Mass Spectrometry (LC/GC), etc. as appropriate. Finally, the claimed yield of BS should be determined accurately. Gravimetric determination of a crude extract or the use of colorimetric methods is generally so inaccurate as to be of little value. Having established a set of guidelines for the search for new producer organisms and new BSs, there is undoubtedly great potential for new discoveries. The key to making useful contributions may lie with the investigation of new previously uninvestigated or poorly investigated environments. Certainly, one such is the marine environment where there are many uninvestigated organisms and where there are environments which have metabolisable substrates available requiring BSs to access them.

# Microbial BSs in the Environment

One of the uses for microbial BSs that have been investigated is in bioremediation applications, either for hydrocarbons or heavy metals, and is coupled to this microbially enhanced oil recovery (MEOR), either from natural deposits or contaminated wastes. In these applications, clearly the purity of the biosurfactant preparation is a minor consideration, providing the preparation has sufficient activity to be effective. The over-riding consideration here will be cost since the existing chemical surfactants and dispersants already in use are relatively cheap. Microbial BSs do have potential in terrestrial and marine bioremediation applications, principally to enhance the biodegradation of polluting hydrocarbons by making them more amenable to the microorganisms present (e.g., Thavasi et al. 2011). However, in many microcosm experiments, the bioaugmentation of the system with hydrocarbonoclastic microorganisms and the addition of extra inorganic nutrients to the system have been shown to produce greater effects than the addition of microbial BSs.

Perhaps something that should be considered in greater depth is the potential effect on the environment that microbial BSs would have if they were to become a major component of many domestic products. If, for example, laundry detergents and household cleaning agents were to contain significant proportions of BS, then this would inevitably find its way into the waste water discharges into the environment. In the opening sentences of many research papers will be found the assertions that microbial BSs are 'environment friendly' and are readily biodegradable. As we have already seen, many microbial BSs are biologically active and have bacteriocidal activity; therefore, to describe them as environment friendly simply because they have been produced through a biological process is illogical. Similarly, if microbial BSs can have an effective role in long term bioremediation applications, then this suggests that they are not necessarily rapidly degraded in the environment. The fate of microbial BSs in the environment and their effects on microbial populations is possibly one area for future research that requires attention.

## The Structure of R&D Projects

In this section, I would like to draw attention to the problems of taking BS technology from TRL 3 (Technology Readiness Level) where proof-of-concept has been established to TRL 9 where an actual system has been thoroughly demonstrated and tested in its operational environment, i.e., in the commercial environment. This transition from laboratory scale to commercial production is a major weakness for much of the applied research that is carried out in research laboratories. The main requirement for the whole process to be completed is a large multidisciplinary team which includes researchers and commercial companies able to deal with the problems of scale-up, formulation into commercial products and full LCA analysis to determine the economic viability of the final product. Luckily, some funding bodies have moved to a funding model that allows such major projects to be completed, often with many partners drawn from various sectors (Baccile et al. 2017). One example of this approach is that adopted by the European Union through its Framework Programmes and more recently Horizon 2020. Such funding has allowed some microbial BS projects to be taken from basic research through to industrial scale at TRL7. This is probably the model that will need to be used in the future if microbial BSs are to be seen in a wide range of commercial products.

## Conclusions

In this chapter, I have tried to throw out ideas about some of the key areas which need to be addressed before microbial BSs will become a major part of all our lives. BSs do clearly have a future in many products and the fact that they can be produced sustainably from renewable feedstocks rather than by chemical processes makes them potentially attractive to many industries. The key to commercial success will be the ability of researchers and industry to work co-operatively to achieve knowledge and technology transfer to produce real products. Perhaps the best summary of the current situation with regard to the incorporation of microbial BSs into high throughput consumer products would be as follows: Microbial BSs are poised to become important components of a wide range of different consumer products, but their full potential has yet to be recognised or exploited.

## References

Baccile, N., F. Babonneau, I.M. Banat, K. Ciesielska, A.-S. Cuvier, B. Devreese, B. Everaert, H. Lydon, R. Marchant, C. Mitchell, S. Roelants, L. Six, E. Theeuwes, G. Tstasos, G. Tsatsou, B. Vanlerberghe, I. Van Bogaert and W. Sotaert. 2017. Development of a cradle-to-grave approach for acetylated acidic sophorolipid biosurfactants. ACS Sustainable Chem. Eng. 5: 1186–1198.

Callaghan, B., H. Lydon, S.L.K.W. Roelants, I.N.A. Van Bogaert, R. Marchant, I.M. Banat and C.A. Mitchell. 2016. Lactonic sophorolipids increase tumour burden in Apcmin+/- mice. PLoS ONE http://dx.doi.org/10.1371/journal.pone.0156845.

Claus, S. and I.N.A. Van Bogaert. 2017. Sophorolipid production by yeasts: a critical review of the literature and suggestions for future research. Appl. Microbiol. Biotechnol. https://doi.org/10.1007/s00253-017-8519-7.

Díaz de Rienzo, M., P. Stevenson, R. Marchant and I.M. Banat. 2016. Antibacterial properties of biosurfactants against selected Gram-positive and -negative bacteria. FEMS Microbiol. Letters 363: 224–231.

Elshikh, M., S. Funston, A.S. Chebbi, R. Marchant and I.M. Banat. 2017. Rhamnolipids from non-pathogenic *Burkholderia thailandensis* E264: Physicochemical characterization, antimicrobial and antibiofilm efficacy against oral–hygiene related pathogens. New Biotechnol. 36: 26–36.

Funston, S.J., K. Tsaousi, T.J. Smyth, M.S. Twigg, R. Marchant and I.M. Banat. 2017. Enhanced rhamnolipid production in *Burkholderia thailandensis* transposon knockout strains deficient in polyhydroxyalkanoate (PHA) synthesis. Appl. Microbiol. Biotechnol. doi: 10.1007/s00253-017-8540-x.

Irorere, V.U., L. Tripathi, R. Marchant, S. McClean and I.M. Banat. 2017. Microbial rhamnolipid production: A critical re-evaluation of the published data and suggested future publication criteria. Appl. Microbiol. Biotechnol. 101: 3941–3951.

Roelants, S.L.K.W., K. Ciesielska, S.L. De Maeseneire, H. Moens, B. Everaert, S. Verweire, Q. Denon, B. Vanlerberghe, I.N.A. Van Bogaert, P. Van der Meeren, B. Devreese and W. Sotaert. 2016. Towards the industrialization of new biosurfactants: biotechnological opportunities for the lactone esterase gene from *Starmerella bombicola*. Biotechnol. Bioeng. 113: 550–559.

Thavasi, R., S. Jayalakshmi and I.M. Banat. 2011. Effect of biosurfactant and fertilizer on biodegradation of crude oil by marine isolates of *Bacillus megaterium*, *Corynebacterium kutscheri* and *Pseudomonas aeruginosa*. Bioresource Technol. 102: 772–778.

# Index

Printed and bound by CPI Group (UK) Ltd, Croydon, CR0 4YY

24/10/2024

01778304-0015